Handbook of Environmental Protection

Handbook of Environmental Protection

Editor: Chuck Lancaster

Callisto Reference
www.callistoreference.com

Callisto Reference,
118-35 Queens Blvd., Suite 400,
Forest Hills, NY 11375, USA

Visit us on the World Wide Web at:
www.callistoreference.com

ISBN: 978-1-63239-995-3 (Hardback)

The publisher's policy is to use permanent paper from mills that operate a sustainable forestry policy. Furthermore, the publisher ensures that the text paper and cover boards used have met acceptable environmental accreditation standards.

Printed in China.

Cataloging-in-Publication Data

Handbook of environmental protection / edited by Chuck Lancaster.
p. cm.
Includes bibliographical references and index.
ISBN 978-1-63239-995-3
1. Environmental protection. 2. Environmental engineering. I. Lancaster, Chuck.
TD170 .H36 2018
363.7--dc23

Table of Contents

Preface

The purpose of the book is to provide a glimpse into the dynamics and to present opinions and studies of some of the scientists engaged in the development of new ideas in the field from very different standpoints. This book will prove useful to students and researchers owing to its high content quality.

Environmental protection is the practice of protecting our environment. Due to a rise in population, the need for environmental protection has increased. Overconsumption of natural resources and pollution have adversely impacted the environment. This has resulted in the imbalance in our ecosystem which can be observed in the form of global warming, natural hazards, ozone depletion, etc. Thus protecting our environment is a major concern for our scientists. The various studies that are constantly contributing towards advancing technologies and evolution of this field are examined in detail. Through this book, we attempt to further enlighten the readers about the new concepts in this field.

At the end, I would like to appreciate all the efforts made by the authors in completing their chapters professionally. I express my deepest gratitude to all of them for contributing to this book by sharing their valuable works. A special thanks to my family and friends for their constant support in this journey.

Editor

Possibilities for achieving the vision of near net zero emission in building sector in Bosnia and Herzegovina

Sandra Martinovic, Rejhana Muhamedagic, Azrudin Husika

Department of Energetics/Faculty of Mechanical Engineering, University of Sarajevo, Sarajevo, Bosnia and Herzegovina

Email address:
s.martinovic@mef.unsa.ba (S. Martinovic), muhamedagic@mef.unsa.ba (R. Muhamedagic), husika@mef.unsa.ba (A. Husika)

Abstract: Increase of greenhouse gases concentration in the atmosphere, mainly carbon dioxide, which leads to global warming, indicates the need to take actions to reduce use of energy from carbon sources. The building sector is the largest consumer of energy, and thereby the largest emitter of greenhouse gases. In this paper the possibilities of achieving "near net zero emission" vision, in the residential sector, by the year 2050, were analysed. The necessary policies and technical energy efficiency measures were analysed, that could be applied in the building sector in Bosnia and Herzegovina by 2050. Large amount of energy is used for space heating and hot water, mainly from fossil fuels, which significantly contributes to air pollution and global warming. Specific energy consumption for space heating in Bosnia and Herzegovina is several times higher than in EU countries with the similar climate conditions. Therefore, it is necessary to reduce energy consumption significantly, ie to increase energy efficiency. It is also necessary to deploy the potential of renewable energy use in buildings and to use buildings as energy producers. In this way, in the long term, energy costs in buildings would be significantly reduced, as well as dependence on energy imports and the need for building of new capacities for energy generation. From other side, employment would be increased while emissions of pollutants and greenhouse gases would be reduced. In this paper, due to the availability of reliable data on energy consumption for the Sarajevo Canton, the necessary policies and measures to promote energy efficiency were analysed, that could be applied in the residential sector in the Sarajevo Canton.

Keywords: Buildings, Energy Efficiency, Renewable Energy Sources, Greenhouse Gases

1. Introduction

Figure 1. *The scheme of Sustainable Development [1]*

Improved consumption in all the segments of human life leads to the enormous deprivation of the natural resources. Consequently, there is a necessity for the reduction of non renewable natural resources consumption, actually the aspiration to the sustainable development.

The basic representation of the symbiosis between man and the nature is the ideology of sustainable development (Figure 1). Sustainable development can be defined in numerous ways. Sometimes, this term is used only to accentuate the importance of continuing and stable economic improvement. It can also define environmental goals. Sustainable development can be defined as the development which is sufficient enough for human needs of the current generations, without causing any problems for the future generations to accomplish their needs [2]. Being sustainable also concerns technology and the capital.

The term "sustainability" makes us imagine numerous scenarios for attaining the sustainable society. Social systems are complex and nonlinear. However, there is a tendency of putting those systems to a certain models, in order to be easily understandable and in order to be able to foresee their functions. Academic tradition is the main reason of it, where the scientists try to make the social laws more similar to the nature ones. On the other hand, the complexity of social systems requires a wider perspective, which includes the entire system and uses the back casting method, based on the principles of sustainability. This could be the chance for achievement of better future. In this process, interdisciplinary approach is the one, which is needed to be applied for better creation of the basic frame, which enables achievement of good results. The changes of the working process, necessary for creation of sustainable society, can be mutually implemented.

2. Research Method

During the last few decades, faster economic and population growth has damaged environment in the form of depletion of ozone layer, climate changes, overuse of fossil fuels and many other consequences, which will surely, in these conditions, be multiplied. Scientists agree that radical measures should be done in order to soothe the negative consequences on environment.

Creation and implementation of radical measures, requires interdisciplinary approach, actually the cooperation of all sectors of the society. The transformation process must start at the level of every individual, reconsidering its own behavior and relation to the nature, its attitudes, as well as its consumption. So far, we have been using the working principle based on the analysis of the problem in the present and an attempt of solving it in the future from the present point of view (forecasting method). However, in the world of constant changes, the mutual perspective or the global vision is something necessary. From the perspective of desired future, finding the way of action in the very present is of crucial importance. This approach is called backcasting method. It is used to describe a future desired scenario, and to value its reason. Objectives are created for a longer period of time, about 50 years [3]. The longer period of time, the greater insecurity and worse control over the possible future scenarios, therefore the vision of future is better defined using the principle of sustainability, unlike specifications for the required period.

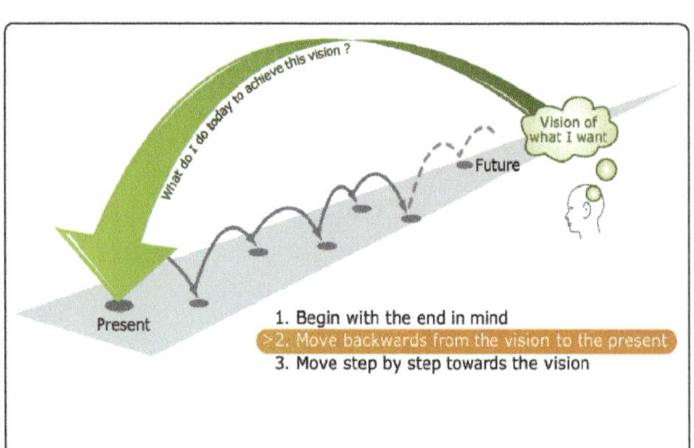

Figure 2. *Applying the Backcasting Method [4]*

The backcasting method is a chance to escape the reality and imagine the vision of future, without any limits of a vision, with realistic possibilities of its realization. This method is usually more effective than the forecasting one, especially in the period of fast and radical changes, because it has no intention of revealing the future, but to analyze more possible future scenarios, and to choose the best variants, as well as to define the steps of realization of the vision (Figure 2).

3. Application of Backcasting Method on Energy Consumption in Buildings

Functioning of cities is based on the constant flows of energy, natural resources, services, people, information, etc. It means it cannot be seen as a simple closed system. All the flows mentioned above are needed for human needs, but it may cause problems such as: air pollution, traffic jams and waste materials. According to the method of trend, it is supposed that cities in the future will face the problems similar to ones of the present, but having in mind that those problems will be more intensive, unless the actions for the development of sustainable cities are implemented.

The concept of the sustainable development requires a good strategic planning, analysis of more factors, including the past, present and vision of the future, creation and realization of the measures, technology development and raising awareness of individuals. In this paper, the concept of the sustainable development will be applied in Sarajevo Canton on energy consumption in buildings. The concept based on the vision of the sustainable cities in the future will be tendered, but since it requires significant changes, the backcasting method will be used. The focus of this work is on reduction of carbon dioxide emission, applying new technologies, sustainable energy resources and energy efficiency [5, 6].

The vision of the energy consumption in housing in Sarajevo Canton (already mentioned in this paper), whose possibility of realization is important to examine, is 40 to 50% less carbon dioxide emission in 2050 in comparison to the present emission.

3.1. Current Situation of Energy Consumption in Housing in Sarajevo Canton

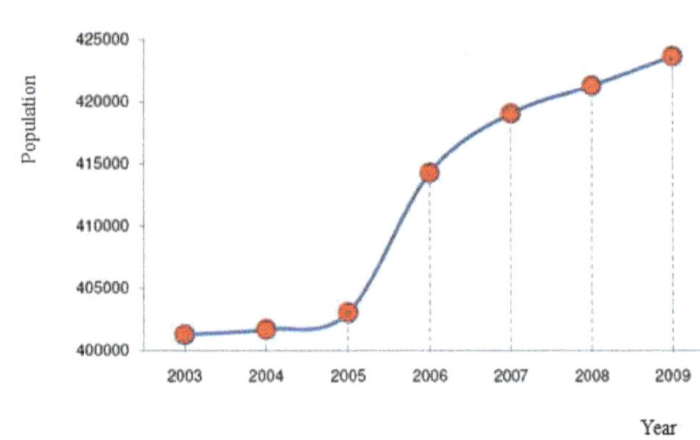

Figure 3. *Population growth in Sarajevo Canton [7]*

According to the available data on consumption of specific energy sources, the analysis of energy consumption has been made for the residential sector in Sarajevo Canton. The first analysis of energy consumption is made for 2008. In the middle of 2008, 421.289 citizens lived in Sarajevo Canton and 304.614 of them lived in the City of Sarajevo. Population in Sarajevo Canton records a growth rate of 1,2 % annually in the period from 2003 to 2010, and from 2003 to 2008, the growth rate was 0,95 % [8] (Figure 3).

Total floor area of residential buildings in Sarajevo Canton in 2008 was 11.354.270 m². According to data for 2008, there were 180.265 residential buildings in Sarajevo [7].

Table 1 shows how many residential buildings make an individual housing (private houses), collective housing (households not connected to the district heating plants), and collective housing (households connected to the district heating plants).

Table 1. *Types of households in the residential sector in Sarajevo Canton*

Households not connected to the district heating plants		
Housing	Number of residential buildings	% out of complete number of residential buildings
Collective housing	29.748	17
Individual housing	105.238	58
Households connected to the district heating plants		
Housing	Number of residential buildings	% out of comp lete number of residential buildings
Collective housing	45.279	25

The structure of energy consumption in residential sector buildings is: 20% for preparation of domestic hot water, 7% for electrical devices, 6% cooking, 5% lighting, 2% cooling, 60 % heating [7].

Figure 4 shows the share of electrical energy consumption in households according to the type of usage. The calculation was based on the available data on total energy consumption in this sector.

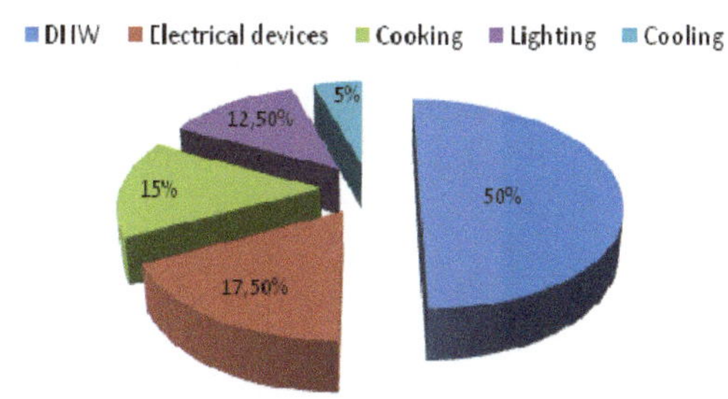

Figure 4. *Electrical energy consumption according to the usage*

In 2008, energy of 1.680,25 GWh was delivered to the residential sector in Sarajevo Canton, which is: electricity of 554,42 GWh or 32,9%, natural gas of 967,95 GWh or 57,6%, solid fuels of 153,37 GWh or 9,12% and oil derivatives of 4,5 GWh or 0,27% [7, 9, 10].

Table 2 shows the energy consumption in 2008 in the residential sector.

Table 2. *Energy consumption in 2008 in the residential sector in Sarajevo Canton [7, 9, 10]*

Residential buildings [MWh]		**Private houses [MWh]**				**District heating system [MWh]**			
Electricity	**Gas**	**Electricity**	**Gas**	**Wood**	**Coal**	**Electricity**	**Gas**	**Light fuel oil**	**Heavy fuel oil**
76.668,6	109.680,8	353.093,3	362.541,4	79.422,5	73.950	124.666,6	495.733,3	2.651	1.850,8

Table 3 shows the estimated carbon dioxide emission from residential building.

Table 3. *Emissions of carbon dioxide in Sarajevo Canton in 2008*

2008	**Coefficient of CO_2 emissions [t/MWh]**	**Mass of CO2 emitted, [t/a]**		
		Residential buildings	**Private houses**	**District heating system**
Gas	0,202	22.155,52	65.513,95	100.138,12
Electricity	0,729	55.891,4	257.405	90.881,95
Light fuel oil	0,281	-	-	739,62
Wood	0	-	0	-
Coal	0,354	-	26.178,3	-
Heavy fuel oil	0,279	-	-	520
Sum		78.046,92	349.097,25	192.279,69

Based on Table 3 total emission of carbon dioxide is calculated on 619.420 t CO_2 in 2008.

4. Measures for Reduction of Carbon Dioxide Emission in Residential Buildings

The last second half of the century has been marked by an enormous consumption of fossil fuels, which is a big problem nowadays. It has let to sudden and significant increase of carbon dioxide concentration in the atmosphere. The problem is that carbon dioxide is a greenhouse gas, and significantly contributes to the process of heating up the Earth, but it should be noted that carbon dioxide is neither the only greenhouse gas, nor the most dangerous one, but it is the widest spread of all. During the 80s of the 20th century, World became aware of the problem which could strike the entire Earth, so they made different laws and measures to slow down the process. Nowadays, the usage of renewable energy is imperative (solar energy, wind, biomass, etc.) in order to achieve a zero emission of greenhouse gases. European Union is very active in emission reduction policy, supporting the usage of renewable energy sources as well as increase of energy efficiency in all areas of human activities. Concerning the usage of energy resources, numerous technical solutions are being implemented, which contribute efficient combustion of fossil fuels with reduced loss and increased level of efficiency. However, at the beginning of the third millennium, with huge technical and technological breakthroughs in many areas, humankind is faced with increasing problems of depletion of fossil fuels, especially minerals, caused by environmental pollution of air, water, soil and faster reduction of forest and agricultural areas. There is an increase in interest for exploration of the natural resources in the conditions of intense technical and economical development, with the aim of attaining sustainability and fulfill the present needs without endangering future generations to fulfill their needs for energy. Aside from reduction of energy consumption by improving energy efficiency in different sectors, it is important to develop more independent, vital and flexible energy system, in which the usage of renewable energy sources could be of a great importance in satisfying needs for energy. In that regard, the usage of alternative and renewable energy sources has been intensified. European commission action plan regarding energy efficiency contains priority measures which include economically viable and energy efficient initiatives, which furthermore include actions in the field of: efficiency of household appliances, energy efficiency in buildings, with an emphasis on promotion of low energy and passive buildings, efficient generation and distribution of energy, introduction of financial mechanisms for energy efficiency as well as for promotion and raising awareness of energy efficiency [11].

Measures for reduction of carbon dioxide emission in residential sector are:

- reduction of heat losses,
- use of solar energy for heating and preparation of domestic hot water,
- use of solar energy for production of electricity,
- use of biomass,
- use of more efficient lighting and household appliances,
- replacement of the conventional boilers with the condensing ones.

5. Assessment of Future Energy Needs

Projections of energy consumption in residential sector in Sarajevo Canton estimated based on expected development, as well as demographic projections, macroeconomic indicators and analysis of expected trends.

In residential sector of Sarajevo Canton, energy sources usually used are: electrical energy, natural gas, liquid fuels and solid fuels (coal and firewood). Energy used for the purpose of satisfying needs for heat in the residential sector is proportional to the residential surface of the space. Energy consumption for the preparation of domestic hot water is primarily dependent on the number of occupants in the household. Energy consumption for other demand is not directly related to the surface, but primarily to purchasing power – living standard (cooling, power of electrical appliances, etc.).

In the following text, we analyze expected trends in the residential area of the average household, according to the number of occupants and individual energy consumption in the residential sector of Sarajevo Canton. Expected energy consumption in the residential sector of Sarajevo Canton until 2050 is analyzed on the basis of two scenarios, which present the projections of the future energy consumption in the residential sector of Sarajevo Canton.

5.1. Scenario I

In this scenario, the development of energy consumption in the residential area is supposed to be on the same level and of the same proportion like in the previous years, following the energy consumption trend from the period between 2002 and 2012. This scenario suggests the development of energy consumption in line with market trends and the consumer's habits, without any national and/or cantonal intervention, relying on increasing of the residential area of the housing sector in Sarajevo Canton, increasing of population and demography. In 2008, 421.289 people lived in 180.265 residential units in Sarajevo Canton. The total area of the residential space was 11.354.270 m^2, which gives an average apartment floor area of 63 m^2.

Assumptions applied in this scenario are:

- Increasing of population in Sarajevo Canton,
- Increasing of GDP/pc,
- Reducing the number of residents per residential units from 2,34 resident/residential unit to 1,7 resident/residential unit,
- Increasing the number of households with the growth rate of 1,7% annualy,
- The same trend in energy consumption as in the period from 2002 to 2012,
- None of measures for reduction of energy consumption were implemented.

On the basis of human migration from 2003, the projection of the number of inhabitants in Sarajevo Canton has been made for 2050 by the method of trend, Table 4.

Table 4. *Population - projection for 2050 in Sarajevo Canton*

Years	2008	2023	2030	2050	Growth rate 2003 - 2050
Number of people	421.289	485.484	518.704	626.680	0,95

Table 5. *Growth rate for particular energy sources in Sarajevo Canton [9, 10]*

	Heavy fuel oil	Electricity	Natural Gas	Light fuel oil	Coal	Wood
	Growth rate, %					
Residential Buildings	-	0,58	-0,64	-	-	-
Private houses	-	2	-0,64	-	1,2	1,2
District heating system for Households	0,5	0,86	0,65	0,5	-	-

On the basis of pathway of consumption of energy sources from 2002 to 2012, the projection of consumption of the fuels in the residential sector has been made by the method of trend for the period until 2050. Table 5 shows growth rate for particular energy sources, used for calculations of the required energy for year 2050.

Energy needs are different in residential sector for the households in the private houses and the ones in the residential buildings (collective housing). Table 6 shows expected energy needs for energy in the residential sector for the individual residential facilities (family houses), collective residential facilities not connected to the district heating system and collective residential facilities connected to the district heating system. Expected needs for energy in 2050 are shown in Table 6.

Table 6. *Energy consumption estimated by the method of trend in 2050 in residential sector of Sarajevo Canton*

Residential buildings [MWh]		Private houses [MWh]				District heating system [MWh]			
Electricity	Gas	Electricity	Gas	Wood	Coal	Electricity	Gas	Light fuel oil	Heavy fuel oil
168.454,8	180.544,2	948.326,4	406.629,7	633.447	416.431,8	282.075,6	1.321.012	6.635,9	4.632,6

It is obvious that the energy needs in residential sector until 2050 will be increased by 61,3% in collective residential units connected to the district heating plants, by 63,8% in private houses and by 46,6% in collective residential units not connected to the district heating plants. Expected increase of population in Sarajevo Canton has also been considered. In case of the same energy mix, the higher increase of energy consumption causes higher increase of carbon dioxide emission.

Table 7 shows the estimated emission of carbon dioxide in the case of increased consumption of energy until 2050.

Table 7. *Emission of carbon dioxide by the method of trend in 2050 in Sarajevo Canton*

Year 2050	Coefficient of CO_2 emissions [t/MWh]	Mass of CO2 emitted, [t/a]		
		Residential buildings	Private houses	District heating system
Gas	0,202	36.469,9	82.139,19	266.844,45
Electricity	0,729	122.803,5	691.329,9	205.633,11
Light fuel oil	0,281	-	-	1.851,42
Wood	0	-	0	-
Coal	0,354	-	87.594,8	-
Heavy fuel oil	0,279	-	-	1.301,76
Sum		159.273,4	861.063,9	475.630,74

Table 7 shows that the total emission of carbon dioxide in 2050, in the case of increased consumption of energy by the method of trend would be 1.495.970 t CO_2, furthermore the increased emission would be 58,6% compared to 2008.

5.2. Scenario II

Vision of the energy consumption in Sarajevo Canton is 50% lower of emitted carbon dioxide in 2050 compared to the present. Both scenarios suppose growth of life standard in Sarajevo Canton, increasing of residential area of the residential sector, as well as the population growth. Unlike the Scenario I, this one suggests development of energy consumption with national and/or cantonal funds, as well as the application of new technical regulations for building of residential facilities; clearly defying the parameters which influence better energy performance of newly built residential facilities. According to the valid legislation in EU countries, the maximum annual energy consumption in buildings is 95 kWh/m^2. Estimation is that an average facility and household in Bosnia and Herzegovina spend twice as much energy then it is usual in EU [12]. Thus, the greatest scope for the increasing of energy efficiency is in housing. It involves a wide range of activities with the final aim of reduction of consumption of all types of energy in the same or similar conditions in the facilities.

Applicable measures for improvement of energy efficiency in housing in Sarajevo Canton are:

- increase of thermal insulation of existing and new buildings (improving of thermal insulation on facade, energy efficient windows etc.),
- increase of efficiency of heating system, cooling and ventilation (energy efficient boilers, insulation of pipes, automatic regulation etc.),
- increase of share of renewable energy sources (solar collectors for domestic hot water system, use of biomass, heat pumps etc.) and
- increase of efficiency of lighting system and electrical appliances.

The current state of energy efficiency in Sarajevo Canton in housing can generally be characterized as unsatisfactory, meaning there are significant potentials for energy savings. The average annual specific heat consumption for space heating for the facilities of collective housing is from 125 kWh/m^2 to 152 kWh/m^2, meaning that those buildings have an energy rating - E, remarked as buildings with unsatisfactory energy efficiency [13]. Concerning individual housing facilities (family houses), the consumption per square meter is about 250 kWh per year, meaning that those buildings have an energy rating - F. According to the energy performance of buildings in Sarajevo Canton and target values recommended by EU, all the existing facilities should be improved in order to have an energy rating - B until 2050. It is of crucial importance to take care that all new buildings have an energy rating A or A+. If renovation of the building (installation of insulation, replacement of the openings, replacement of boilers, etc.) reduce energy needs and improve its energy rating from E into B, the annual energy savings would be 70% per building for collective housing, and 82% per facility for individual housing.

The efficiency of heating system can be increased by installing of high - efficient condensing boilers, which would lead to the significant savings in consumption of natural gas because the projected reserve on the surface of heating bodies enables satisfying of thermal comfort, even at lower temperatures of supply pipeline. Modern boilers, under optimum operating conditions have energy efficiency over 90%. In order to increase the efficiency, condensing boilers use latent heat of condensation of water vapor in flue gases, and as a result of this, during the warmer periods of the heating season, they work with efficiency greater by 10 - 15% than the traditional boilers [14]. Improvement of efficiency of heating system can be attained with the installation of thermostatic valves acquiring savings in energy consumption from 6 - 10%. Also, the heat pump can make 10% of electricity consumption, therefore, the installing of high - efficient heat pump, as well as the optimization of pump system, is an important element for electricity savings, which enables the savings of 80% [15].

Since the potential of renewable energy sources in Bosnia and Herzegovina is insufficiently used, significant savings in fossil fuel consumption are possible to attain, as well as the reduction of carbon dioxide emission by increasing share of renewable energy sources in a total consumption. Savings in consumption of non - renewable energy sources can be attained by 29% by installation of solar panels in the facilities of individual housing for the domestic hot water preparation, with the average annual efficiency of 50% per year [16]. Also, savings in electricity consumption can be attained by 80% by installing heat pumps for cooling and domestic hot water preparation in the facilities of collective housing. Heat pumps use electricity to divert the heat from the environment to the facility, or to divert the residue of heat from the facility to the environment. Such diverting of the heat requires significantly less energy than other solutions for heating or cooling.

Bosnia and Herzegovina has a remarkable potential of biomass. About 50% of the territory is covered by forests. Biomass made in agriculture should not be ignored either. However, the share of biomass in the total energy supply is low at the moment. The potentials of residual wood and wood waste are not used, which could ensure the thermal energy for all the facilities of individual housing in Sarajevo Canton. Using the biomass as a fuel for heating of the facilities of individual housing, it is significantly possible to reduce the emission of carbon dioxide in residential sector.

Out of 12,5% of total electrical energy in households is consumed on lighting. Most efficient measure for improving lighting indoor comfort is to use daylight as much as possible, so, equipping the apartment, the proper selection and arrangement of lighting in functional, as well as in decorative sense, is as important as the selection of furniture and wall paintings [15]. Since daylight is usually insufficient, it is replaced by artificial lighting. According to the EU regulation on lighting products (EC Regulation 244/2009), it is estimated that incandescent light bulbs will not be produced until 2016, and all the light bulbs will be replaced by energy saving ones. Using the energy saving bulbs, the electricity consumption for lighting can be reduced even by 80% [9]. Electrical devices have a share of about 17,5% of total electricity consumption in household, and together with appliances, that share is 32,5%. Using more energy efficient appliances, expenses could be reduces to only 10% of nominal and turning off the appliance instead having it in standby mode is also a way for additional saving. Most of electrical energy used for household appliances (fridge, freezers, washing - machines, drying machines, cookers, etc.) shows that a great number of old and less energy efficient electrical appliances are used in Bosnia and Herzegovina. Energy efficient appliances use less energy (up to 60%) compared to the standard ones, so the basic recommendation when purchasing is to strictly pay attention to that, where is economically justified to buy appliances of energy class A or perhaps B. Using the modern energy efficient appliances saves a lot, e.g.: the modern fridge needs up to 55% less electricity than the model from 1998. No matter what appliance it is: fridge, washing - machine or dishwasher, the mutual characteristic is that modern appliances use less electrical energy than 10 years ago: combination fridge - freezer up to 60%, dishwasher up to 40%, and washing - machine up to 35% [15]. Applying these measures of increasing energy efficiency, it is possible to reduce the energy consumption in residential sector in Sarajevo Canton. Table 8 shows the expected needs for energy in 2050.

Table 8. *Energy consumption in 2050 in residential sector in Sarajevo Canton, after applying energy efficiency measures*

Residential buildings [MWh]		Private houses [MWh]				District heating system [MWh]			
Electricity	**Gas**	**Electricity**	**Gas**	**Wood**	**Coal**	**Electricity**	**Gas**	**Light fuel oil**	**Heavy fuel oil**
58.264	80.304	238.212,8	26.419,3	155.911,6	0	92.297,7	270.274,8	2.173	2.173

Table 9. *Emission of carbon dioxide for year 2050 in Sarajevo Canton, after applying energy efficiency measures*

2050	Coefficient of CO_2 emissions [t/MWh]	Mass of CO_2 emitted, [t/a]		
		Residential buildings	Private houses	District heating system
Gas	0,202	16.221,2	5.336,6	54.595,5
Electricity	0,729	42.474,4	173.657	67.285
Light fuel oil	0,281	-	-	610,6
Wood	0	-	0	-
Coal	0,354	-	0	-
Heavy fuel oil	0,279	-	-	599,7
Sum		58.695,6	178.993,6	123.090,8

Applying the measures of energy efficiency, the consumption of total energy in 2050 would be 45% reduced, compared to 2008, and 78,85 % reduced, compared to energy needs expected by the method of trend for 2050. So, reduced energy needs would have reduced carbon dioxide emission as a result. Table 9 shows estimated carbon dioxide emission in the case of application already described measures for energy efficiency until 2050. Table 9 shows that the total carbon dioxide emission in 2050, in Scenario II, would be 360.780 t CO_2. Reduction of emission of carbon dioxide would be 42% compared to the base year 2008. In comparison with Scenario I, reduction of emission would be 76%, which can be seen in Figure 5. Using the assumption that the average apartment floor area is 63 m^2, with the possible increasing of residential units with growth rate of 1,7%, according to Scenario II, it is estimated that the annual energy consumption would be 72,8% lower compared to 2008, and compared to 2050 in Scenario I, the energy consumption would be lower for 78,8%, which would be 40,2 kWh/m^2, Figure 6.

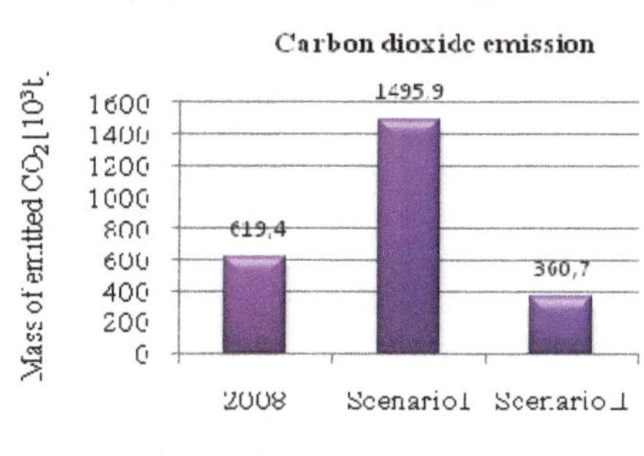

Figure 5. *Emission of carbon dioxide for the analyzed scenarios of the energy consumption in residential sector of Sarajevo Canton*

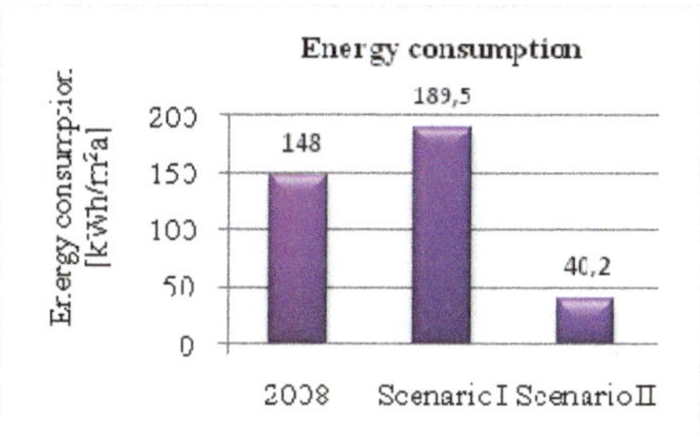

Figure 6. *Energy consumption for the analyzed scenarios in residential sector of Sarajevo Canton*

In case of implementation of Scenario II, energy consumption in residential sector in Sarajevo Canton will be lower for 3.449 GWh in 2050 than in case of Scenario I. Bosnia and Herzegovina does not have any commitment to reduce greenhouse gases emission, at least so far. Therefore, the primary driving force for implementation of Scenario II should not be achievement of certain reduction of emission, but, first of all, reduction of energy costs for buildings. Buildings in Sarajevo use mainly natural gas for heating which is imported and very expensive. Annual cost for natural gas is almost 10% of the cantonal budget. Therefore, any savings would improve overall economy in the Canton. Another benefit of implementation of Scenario II is creation of local jobs. Taking into account that 0,38 job-year is created per GWh of energy saved during implementation of energy efficiency measures [17], it can be calculated that about 65.000 job-year will be created, assuming 20 years of lifetime of energy efficiency measures or about 1.870 sustainable jobs.

6. Conclusion

Possible reduction of carbon dioxide emission in residential sector in Bosnia and Herzegovina, with the aim of approaching to "near net zero emission" until 2050, is analyzed in this paper. Two scenarios have been analyzed, applying different assumptions for future energy consumption in residential sector in Sarajevo Canton. Scenario I involves a development of energy consumption on the same level and of the same proportions as in the previous years, following the trend of energy consumption in the period between 2002 and 2012. It was assumed that the population growth rate would be 0,95% per year, and non measures for reducing of energy consumptions were provided. As the base year 2008 is taken. Measures for improvement of energy efficiency were described in Scenario II, as well as the assumption of increasing of population, mentioned in Scenario I. In Scenario I, the emission of carbon dioxide was analyzed to be higher by 58,6% for 2050 compared to 2008 and energy consumption per square meter would be higher by 21,9%. Scenario II shows that the emission of carbon dioxide, compared to the base year would be reduced by 42% and energy consumption per square meter would be reduced by 72,8%. By applications of measures for improvement of energy efficiency while compared to the emission of carbon dioxide in 2050 in Scenario I, it would be reduced by 76% and energy consumption would be reduced by 78,8%. The results indicate feasibility of achievement of realization of "near net zero emission" vision, establishing appropriate policy and applying technical measures for improvement of energy efficiency, as well as using the potentials of renewable energy sources.

References

[1] M. Joshi, S. Ravindranath, G. K. Jain, K. Nazareth, "Sustainable Development: An Introduction", Internship Series, Volume - I, Centre for Environment Education, 2007.

[2] M. Lucic, M. Mack, "qLife", Cotrugli BS, Quantum21.net, Rijeka, No 2/Vol 1/Spring 2009, pp. 5-9, pp. 81-91.

[3] A. Miola, "Backcasting approach for sustainable mobility", JRC Scientific and Technical Reports, European Commission Joint Research Centre, Institute for Environment and Sustainability, European Communities, 2008.

[4] http://www.naturalstep.org/en/backcasting, (Accessed on the 15th of July 2014).

[5] A. Phdungsilp, "Futures studies' backcasting method used for strategic sustainable city planning", ELSEVIER, Futures 43, 2011, pp. 707–714.

[6] M. Höjer, A. Gullberg, R. Pettersson, „Backcasting images of the future city—Time and space for sustainable development in Stockholm", ELSEVIER, Technological Forecasting & Social Change 78, 2011, pp. 819–834.

[7] CETEOR Ltd., "Study on optimal energy supply in Sarajevo Canton", vol 1, Goverment of Sarajevo Canton, 2010.

[8] Institute for development planning of Sarajevo Canton, "Demographic analysis of Sarajevo Canton by city areas from 2003 – 2010", Goverment of Sarajevo Canton , Sarajevo, 2011.

[9] HVAC DESIGN Ltd., REGEA, "Action plan on sustainable energy development of Sarajevo city", Goverment of Sarajevo Canton, Sarajevo, 2011.

[10] Ministry of Agriculture, "Balance of energy needs of Sarajevo Canton", Goverment of Sarajevo Canton, Sarajevo, 2008-2013.

[11] D. Kolic, T. Simunovic, "Economic and ecological characteristics of energy efficient construction", University of Zagreb, Faculty of civil engineering, Zagreb, 2009.

[12] S. Cenic, M. Dakic, A. Husika, "Unlocking the future - Sustainable energy in Bosnia and Herzegovina: Challenges and perspectives", Heinrich Böll Foundation, Sarajevo, 2013.

[13] REIC (Regional Education and Information Centre for Dustainable Development in South-East Europe), "Improving energy efficiency of residential buildings in Sarajevo Canton", Center for energy efficiency, Sarajevo, 2009, pp. 70-200.

[14] http://www.computherm-fabo.rs/download/Usteda_energije_i_termicka_ugodnost.pdf, (Accessed on the 5th of July 2014).

[15] S. Maksumic, "Analysis of possible measures for reduction of electricity consumption", Training material for energy certification of buildings, CETEOR Ltd., 2011.

[16] http://www.centrala.org.rs/knjiga%20liber%20perpetuum/02Sunceva_energja_Solar_Energy.pdf, (Accessed on the 12th of July 2014).

[17] M. Ghani-Eneland, M. Renner, A. Chawla, "Low Carbon Jobs for Europe: Current Opportunities and Future Prospects". Gland, Switzerland: World Wide Fund for Nature (WWF), 2009.

2

Sensitive Alpine Plant Communities to the Global Environmental Changes (Kazbegi Region, the Central Great Caucasus)

Otar Abdaladze[1], Gia Nakhutsrishvili[2], Ketevan Batsatsashvili[1], Khatuna Gigauri[1], Tamar Jolokhava[1], George Mikeladze[2]

[1]Alpine Ecosystems Research Program, Institute of Ecology, Ilia State University, Tbilisi, Georgia
[2]Department of Plant Systematic, Institute of Botany, Ilia State University, Tbilisi, Georgia

Email address:
alpine_ecology@iliauni.edu.ge (O. Abdaladze)

Abstract: Sensitive plant communities are complexes of species particularly susceptible to global environmental changes (climate, land use, etc.). In the temperate zone alpine areas are considered as the most important "hot spots" in this respect. In the Central Great Caucasus, which is the traditional alpine vegetation monitoring site in the Caucasus, on the basis of 50-years long (1964-2014) phytosociological and ecological studies the most sensitive plant communities were distinguished: 1) Treeline ecotone communities, including: (a) Evergreen prostrate shrubbery dominated by *Rhododendron caucasicum*, (b) Dwarf semi-shrubbery dominated by *Dryas caucasica* and (c)Thermo-hygrophilous subalpine tall herbaceous vegetation dominated by *Heracleum sosnowskyi*; 2) Subalpine broad-leaved mesophilous meadows dominated by *Anemonastrum fasciculatum, Geranium ruprechtii, Betonica macrantha* and *Trollius ranunculinus*; 3) Alpine carpet-like meadows ("Alpine carpets") consist of *Campanula biebersteiniana, Veronica gentianoides, Taraxacum porphyrantum, Sibbaldia semiglabra*, etc.; 4) Snow-bed vegetation (*Galanthus platyphyllus, Fritillaria latifolia*, etc.), and 5) Subnival/nival vegetation patches formed by 2-5(7) species (*Cerastium kasbek, Alopecurus dasyanthus, Tripleurospermum subnivale, Saxifraga sibirica, S. flagellaris, Delphinium caucasicum, Nepeta supina, Pseudovesicaria digitata, Symphyoloma graveolens*, etc.). Totally habitats of these plant communities cover about 1/3 of Kazbegi region area. Temperature rise, decrease in precipitation will lead to abrupt decrease of already small areas covered by Tertiary's relict tall herbaceous vegetation; elimination of a number of highly sensitive plant species including: relic, rare, endemic and critically endangered; disappearance of alpine snow-bed species. On account of early snow thawing or belated snowfall in autumn chionophyte plants (elfin, prostrate and dwarf shrubs and forbs in the alpine zone) sensitive to low moisture content, few and short-term snow cover and high temperature will lose. Probably sharp changes should be expected in subnival/nival zones, first of all, related to abrupt decrease in the glacier areas and subsequent increase in the distribution range of many species. According to the scenario suggested for the South Caucasus, which is based on predicted 3.9°C rise of temperature and decrease of precipitation by 9-13% during the century, it is expected that further climate warming may significantly change the vegetation and consequently the landscapes in the region. The vegetation may become similar to that of dry gorges of the Rocky Ridge of the Great Caucasus, which are situated 10-15 km to the north of the Kazbegi region.

Keywords: Caucasus, Climate Change, Sensitive Plant Communities, Alpine Plant Diversity

1. Introduction

The world climate continues steady warming. Global climate change and its ecological perspectives are the topical ecological problems of modernity [1]. The climate change effect will occur more or less in all ecosystems of the world, but most acute and rapid changes will expected in a highly sensitive alpine environment. The sensitivity of alpine species and plant communities are determined by low temperature conditions of alpine habitats [2-6]. Because of this, alpine environments of temperate zone are considered as the most important "hot spots" and alpine vegetation is viewed as the most noticeable bio-indicator [7-9]. Changing of low temperatures limiting effects will have a sharp impact on the diversity of alpine vegetation. Species vertical (along the altitudinal gradient) and horizontal (along the geographic sites) migrations will be the most prominent effect due to climate change [10, 11]. As a result contemporary structure of vegetation, pattern of its spatial distribution formed during centuries will be disturbed, new plant communities will gradually develop, some species will disappear, and others will occupy new more or less suitable ecological niches. The mosaic structure of micro-relief in high mountains and hence highly heterogenous microclimate create diverse ecological niches, which will be very important for local plants to find shelters during vegetation transformation caused by climate change [12-14].

Sensitivity of alpine species is strongly dependend on plant living form, development rithm and other ecological peculiarities which determine vitality, competition, resistance, etc. [2, 9, 15-18].

Expacted transformation of alpine vegetation threatens populations of species adapted to very specific ecological conditions; among such species are many rare, relict and endemic taxa. For this reason biodiversity degradation will become a real problem. These factors determine the importance of identification and ecological assessment of sensitive species and plant communities.

The goal of this study was to identify alpine species and plant communities sensitive to the global climate change in the Central Great Caucasus.

2. Study Area & Methods

Studies were conducted in the Kazbegi region, which is located (N 42°39'; E 44°37') on the extreme eastern, highest and geomorphologically the most complex central part of the Great Caucasus range (Fig. 1).

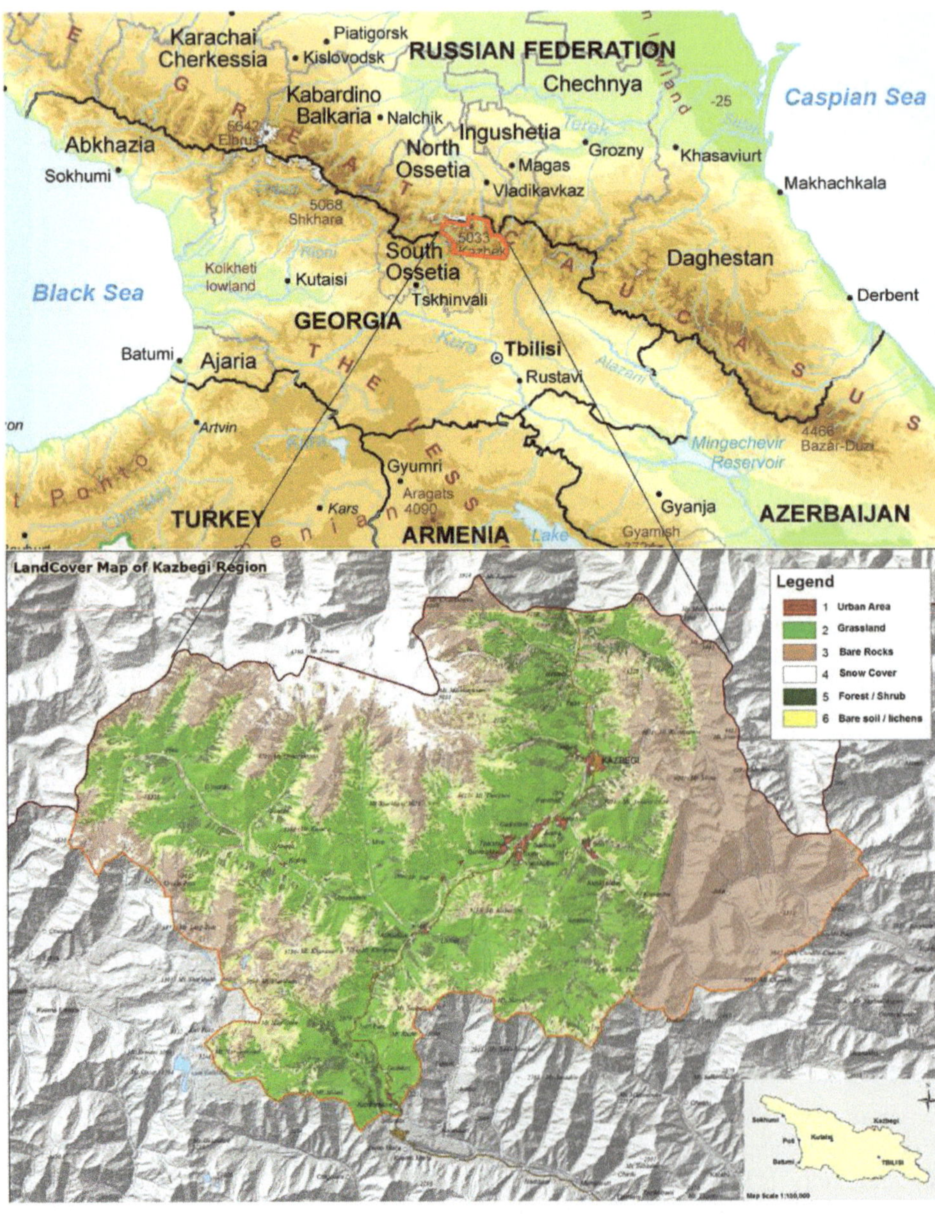

Figure 1. Location and main types of vegetation of the Kazbegi region (the Central Great Caucasus)

The study area is of volcanic origin. Topography is formed by Jurassic rocks, Paleozoic and even older granites, younger lava and moraines. Mountain massifs of the Kazbegi volcanic area are overlain by Quaternary glacial or river deposits and rock falls as well as major accumulations of calcareous tuffs and travertines. Glacial deposits occur in many places. The relief of the Kazbegi region is formed by: ascending, bare, sharp ridges; isolated peaks; very steep rocky slopes; narrow gorges; and caves of erosion-tectonic origin. Ninety-nine big and small glaciers with different exposure and morphology are present in the region. The lowest, average and highest elevations of the region are 1210 m a.s.l., 2850 m a.s.l. and 5033 m a.s.l. (Mt. Kazbegi). Area of Kazbegi region is 1081.7 km2 [19, 20]. The dominant soil types are: leptic, folic, alumic, humic umbrisols [21].

The following factors determine the climate in the region: high elevations, complex topography, and location on north-facing macro-slope of the Great Caucasus. Tab. 1 represents some main characteristics of climate of the region. The climate type characteristic to upper forest-subalpine zones is moderately dry with short and cool summer and relatively cold winter. The climate type of subalpine-alpine zones is moderately moist, with short and cool summer and relatively dry and cold winter. The climate type of alpine-subnival zones is moderately dry and characterized by severe and long winter and absence of real summer. The climate type of subnival-nival zones is moderately moist, without summer and characterized by the permanent snow cover and ice [20, 22].

The Kazbegi region is characterized by rich flora. It contains 1112 vascular plant species, which makes up 27% of the total number of plants (about 4100 species; [23]) recorded in Georgia. The study area is characterized by a high richness of Caucasus endemic species (26%) and genera (6 out of 11) [18, 24-27], as well as high diversity of plant communities [20, 28].

The altitudinal zonation of the vegetation is well pronounced. Mountain forest fragments, shrubberies, rock and scree vegetation is typical in the mountain-forest zone (up to 1800/1850 m a.s.l.). The alpine vegetation can be divided into four zones: subalpine (1800/1850– 2450/2500 m a.s.l.), alpine (2450/2500–2950/3000 m a.s.l.), subnival (2950/3000–3650/3700 m a.s.l.) and nival (> 3700 m a.s.l.) [18, 20, 22].

In the subalpine zone birch forests and elfin woodland (*Betula litwinowii*), shrub communities (*Rhododendron caucasicum*), herbaceous and tragacanthic (*Astragalus denudatus*) vegetation are typical remnants of natural vegetation types. However, vast grasslands, pastures and hay meadows characterized by e.g. *Festuca ovina, F. varia* subsp. *woronowii, Hordeum violaceum*, and *Bromopsis variegata* dominate in the subalpine landscape. The alpine zone is mainly characterized by grazed grasslands, which cover mountain ridges and south-facing steep slopes (dominated by *Festuca varia* subsp. *woronowii*), while the north-facing slopes are mainly covered with shrubs (*Rhododendron caucasicum*). Carpet-like meadows (where dicots mostly dominate), rock and scree vegetation are also well pronounced in the alpine zone. In the subnival zone vegetation is present as patches of few (2-5) species. In such micro-coenoses (nanocoenoses) cushion-like, dense-tussock and rosette life forms are predominate. In the nival zone only two species (*Cerastium kasbek, Alopecurus dasyanthus*) are recorded and they reach 4000 m a.s.l. [18, 20, 24, 25, 28]. In the Kazbegi region alpine vegetation is under long-term human impact [17].

Plant species were identified according to the following sources [29, 30]. Critical samples were compared with samples in the Herbarium of the Tbilisi Institute of Botany and with collection of Stephantsminda Alpine Ecology Institute of Ilia State University. Botanical nomenclature follows to [23, 30]. Phytosociological surveys have been carried out according to the cover-scale method [31, 32] during 1961-2014. The altitudinal differentiation of the vegetation as well as classification and nomenclature of plant communities are given according to [18]. ArcGIS Desktop version 9.3 (ESRI Inc., Redlands, CA) was used to calculate of areas covered by sensitive plant communities.

Table 1. *Climate characteristics of the Kazbegi region (according to: Nakhutsrishvili, 2003; Nakhutsrishvili et al., 2005)*

Vegetation zone/ Alttude (m a.s.l.)	Temperature of worm (July-August) months (°C)		Temperature of cold (Janary-February) months (°C)		Duration of snow cover (month)	Max. depth of snow cover (cm)	Average annual presipitation (mm)	Other information
	Ave.	Max.	Min.	Ave.				
Upper forest; Subalpine/ 1500–1900 m	14–18	33	–28	–5	3–5	≤ 80	780–800	In summer air humidity two times higher than in winter; in August number of clear, cloudless and hot days are 70%; Drought is often
Subalpine; Alpine/ 1900-2600 m	10-14	33	–30	–11	5–7	115–120	1000–1500	Number of fogy days: 135/yr; Winds are often
Alpine; Subnival/ 2600-3400 m	10	31	–30	–12	7–8	≤ 200	800–1000	In summer morning frosts and winds are often; Weather is strongly change during a day
Subnival; Nival/ > 3400 m	2.2	15	–33	–14	12	> 200	1000–1200	Number of windstorms: 12/month; Number of snowy days: 170/year

3. Results & Discussions

In the alpine landscape of temperate mountains upper timberline is the most noticeable and well marked phenomenon. The treeline ecotone has various important ecological functions, such as: protection against avalanches, landslides, mud-flows, debris-flows and soil erosion; regulation of water balance in lowlands; creation of natural reservoir of clean water; maintenance high species diversity, etc. [33, 34]. The treeline plant communities of the Kazbegi region are distinguished by exceptional diversity, which is expressed in the contrasting species composition and caused by various ecological conditions [22, 35]. High degree of species richness is mainly due to peculiar structural properties of plant communities, rather than ecological and human factors [28]. In the treeline ecotone of the Kazbegi region the following ecosystems should be referred to as highly sensitive: a) Evergreen prostrate shrubbery dominated by *Rhododendron caucasicum*; b) Dwarf semi-shrub community dominated by *Dryas caucasica*; c) Thermo-hygrophilous subalpine tall herbaceous vegetation (Megaphorbia); d) Subalpine broad-leaved meadows dominated by *Anemonastrum fasciculatum, Trollyus ranunculinus, Geranium ruprechtii* and *Betonica macrantha.*

The sensitivity of above mentioned plant communities is caused by: high slope degree (35-750), not well-formed and mobile soil cover, frequent solifluction, insignificant and/or short-term snow cover, etc. In Georgia droughts, high air and soil temperatures of the recent years [36] have particularly affected chionophils (*Daphne glomerata, Rhododendron caucasicum, Vaccinium myrtillus, V. vitis-idaea, Empetrum caucasicum*) widely presented on north-facing slopes in evergreen *Rhododendron* shrubbery. Withering of *Rh. caucasicum* leaves and noticeable decline of *Daphne glomerata* are observed in the treeline ecotone. Some other factors also invoke the significant suppression of *Rh. caucasicum* vitality. For example, after snowless winter in spring insufficient snow thaw rapidly at the edges of *Rhododendron* thickets and high solar radiation and warm air provoke photoinhibition and "winter drought" phenomena (Fig. 2; [16]). Mass withering of *Rh. caucasicum* will have strong negative impact on elfin shrubs and rare stenotopic herbs, which grow within the dense thickets in very specific ecological conditions. It will also break important facilitation relationships with main forest forming species in the region (Betula litwinowii). Dense thickets of *Rh. caucasicum* promotes establishment of birch seedlings in the treeline ecotone [37, 38].

The area occupied by dwarf semi-shrub *Dryas caucasica* is not large and covers skeletal and stony calcareous soils on steep (20-500) slopes with North and North-West aspects between 2000-2600 m a.s.l.. Normally it grows on slopes with dense snow cover, often in the avalanche places. At present its area is significantly reduced. In the lower boundary of its distribution (2000-2200 m a.s.l.) only single and suppressed individuals can be found.

Figure 2. *Damaged Rhododendron caucasicum thickets due to winter drought (a) and damaged leaves due to high insolation (b) in the treeline ecotone of the Kazbegi region*

Our studies have shown, that the habitat of special natural phenomenon represented by stenotopic Tertiary relict thermo-hygrophilous subalpine tall herbaceous vegetation has a specific character: a) variable slope angle (between 1-20 and 600) and all available exposures; b) optimal air (15-21°C) and soil (10-14°C) temperatures without strong daily fluctuations; c) high level of soil moisture (33-62%) and air humidity (58-89%), including frequent horizontal precipitation (fog); d) low or moderate illumination (800-1300 μmol photons•m^{-2}•s^{-1}); e) rich soils usually with near neutral pH (5-7), etc. This type of vegetation is composed of tall (1.8-2.5 m, rarely >3 m) forbs (chiefly dicots) and grasses. They are characterized by rosetteless shoots, short top roots and rhizomes. Stratification is not observed in the canopy. Species richness of an individual community is not high (from 13 up to 33). Soil pH measurements revealed that tall herbaceous vegetation is confined to weak acid soils. It is known that in such type of soils humus horizon is well developed and mineralization is rather high [39]. It is also mentioned that the weak acid reaction of the soil helps to increase respiration of root system [40]. Soil temperature and moisture also influence CO_2 output, although effect of these factors depends on vegetation type [41, 42]. Temperature increase may dry out soil upper layer, change soil pH and after certain time decrease mineralization rate. At the same time, if pool of nutrients will not correspond to excess CO_2, flower and fruit development will decrease [16]. These processes may create some threats to tall herbaceous vegetation, which is strongly adapted to the certain environmental conditions. Tall herbaceous vegetation is one of the main bio-indicators for climate warming. It is very sensitive to the fluctuations of temperature and humidity. Recent droughts have had negative influence on the development of such species as: *Senecio rhombifolius, S. propinquus, S. othonnae, Aconitum nasutum, A. orientale, Swertia iberica, Gadelia lactiflora, Dolychorhiza renifolia, Heracleum sosnowskyi, H. roseum, Cicerbita racemosa, Polemonium caucasicum, Delphinium flexuosum, Aquilegia caucasica, Asyneuma campanuloides, Doronicum macrophyllum, Inula helenium, I. magnifica, Lilium monadelphum subsp. georgicum, Valeriana tiliifolia, Cephalaria gigantea, Calamagrostis arundinacea*, etc. Totally 85 tall herbaceous species are mentioned in the Central Caucasus [18]. In particular, plants will not grow properly; undeveloped inflorescence will be observed; flowers will be open permanently; early senescence and even complete wilting of leaves will occur [20, 28].

Current climate warming will induce drastic changes in subalpine mesophilous broad-leaved meadows. Some of these meadows are associated with cool habitats above timberline, and some of them – with secondary meadows in the forest zone, on gentle or slightly sloping hillsides ($\leq 10°$) of both subalpine and alpine zones. Species composition in the plant communities will change due to soil drying, and populations of such species as: *Anemonastrum fasciculatum, Trollyus ranunculinus, Geranium ruprechtii* and *Betonica macrantha* will decline, or even some taxa will disappear.

Besides above mentioned sensitive treeline ecotone habitats there are some other ones as well. These are: a) Snow-bed patchy vegetation; b) "Alpine carpets" and c) Subnival plant communities. The snow-bed vegetation mostly occupies shady and wet cirques and small relief depressions with long-term snow cover and cool microclimate. These communities strongly depend on snowmelt water. Especially stress-dominated situations are observed in early spring flowering species (*Galanthus platyphyllus, Fritillaria latifolia, F. Lutea, Ficaria calthifolia, P. auriculata*). For example, in the habitats of early spring geophyte (*G. platyphyllus*) the highest level of temperature and moisture stress probability was observed in the areas completely clear from snow cover, where plants are in the phase of fruiting. Individuals which start their flowering phase at the edges of snow-beds were in somewhat better hydro-thermal conditions than individuals in the snow cover.

Snow cover reduction caused by climate warming and early snowmelt will drastically change the ecological situation. Because of this the snow-bed plant communities could be considered as highly sensitive, as well as carpet-like alpine meadows ("Alpine carpets") occurimg in the upper part of the alpine zone. These small plant communities are found among large stones and moraincs. They are predominantly composed of *Campanula biebersteiniana, Veronica gentianoides, Taraxacum porphyrantum, Gnaphalium supinum, Sibbaldia semiglabra, Poa alpina, Pedicularis crassirostris*. "Alpine carpets" which occur around the permanent snow spots and at the edges of glaciers contain the following species: *Ranunculus oreophilus var. pumilus, R. baidarae, Primula algida, Gentiana djimilensis* (= *G. pyrenaica*), *G. angulosa, G. nivalis, Corydalis emanueli, Minuartia aizoides, Cerastium cerastoides*, etc. "Alpine carpets" differ from true alpine meadows by their floristic composition and ecological peculiarities. Due to the lasting snow cover, they are characterized by short vegetative period. Herb cover does not exceed 3-4 cm, and 80% of aboveground phytomass is accumulated near the soil surface. Besides the climax "alpine carpets", secondary ones, enriched mainly by *Alchemilla* species, are present widely. Their development is caused by overgrazing. They occupy the most exploited alpine pastures [18, 22, 28].

At the edge of alpine and subnival zone climatic and edaphic conditions become more severe and only few species are adapted to them [1, 4, 6, 15]. In the Central Great Caucasus in the subnival zone vegetation is represented by individuals, or by small patches, which consist of 2-5(7) species. These are mostly cushion, trellis-like, tussock and rosette living form plants [26]. These plant ultra-microgroupings or micro-coenoses are determined as nanocoenoses [22]. The "hosting plants" of these nanocoenoses create microclimate, which use some other species [43, 44] among which are many rare, relict and endemic plants [18, 25, 27]. The majority of subnival/nival species are chionophytes: they need deep and long-term snow cover for successful wintering. Probably sharp climatic changes in subnival/nival zones enforce plants to change habitats: move upwards, or even change slope aspect. During these processes a number of species might lose their habitats and completely disappear, populations of some others will be depressed and decreased, and vegetation patches will change their composition and areas.

A number of studies in the worldwide (GLORIA Project first monitoring cycle – 2001-2008) has shown that area of some alpine species, distribution character and vitality were changed due to of climate change [45]. According to these study vascular plant species number increased on 45 GLORIA target regions (especially in the boreal-to temperate mountains) and decreased on 10 summits. Such alteration is a result of warmer conditions; the combination of rising summer temperatures and stable to decreasing precipitation, which was recently documented for southern Europe for the past decades [46, 47]. In the Central Great Caucasus GLORIA studies have shown insignificant increase in the abundance of resistant species individuals. This was recorded in parallel with an insignificant decrease in abundance of sensitive species. In subnival zone the changes are more visible, as a number of alpine species can be found at unusually high altitudes. In spite of this species migration processes from lower altitudes was not observed yet. Therefore, all of these changes must be considered as local fluctuations, and not as a succession phenomenon [48-52].

According to the scenario suggested for the South Caucasus and particularly the Central Great Caucasus which is based on predicted 3.9°C rise of temperature and precipitation decrease by 9-13% during a century [53], it is expected that further climate warming may significantly change the vegetation and consequently the landscapes in the region. Vegetation change models based on these data predict subalpine zone rise by 400-500 m. Subnival/nival zones will also rise by 500-600 m. This will be problematic to species occurring on plane areas at high elevation. These species are adapted to wintering under deep and long-term snow cover. It is doubtful that they can survive on steep slopes of the peaks. At the same time low competitiveness of subnival/nival plants may cause their suppression by expansion of subalpine and alpine plants. The worst predictions are plausible for such rare endemic species of monotypic genera as *Symphyoloma graveolens* and *Pseudovesicaria digitata*. The vegetation of Kazbegi region will resemble to that of dry gorges of the Rocky Ridge of the Great Caucasus, which are situated 10-15 km north of the Kazbegi region, in Russian Federation.

All above mentioned types of alpine vegetation are rather sensitive not only to the climate change, but also to damage by grazing, trampling and other human impacts [17, 18, 20]. Besides, they are characterized by low ability for natural recovery. The total area covered with habitats of sensitive plant communities is rather significant and reaches about 1/3 of the Kazbegi region area (Fig. 3).

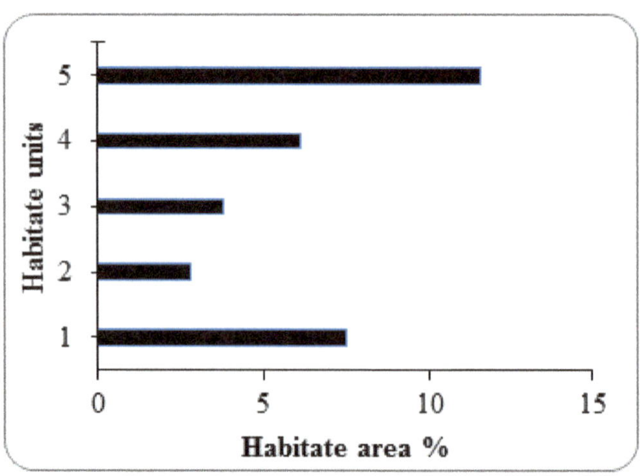

Figure 3. *Proportion of sensitive plant community habitats from the total area of the Kazbegi region (1 – Rhododendretum caucasici and Betuletum litwinowii together; 2 - Dryetum caucasici; 3 - Broadleaf mesophilous meadows; 4 – Tall herbaceous vegetation and snowbed vegetation together; 5 – Subnival/Nival vegetation patches*

Transformation of sensitive alpine vegetation due to of climate global worming will drastically change phytolandscapes in the Central Great Caucasus.

4. Conclusions

In the Kazbegi region (the Central Great Caucasus) on the basis of long-term phytosociological and ecological studies the most sensitive plant communities were distinguished: 1) Alpine treeline ecotone communities: (a) Prostrate evergreen shrubbery dominated by *Rhododendron caucasicum*, (b) dwarf-shrub community dominated by *Dryas caucasica*; (c) Tertiary's relict thermo-hygrophilous subalpine tall herbaceous vegetation dominated by *Heracleum sosnowskyi, Aconitum nasutum, Senecio rhombifolius*, etc.; 2) Subalpine broad-leaved mesophilous meadows dominated by *Anemonastrum fasciculatum, Trollyus ranunculinus, Geranium ruprechtii* and *Betonica macrantha*; 3) Alpine carpet-like meadows ("Alpine carpets") represented by *Campanula biebersteiniana, Veronica gentianoides, Corydalis emanueli, Taraxacum porphyrantum, Sibbaldia semiglabra, Poa alpina*, etc.; 4)Snow-bed vegetation (*Galanthus platyphyllus, Fritillaria latifolia, Primula auriculata*); 5) Subnival/nival vegetation patches consist of few (2-5) species: *Delphinium caucasicum, Cerastium kasbek, Saxifraga flagellaris, S. Sibirica, Alopecurus glacialis, A. Dasyanthus, Nepeta supina, Tripleurospermum subnivale, Pseudovesicaria digitata, Symphyoloma graveolens.*

Further climate warming may significantly change the vegetation and consequently the phytolandscapes in the Central Great Caucasus.

Aknowledgements

The authors would like to thank all the people of Ilia State University and the Institute of Botany who helped in organizing the field work results of which are presented in this paper.

References

[1] Ch. Korner, "Mountain ecosystems in a changing environment", Eco. mont - Journal on Protected Mountain Areas Research, vol. 6, 2014, pp. 71-77.

[2] W. Larcher, K. Kainmuller, and J. Wagner, "Survival types of high mountain plants under extreme temperatures", Flora, 2008, pp. 1-22.

[3] Ch. Korner, "Mountain vegetation under environmental change", in R. Jandl, A. Borsdorf, H. van Miegroet, R. Lackner, R. Psenner, Eds. Alpine Space - Man and Environment. vol. 7: Global Change and Sustainable Development in Mountain Regions. Innsbruck Univ. Press, Innsbruck, 2009, pp. 25-30.

[4] Ch. Korner, "Alpine ecosystems", in S. A. Levin, Ed., Encyclopedia of Biodiversity, Elsevier, 2013, pp. 148-157.

[5] L. Nagy, and G. Grabherr, The Biology of Alpine Habitats, Oxford: Oxford University Press, 2009.

[6] W. Larcher, "Bioclimatic temperatures in the High Alps", in C. Lutz, Ed. Plants in Alpine Regions: Cell physiology of adaptation and survival strategies. Springer, 2012, pp. 21-27.

[7] G.. Grabherr, M. Gottfried, A. Gruber, and H. Pauli, "Patterns and current changes in alpine plant diversity", in F. S. Chapin, Ch. Korner, Eds. Arctic and Alpine Biodiversity: Patterns, Causes and Ecosystem Consequences. Springer, Berlin-Heidelberg-New York, 1995, pp. 167–181.

[8] G.. Grabherr, M. Gottfried, and H. Pauli, "GLORIA: A Global Observation Research Initiative in Alpine environments", Mt. Res. Dev., 20, 2000, pp. 190-191.

[9] Ch. Korner, Alpine Plant Life: Functional Plant Ecology of High Mountain Ecosystems, Springer, Berlin-Heidelberg, 2003, 333 p.

[10] M. Gottfried, H. Pauli, K. Reiter, and G. Grabherr, "Potential effects of climate change on alpine and nival plants in the Alps", in: Ch. Korner, and E. M. Spehn, Eds., Mountain Biodiversity - A Global Assessment, London-New York, Parthenon Publ., 2002.

[11] C. M. van de Ven, S. B. Weiss, and W. G. Ernst, "Plant species distributions under present conditions and forecasted for warmer climates in an arid mountain range", Earth Interact. vol. 11, 2007, pp. 1-33.

[12] D. Scherrer, S. Schmid, and Ch. Korner, "Elevational species shifts in a warmer climate are overestimated when based on weather station data", Inter. J. of Biometeorology, vol. 55, 2010, pp. 645-654.

[13] D. Scherrer, and Ch. Korner, "Infra-red thermometry of alpine landscapes challenges climatic warming projections". Global Change Biology, vol. 16, 2010, pp. 2602-2613.

[14] D. Scherrer, and Ch. Korner, "Topographically controlled thermal-habitat differentiation buffers alpine plant diversity against climate warming". Journal of Biogeography, vol. 38, 2011, pp. 406-416.

[15] W. Larcher, "Klimastress im Gebirge – Adaptationstraining und Selektionsfilter fur Pflanzen", Rheinisch-Westf. Akad. Wiss. Vortr, vol. 291, 1980, pp. 49–78.

[16] W. Larcher, Physiological Plant Ecology: Ecophysiology and Stress Physiology of Functional Groups", Springer, Berlin-Heidelberg, 2003, 513 p.

[17] G.. Nakhutsrishvili, M. Akhalkati, and O. Abdaladze, "Main treats to the mountain biodiversity in Georgia (the Caucasus)", Mountain Forum Biull., IX, vol. 2, 2009, pp. 18-19.

[18] G.. Nakhutsrishvili, The Vegetation of Georgia (South Caucasus), Springer, Berlin-Heidelberg, 2013, 235 p.

[19] L. Maruashvili, Geomorphology of Georgia (Geoporfologya Gruzii), Metsniereba, Tbilisi, 1971, 346 p. (In Russian).

[20] G.. Nakhutsrishvili, O. Abdaladze, and A. Kikodze, Khevi: Kazbegi Region, Tbilisi, 2005, 54 p.

[21] World Reference Base for Soil Resources, "A framework for international classification, correlation and communication", Food & Agriculture Organization of the UN, 2006, 128 p.

[22] G.. Nakhutsrishvili, "High- mountain vegetation of the Caucasus region", in L. Nagy, G. Grabherr, Ch. Korner, and D. B. A. E. Thompson, Eds., Alpine Biodiversity in Europe. Springer, Berlin-Heidelberg, 2003, pp. 93-103.

[23] R. Gagnidze, Vascular Plants of Georgia: A Nomenclatural Checklist, Universal, Tbilisi, 2005, 247 p.

[24] A. L. Kharadze, "On the subnival belt of the Greater Caucasus" (Subnivalnyi poias Bolshogo Kavkaza), Inst of Bot. Acad. of Science GSSR, vol. 25, 1965, pp. 103-104.

[25] M. F. Sakhokia, and E. Khutsishvili, (1975): Synopsis of the Flora of Higher Plants of Khevi", Metsniereba, Tbilisi, 205p.

[26] G.. Nakhutsrishvili, and R. I. Gagnidze, "Die Subnivale und Nivale Hohgebirgsvegetation des Kaukasus", Phytocoenosis, vol. 11, 1999, pp. 173-182.

[27] G.. Schatz, T. Shulkina, and J. Solomon, Red List of Endemic Plants of the Caucasus Region, Saint Louis, MO, USA, Missouri Botanical Garden Press, 2013.

[28] G.. Nakhutsrishvili, O. Abdaladze, and M. Akhalkatsi, „Biotope types of the treeline of the Central Greater Caucasus", in: D. Gafta, and J. Akeroyd, eds., Nature Conservation: Concepts and Practice, Springer, Berlin-Heidelberg, 2006, pp. 211-225.

[29] Identification Guide of Plants of Georgia (Sakartvelos Mtsenareta Sarkvevi), vol. I-II, Metsniereba, Tbilisi, 1964/1969. (in Georgian).

[30] Flora of Georgia (Sakartvelos Flora), v. I-XII, Metsniereba, Tbilisi, 1971-2013 (in Georgian).

[31] J. Braun-Blanquet, Pflanzensoziologie. Grundzuge der Vegetationskunde. 3. Aufl. Springer, Wien, 1964.

[32] E. van der Maarel, "Transformation of cover-abundance values for appropriate numerical treatment – alternatives to the proposals by Podani", J. Veg. Sci. vol. 18, 2007, pp. 767–770.

[33] Ch. Korner, "A re-assessment of high elevation treeline positions and their explanation". Oecologia. vol. 115, 1998, pp. 445-459.

[34] Ch. Korner, Alpine Treelines: Functional Ecology of the Global High Elevation Tree Limits, Springer, Heidelberg, 2012, 220 p.

[35] A. G. Dolukhanov, "Timberline and subalpine belt in Caucasus mountains, USSR". Alpine Arctic Research, vol. 10, 1978, pp. 409-422.

[36] E. S. Elizbarashvili, T. F. Urushadze, M. E. Elizbarashvili, S. E. Elizbarashvili, and M. K. Schaefer, "Temperature regime of some soil types in Georgia", Eurasian Soil Sci. vol. 43, 2010, pp. 427–435.

[37] M. Akhalkatsi, O. Abdaladze, G. Nakhutsrishvili, and W. K. Smith, „Facilitation of seedling microsites by Rhododendron caucasicum extends the Betula litwinowii alpine treeline, Caucasus mountains, Republic of Georgia", Arctic Antarctic Alpine Research, vol. 38, no. 4, 2006, pp. 481-488.

[38] N. M. Hughes, D. M. Johnson, M. Akhalkatsi, and O. Abdaladze, „Characterizing Betula litwinowii seedling microsites at the alpine-treeline ecotone, Central Greater Caucasus mountains, Georgia", Arctic Antarctic Alpine Research, vol. 41 (1), 2009, pp. 112-118.

[39] T. F. Urushadze, (1997): Basic Soils of Georgia (Sakartvelos Dziritadi Niadagebi), Metsniereba, Tbilisi (in Georgian).

[40] S. Reth, M. Reichstein, and E. Falge, "The effect of soil water content, soil temperature, soil pH-value and the root mass on soil CO2 efflux – A modified model", Plant & Soil., vol. 268, 2005, pp. 21-33.

[41] X. F. Wen, G. R. Yu, X. M. Sun, Q. K. Li, Y. F. Liu et al., "Soil moisture effect on the temperature dependence of ecosystem respiration in a subtropical Pinus plantation of Southeastern China". Agric for Meteorol. Vol. 137, 2006, pp. 166-175.

[42] W. Y. Kao, and K. W. Chang, "Soil CO2 efflux from a mountainous forest-grassland ecosystem in central Taiwan". Botanical Studies, vol. 50, 2009, pp. 337-342.

[43] Z. J. Kikvidze, and G. Nakhutsrishvili, "Facilitation in the subnival vegetation patches". J. Veg. Sci., vol. 9, 1998, pp. 222-226.

[44] R. M. Callaway, R. W. Brooker, P. Choler, Z. J. Kikvidze, Ch. J. Lortie et al., "Positive interactions among alpine plants increase with stress". Nature, vol. 417, 2002, pp. 844-848.

[45] H. Pauli, M. Gottfried, S. Dullinger, O. Abdaladze, M. Akhalkatsi et al., "Recent plant diversity changes on Europe's mountain summits". Science, vol. 336, no. 6070, 2012, pp. 353-355.

[46] A. Mariotti, N. Zeng, J. H. Yoon, V. Artale, A. Navara et al., "Mediterranean water cycle changes: transition to drier 21st century conditions in observations and CMIP3 simulation", Environm. Res. Lett. Vol.3, 2008, 044001. (doi:10.1088/1748-9326/3/4/044001).

[47] S. Del Rio, L. Herrero, R. Fraile, A. Penas, "Spatial distribution of recent rainfall trends in Spain (1961- 2006)", Int. J. Climatol., vol. 31, 2011, pp. 656-667.

[48] Kh. Gigauri, M. Akhalkatsi, G. Nakhutsrishvili, O. Abdaladze, "First Signs of weak thermophilisation on the GLORIA-EUROPE Target Region in the Central Caucasus GE-SAK)", (Termopilizaciis Pirveli Sustad Gamokhatuli Nishnebi Centraluri Kavkasionis GLORIA-EUROPE-is Samonitoringo Qselis Sakvlev Mtsvervalebze - GE-SAK), in G. Nakhutsrishvili, ed., Biodiversity of Georgia. Tbilisi, Georg. Nat. Acad. of Sci. Publ, 2011, pp. 19-22 (in Georgian with English summary).

[49] Kh. Gigauri, M. Akhalkatsi, G. Nakhutsrishvili, O. Abdaladze, "Monitoring of vascular plant diversity in a changing climate in the alpine zone of the Central Greater Caucasus", Turk. J. Bot., vol. 37(6), 2013, pp. 1104-1114.

[50] Kh. Gigauri, O. Abdaladze, G. Nakhutsrishvili, M. Akhalkatsi (2014): „Vascular plant diversity and climate change in the alpine zone of the Central Greater Caucasus", International Journal of Ecosystems and Ecology Science. Vol. 4 (4), 2014, pp. 573-589.

[51] B. Erschbamer, M. Mallaun, P. Unterluggauer, O. Abdaladze, M. Akhalkatsi M et al., (2010): "Plant diversity along altitudinal gradients in the Central Alps (South Tyrol, Italy) and in the Central Greater Caucasus (Kazbegi Region, Georgia)", Tuexenia, vol. 30, 2010, pp. 11–29.

[52] B. Erschbamer, M. Mallaun, P. Unterlugauer, G. Nakhutsrishvili G, M. Akhalkatsi M et al., „Plant diversity along altitudinal gradients in the Central Alps and in the Central Caucasus: the project GLORIA in south Tyrol (Italy) and in the Kazbegi region (Georgia)", in F. Pedrotti, ed., Colloques Phytosociologiques, XXIX, Camerino, Italy, 2013, pp. 167-188.

[53] Georgia's Second National Communication Under the UN Framework Convention on Climate Change", Tbilisi, 2009, 230 p.

Changes of Georgian Mountainous Rivers Water Flows, Problems and Recommendations

Tsisana Basilashvili

Institute of Hydrometeorology, Georgian Technical University, Tbilisi, Georgia

Email address:
jarjinio@mail.ru

Abstract: As a result of Global warming the Caucasus glaciers increasingly start melting, which in its turn leads to the higher water flows and disasters. While in drier regions evaporation increases, the water levels in the rivers decrease and the crops dwindle. In the long run, some of the glaciers will disappear, which will have a negative impact on water resources, water supply will lower and the country's economic development will be halted. For increasing water resources, reforestation is a solution as forests play a vital and regulating role for superficial and underground water supplies. Forests enhance the quality of water and increase its supply and discharge. Forests strengthen the soil and prevent erosion, landslides, high water flows, floods and avalanches. Thus, the more forests the less negative consequences. In agricultural lands, forests are needed to protect vegetation and soil from emaciating and drying. In the dry climate regions water reservoirs must be built in the terrains which permit to do so, which will resist to serious water flows, mitigate the consequences of disaster and supply water to farms and population during the dry spells. For protecting river waters from the elements, for their rational utilization and effective and safe exploitation having forecasts of water flows is vital. In East Georgia work to increase rainfall artificially should be restored, which in its turn will protect the region from hail. For obtaining technical water technologies for turning salty sea water into fresh one should be utilized.

Keywords: multi-year dynamics, water resources, forest

1. Introduction

The world is constantly changing. Over the 4,5 billion years, the earth has undergone multiple changes. Humankind met the XXI century with a seriously complicated ecological condition. Now, we are extremely worried about the climate change caused by global warming. Besides its positive feedback, modern technologies also created the self-destruction possibility for live nature [1].

Temperature growth on earth was caused not only by the man-made factor (carbon dioxide emission into the atmosphere), but by activation of the ongoing processes on the sun's surface and depletion of the ozone layer as well (the earth shield absorbing UV radiation from the sun). Climate change caused by such "greenhouse effect" is a really pressing problem for the world and the most important and potential danger to our environment. Ongoing climate fluctuations contribute to strengthening of climate contrast range, imbalance of humidity distribution on earth and the existing created situation in this regard [2].

Fresh water resources are one of the most complex issues nowadays that deserve appropriate attention from our side. Life on earth totally depends on these resources as well as each country's economy and development. Nature protection, rational use and reproduction of its resources is the basis for sustainable development of any country's economy. The climate-related crisis and anomalies caused by global warming and happening on earth (in real space and time) exert huge influence on natural resources, particularly on water and land resources. Against this background, the risk of development of any country is daily increasing, mainly caused by reduction of the volume of water resources, their quality deterioration and frequent natural disasters [3-4].

2. Study Area

Diverse landscape, relief and climate are characteristic to Georgia, starting from the moist subtropical zone of the Black Sea, eastern part and ending with the highland alpine Nival climate on Great Caucasus. Mountains occupy nearly 85% of Georgia's territory (fig. 1). Among other risks, floods/flash flows and water erosion processes are on top of the list in Georgia, as well as the sea coastline wash off, landslides, mudflows and snow/glacier avalanches. In this regard, considering the scale of the development of similar processes and their negative consequences, Georgia is an outstanding region among other highland countries. Floods and flash flows are typical for all rivers in the mountains and lowlands here as they are conditioned by the density of hydrographic network, the specificity of the relief and climatic-meteorological regime. Density of rivers' network is average 0,85 km/km^2. Regarding the distribution of atmospheric precipitations, this coefficient is 1,07 km/km^2 in West due to moist, while it is 0,68km/km^2 in dry East [5].

Danger of floods is especially big in foothills and lowland relief. As a result of massive precipitations, rivers overflow and flood territories, settlements, cities, roads, hundreds of arable get lost and often we face victims as well. Similar processes were described in the ancient historical sources in Georgia [6-7]. The fact should be noted that earlier, the population used to reside mainly in the mountains, while the flooded rivers caused no danger to lowlands. Even more, flooding fertilized and enriched soils that later resulted in crop productivity. Though, in parallel to the population's propagation, people started organizing their living habitats near rivers, thus increasing the risk of danger. The 2/3 of the recorded floods/flash flows falls at West Georgia rivers due to the fact that West Georgia is characterized with prolonged and heavy rains. Besides that, present eustasy – uniform worldwide change in sea level caused by water thermal expansion, imbalance of fresh waters between the land and the ocean in favour of the ocean, seriously threatens Black sea coastline, where the sea level has raised by 20 cm and this is a continuous process with unfavorable forecast. This vivid example is figure 2, that shows the flooding result happened on 1987 31 January in river Rioni lower element near the Black Sea caused by heavy showers and intense melting of snow avalanche [7].

In order to study the impact of the ongoing climate change caused by global warming, research on the change of dynamics of quantities of Hydrometeorological processes shall take a leading position. With this purpose, we discussed analysis of the rows of multi-year observations made in the Hydrometeorological network on the territory of Georgia. In this regard, rather important is the qualitative and the quantitative definition of multi-year dynamics of these elements. This implies assessment of trends in an equation reflecting their rectilinear approximation:

$$T = a N + b, \qquad (1)$$

where T – is a trend – averaged straight line reflecting the tendency of multi-year change of the element at research; N - is an ordinal number of observation data of this element starting from their initial year, for which N = 1, for each following (i) year N = 1 + i; a – is a coefficient of the equation, its mark (+ or -) indicates at direction of change of the element at research: positive (+) mark expresses its growth, i.e. ascending tendency, while negative (-) mark shows decline of the element at research, i.e. descending tendency. a itself - quantitative meaning of the parameter defines intensity of the change of the element at research, i.e. speed; b – is an equation constant.

Table 1 gives a and b parameters of the equation received from approximation of the trends reflecting the multi-year change (60-70 years) until 2010 on meteorological stations operating in West and East Georgia, atmospheric precipitations and air temperature. As the table data show, due to global warming, climate in separate regions of Georgia is getting colder rather than warmer. For example, in some regions of West Georgia, the weather gets colder, while in East Georgia, it gets warmer and air temperature growth is expected.

Table 1. *Parameters (a, b) of atmospheric precipitations (R) and air temperature trends (θ).*

River	Meteo Station	Height	a_R	b_R	a_θ	b_θ
West Georgia						
Bzifi	Gagr. Ridge	1644	1,925	1623	-0,005	6,02
Kodori	Lata	299	0,259	1859	-0,013	12,45
Enguri	Mestia	1441	-1,378	998	-0,004	5,90
Enguri	Khaishi	730	-	-	-0,011	10,89
Rioni	Mamisoni	2854	-	-	0,008	-2,27
Rioni	Lentekhi	760	2,637	1211	0,004	9,39
Rioni	Ambrolauri	544	1,473	1008	-0,010	11,50
Rioni	Kutaisi	114	-1,421	1499	-0,013	0,01
Kvirila	Saqara	148	0,179	1255	0,004	14,02
Supsa	Bakhmaro	1926	-3,974	1630	0,006	4,00
Acharisck.	Khulo	923			0,002	10,40
Acharisck.	Qeda	256	5,517	1541	-0,010	12,96
East Georgia						
Aragvi	Kazbegi	3653	-4,613	1865	-	-
Aragvi	Gudauri	2194	-1,934	1579	0,012	2,13
Aragvi	Barisakho	1325	-2,466	1150	0,001	6,96
Aragvi	Pasanauri	1070	-1,358	928	0,007	7,95
Alazani	Telavi	568	-1,183	831	0,005	12,08
Alazani	Lagodekhi	362	-2,318	1086	0,767	30,10

All the discussed meteo-points in East Georgia show significant reduction of atmospheric precipitations, while their reduction is observed in separate places of West Georgia (Mestia, Kutaisi, Bakhmaro); the rest stations in dicate at increase of precipitations.

In order to study the change in river runoff, we discussed the 50-70 year observation rows of average annual discharge on rivers of Georgia (Q, m^3/sec), also their annual maximal discharge (Q_m, m^3/sec), as they represent the main source of natural disaster developments. Table 2 provides the research results. Based on the analysis, in multi-year dynamics of average annual discharge, decrease tendency is observed in these river basins, where there are no glaciers and eternal snow coverage. There are exceptions in this regard; e.g. river Adjaristskhali, with no Nival zone, water average annual discharge is increased. The reason is that according to Table 1, the highest intensity of growth of atmospheric precipitations (R, mm) is observed in this river basin (m/s Keda), which is depicted as follows:

$$T_R = 5{,}517\ N + 1541, \qquad (2)$$

Descending (decreasing) trend of air temperature (θ°C) is reflected as the following equation:

$$T_\theta = -\ 0{,}01\ N + 12{,}96. \qquad (3)$$

Closeness to the Black Sea caused this fact in the river Ajaristskhali basin, where cold, moist air masses are intensively (compared to an earlier period) intruding.

Dynamics of rivers' maximal discharge sharply outlines the growth tendency, if the glaciers and melted permanent snow waters participate in their enrichment. Growth of maximal discharge in the river Kvirila basin is an exception in this regard, notwithstanding the absence of glaciers here. Pursuant to the data from Table 1, atmospheric precipitations significantly increase in this river basin (m/s Saqara) due to intrusion of Black Sea moist air masses.

From the river basins discussed in the table 2, river Rioni is distinguished by the highest intensity of water discharge growth and feeds (partially) from the glacier and melted permanent snow waters. Counting of observation years on river Rioni (at village Saqochakidze) begins from 1928, for which N = 1. The trend relevant to the dynamics of average annual discharge is given as following equation:

$$T_Q = 1{,}303\ N + 372, \qquad (4)$$

The biggest intensity of the annual discharge drop is observed on river Chorokhi, the trend of the change in the nearby territory of its mouth (near village Erge) can be read as the following equation:

$$TQ = -\ 0{,}458\ N + 290. \qquad (5)$$

Ajaristskhali is characterized by the biggest intensity of river water maximal discharge drop ($a_{Qm} = -2{,}95$), while the biggest intensity of growth ($a_{Qm} = 35{,}5$) is observed at village Saqochakidze, river Rioni. Multi-year dynamics of maximal discharge of these rivers with relevant trends is given in the fig. 3.

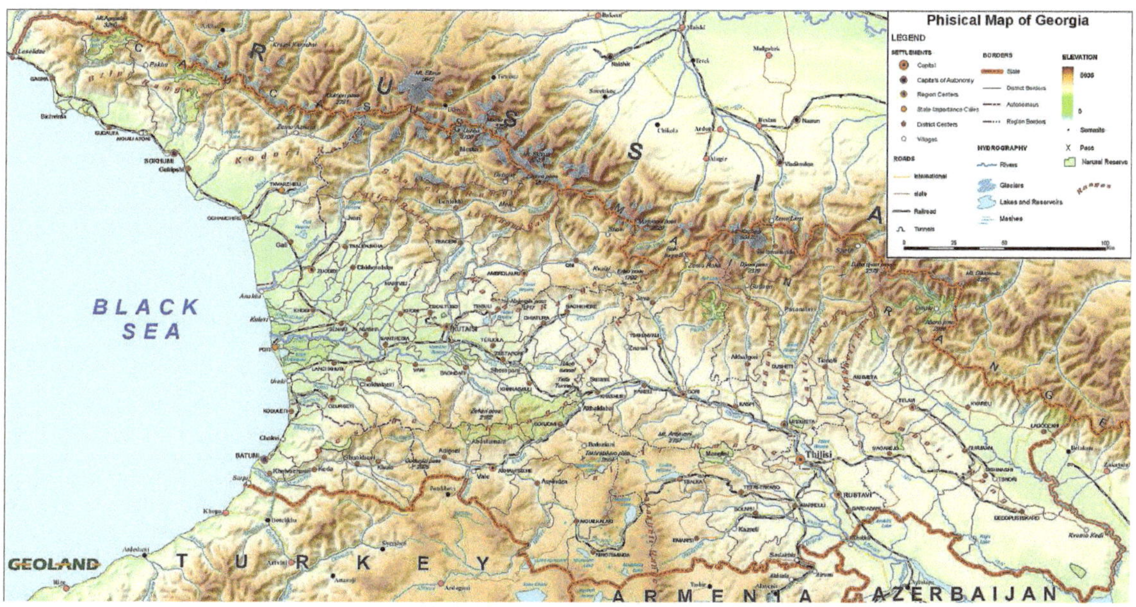

Figure 1. *Map of Georgia.*

Figure 2. *Catastrophic flooding on River Rioni on 31 of January, 1987 at vil. Chaladidi.*

Table 2. *Trend parameters of average (Q) and maximal (Qm) water discharge of Georgia's rivers.*

River	Place	Area km²	Height m.	Average		Maximal	
				a_Q	b_Q	a_{Qm}	b_{Qm}
Kodori	Lata	1480	1920	0,357	81,2	5,92	310
Enguri	Khaishi	2780	2320	0,960	89,4	8,50	319
Rioni	Saqochakidze	13300	2620	1,303	372	35,5	828
Kvirila	Zestafoni	2490	960	-0,028	61,6	0,83	504
Chorokhi	Erge	22000	-	-0,458	290	-0,80	1369
Acharisck.	Qeda	1360	1470	0,118	43,0	-2,95	391
Mtkvari	Tbilisi	21100	-	-	-	-0,76	1181
D. Liakhvi	Kekhvi	924	2100	0,114	24,4	0,85	115
P. Liakhvi	Vanati	422	1940	-0,014	10,3	-0,43	62,1
Ksani	Korinta	461	1830	-0,070	11,2	-0,50	72,2
Alazani	Birkiani	282	2200	-0,002	14,0	-1,08	98,4
Alazani	Shaqriani	2190	1210	0,050	43,8	-1,18	348
Aragvi	Jinvali	1900	1890	0,028	44,1	-	-
T. Aragvi	Mleta	107	2620	0,008	5,08	-	-
T. Aragvi	Pasanauri	335	2189	0,003	12,0	0,56	45,6
Sh. Aragvi	Shesartavi	240	2020	-0,008	7,84	-	-
F. Aragvi	Magaroskari	945	1960	0,052	17,6	-	--
Khadiskhevi	Tskere	18,8	2500	0,005	0,87	-	-
Iori	Lelovani	494	1640	0,934	12,7	-	-

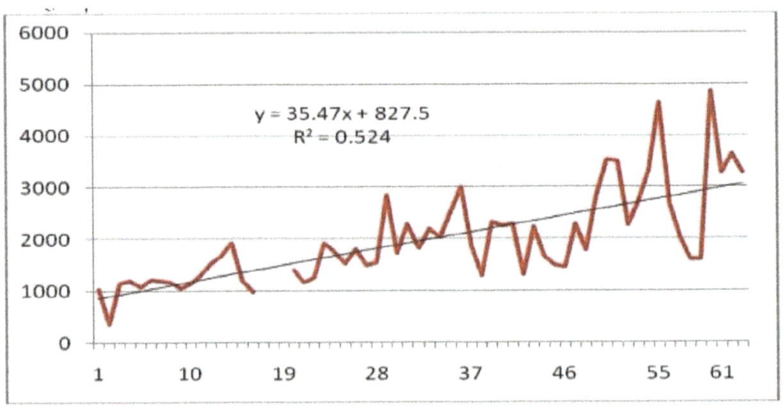

Q_m, m³/s A

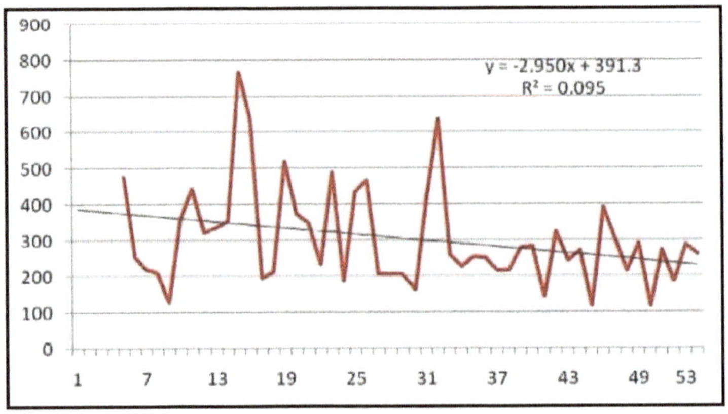

Q_m, m³/s B

Figure 3. *Dynamics of maximal water discharge (Q_m, m³/s) on river Rioni (A) and Ajaristskhali (B).*

3. Results and Discussion

In the XXI century, evaporation will also rise in parallel to air temperature growth due to global warming [8]. Accordingly, liquid precipitations will increase and snow coverage area decrease, glacier melting will become intensive, underground water levels will lower as well as the runoff in river water scarcity period, while runoff in water abundancy period will go up and floods start to occur frequently. Expected climate change will result in droughts, floods, water deficit during on one hand and increased flooding-inflicted damage, on the other.

Humankind has been struggling against natural disasters for many years already, though has not found reliable tools yet and the floods happening everywhere around (including Georgia) with its accompanying tragic results confirm that. Science and technology are unable to deal with the causing factors of ongoing global climate processes and is forced to be limited only with searching for different tools to eliminate the consequences. In order to reduce the negative impact of floods and flash flows, we developed specific recommendations based on the analysis.

Each year, as soon as flooding ends, to ensure smooth ending of the next year flooding period, riverbed conductivity and admissible depths of river ravines shall be maintained. For that shall be conducted: cleansing of riverbeds, protecting the banks with dams, dikes and bulwarks; shall also be constructed roundabout (release) spillway channels. Besides, slopes of river ravines shall be terraced and different agromelioration measures conducted: across ploughing, selection of plant varieties, etc.

Special attention shall be paid to renewal and expansion of forest on the territories of river basins, as it protects and strengthens the soil on mountain slopes, hinders the process of erosion, mudflows, landslide and snow avalanche formation. So, the more wider and expanded the forest and trees are, the less originate the negative processes around.

Forest is one of the most important factors for underground and surface water regulation. It improves quality of underground waters, increases the points (sources) of their release and their discharge. Most importantly, forest stabilizes river water regime and reduces their maximal peak. Forest is the main source of oxygen and is of crucial importance for air purification and human health improvement. It provides favorable habitat for fauna and avifauna propagation. Useful plant varieties shall be selected and selective cutting shall be organized, while windbreak lines shall be placed in agricultural arable to protect plants/soil from losing moist, also hamper the process of soil drying out and its depletion.

Besides that, the reservoir is one of the most effective tool to regulate river waters. It helps to retain huge water masses during flooding, while in water scarcity, it is the main water resource that could be used for water supply of population and farming, to get electricity, develop fishery and most importantly, for hydromelioration, especially in dry climates [9]. In the conditions of favorable and geologically sustainable relief of rivers' ravines, construction of small size reservoirs would be very useful. They will be the most reliable and effective constructions against floods and flash flows. River ravines or dried lake castles could be used for that.

When constructing the reservoir, based on the accurate hydrological calculations, controller's time-tables shall also be developed in a way that will make possible to empty reservoirs in case of emergency, before the flood occurs, i.e. step by step (before the peak) instead of concurrently that could intensify flash flows below. Timely unloads of reservoir and retainment of huge masses of flood will avert damage and victims. This is the reason why the network of small reservoirs shall be expanded more in future. Their construction truly needs big funds, though sometimes flooding of a small mountainous river can cause more damage than those funds spent on the reservoir construction. In the conditions of its proper exploitation, the reservoir will protect the environment and population from natural disasters in the future many times. It will also improve the economy of the country.

The most acceptable and easy method to protect riverside settlements from water-related disasters, is to determine the borders of floods/flash flows and announce this territory as a dangerous zone, where construction of residential and other premises will be prohibited as well as all economical and domestic activities.

4. Conclusions

With the aim to maintain ecological sustainability of the environment, ensure population security, avoid flooding of water objects (hydroelectric stations, water supply and irrigation systems), railway and motor roads, agricultural arable, also protect them from other dangers and reduce the damage, most important is to forecast floods and their maximal discharge [10]. Forecasts are also required for planning the rational use of water resources and effective/safe exploitations of water reservoirs.

Development of only forecasts cannot stop natural disasters, it is true, but can play a big role in the reduction of their negative impact. Long-term forecast (1-3 months) of floods' maximal discharge enables to timely carry out all the preventive measures in case when the expected water is at its peak, protect all the objects at risk and avoid damage/victims. So, forecast of maximal river discharge, even in terms of consultations at the initial stage, shall be rapidly delivered to the relevant organizations to enable them to timely warn the population and if required, conduct evacuation and ensure security of material values. In parallel to this, reservoirs shall be gradually unloaded to receive high water flow.

Population living in highland areas is the initial provider of information re dangerous natural disasters occurring in river ravines. The first signs appear in the form of a terrible buz, or a whirlwind of dust, river water levels can termination of river-flow in case of ravine blockage, etc. Upon the receipt of similar notifications, an alarm shall be rung and all the relevant organizations mobilized in order to provide with all the required preventive measures.

References

[1] L. Douglas and A.Johnson, "Laurence Levis Land Degradation: Creation and Destruction" 2007, 303 p.

[2] G. Metrevely, N. Tsivtsivadze, M. Goginava, R. Diasamidze and M. Beridze" New approaches of the World Ocean level series analysis" Advanced Problems in Geophysics, Tbilisi, 2014, pp. 226–229.

[3] O.P. Sosnovskaya, "Evaluation of Fresh Water Ecosystems State", Geography: Science and Education Problems, St-Petersburg, 2011, pp. 140-142.

[4] Wei-Hua Guo, Bo Li, Xin-Shi Zhang and Ren-Qing Wang, "Water Balance in Soil-Plant-Atmosphere-Continum (Spac) and Garagana Intermedia", Environment and Ecology in the Mediterranean Region, Cambridge Scholars Publishing, UK, 2012, pp. 261-270.

[5] "Water Resources of TransCaucasia", edited by G. Svanidze and V. Tsomaia, Hydrometeoizdat, Leningrad, 1988, 264 p.

[6] G. Svanidze and G. Khmaladze, "Freshets and Floods", Hazardous Hydrometeorological Phenomena in Caucasus, Hydrometeoizdat, Leningrad, 1983, pp. 191-210.

[7] Ts. Basilashvili, M. Salukvadze, V. Tsomaia and G. Kherkheulidze, "Catastrophic of Flooding, Mudflow and Avalanches in Georgia and their Safety", Publishing House "Technical University", Tbilisi, 2012, 244 p.

[8] Georgia's Second National Communication to the UNFCCC, Tbilisi, 2009, 230 p.

[9] Ts. Basilashvili, J. Tabatadze and M. Janelidze, "River Water Regulation under Modern Climate Change Conditions", Environment and Ecology in the Mediterranean Region, Cambridge Scholars Publishing, UK, 2012, pp. 347-352.

[10] Ts. Basilashvili, "Multifactorial Statistical Methodology for Forecasting Floods-High Water Flows", Publishing House "Technical University", Tbilisi, 2013, 180 p.

4

Natural-Geographical Zoning and Geo-Ecological Problems of Georgia's Black Sea Coast

Melor Alpenidze[1], Zurab Seperteladze[2], Eter Davitaia[2], George Gaprindashvili[3], Tamar Aleksidze[2]

[1]Sokhumi State University, Department of Geography, Tbilisi, Georgia
[2]Ivane Javakhishvili Tbilisi State University, Department of Geography,Tbilisi, Georgia
[3]National Environmental Agency, Department of Geology, Tbilisi, Georgia

Email address:
Melor07@mail.ru (M. Alpenidze), zura_sep@mail.ru (Z. Seperteladze), gaprindashvili.george@gmail.com (G. Gaprindashvili)

Abstract: The region's natural geographic zones meant detection and mapping of Geo-complexes of Identified taxonomic rank. Geo-complexes - Individual territorial units differ from one another by natural conditions, landscapes hypsometry location and their spatial distribution, geo-ecological problems, shape and scale of anthropogenic impacts and extent of the character and quality of agriculture. Interest of Geo-ecological problem of the Region is caused by Black Sea (within Georgia) coast nature use process, utilization of the dynamic equilibrium potential faults. Negative changes in the functioning of natural systems and the disorder lead to unwanted changes. Possible solution to the problem of ecological science and environmental components geo-complexes scientific studies based on geographical and ecological aspects of the current state estimates, analysis and forecasts. Research object: Black Sea coastal zone – the eve of the land and sea include three geomorphological elements: coastline, beach and underwater slope. It is obvious that the acute deficit of sediment to the main geo-ecological problem even worse by strong anthropogenic pressure, which in turn is connected with the region's natural resource potential (Resort-recreational, Balneology, mineral). Studies found that the magnitude of the loads exceeded the scale of anthropogenic and natural coastline of the permitted limits, contradicts the condition of the natural environment, focused on the conflict situation and the creation of an entirely new type of anthropogenic landscapes (aquatic, technogenic, recreational) towards development. Sea landscapes of natural conditions and ecological problems in the development of mixed-mode based on the detailed study was made possibility of natural-geographical zoning.

Keywords: Geo-Ecology, Zoning, Landscape, Nature Use

1. Introduction

Within the natural-geographical zoning of research coastal region, authors cover revealing, mapping and complex characterization of particular features of some specific taxonomic geographical complexes. Geographical complexes revealed as a result of the research, are individual territorial units and differ from each other as with the features characterized for natural conditions, and with hypsometrical location of landscapes and spatial disposition, as well as with geo-ecological problems, forms and scales of anthropogenic effects, character and quality of economic coverage of the regions.

Interest in geo-ecological problems of the region stand on rational (reasonable) coverage of Black Sea coast and solving the practical tasks of nature management. Namely, with the expecting collapses of natural dynamic balance connected with the long-term economic coverage of coast. Negative remakes of similar natural systems simple prove the collapse of their functioning and improper development [1]. Researches of separate components of coast and scientific results, including evaluations of modern day status of geographic - ecological aspects, analysis and prognosis, allow us to solve this geographical problem.

In the course of intensive coverage of environment of Black sea coast of Georgia, coastal system (human being - natural - economy) influenced rough negative changes. It incurred collapse of dynamic balance in nature, destroying of its functioning and multiple negative results: dynamic collapse of the "seashore" system connected with the shortage of deposits; falling of coastal landscapes under rough anthropogenic press by intensive coverage of coastal natural potential. Aim of this study is to provide natural-geographical zoning of coast and to research specific geo-ecological problems - collapse of natural system incurred by anthropogenic Influence of the coast and degradation of shores. By fulfilling the given aim, major task set by the authors about the coastal environment, is to be solved.

Major part of the Georgian population, including resort and economic infrastructure locates within 5-10 km zone of the Black Sea coast, being set as the major priority region [2]. Thus, the task set for rational coverage of natural resources, aims to elaborate measures of keeping modern condition, stability of the coast, including its reasonable coverage.

Modern coast is formed based on some factors, including tectonic movement [3] of crust and changes incurred by sea wave storming [4], changing of sea shore line and movement of alluvial accumulation, melting of glaciers and increasing of energetic level of sea as a result of global climate changes, etc. Thereby, long term and chaotic manner of economic coverage within the Georgian coast roughly weakened geo-ecologic stability (abrasion, slipping, erosion) of the coast.

Thus, intensive washing of the beaches and degradation of beach, turned into irreversible process [5], long term and chaotic manner of economic coverage of the coast roughly weakened geo-ecologic stability of the coast. Thereby, for developing social - economic growth of Georgia and sustainable development of the region, increasing of marine goods turnover is very important, meaning building of ports for deep displacement cargo ships. If this is accomplished, coast will be largely loaded (widening of port infrastructure, deepening of underwater coast, termination of deposit flows along the shore) and corresponding negative geo-ecological changes, which falls under the competence of coastal researches, including us [6]. Geo-ecological problems of the coast are neatly connected with the surface and ground water regime of the region, including global changes of current climate. For softening influence of possible climate changes on water resources, authors think it appropriate to provide refilling, keeping and regulatory measures of water resources.

2. Study Area & Methods

Black Sea coast of Georgia includes coast from the River Psou and the River Chorokhi outfalls (315 km). Considering local physical - geographical aspects, coast was divided into four parts: 1. Northern part (Delta of the River Psou - Eastern periphery of Eshera massif - 102 km); 2. North-west part (outfalls of the River Gumista and Inguri - 103 km); 3. West (Outfalls of the River Inguri and Natanebi - 57 km); 4. South-east (Outfalls of the River Natanebi and Chorokhi - 53 km). Coastal line of the first one is directed from north-west to south east, generally and is characterized with significant vertical separations of underwater coastal slope. There is a slim shelf line with lots of deep and steep underwater canyons (Bichvinta) [7], where pretty wide [8] shallow line is developed. Low and hilly line of the sea shore (Gantiadi - Gagra) create river and sea terraces, where in between branches of rocky hills create high (up to 100 m) cliffs (South tip of Arabika massive, Hilly valleys of Miusera, Akhali Atoni and Eshera). Coastal plain and vast deltas of the rivers are occupied by modern settlements (Gantiadi, Gagra, Bichvinta, Gudauta, Akhali Atoni).

On the underwater coastal slope of the second part, there are intruding outflows of deep underwater canyons (Sokhumi, Kelasura, Kodori, Inguri). Often swallow part of the beach. Low underwater coastal slope is characterized with wide shelf, though pretty deep intrusion of the underwater canyons (Inguri, Poti) is not rare. Sea shore is mainly low (Rioni Delta) and bogged up, separated by sea dunes. Relief shows sea generation [9] signs. Modern form of connate lakes (Paliastomi, Imnati, Bebesiri, Maracha) refers to the existence of the old lagoons.

Underwater slope of the far south - east coast is distinguished with many deep canyons (Batumi, Chorokhi), incurring signs of modern dynamics of the coast (swallowing of deposits) [10]. Coastal line mainly is low (Chorokhi Delta) and hilly plain, though in the areas of Tsikhisdziri, Green Cape and Sarpi high hills form sharply expressed steep cliffs near the shore line.

For researching geo-ecological signs of Black Sea coasts and region, as well as natural - geographical zoning, we have used the following research methods:

2.1. Space and Time Method

Provides system comprehension of micro and nano-forms of the sea coast, understanding of their disposition and spatial changes of numeric-qualitative indexes. For revealing artificial changing factors of coastal parameters (alluvium, beach, sea shore, shore flow) we have used the results of our observations [11, 12], supporting research of numeric features and time changes (separating and reforming of whole), regular changing of quality signs (collapsing of balance, lower washings), comprehension of development tendency of morpho-dynamic signs, cause and effect of occurrences and their interconnection.

2.2. Deductive Methods

Stands on generalization of the occurrences based on the information (changes of coasts on the shores of Sochi, Crimea, Japan, USA) about general morpho-dynamic regularity of coasts [13,14];

2.3. Multifactorial (Complex) Method

This method allowed to reveal geo-ecological problems of economic coverage, forms and scales of anthropogenic influence of separate components and natural - territorial complexes of the coast. Presently, coast is transformed and partitioned, having own signs and features: 1. Feeding source; 2. Granulometry of alluvium and beach; 3. Sea shore and shore line; 4. Underwater coastal slope; 5. Sea regimen; 6. Anthropogenic factor. As a result of common signs of these objects, we receive the following: analog - objects - 2 from 6 with common sign; one type objects - 4 from 6 with common sign; similar objects 4 from 6 with common sign;

2.4. Method of Merged Analysis and Zoning of Components

Is used for natural - geographical zoning and it is based on landscape - genetic principle of the object research and includes the following: research of the coast components (beach, sea edge, underwater slope, sea area, coast), factors (waves, alluvium, relief), processes (litho and morpho-dynamics) and ecologic condition; researching of anthropogenic loads (construction of power stations and ports, withdrawing of inert materials); revealing of geo-ecologic consequences (beach washing, abrasion of shores, lowering of attraction); marking of taxonomic geo-complexes and geo-ecological evaluation.

2.5. Geo-Informative Method

This method allows to acquire and process data (geo-data) about the objects placed in the space, to determine occurrence and development tendency of geo-ecological problems, to process graphical data and to elaborate geo-ecologic and environment programs;

2.6. Modeling and Balance Methods

Allowed us to provide graphic expression of natural processes and cartographic modeling, when it is possible to record modern form and expecting changes of coastal fragments via comparison of the initial and repeated planning. Balance method stands on verifying of correlation of deposit inflow-outflow and is used for defining the dynamics of beaches.

3. Results & Discussions

Considering of multifactorial nature of coastal processes, research was carried out in two stages. First stage covered the research of general signs and development aspects of coastal evolution. It was proved that in earlier stage of the coast, according to the partition of the shore line and nature of beach feeding, only four dynamic system were marked out (1. Mzimta - Psou - Miusera; 2. Miusera - Kodori; 3. Kodori - Rioni; 4. Rioni - Chorokhi).

Based on the petrographic analysis of the coastal constructing material and feeding sources, it was turned out that research coast line [24, 25] has been partitioned in two nine independent dynamic system (1. Psou - Bzyb, 2. Bzyb - Bichvinta; 3. Bichvinta - Hiphsta; 4. Hiphsta - Gumista; 5. Gumista - Besleti; 6. Besleti - Kodori; 7. Kodori - Inguri; 8. Inguri - Natanebi; 9. Natanebi - Sarpi) It has become clear that differentiation of dynamic systems relevant to the earlier stage is conditioned by natural occurrences of coastal evolution (growing of deltas, approaching of capes to the shelf edge, formation of gulfs).

Detailed research of coastal processes was carried out on key research spaces (coasts) on the second stage, which was selected as follows: 1. Marking of most typical and characterizing fragment (Coast of Gumista delta) and generalization of research results (Morpho-dynamic and geo-ecological signs) on similar objects (Deltas of Bzyb, Kodori and Chorokhi) via extrapolation; 2. Marking of sharply significant fragment (Middle Kolchida or Kolcheti) and spreading of the research results to the entire perimeter of low coast of Kolchida. 3. Marking of less atypical key space (Gudauta bank) and using of research results for theoretical and practical purposes; 4. Analyzing and generalizing of geo-ecological results (lower washings) on key parts of damaged coasts via anthropogenic mode.

First key coast falls between the outfalls of the River Gumista and the River Kelasura (14 km) and includes the delta of the River Gumista, perimeter of Sokhumi Cape and Gulf, seashore accumulative plain.

For the purposes of projecting commercial port and urban infrastructure of Sokhumi, searching and analyzing of wells in coastal line and bottom of water area; comparison of planning materials of the coast (1896, 1905, 1932, 1947, 1956, 1980, 1990 years) and analysis of results; observations on underwater slope of Sokhumi cape and the canyons of the River Gumista; marking of the coastal development phases; studying and analyzing of micro relief of Sokhumi (Sea generation embankments); analysis of the effect of "external blocking" in Sokhumi Gulf; elaborating measures for regulating geo - ecological condition.

Spreading of the turf (12, 13, 20) between the sea stone and sand layers by the wells of bottom and shore of the Sokhumi Gulf coast is proved. It also made clear that, Gulf area was occupied by vast lagoon in Holocene, which was separated from the Sea with large embankment. Research results of detailed plans of sea generation embankments of Gumista Delta (1932, 1947, 1956 years) showed the growth of accumulative cape of Sokhumi (Fig. 1).

Approaching of distal part of the Cape to the shelf edge, should have incurred movement of sea shore materials towards the steep slope inclination of the shelf.

We have to look for the facts of falling of deposit in deeper layers and limitation of cape growth, in front of the cape - underwater coastal slope: underwater observations and planning (Fig. 2) showed [14] us sharp critical inclination of the slope (20-250). Potholes incurred by movement of pebble-stone deposit and sludge suspension are seen on the underwater slope, made of sludge filling and beach material, with small "crowns".

Space of Gumista Delta is about 18-20 km^2. Comparably wide line of Delta falls within accumulative cape of Sokhumi. On the stage of coastal formation, drying of the lagoon formed in delta isolated from water area by embankment, helped to form separate marchland fragments and connate lakes [13]. Turf layers accrued in bogged lines of lows formed between the small generations. Presently, major part is drained and modern city was constructed instead. Small nameless connate lake and marchland fragments can be seen in proximal part of the Cape.

Unlike similar perimeters, on stabile coast of Sokhumi Gulf, really significant changes were revealed in 40s of the last century. In 1942, military ship (Tanker "EMBA") and reinforcement dock were cast ashore on shallow slope of Sokhumi Gulf. This fact, as it was theoretically expected caused external blocking effect [12] of the shore. As a result of strong diffraction of waves, ships caused active deposition of beach materials in front for the ships and for 50 years, in "wave shadow" beach width was increased by 155 m) and was spread over 43 thousand m2 and deposited pebble stone reached 542 thousand m3.

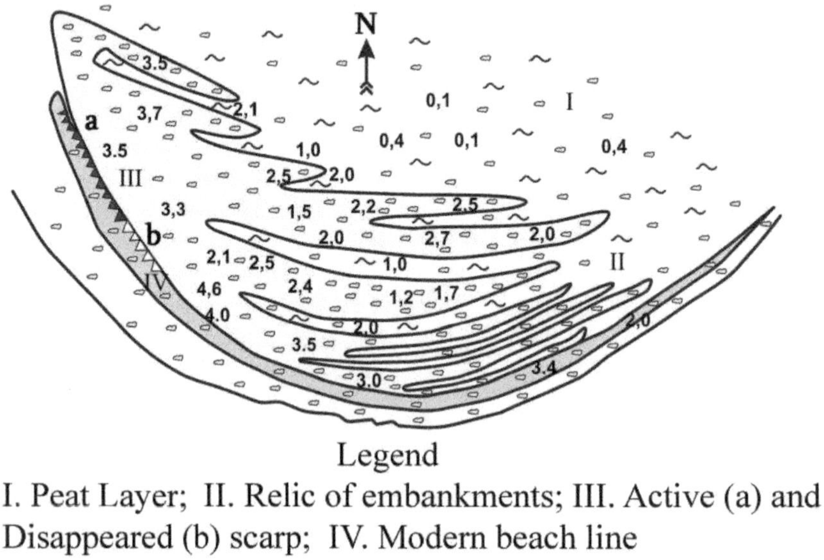

Fig. 1. Sukhumi Cape growth tendency.

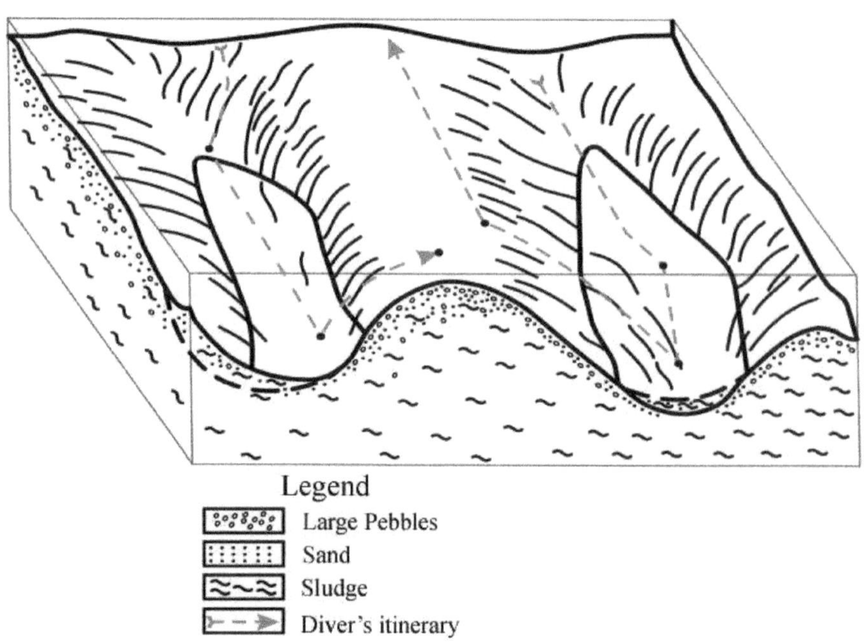

Fig. 2. Block diagram of Sukhumi cape underwater slope.

Growth of beach around ship accident area, incurred deposit shortage on the shore and active collapse of seashore fragment of the city (145 thousand m^2). Changing of shore line incurred formation of negative geo-ecological condition - intensive abrasion of surrounding shore line [12]. Protecting of city boulevard line and constructing of 2 km coast-protecting wall costed 3,8 million USD.

Pretty wide and stabile beach located at Sokhumi Gulf protected the shore line very well. Though, slow decreasing of sea edge (0,3 meter per year) was seen in historical period as well. According to archeological remains, decreasing of shore line in the area of Sokhumi Gulf (II-XIX cc) was about 520-550 m [15]. 40 m bogged line of antique fortress (Dioskuria) was actively collapsing a century ago and finally it was completely swallowed up by the Sea. Similar occurrences are recorded in other fragments of research shore as well.

Thus, common aspects of the research coast - deltas of the

Rivers Bzyb, Kodori and Chorokhi, are characterized with genetic, morphologic and dynamics signs and with spreading of similar occurrences, showing tendency of similar natural development.

Second key coast includes Middle Kolkheti sea area and mainly outflow of the River Inguri and perimeter of its surrounding water area. It is between Ganmukhuri and Kulevi, occupying 20-22 km shore. Shore line is pretty vast plain area, which is quite transformed in terms of anthropogenic loads, causing and forming new type aquatic, induced and residential anthropogenic landscapes, having passed natural, natural - anthropogenic and other modification stages. Kolchida sea line and surrounding fragments are mugged (Phichori - Kvishona, Ganmukhuri and Nakargali, Tikori - Churia) and are separated by sea generation embankments. Natural conditions and limited resources of Kolchida Region, in general do not give perspective of sustainable development. Its major part is occupied by marchlands, only small part of soil (25-27%) is valid for cultivation. In terms of nature management, it is decreased by 18-20 %, annual growth of drained (secondary marching) soils are low, as well as perspective of recreational resources. Based on the low convenience of weather conditions (strong winds, high level of moisture), Kolchida sea area is less attractive for resort-tourist opportunities. More attractive is far peripheral parts of Kolchida coast (Kobuleti, Ochamchire).

Coast line is characterized with large scales of anthropogenic influence. Their transferring into natural regimen is sufficiently difficult, economically ineffective, ecologically dangerous matter of activity. Though, scientific research of perspective field and sustainable development of the region responses to time requirements. Thus, the most important perspective for sustainable development is deemed reasonable planning and realization of port infrastructure in coast line. After occupation of the Black Sea area of Georgia (64%), perspective area of marine transport infrastructure and railway of the country was sharply decreased. Ports of Poti and Batumi cannot satisfy increased goods turnover. Thus, constructions of new ports are being discussed. Long-term practice of port constructions in other parts of the world (including seas of Georgia) (Gagra, Ochamchire, Kulevi, Poti, Batumi) showed that exploitation of capital buildings within labile, dynamic and vulnerable coastal zone often gives negative result. Major reasons include geologic, morpho-dynamic and landscape - ecologic aspects: 1. Possibility to pan the port with siltstone and fine sand; 2. Weak partitioning of coastal line and surrounding land; 3. Formation of shore with alluvial material and spreading of deposit flows along the shore; 4. Spatial limitation of port infrastructure by marchland fragments.

Notwithstanding the above said, relief of underwater slope of the research coast often is characterized with difficult vertical partitions and deep underwater canyons are intruded in sea shelf and coastal zone. This fact provides successful conditions for port constructing. Comparably successful conditions are in the area of old outfall of the River Inguri. Here, beginnings of the underwater canyons are in 150-200 m distance from coast

Line and create deep natural entrance, defining low costs of the port construction and economic effect of exploitation (passive mobility). Port infrastructure shall have positive geo-ecologic aspects as well. It is known that, as a result of blocking the River Inguri water flow and solid eduction (277,5 thousand m^3 per year) by the Hydro Power Station, decreased the volume of beach-formation fraction (165,0 thousand m3/ per year) in 4,5 - 5,0 times [16] and presently it is 29-30 thousand m^3/ per year. Sustainability of the coast was more or less kept by the river deposit and flow deposit along the shore (20 thousand m^3 per year). Negative geo-ecological consequences of sharp shortage of deposit in the area of the River Enguri outflow have no positive perspective in the future: decreasing of the coastal line of Gagidi, Ganmukhuri, Anaklia, Tikori is between 1-3 - 4-5 [3].

Thereby, it is known that the largest part of the deposit was swallowed by the underwater canyon directly located at the outflow of the River Inguri [17]. For avoiding this negative consequence [18], project covers stabilization of the opposite part of the head of Inguri canyon, constructing of long, deep mole (10m) near canyon head, blocking of deposit flow path. Task aims both coast-protection and construction of port for deep displacement ships. Head off the deep underwater canyon located at the outflow of the River Inguri can host deep displacement ships. On the lower part of the dewatered canal of the River Inguri, it is planned to create port basin. If this artificial basin is reasonably managed, it will have great perspective: realization of 400 mln m^3 pebble-stone in coast protection, received from processing of bottom material; arranging of sport, recreation, tourist infrastructure. Based on the historical practice of nature and culture comprehension, following complex research of the region, authors presented number of measures for elaborating projects aiming to manage natural conditions and resources reasonably:

1. Because of close location of coastal line and mineral water resources (Menji, Tsaishi), solving of conditions (distance location of natural resources and historical heritage) hindering recreation nature management of the region: a) to use Zugdidi - Mestia highway and building of airport in Mestia; b) to construct Ingiri (Zugdidi surroundings) airport and Anaklia port;

2. To create transport communication and relevant infrastructure: a) Ganmukhuri - Batumi highway along the sea; b) to connect Ganmukhuri - Anaklia - Tikori - Churia; c) to improve Svaneti highway; d) to arrange ropeway "Mountain - Sea - Seashore";

3. While elaborating resort - recreational, sanitary - hygienic projects of Anaklia coast, natural factors should be considered at first place, namely: a) considering of modern tectonics in transport network, storehouse construction; b) planning of beaches, recreational parks, squares and prognosis of sea level rising tendency; d) arranging of sea complexes, studying of biotic components incurred as a result of global weather changes, revealing of negative signs of ecosystem, studying of regulation of phytoplankton and

water ecosystem management; e) regulating of positive (comfort condition) and negative (temperature, moisture) consequences (changing of water balance, plants, secondary bogging) of marchland draining.

Third key coast (Gudauta Bank perimeter) is less typical key space in research region, thus research aims to accomplish its results as theoretically, as well as in practice. Key space is sharply atypical fragment. Studying of the bank on the second stage of the coast research [19] was conditioned for its special condition. Slope of the bank is constructed with meiotic conglomerate. It is the paleo-delta of the River Bzyb, proved by the reef of the underwater slope. Underwater (diving) observations of the Gudauta Bank bottom [19] showed that, it is conglomerate plate, slightly sloped (0,01- 0,001), covered by break-stone and grey sand. In 200 m, there are medium pebble-stones and boulder between the large stones. Often ruins host water plants (Cystosiera barbata) and are covered by thick serpulidae, crust or moss. In between there are left folds of Ostrea edulis, which do not inhabit here anymore, because of predatory nature of Rapana (Rapana venosa). Carbonate plates are inhabited with (Petricola lithophaga, Barnea candida, Pholas dactylus).

Studying of bank showed us that, ruins of crushed stones are spread through the bottom. On the shore, 5-7 m beach materials contain pebble-stones bored by boring mollusks, including them in the holes of limestone plates (35-40 cm) mollusk valves are proved. In 13-15 meters, water plants - Zostera, Cystoseira, Phyllophora etc can be discovered. After each 3-4 magnitude storm, you can sea dead water plant deposits on the shore. It is clear that water plants and break-stone can be seen as on the bottom of the bank, as well as on the beach surface. Authors raise quite a natural question, whether if water plants play some role in dynamics of pebble-stone. Observation of storm waves showed that by wave force, floating ability of water plants are increasing. Observation proved that water plants are better attached with their roots to the break-stone, which are located in 5 meters deeper. For high floating abilities, stones attached to their roots are moving towards the shore. On small depth (3-4m), mechanical action of the waves is increasing and boulder removes from the roots, though storm waves cast them ashore.

Thus, mobility of pebble stone deposits from bottom to the shore and feeding of the beach is proved. As the occurrence is very important, volume of the feeding source was determined via balance method. For these

Purposes, within the coastal zone of key space (6 km) inflows and outflows of deposits were calculated. Annual inflow: 1. Alluvium of the rivers (6000 m^3) and 2. Bottom material (4000 m^3). Outflow: 1. Accumulation on the shore (5000 m^3) and 2. Grinding of pebble-stones (5000 m^3).

Balance report shows: on the coast of the kea space, keeping of balance between the volumes of natural inflow of the beach materials from the bottom, its inflow and grinding.

Inflowing of the deposit from underwater slope to the beach is pretty familiar process. It is recorded on the shores of Tarkhankut, Absheron and Estonia. Thus, outflow of the deposit from the bottom is proved process.

It is important for both theoretical and practical point of view: existence system interaction of underwater and surface beach materials, turned wave power in favor of human being (artificial beaches). This allowed to extract deposit from small depth (4-5m) (coast protection of Kobuleti, Poti, Sokhumi).

Fourth key coastal parts includes shores damaged anthropogenically, showing negative geo-ecological consequences (lower washings, abrasion of major shores). On the key parts of the coasts under economic pressure, negative geo-ecological consequences are conditioned by the construction processes of port infrastructure and improper exploitation, unreasonable coast-protection "concrete policy", intensive extracting of beach materials from the sea shores. Negative consequences of economic activities on the shores are spread generally. World practice showed that [21, 24] geo-ecological aspects developed as a consequence of strong anthropogenic pressure affecting the shores of the seas and oceans (Japan, USA, Russia) shows undesirable signs. For clarification, we shall discuss some of them:

1. Construction of port in Old Gagra, within 6-7 km perimeter, slowing of "lower washing" was followed by dyke construction and formation of new washing parts; for the purposes of railway contour, to the west of Gagra Gulf (surroundings of the village Grebeshok), deposit flow along the shore was cut by dyke series (32 dykes and 3 wave breaker). Later, dyke series was filled with deposit material (30 thousand m^3/ per year) [22]. Notwithstanding this, "lower washing" was spread over 5-6 km; similar occurrence was recorded in the surroundings of Gagra, Gudauta, Akhali Atoni, Ochamchire. Stopping of local abrasion via reinforcement dyke series and shaped constructions (tetrapods, dipods), including wave breakers, was followed by formation of new sections of lower washings";

2. Deposit flow (69 km) of the Kodori Dynamic system (seashore perimeter of the River Kodori and the River Inguri), was cut by the port mole near Ochamchire. Thus, shore was decreased with 250 in some sections as a result of "lower washings". Arrangement of coast-protection walls and dyke series for protecting the city were followed by formation of "lower washing" sections. In the north area of the port mole, pretty wide beach was formed on the left section of the River Mokvi outflow. Similar occurrences were recorded within the areas of Poti and Batumi ports. Especially, strong negative geo-ecological consequences were recorded in terms of construction and exploitation of Poti Port. On the north - Nabada section of Port, 200-250 m sand beach was formed and in the south, in surroundings of the great island, coast line was decreased in 800 m;

3. Notwithstanding that alluvium was proved as the major feeding source of the beach, because of the construction of the Inguri Power Station (1972), major part of water flow was transferred into the canal of the River Eristskali; later was used as channel and its intrusion in coastal water area formed "dyke effect" in the south of Ochamchire (Gali

Municipality, causing active abrasion of the major coast. On the right zone of the "outflow" of the River Eristskhali wide (80-110m) beach was formed, and coastal line was decreased in 12-15m/yy on the left section. Power of deposit flow along the show fell down 2 times and actively decreases. Because of rough decreasing of the beach feeding material (4-5 times), Kolchida coast decreases in 2-5m/ per year [16];

4. Constructing of power station on the River Chorokhi on the territory of Turkey, provides negative geo-ecological affect on the coast of Achara - Guria, putting it under danger [23]. Village Adlia (Achara) is especially under danger, where beach washing and decreasing of the shores is more than 5-7 meters annually.

Thus, reviewed material and its analysis shows that, in terms of coverage and exploitation of coastal region, they had gone under strong and impressive changes. Based on the clear image of the changes and differentiation, geo-ecological features of the coasts, natural conditions, landscape development and detailed research of geo-ecological problems, provided natural - geographical zoning. Taxonomic zoning units are as follows (Fig. 3):

Region - sharply expressed geotectonic unit, with different modern tectonic, with natural conditions and resources, homogenous geo-ecological problems;

Subregion - regional geomorphologic unit, with non-homogenous natural factors (relief, air, hydrography), similarity of geo-ecological processes;

District - comparably local territorial unit, with comparably similar natural conditions and geodynamic processes;

Subdistrict - low regional unit, with homogenous relief and similarity of natural processes;

Area - local territorial unit with micro relief forms and priority development of one of the elemental processes.

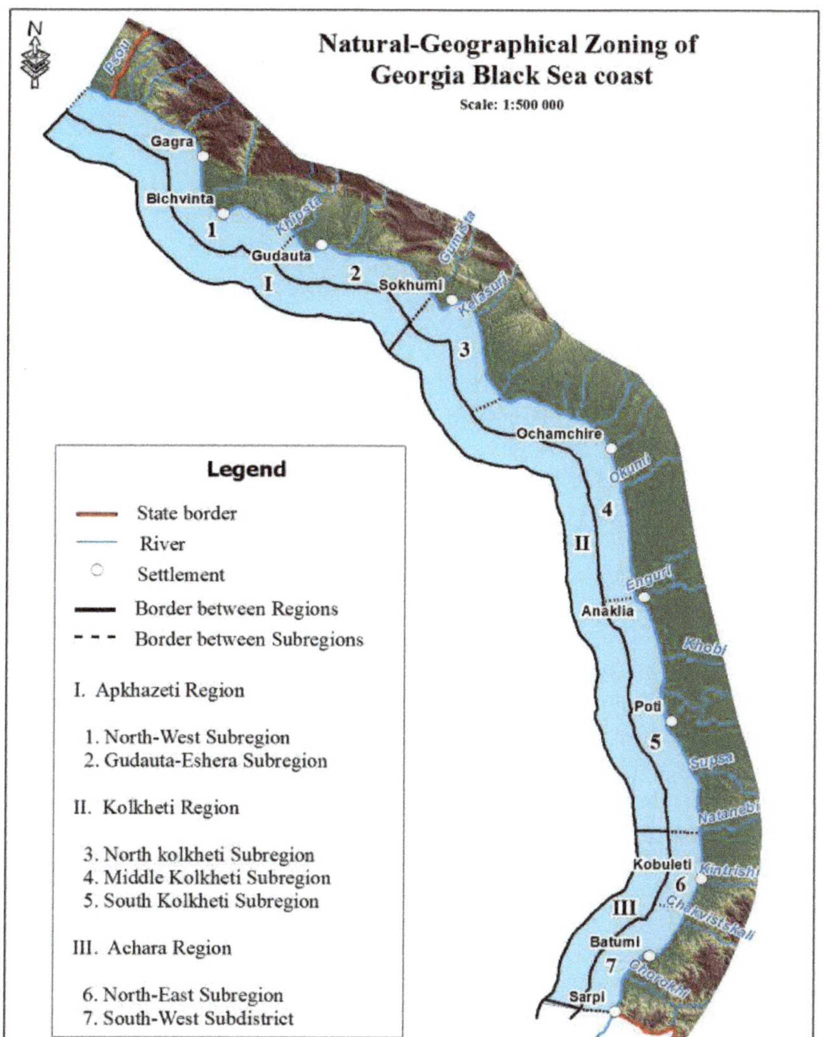

Fig. 3. *Natural-Geographical Zoning of Georgia Black Sea coast (Within the Subregions it is Separated: 19 districts, 42 subdistricts, 103 area (because of research scale it was impossible mapping of relatively thin taxonomic units - Districts, Subdistricts, Area).*

4. Conclusion

1. Environment object regulation and management measures, including some economical problems - elaboration and realization of reasonable projects of port infrastructure (Anaklia Port) are offered for the optimization of sanitary

pollution of dewatered sections (lower flow of Inguri), as a result of construction of power stations, remaking of river canals and extracting of inert materials from the beach, negative geo-ecological consequences as a result of strong anthropogenic pressure of the low and vast plain and underwater coastal slope of Kolchida.

2. Based on comparing of common and differentiating morphological - dynamic features of the coasts (feeding source, granulometry of deposit, coastal relief, underwater slope, sea regimen) the following can be marked out:

A). More than four from six with common signs i.e. Similar objects: according to the genetic - tectonic and morphologic similarities, they are grouped in 2 parts:

A) Coasts of accumulative plains of alluvium - delta generation (Psou-Mzimta Delta - Bzyb Delta - Sokhumi Cape - Kodori Delta - Rioni Delta - Chorokhi Delta);

B) Coasts of hilly seashore massifs (Tkhemi - Miusera - Eshera - Green Cape - Tsikhisdziri - Gonio - Sarpi);

B) Four from six with common signs i.e. One type objects: plane and narrow plains sea shore parameters (Gantiadi - Gagra - Bombora Roadstead - Bichvinta Bay - Gudauta Bay - Kindgi-Kobuleti-Kakhaberi);

C) Two from six with common signs i.e. Analog - objects: low coasts of vast and plane underwater slope, more or less showing bottom feeding: Gudauta - Ochamchire - Middle Kolchida;

3. Natural inflow of materials (break stone), attached to the roots of the water plants, formed as a result of biological drying (boring of underwater stones by boring mollusks) in underwater bottom constructed with meiotic conglomerate of Gudauta Bank and its role in feeding of the beach is proved. Launching it into practice (filling of beach shortage with artificial inflow of deposit) is pretty attractive activity, being economically effective, based on ecologically safe technology.

Acknowledgements

This research would not have been possible without the support of many people and also without using many scientific researches done by the staff of Sokhumi State University and the Department of Geography (Faculty of Exact and Natural Sciences) of Ivane Javakhishvili Tbilisi State University. Authors of the work express their gratitude to all them.

References

[1] M. Alpenidze, Z. Seperteladze and E. Davitaia (2013) Landscape and environmental problems nature manajement Black Sea Coast Kolkhida (the effects, perspective). I International Conference “Kolkheti Lowland Water Ecosystems – Protection and Efficient Use”, Workbook, pp. 132-135.

[2] M. Alpenidze and Z. Lomtatidze Black Sea: Abiotioc and Biotic process dynamic. Abiotic Factors. Part 1. Apkhazeti National Academy of Sciences, Tbilisi, 2011, pp. 235.

[3] N. Dunaev The contribution of recent tectonics in the state of the sea coast. In the coll. Seashores - evolution, ecology, economy. Proceedings of XXIV International Coast Conference. Tuapse, 1-6 X, 2012, pp. 152-155.

[4] N. Fokina. Anthropogenic factors abrasion processes. In: Construction and technological safety. Vol. 15-16, Simferopol, 2006, pp. 138-142.

[5] Kolkhida lowland: Scientific background and development. Moscow, 1990, -248 p.

[6] M. Alpenidze and G. Lominadze. Where and How should built Port in Anaklia "Georgia and World" # 21 (256), 11-17 June, Tbilisi, 2014, pp. 18-19http://www.geworld.ge/View.php?Artid=5966&lang=ge

[7] V. Zenkovich. Research of the coastal zone of the Black Sea within Georgia. Proc: Natural foundations coastal protection. // under. Ed. V. P Zenkovich, E.I Ignatov, S.A. Lukyanova. "Nauka", Moscow, 1987, pp. 45-51

[8] M. Alpenidze, Z. Seperteladze and E. Davitaia Black Sea rational nature landscape-ecological research and regulation-management of anthropogenic processes (on example of Georgia). Collected papers: "Current Problems of Mountainous Regions Geography", Tbilisi, 2008, pp. 329-341.

[9] M. Alpenidze and E. Davitaia Kolkheti Black Sea rational nature use regional and geographical issues. "Science and Modernity", Tbilisi, Science, 2003, pp. 135-142.

[10] G. Russo and S. Khorava. Coastal zone protection with artificial beaches (Kobulety city, Georgia) http://coruna.coastdyn.ru/iccal/pub/ russo.pdf;

[11] M. Alpenidze. Latest developments of Sukhumi district coastline. Msg. ANGSSR, Volume 90, № 2, Tbilisi, 1978, pp. 397-399.

[12] M. Alpenidze. Features of Sukhumi district. Geomorphology, Moscow, № 3, 1983, pp. 56-61.

[13] V. Menshikov, V. Peshkov and M. Alpenidze Some structural features of submarine canyons in the area of Cape Sukhumi. Msg. Academy of Sciences of the Georgian SSR, 95, № 3, Tbilisi, 1979, pp. 637-640.

[14] I. Voronov about dynamics of the coast of Sukhumi Bay in historic times. - In Proc. Of young scientists - historians of Abkhazia. Sukhumi, “Alashara”, 1974, pp. 24-38.

[15] Sh. Jaoshvili Black Sea Rivers. European Agency for Environmental Protection. Tbilisi, 2003, - 186 p.

[16] G. Lominadze, N. Megreli and G. Russo. Change of Black Sea coastal area (central part of Kolkhida, 1976-2002) under influence of man-induced factors. Change of Environment at the Turn of the Millennium. Proceedings of International Internet Conference. Tbilisi-Moscow, 2006, pp. 133-139.

[17] V. Zenkovich. Expected changes in the north sea coast of Kolkheti. Academy of Sciences of the Georgian SSR 83, №3, Tbilisi, 1976, pp. 637-640.

[18] M. Alpenidze. Bottom power long shore sediment. Geomorphology, № 2, Moscow, 1985, pp. 65-70.

[19] M. Alpenidze. Formation and development of Sukhumi alluvial plain. Geography and modern environmental problems // Proceedings E-Conference (25 May - 25 June), Tbilisi: SGU, 2014, pp. 75-91.

[20] V. Zenkovich. From foreign experience marine coastal protection. Proc.: Natural foundations coastal protection. "Nauka", Moscow, 1987, pp. 149-153.

[21] A. Peshkov. Artificial pebble beaches in the marine coastal protection. USSR Academy of Sciences. The Commission on the World Ocean. Nauka, Moscow, 1987, pp. 165-171.

[22] Sh. Jaoshvili What may happen to the Adjarian Coast after construction of the Dams on the Chorokhi river. 2000, http://www.cenn.org/ Magazine1/Magazine3/whatmayhappenrus.html

[23] M. Orlova. Sea coast of the Crimea as a resource for recreation (on example of the coast of Zap. Crimea). Author. Dissertation of candidate of Geographical Science, Moscow, 2010, 26 p.

[24] A. Kiknadze. Application of the method to the lithological study of the dynamics of the coastal zone (on example of the Black Sea coast). Dissertation of candidate of Geographical Science, Moscow State University, 1970, -17 p.

[25] A. Kiknadze. Dynamical systems and sediment budget along the Black Sea coast of Georgia. Proc.: Man and the environment, Sukhumi, "Alashara", 1977, pp. 59-67.

5

Alga Species Diversity of the Natural Park “Borovoye”

Satbay Nurashov, Elmira Sametova

Institute of Botany & Phytointroduction Ministry of Education and Science, laboratory of Mycology and Algology, Almaty, Republic Kazakhstan

Email address:
nurashs@mail.ru (S. B. Nurashov), elyasam@mail.ru (E. Sametova)

Abstract: The work was carried out within the project “Algae-flora of lake systems of Kokshetau-Borovoye, development of the recommendations about fighting against species of micro- and macro-algae causing oozing and “blossoming” of ponds in mass people recreation zones”. Sample collecting was made during the expedition trips at summer season from the lakes Akkol, Lebedinoye and from the several region lakes such as Borovoye, Schuchiye, Bolshoye and Maloye Chebachye, Lebiyazhiye, Maybalyk, Katarkol and Shalkar. As a result of cameral processing and preliminary identification of the collected material more than 231 alga species were defined: Bacillariophyta – 151; *Chlorophyta* – 46; *Cyanophyta* – 16, *Dinophyta* – 4; *Charophyta* – 6; *Euglenophyta* – 7 and *Chrysophyta* – 1. Alga species and their distribution by systematic groups were determined. They belong to 71 genera, 39 families, 24 orders, 11 classes and 7 departments. The greatest species variety was belonged to *Bacillariophyta* and *Chlorophyta* alga species. Often seen, dominating and causing water “blossoming” alga species of all departments were revealed in lakes Akkol and Bolshoye Chebachye. Mass development was demonstrated by the species *Anabaena* flos-aquae Breb. ex Born. et Flah. and species of the genus *Oscillatoria* Vauch. ex Gom. On the stones there were revealed mucilaginous yellow-brown fouling of *Bacillariophyta* algae such as *Gomphonema*, *Synedra* and *Melosira*. Reasons of mass development of the activators of blossoming and oozing of the ponds were revealed. They were caused by: increase of the content of nitrates, phosphates and other nutrients in the water; washouting the non-organic fertilizers from the nearest hills and fields; sedimentary waters coming into the ponds. The abstract and the biological description of the alga species (231 species of 7 departments) revealed in the lakes of the studied territories were made. Among found algae the perspective species of *Chlorophyta* and *Charophyta* algae for purification of the polluted waters were marked out.

Keywords: Natural Park, Lake, Algae, Kazakhstan

1. Introduction

Anthropogenic disturbance of the natural balance of water bodies some algae species can cause processes affecting the life of reservoirs, reducing the quality of water which becomes unsuitable for aquatic organisms and economic needs of people.

Due to intense anthropogenic impactat the “Borovoye” lake ecosystems the following events take place: the accelerated sedimentation of bottom of the reservoirs, water quality deterioration, unavailability of water bodies for recreational, fishing and other useful purposes. A sharp increase inload, especially in the areas of mass rest of the people leads to an increase in revenues in ponds of various organic wastes that greatly exceeds the ability of aquatic ecosystems to cleanse itself. This, inturn, contributes to the rapid growth of many harmful toxic microalgae and macrophytes. Settling on the bottom of water bodies, they accelerate the process of anaerobic decay; create conditions for rapid growth of benthic organisms – micro- and macro algae and higher aquatic macrophytes that is the root cause of the sedimentation and water-bloom.

The State National Natural Park "Borovoye"is located in the Akmola area and it was established in August, 2000. The territory of the SNNP "Borovoye" is a part of Kokshetau steppe, forest-steppe and ridged small mound upland. The climate is sharp-continental with hot summers and cold winters. Modern structure of landscapes within the park is represented by steppe, lake, forest, forest-steppe landscapes and landscapes of lowlands.

The State National Natural Park "Borovoye"is located in the Akmola area and it was established in August, 2000. The territory of the SNNP "Borovoye" is a part of Kokshetau steppe, forest-steppe and ridged small mound upland. The climate is sharp-continental with hot summers and cold winters. Modern structure of landscapes within the park is represented by steppe, lake, forest, forest-steppe landscapes and landscapes of lowlands.

The Kokshetau-Borovoye lake systems are located in the North of Kazakhstan, between Astana and Kokshetau cities. The highest point of the territory is Kokshetau mountain (Blue Mountain). In the territory of the Natural Park such large lakes as Shchuchiye, Borovoye, Bolshoye and Maloye Chebachiye and Katarkol were studied. In addition. such small lakes as Svetloye, Karasiye, Gornoye and Lebedinoye were also studied.

The aims of this study were: - to determine the species composition of algal flora and vascular aquatic plants of the lake system, including water-bloom pathogens and sedimentation of water bodies in the areas of mass recreation; - to study biology and ecology of these pathogens and the conditions for development and mass distribution of the species.

This work was directed to studying the algal flora in the Kokshetau-Borovoye lake systems due to the deterioration of water quality in water bodies of the region.The following tasks were formulated: - collecting the material-samples of algae from ponds in the Kokshetau-Borovoye lake systems; - establishing the algae species; - determination the cause of the mass development of water-bloom pathogens and sedimentation of water bodies in recreational areas in the region. As well as identifying the dominant and causing water-bloom species of algae; making recommendations for the conservation of water bodies in the State National Natural Park (SNNP) "Borovoye".

2. Study Area and Methods

During 2012-2014 algae from the water bodies of the SNNP "Borovoye" were the objects of our researches. Expedition trips by means of the route inspection and permanently in the advance planned places and during different seasons of the year were the methods of our researches. While collecting algae the weather conditions, air and water temperatures were noted, transparency of water was determined by Secchi disk, chromaticity – by pH-universal indicator paper. The standard classical hydrobotany and algology methods were used. Phytoplankton tests for the qualitative analysis were selected by the Apstein planktonic network made of silk gas No. 76, 45 cm diameter using methods of M.M. Gollerbakh and V.N. Polyansky, N. P. Masyuk, etc. [1, 2].

Samples of benthos and fouling from stones and boulders were taken by the scalpel; floating "flat cakes" were gathered with hands or rakes. In winter the samples were gathered from under the ice, for this purpose the drill and bathometer were used. Collected material was fixed in places of collecting by 4%-formalin solution. While collecting the material data of occurrence of macrophytes, their quantity and forms of colonies of characteristic algae, color and nature of the samples were registered. Processing the material was made only after the end of field researches. Cameral processing of the samples was carried out in two steps, at first blue-green, green, euglenophyta, chrysophita and dinophyta algae were determined, then part of the samples were cleared from various mixtures and the identification of the diatoms was started. Specimens of diatom species were made by a calcinating method. The material washed from formalin was applied to the cover glass and calcinated on the electric stove. For definition of the diatoms the constant specimens were prepared. Cleaning the diatom frustules from organic substances was carried out by boiling in strong acids.

Observation over the morphology of algae was carried out by means of the light microscopeMBI-3 and by the binocular microscope Motic BA 400 with the digital camera and the computer, the sizes of cells were measured with the eye-piece–micrometer, processing and definition of material were carried out according to the general accepted techniques of the algology and hydrobiology researches [3-18].

3. Results and Discussion

At studying the algal flora in the Borovoye lake systems during the summer, autumn and winter periods for clarification of specific structure and regularity of distribution of algae more than 200 samples of micro algae (plankton, benthos and fouling) were collected by us. In all water bodies the existence of the vascular water plants and chara algae were revealed, in some lakes their mass development was observed.

As a result of processing the assembled alga samples from the studied water bodies 231 species and varieties were revealed: - 151 diatom species; - 46 green alga species; - 16 blue-green alga species; - 7 euglenophyta species; - 4 dinophyta species; - 1 chrysophyta species and 6 chara alga species. The found species belong to 71 genera, 39 families, 24 orders, 11 classes and 7 departments (Table 1).

The diatoms differed in specific wealth (*Bacillariophyta*). The following genera were especially numerous: *Navicula* Bory – 19 species; *Nitzschia* Hass. – 12; *Pinnularia* Ehr.-10; *Gomphonema* Ag. – 9; *Cymbella* Ag. – 8; *Eunotia* Ehr. – 7, *Fragilaria* Lyngb. – 7, *Surirella* Turp. – 7. The most often met species were: *Rhoicospheinia curvata* (Kutz.) Grun, *Cymbella ventricosa* Kutz., *Gomphonema acuminatum* Ehr., *G. constrictum* Ehr., *G. olivaceum* (Lyngb.) Kutz., *Cocconeis placentula* Ehr., *Caloneis silicula* (Ehr.) Cl., *Navicula cuspidata* Kutz., *N. bacillum* Ehr., *N. criptocephala* Kutz., *N. radiosa* Kutz., *Gyrosigma spenceri* (W. Sm.) Cl., Amphora ovalis Kutz., *Nitzschia hungarica* Grun., *N. tryblionella* Hautzsch., *N. sigma* (Kutz.) W. Sm., *N. sigmoidea* (Ehr.) W. Sm., *Epithemia zebra* (Ehr.) Kutz., *Rhopalodia gibba* (Ehr.) O. Mull., *Surirella ovalis* Breb. and *Cymatopleura solea* (Breb.) Sm.

Green algae, (*Chlorophyta*), by number they were at the 2nd place (46 species or 19.91 % of the total algae species). There were met some species from the orders *Desmidiales* and *Chloroococcales*. The largest number of species was noted in the genera *Cosmarium* Corda., *Scenedesmus* Meyen., *Ankistrodesmus* Corda., *Staurastrum* Meyen. and *Pediastrum* Meyen.On the shores there were met such species as:

Cladophora glomerata (L.) Kutz. and *Ulothrix zonata* Kutz.

In the lakes Borovoye and Bolshoye Chebachiye there were found Chlorococcophyceae species of such genera as Scenedesmus Meyen., *Ankistrodesmus* Corda. And *Pediastrum* Meyen. Especially often there were met such species as: *Scenedesmus acuminatus* (Lagerh.) Chod., and *S. quadricauda* (Turp.) Breb., *S. obliquus* (Turp.) Kuetz.,S. acutus Meyen., *S. falcatus* Chodat., *S. ellipticus* Corda., *Ankistrodesmus fusioformis* Corda., *A. falcatus* (Corda) Ralfs., *A. gracilis* (Reinsch) Korsch., *Pediastrum boryanum* (Turp.) Menegh., *P. duplex* Meyen., *P. angulosum* (Ehr.) Menegh. and *P. tetras* (Ehr.) Ralfs. (Table 1).

Table 1. *Taxonomical structure of sea algae in the lake systems of the SNNP "Borovoy".*

Division	Number					
	classes	orders	families	genera	species	% from the total number of species
Bacillariophyta	3	13	20	33	151	65.37
Chlorophyta	2	4	9	20	46	19.91
Cyanophyta	2	3	6	9	16	6.93
Euglenophyta	1	1	1	3	7	3.03
Dinophyta	1	1	1	3	4	0.73
Chrysophyta	1	1	1	1	1	0.43
Charophyta	1	1	1	2	6	2.60
Total:	11	24	39	71	231	100

Green algae, (*Chlorophyta*), by number they were at the 2nd place (46 species or 19.91 % of the total algae species). There were met some species from the orders *Desmidiales* and *Chloroococcales*. The largest number of species was noted in the genera *Cosmarium* Corda., *Scenedesmus* Meyen., *Ankistrodesmus* Corda., *Staurastrum* Meyen. and *Pediastrum* Meyen.On the shores there were met such species as: *Cladophora glomerata* (L.) Kutz. and *Ulothrix zonata* Kutz.

In the lakes Borovoye and Bolshoye Chebachiye there were found Chlorococcophyceae species of such genera as Scenedesmus Meyen., *Ankistrodesmus* Corda. And *Pediastrum* Meyen. Especially often there were met such species as: *Scenedesmus acuminatus* (Lagerh.) Chod., and *S. quadricauda* (Turp.) Breb., *S. obliquus* (Turp.) Kuetz.,S. acutus Meyen., *S. falcatus* These alga species developing in large numbers, causing water-bloom can become the reason of fish death and poisonings.

In the studiedwater bodies there were found two families of the order *Desmidiales* (*Desmidiaceae*) of the 7 genera, 19 species and 2 forms. We often met *Desmidiaceae* from the genera *Cosmarium* Corda and *Staurastrum* Meyen. From the genus *Cosmarium* Corda there were found: *Cosmarium regnesii* Reinsch., *C. angulosum* Breb,. *C. impressulum* Elfv., *C subimpressulum* Borge., *C. gayanum* De Toni., *C. gostyniense* (Racib.) Gronbl., *C. venustum* (Breb.) Arch., *C. venustum* var. hypohexagonum West., *C. botrytis* Menegh., *C. retusum* var. in aequal ipellicum W. et G.S. West. And *C. granatum* Breb.

From the *Dinophyta* there were found four species – *Ceratium hirundinella* (O. F. Mull.) Bergh., *Peridiniopsis penardii* (Lemm.) Bourr. *Peridinium aciculiferum* Lemm. and *Peridinium lomnickii* Wolosz.

In the lakes Bolshoye Chebachiye and Borovoye, Maloye Chebachiye, Katarkol and Shchuchiye among vascular aquatic plants there were met Chara algae – *Charatomentosa* L., *Characanescens* Desv.et Lois. *C. kirghisorum* Lessingemend. Hollerb and *Nitella confervacea* A. Br. On these Chara species there were found periphyton species such as: *Cocconeis placentula* Ehr., *Cymbella ventricosa* Kutz. and *Rhoicospheinia curvata* (Kutz.) Grun.

The lakes Borovoye, Lebiazhiye and Shchuchiye are freshwater ponds, and the lakes Bolshoye Chebachiye, Maloye Chebachiye, Maibalyk and Katarkol are brackish. The exsicating lake Akkol was water-logged, on its shore there were found fouling of blue-green algae especially often there were met species of the *Oscillatoria* Vauch. ex Gom genus. In the estern part of the lake Shchuchiye there were met vascular aquatic plants. They formed tangles made of macereed, water spike, bur-reed and chara alga species such as *Charatomentosa* L. and *Characanescens* Desv. et Lois. Also these species were noted in the lakes Katarkol, Maloye Chebachiye and Lebiazhiye. In the lake Chebachiye there was noted water-bloom of blue-green algae, among them especially dominant was *Anabaena flos-aquae* Breb. ex Born. et Flah. This toxic species was the cause of water-bloom of the reservoir. Also on the stones there was found limy yellow-brown fouling of diatoms species of the genera *Gomphonema* and *Synedra.* By the composition of algal flora the lake Borovoye, Bolshoye Chebachiye, Shchuchiye and Katarkol are moderately polluted lakes, and the lake Maloye Chebacheis more or less clean.

From the *Euglenophyta* species we met species from the genera *Euglena*, *Leposinclis* and *Trahelomonas*.

From the *Chrysophyta* species in the lake Borovoye there were found only one species *Dinobryon divergens* O.E. Imhof.

In the study of yellow-brown patina of the coastal stones it was found that the dominant species were species of the genera *Gomphonema*, *Synedra* from the diatoms. Also, we have identified species of algae causing water bloom and sedimentation of water bodies (Lakes Akkol and Bolshooye Chebachiye), blue-green algae: *Anabaena flos-aquae* Breb. ex Born. et Flah., *Anabaena constricta* (Szaf.) Geitl., *Microcystis aeruginosa* (Kutz.) Kutz., *Oscillatoria brevis* (Kutz.) Gom. From green algae: *Scenedesmus acuminatus* (Lagerh.) Chod., *S. quadricauda* Chod., *S. obliiqus* (Turp.); diatomas: *Asterionella formosa* Hass., *Fragilaria crotonensis* Kitt.

It was noted that when there were water-bloom processes in the water bodies mainly one or several microalgae species dominated, usually confined to such eutrophic environments. In such conditions, other species stop their development or precipitate. When the water bloom-season elapses, these species again begin their development.

There were alga species which were not found in summer time, but were found in winter samples in the lake Bolshoye Chebachiye (*Fragilaria leptostauron* (Ehr.) Hust., *Amphiprora paludosa* W. Sm., *Mastogloia grevillei* W. Sm., *Rhopalodia parallela* (Grun.) O. Mull., *Navicula binodis* Ehr., *Campylodiscus noricus* Ehr). They belong to the benthos species; their sizes are in several times higher than the planktonic species. Among these benthos species the plankton species were found in small quantities. While view in gunder a microscope it was clear that almost all species had the protoplast of green-yellow-brown color. It means that all of the detected species under the ice are located in optimum conditions.

Almost in all studied water bodies the tangles of the vascular water plants which participate in the process of purification of the muddy waters were found, lots of planktonic algae settle on the bottom of the water bodies and symbiotic epiphytes grow on them. Lots of vascular water plants from the genera *Potamogeton*, *Najas*, *Sparganium*, *Scirpus*, and *Typha* were found.

As a result of researches of the water bodies of the SNNP "Borovoye" the recommendations about the fight against sedimentation and water-bloom in recreational zones were made. Strict observance of sanitary and preventive norms in the recreational territories aimed at restriction of receipt of organic and other waste in reservoirs is necessary. The organization and systematic carrying out the ecological-educational actions in places of mass concentration of vacationers and the organization of works on improvement of the territory of other neighboring lakes to remove overloads from the main reservoirs – Borovoye and Bolshoye Chebachiye are recommended.

4. Conclusion

As a result of the researches of the reservoirs of the SNPP "Borovoye" 231 species and varieties of algae relating to 71 genera, 39 families, 24 orders, 11 classes and 7 departments were revealed.

The greatest specific variety was presented by diatom algae – 151 species. The second place was taken by Green algae – 46 species. Blue-green algae – 16 species; Euglenophyta – 7, Chara algae – 6, Dinophyta – 4 species.

In water bodies of the SNNP "Borovoye" during the whole vegetative period the diatom algae was dominant, it was noted that their specific quantity decreased in summer, in comparison with spring and autumn species. The quantity of species of green algae, on the contrary, in the summer increased and decreased in spring and in autumn.

Species of algae causing the water-bloom and sedimentation in the lakes Akkol and Bolshoye Chebachye were revealed, such as blue-green algae: *Anabaena flos-aquae* Breb. ex Born. et Flah., *Anabaena constricta* (Szaf.) Geitl., *Microcystis aeruginosa* (Kutz.) Kutz. and *Oscillatoria brevis* (Kutz.) Gom.

It was noted that the ecological structure of the alga flora of the water bodies of the SNNP "Borovoye" was formed by planktonic (80 species), benthophytes (125), fouling (47) and epiphytes (27) species.

References

[1] Hollerbach , MM, Polyansky VI (1951). Pres - novodnye algae and their study . Define - Tel freshwater algae USSR. The general part. " Soviet science", M.,199 p.

[2] Masuk NP, Kondratieff NV, Wasser SP (1989). Algae . Spravochnik.Kiev, 608 p.

[3] To freshwater algae USSR in 14 editions. (1951; 1953; 1982; 1983). M.

[4] To freshwater algae Ukrainian SSR. (1960). Vol. 11. Diatoms. Kiev. 412.

[5] Lothar Kalbe. (1980). Kieselalgen in Binnengewassern. Diatomeen. Wittenberg Lutherstadt, 206 p.

[6] Round F.E., Crawford R.M., Mann D.G. (1990). The Diatoms. Biology and morphology of the genera. Cambridge Univ. Press. 747 p.

[7] P. Carenko Short determinant Chlorococcales algae Ukrainian SSR. (1990). Kiev. 198.

[8] Ergashev AE Qualifier Protococcales algae Central Asia. (1979). Vol. 1-2. Tashkent. 344 p., 384 p.

[9] Genkal SI, Kulikov MS, TM Mikheev, Kuznetsova IV, EV Lukyanov Diatoms plankton Svislach River and its reservoirs. M .: Science World, 2013. 236 p.

[10] Genkal SI Trifonov IS Diatoms of the plankton of Lake Ladoga and its basin reservoirs. Ed.: JSC "Rybinsk Printing House", 2009. 72 p.

[11] Barinova S.S., Medvedova L.A., Anissimova O.V. Biodiversity of algae-environmental indicators. Tel Aviv, 2006. 498 p.

[12] Barinova S.S., Medvedova L.A., Anissimova O.V. Diversity of algal indicators in Environmental Assessment. Tel Aviv: PiliesStudio, 2006, 498 p.

[13] Popovskaya GI Genkal SI, Likhoshvai EV Diatoms of the plankton of Lake Baikal: Atlas determinant. Novosibirsk: Nauka, 2002, 168 p.

[14] Dorofeyuk N.I., Kulikovskiy M.S. Diatoms of Mongolia. Moscow, 2012, 366 p.

[15] Lange – Bertalot. Diatomin im Subwasser- Benthos von Mitteleuropa. A.R.G. Gantner Verlag K.G. 908 p.

[16] Krammer K., Lange-Bertalot H. Bacillariophyceae. 1. Teil: Naviculaceae // SuBwasserflora von Mitteleuropa. 1986. Bd. 2/1. 876 s.

[17] Krammer K., Lange-Bertalot H. Bacillariophyceae. 2. Teil: Bacillariaceae, Epithemiaceae, Surirellaceae. SuBwasserflora von Mitteleuropa. 1988. Bd. 2/2. 536 p.

[18] Krammer K., Lange-Bertalot H. Bacillariophyceae. 3. Teil: Centrales, Fragilariaceae, Eunotiaceae. SuBwasserflora von Mitteleuropa. 1991a. Bd. 2/3. 576 p.

6

Assessment of some heavy metals distribution and their possible human health risks: A case study of parts of Langtang south area, middle Benue trough, Nigeria

Lar Uriah[1], Yenne Ezekiel[1], Ozoji Toochukwu[2], Jibo Jesse[1]

[1]Department of Geology, University of Jos, P.M.B. 3084, Jos, Nigeria
[2]Department of Geology and Mining, Ibrahim Babangida University Lappai, Niger State, Nigeria

Email address:
ualexanderlar@yahoo.co.uk (L. Uriah), ezekielyenne@yahoo.com (Y. Ezekiel), ozojitoochukwu@gmail.com (O. Tochukwu), jessejibo@gmail.com (J. Jesse)

Abstract: An investigation on the distribution of heavy metals in some parts of Langtang South area located within the middle Benue Trough of Nigeria was carried out with the aim of determining their concentration levels for principally environmental assessment and for future mineral exploration studies. The area is characterized by four (4) lithologic formations; Asu River Group, Awe formation, Keana formation and Lafia formation. The area is endowed with abundant Pb-ZnS and $BaSO_4$ mineralizations. Geochemical analysis of soil samples collected from the area revealed the following heavy metal concentrations; Co = 0.022 - 30.09ppm; Cr = up to 21.64ppm; Ni = 0.02 - 50.67ppm; Cu = 0.39 - 63.77ppm; Zn = 0.04 - 3468ppm; As = up to 465.7ppm; Pb = 0.012 - 9322ppm, Cd = up to 17.74ppm. It is also observed that Co, Cr, Zn, As, Cd, Ni and Cu have relatively low spatial distribution pattern in Agri and Bolya villages as well as its surroundings while high spatial distribution pattern of these elements are observed in Jigawa ba da Goshi. However, Pb is observed to have high spatial distribution pattern in almost all northern parts of the mapped area including Agri, Jigawa ba da Goshi and Bolya villages. The concentrations of heavy metals in water samples decreases in the order Cu(223μg/l) > As(75.3μg/l) > Co(28μg/l) > Ni(27.8μg/l) > Pb(14.3μg/l) > Cr(6.8μg/l) > Zn(2.008μg/l) > Cd(1μg/l) and these are greater than the recommended no-effect values in water. The calculated daily intake of metal showed that the heavy metals portent potential human health risks in decreasing order of Cu> As >Cd. Health risk index found for these heavy metals was less than one suggesting that the consumption of such waters is safe and therefore pose no human health risks. However, the cumulative intake of such heavy metals over a long period of time may cause serious human health problems.

Keywords: Heavy Metals, Health Risks, Middle Benue Trough, Langtang South

1. Introduction

Many people are oblivious of the danger posed as a result of their interactions with the environment. The environment being the storehouse of many metals release them through a range of media including stream sediments, soils and waters [1]; [2]. These metals find their way into natural water bodies, or are remobilized into soils [3]. They originated from both natural and anthropogenic sources containing heavy metals which accumulate in the media and subsequently enter the human body through several pathways including food and water intake, dermal contact and inhalation [4], [5], [6]. It is imperative to understand that once they enter into the system, they are deposited in bones and fat tissues, overlapping noble minerals and cause several diseases and to some extent death [7]. They are often problematic environmental pollutants, with well-known toxic effects on living systems [8]; [9] especially when they are more than the standard required for consumption in the area. High concentrations of heavy metals such as such as Cd, Cr, Cu, Mn, Ni, Pb and Zn causes wide range diseases in humans such as abdominal pain, hypertension, cardiovascular diseases, immune dysfunction, liver diseases, anorexia, and kidney related disorders, as well as various kinds of cancers due to excessive intake in contaminated food and drinking water.

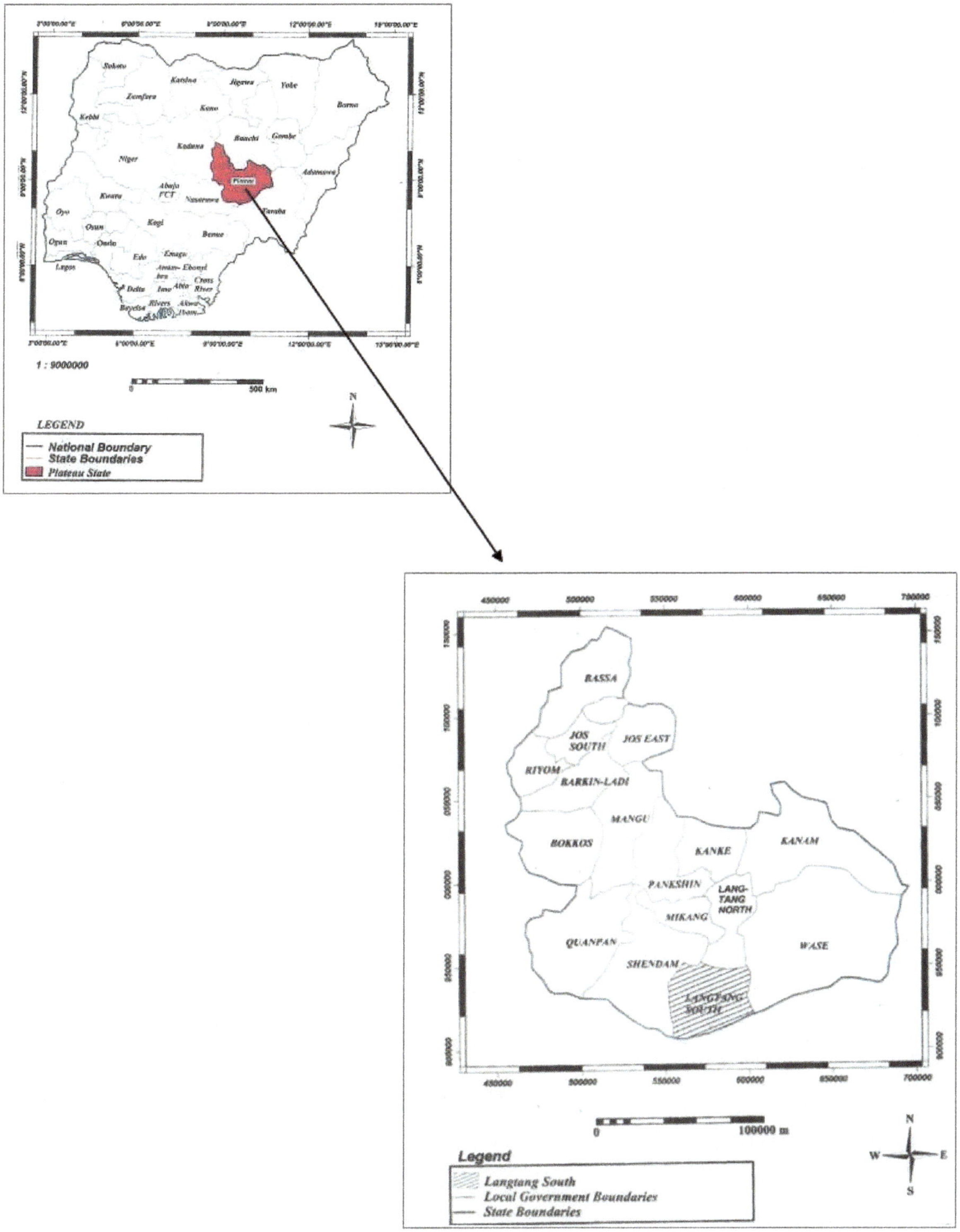

Figure 1. *Location of the study area*

They study area is faced with the scarcity and inadequacy in portable water supply, hence, surface and shallow water have become the most available source that is directly used for human consumption and domestic uses and as such increases the chance of heavy metal intake. Also, the polluted soil/stream sediments are use for the cultivation of crops and since the toxic metals are non-biodegradable in nature [10], they are transferred and remobilized through the food crops and directly consumed in the body. Therefore, this study is aimed at establishing and assessing the extent of the distribution of heavy metals and evaluating the health risks indexes in the area since the populace are very vulnerable to the direct exposure of heavy metals in the soils, stream sediments and water.

2. The Study Area

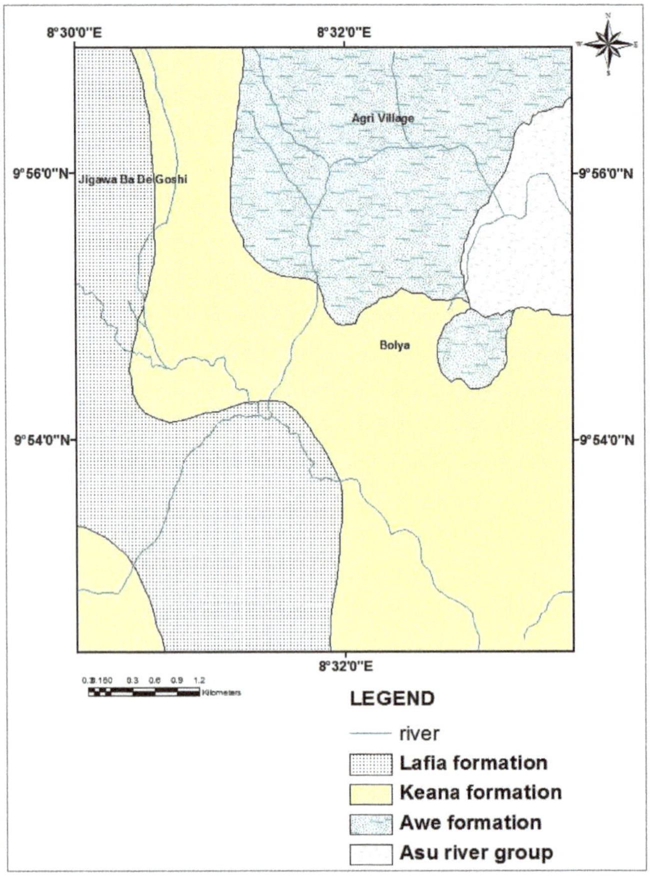

Figure 2. The geology of the study area

The study area is located within Shendam sheet 212 SE and covers an area of 54km^2 on Latitude 9^0 52' 28" to 9^0 57' North and Longitude 8^0 30' to 8^0 33' 42"East (Figure 1). The area is accessible through Mabudi major and minor roads like the Timya to Yamini minor, Jigawa ba de Goshi, Taka lafiya etc. The area is characterized by Guinea savannah with variety of tall as well as short grasses and trees [11]. Agricultural activity is the major occupation of the area with animal rearing and crop productions. The climatic conditions of the area is very typical of the tropics with two distinct rainy seasons; dry and wet seasons with relatively high temperature ranging from 30^{0}C to 33^{0}C witness especially during the dry season. The topography is relatively low with gentle slope towards the valley in the central portion. The drainage pattern is dendritic with major river (Gamakai) flowing in a south east direction which is been fed by several tributaries. However, the rivers are seasonal which prompted the construction of an artificial dam in the area. Surface water is usually polluted due to the presence of cattle and dust from trees leading to contamination of the water. Regrettably, it's this water source that becomes the major water supply to the populace during the dry season. The settlement pattern is characterized and controlled by the availability of the seasonal water bodies and soil fertility forming a nucleated or dispersed settlement. The soil is invariably sandy, loose and coarse grained and often very thick.

3. The Geology of the Study Area

The study area is located within the Middle Benue Trough i.e. at the middle portion of the Nigerian Benue Trough. The Benue Trough itself is a rift basin in Central West Africa that extends NNE–SSW for about 800km in length and 150km in width with Cretaceous-Tertiary Sediments [12]. The Middle Benue Trough is characterized by six (6) lithogenic formations though four (4) are observed in the study area (Figure 2); Asu River Group, Awe formation, Keana formation, Ezeaku formation, Agwu formation and Lafia formation [13]. However, the volcanics are seldom observed in most parts of the basin but extrude especially at the north-eastern parts. The Asu River formation is Mid-late Albian in age and composed of dark micaceous siltstone, fissile shales, mudstone and subordinate clays. They are the oldest known marine geological formation within the basin. The Awe formation is Cenomanian to Turanian in age and overlies the Asu River formation. It is composed of calcareous sandstone, carbonaceous shales and clays. The Keana formation is Campanian to Maastrichian and overlies the Asu River formation. Its lithology consists of whitish coarse felspathic gritty sandstone which is massive and heavy as a result of water current. Ezeaku formation is Turonian and overlies the Keana formation. It is composed of calcareous shales, micaceous and friable sandstones. It inter-fingers in the transition environment. Agwu formation is Conanian in age and overlies the Ezeaku formation. It is composed of limestone, shales and coal seams and observed around the Lafia-Obi area with thickness covers of not more than 700m [11]. The Lafia formation is Maastrichian in age and deposited within the continental (fluvitile environment). It overlies unconformably on the Agwu formation. It is the youngest formation in the Middle Benue Trough and consists of ferruginized sandstones, red loose sands, flaggy mudstones, clays and clay stone.

4. Geomorphology, Soil and Land Use

The study area is characterized by undulating landscape of low relief which is between 250-300m. Other topographic features include valleys, tributaries and undulating hills. The drainage pattern in the study area is dendritic. The major river Gamakai drains the area in a NNE-SWW flow direction. Human activities such as farming and overgrazing in this area are predominantly dependant on the existence of these tributaries. The area of study is predominantly composed of lowlands with few denudated low lying conical hills which area characterized by flat tops. The soil varies from place to place both in textures and colour depending on the underlying strata. They are invariably sandy, loose and coarse grained and often the covering is thick in some areas, the sand are thick enough to be quarried for construction purposes. Laterization is also observed along most river channels.

5. Materials and Methods

The topographic map of the study area (Shendam sheet 212) was digitized in ArcGIS and gridded for field reconnaissance survey. This reconnaissance survey was to find out exposures and structures and also locate the rivers, streams channels. Stream sediments, soil and water samples were obtained along river channels, farmlands, hand dug wells, stagnant ponds etc. where about 4.5g weight samples of stream sediments and soils were collected at a depth of between 10-15cm for fresh samples and kept in a clear polyethene bags. The samples were then prepared by drying in an oven to remove moisture content and pulverized to powder using an agate mortar and sieve to a size of 80micrometer. About 0.1g of the fines was later dissolved by adding aqua regia to the mixture of hydrochloric acid (HCl) and nitric acid (HNO_3) in a ratio of 3:1 and then heated on a sand bath for about 5-6hours after which deoxidized water in a 100ml volumetric flask is added, filtered out and then loaded for analyses in ICP-OES. Also, the water samples were collected from hand dug wells, stagnant ponds and river channels; and acidified with nitric acid (HNO_3) to preserve the constituents of the water. This was then filtered using filtering cone into a sterilized bottle and then loaded for analyses also in ICP-OES. The analytical techniques employed for all the prepared samples collected from the study area was carried out in the geochemical laboratory of the Department of Geology and Mining, University of Jos. The heavy metals obtained were statistical analyzed in ArcGIS and Microsoft excels to quantify and categorize each element. Subsequently, concentrations of the heavy metals were compared with World Health Organization (WHO) and average crustal abundance standards.

In assessing health risks, the health risk indicators parameters such as chronic daily intakes (CDIs) and health risk indices (HRIs) of metals were calculated for both adults and children from water samples since scarcity of water is eminent and consumption of it is without any treatment and purification. Thus, if the HRI value is less than one, it is considered to be safe for the consumers otherwise it indicate health risks [6], [14].

Therefore, the equation below was used for estimating the chronic daily intake, CDI of the study area from the water sample according to [15]

$$CDI = (C_m \times D_w)/ W_b \quad (1)$$

Where, C_m (μg/L) is the heavy metal concentration in water

D_w (L/day) is the average daily intake of water

W_b (kg) is the average body weights

A health risk index of the heavy metals was also estimated by Equation (2) [10]; [16].

$$HRI = CDI/ R_fD \quad (2)$$

Where, R_fD is the oral toxicity reference dose (μg/(kg·day))

6. Results

The results obtained from the geochemical analyses for heavy metals of stream sediments, soils and water samples are presented in table 1. Figures 3a and b, 4a and b, 5a and b, 6a and b, 7a and b, 8a and b, 9a and b and; 10a and b show the relative distribution of the concentrations of the analyzed heavy metals in the samples with their respective prediction pattern map. Comparison of the heavy metals in water samples with WHO permissible values [17] (Table 1) are presented thus in figures 11, 12, 13, 14, 15, 16, 17 and 18. Figures 19, 20, 21, 22, 23, 24, 25 and 26 give the comparison between the various concentrations of the heavy metals in the soil and stream sediment samples with that of average crustal abundance of the metals (Table 2). Table 3 present calculated estimates of the mean concentrations of the heavy metals, chronic daily intake and the health risks indexes of water consumption the study area.

Table 2. *WHO guidelines for drinking-water quality (2008) and International Standard for Abundance of element in average crustal rocks (Green and Taylor, 1978)*

Element (symbols)	Standard health guideline by the WHO in mg/l	Average Abundance
Arsenic, As	0.01	2ppm
Calcium, Ca	15	3.3%
Cadmium, Cd	0003	0.1ppm
Cobalt, Co	0.001	25ppm
Copper, Cu	0.10	50ppm
Chromium, Cr	0.02	100ppm
Iron, Fe	4.4	4.65ppm
Magnesium, Mg	2.7	1.7%
Nickel, Ni	0.01	75ppm
Lead, Pb	0.01	10ppm
Zinc, Zn	0.2	10ppm

Table 1. *Geochemical Analyses of heavy metals in the study area*

S/N	Sample name	Sample type	Long. (E)	Lat. (N)	Co (ppm)	Cr (ppm)	Ni (ppm)	Cu (ppm)	Zn (ppm)	As (ppm)	Pb (ppm)	Cd (ppm)
1	TOB 1	Stream sediments	8.510281	9.944008	19.9	21.64	36.68	63.77	09.02	168.7	111.6	17.74
3	TOB 4	Stream sediments	8.529783	9.921025	30.09	13.25	50.67	45.04	2781	321.9	118.3	<DL
4	TOB 5	Stream sediments	8.524253	9.904	12.88	<DL	28.92	21.19	3468	465.7	64.27	<DL
6	TOB 12	Stream sediments	8.506717	9.916683	22.67	<DL	23.73	33.31	2602	156.2	73.03	<DL
8	JJM3	Soil	8.525	9.880556	9.866	<DL	16.76	30.14	2918	113.7	84.89	0.178
9	JJM4	Stream sediments	8.541194	9.947619	2.82	<DL	4.74	10.25	2077	<DL	9322	<DL
10	JJM6	Stream sediments	8.539611	9.945936	1.466	<DL	19.13	42.22	1735	<DL	54.62	<DL

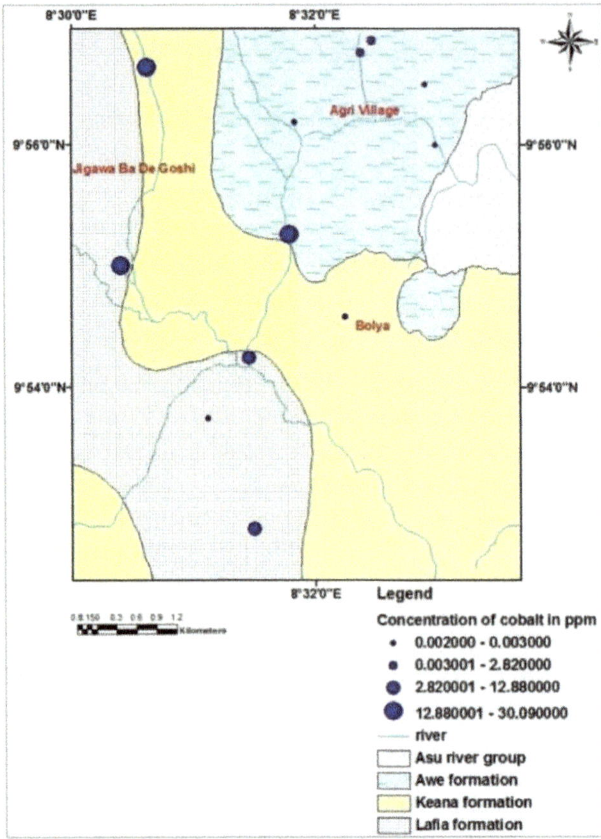

Fig 3a. *Cobalt concentration map of the study area*

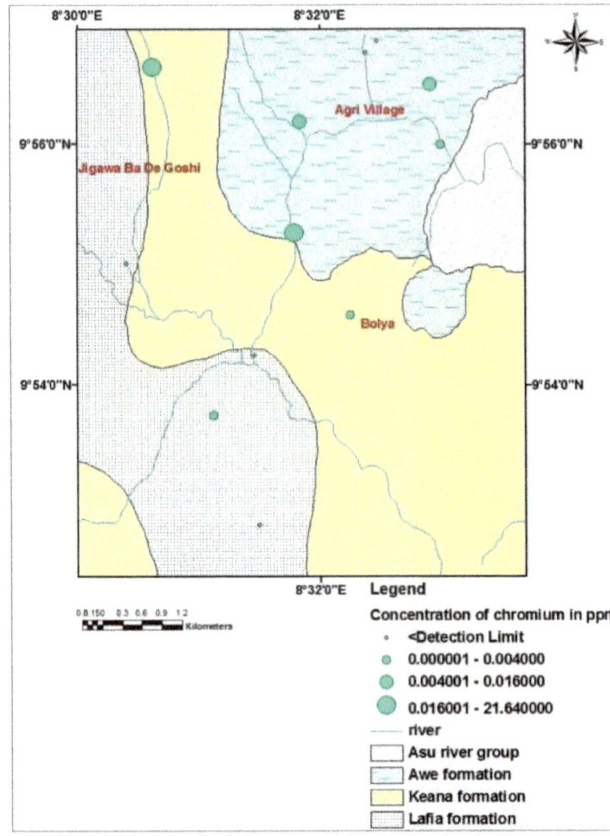

Fig 4a. *Chromium concentration map of the study area*

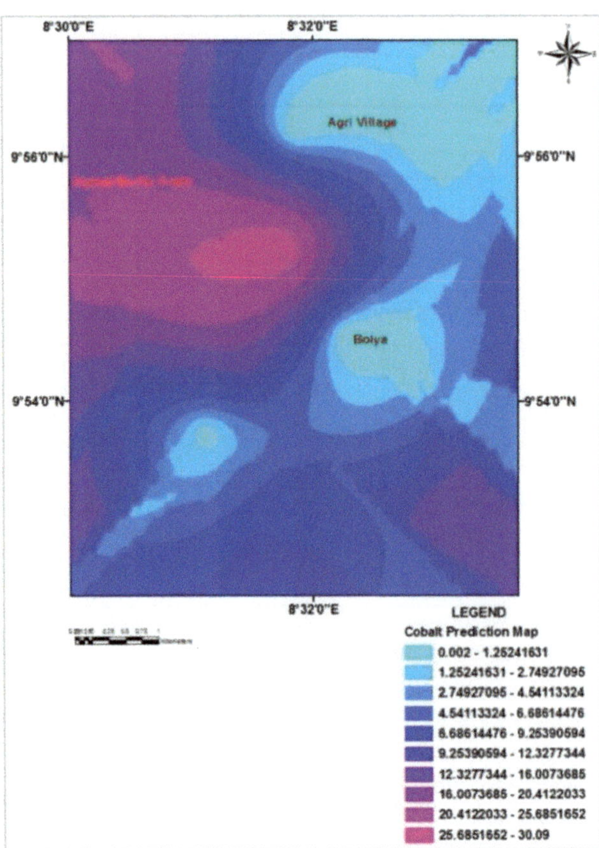

Fig 3b. *Cobalt prediction map of the study area*

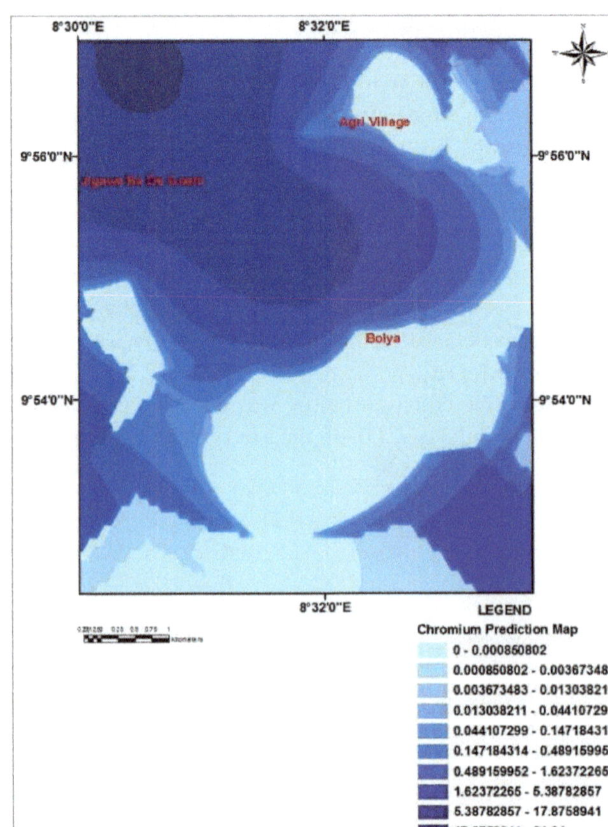

Fig 4b. *Chromium prediction map of the study area*

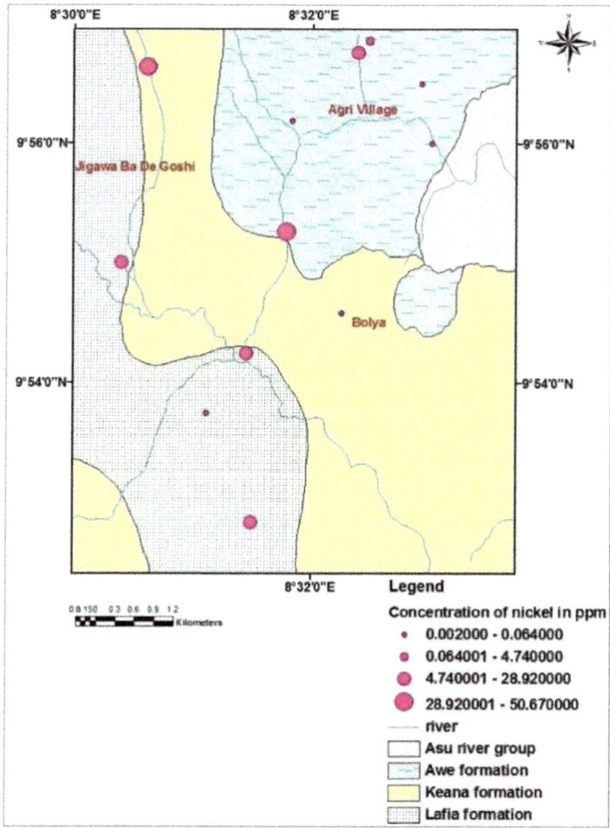

Fig 5a. Nickel concentration map of the study area

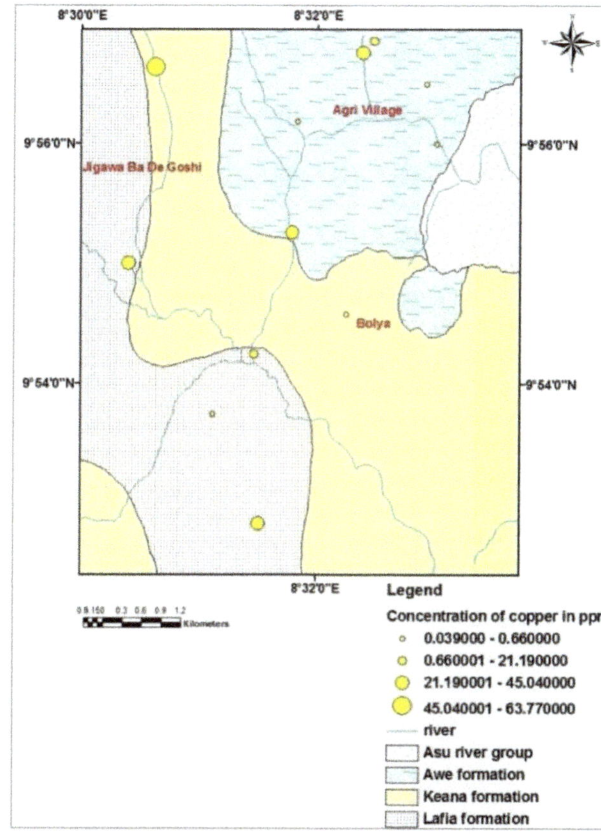

Fig 6a. Copper concentration map of the study area

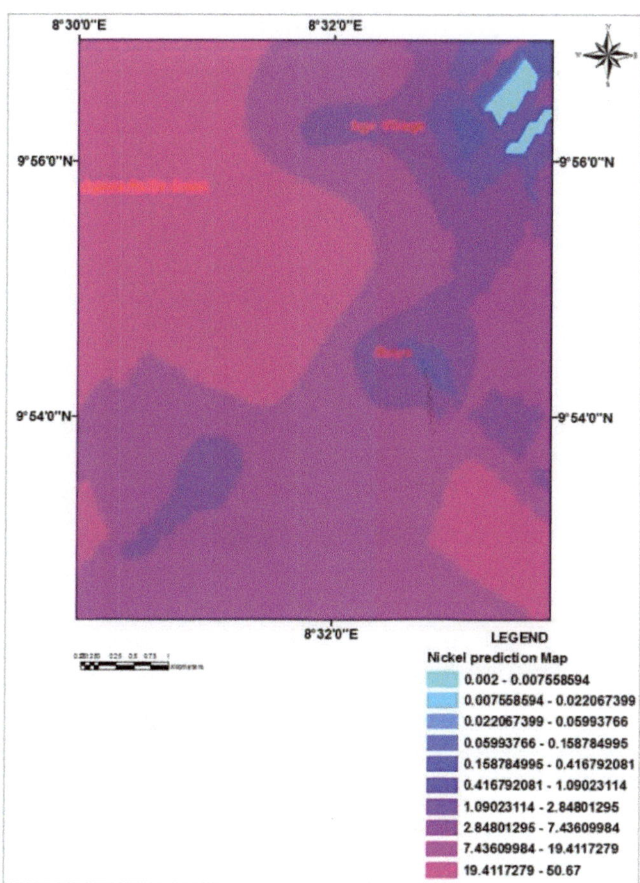

Fig 5b. Nickel prediction map of the study area

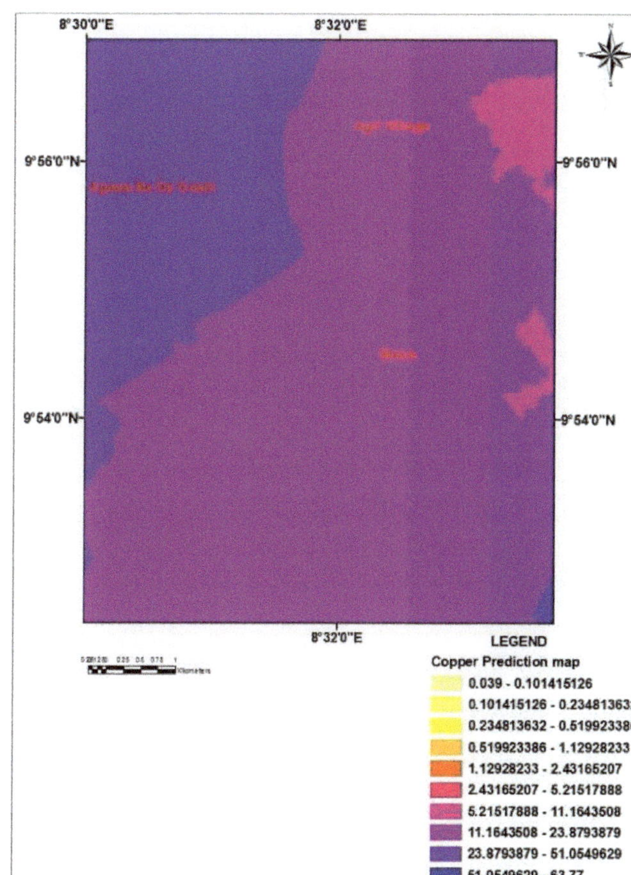

Fig 6b. Copper prediction map of the study area

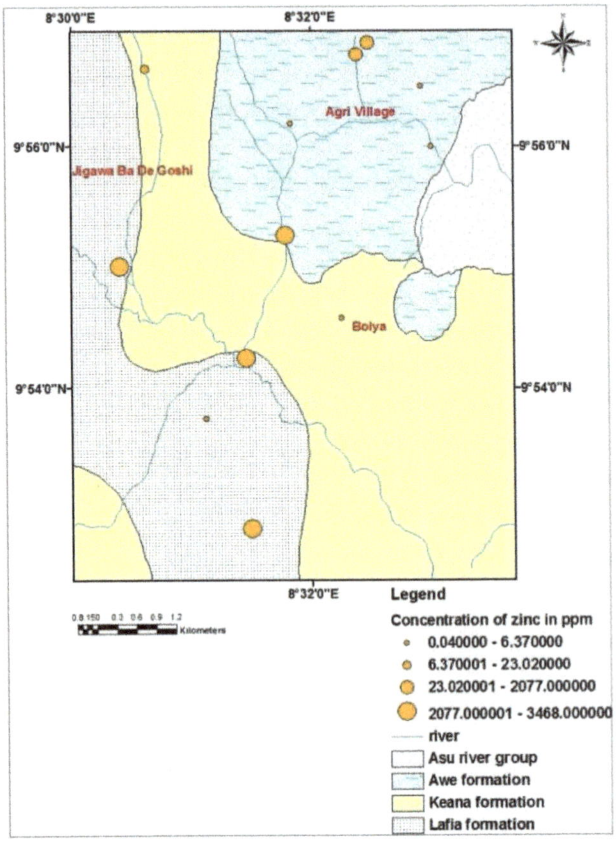

Fig 7a. *Zinc concentration map of the study area*

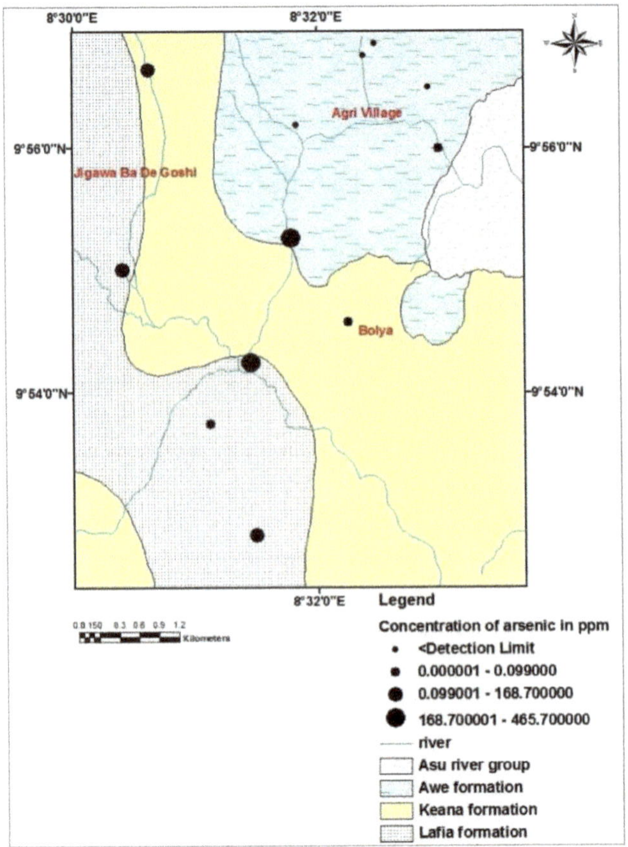

Fig 8a. *Arsenic concentration map of the study area*

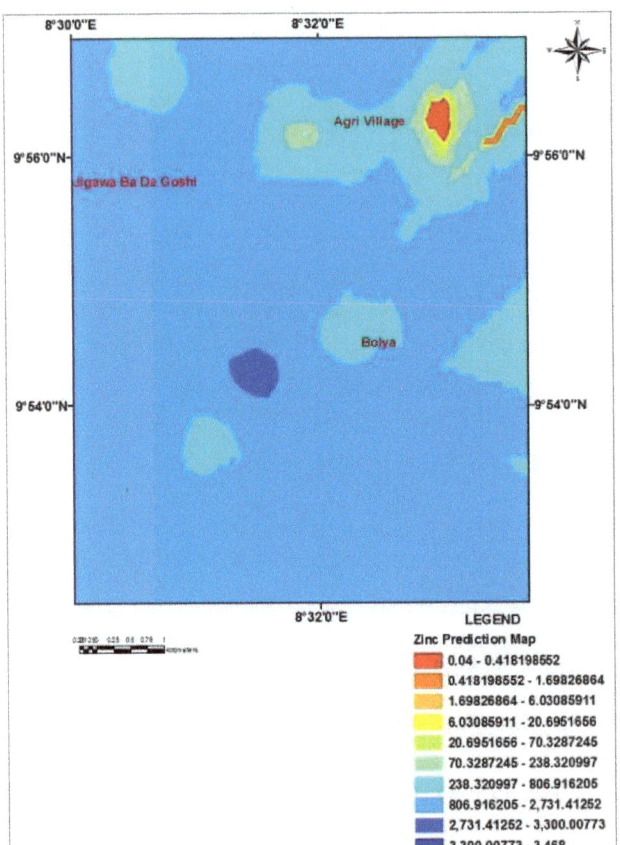

Fig 7b. *Zinc concentration map of the study area*

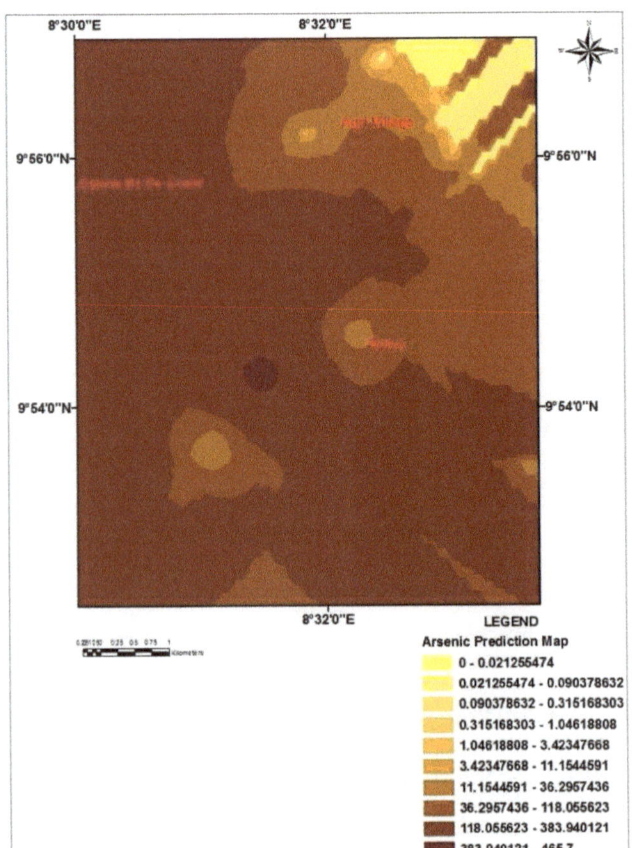

Fig 8b. *Arsenic prediction map of the study area*

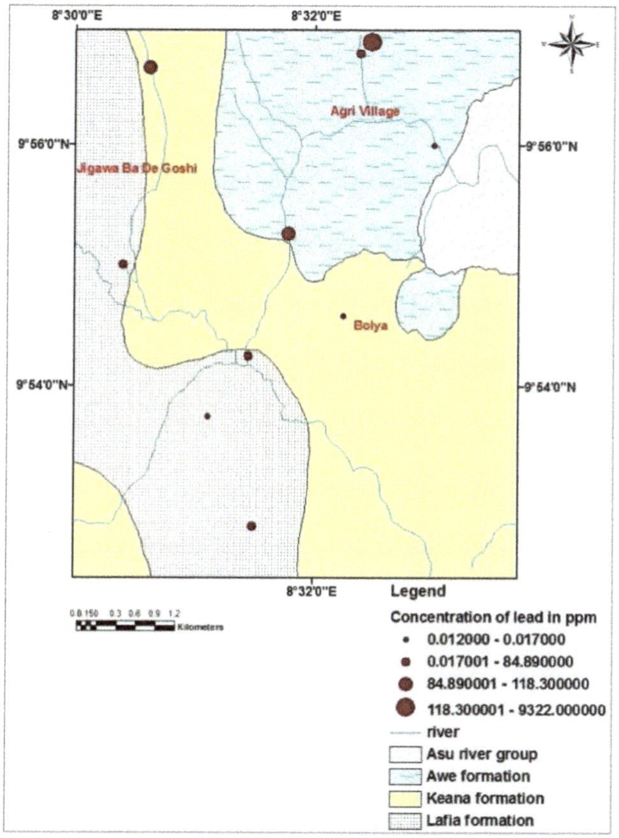

Fig 9a. Lead concentration map of the study area

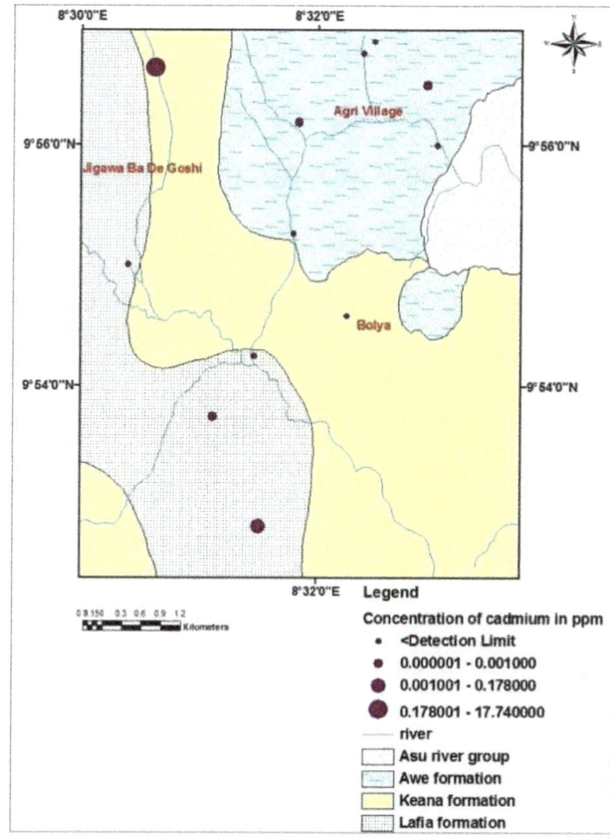

Fig 10a. Cadmium concentration map of the study area

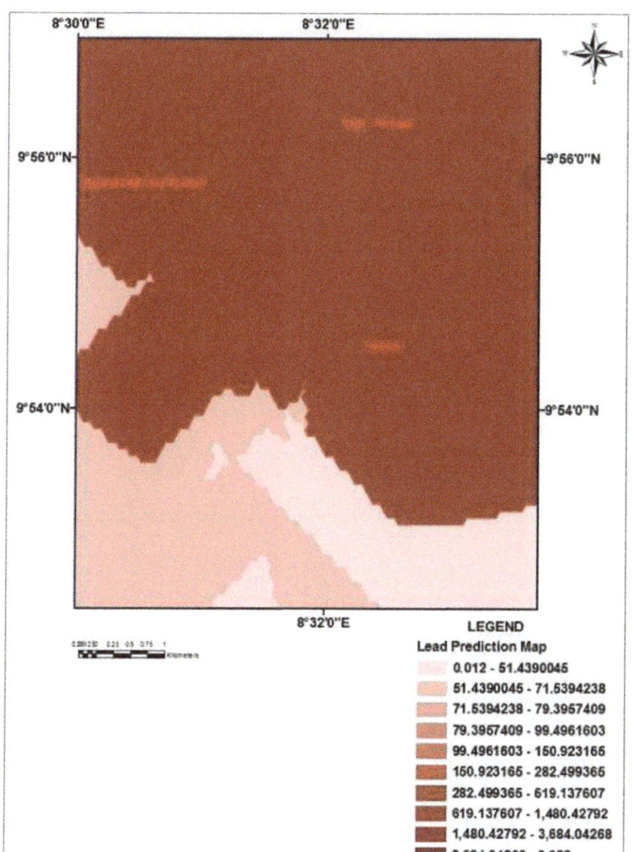

Fig 9b. Lead prediction map of the study area

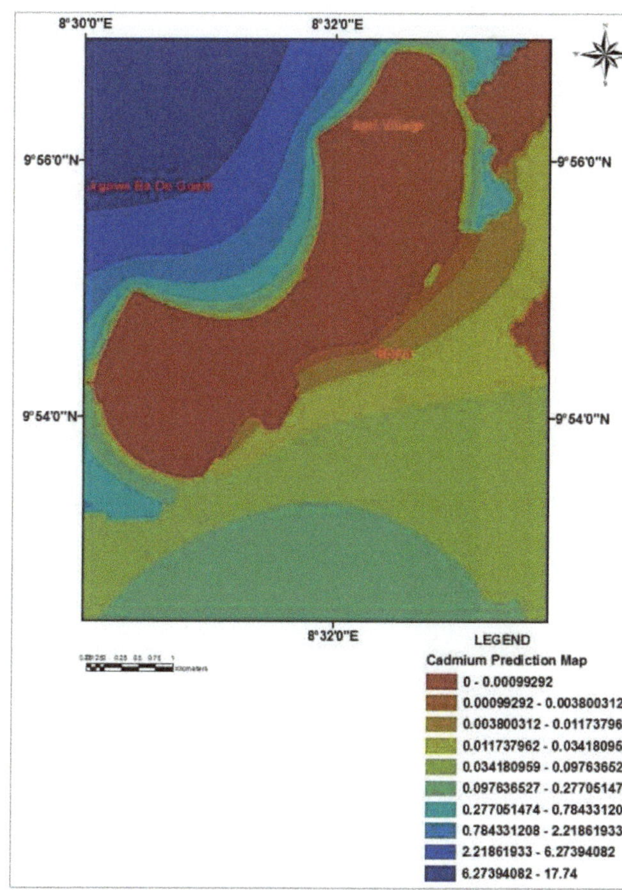

Fig 10b. Cadmium prediction map of the study area

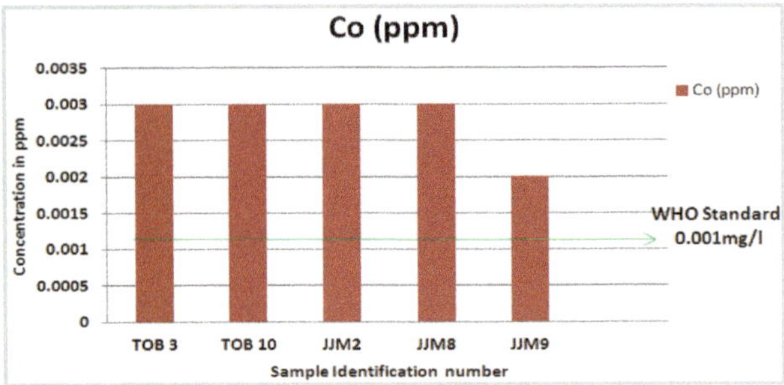

Fig 11. Comparing cobalt concentrations in water with WHO standard

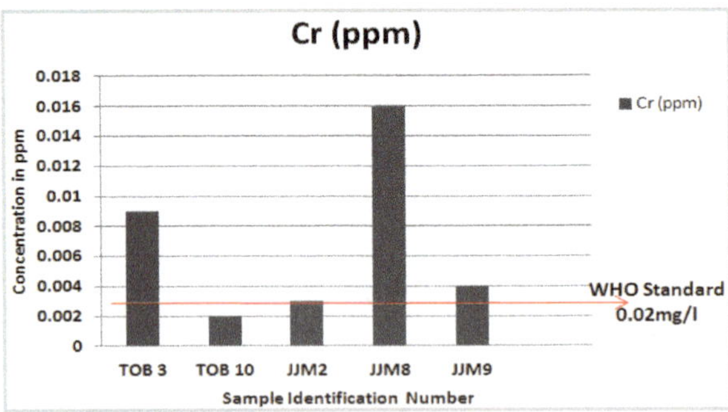

Fig 12. Comparing chromium concentrations in water with WHO standard

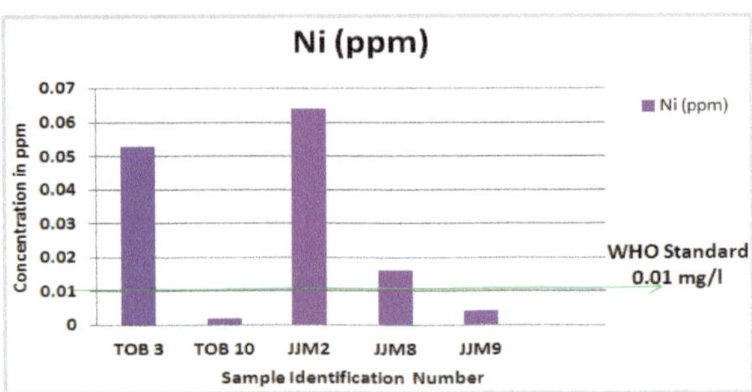

Fig 13. Comparing nickel concentrations in water with WHO standard

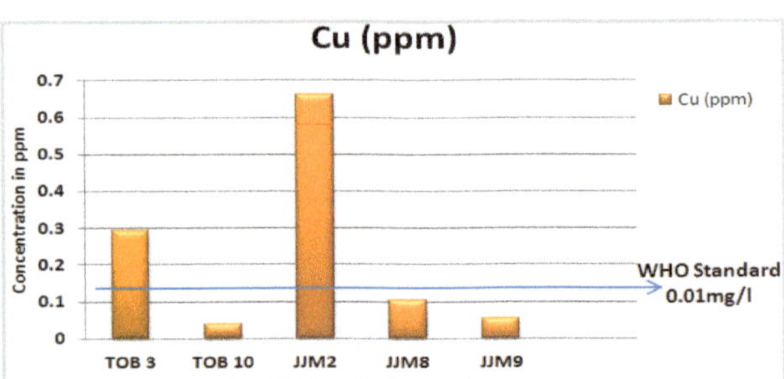

Fig 14. Comparing nickel concentrations in water with WHO standard

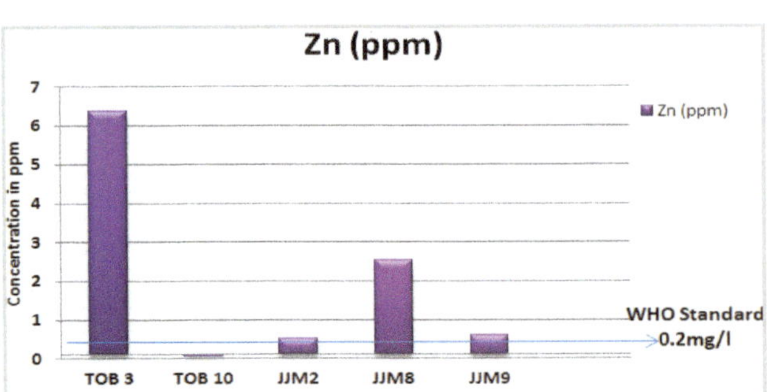

Fig 15. Comparing zinc concentrations in water with WHO standard

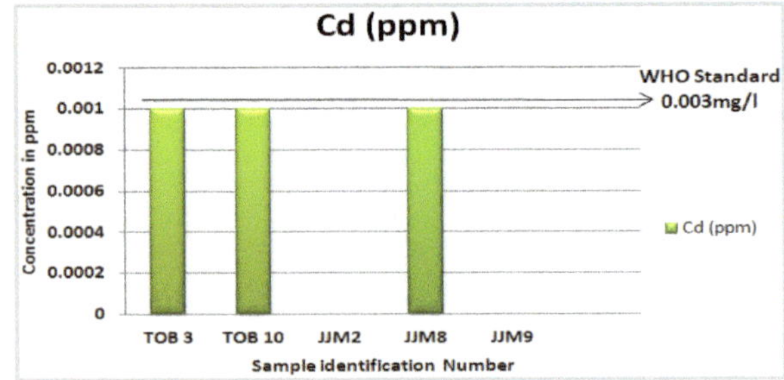

Fig 16. Comparing cadmium concentrations in water with WHO standard

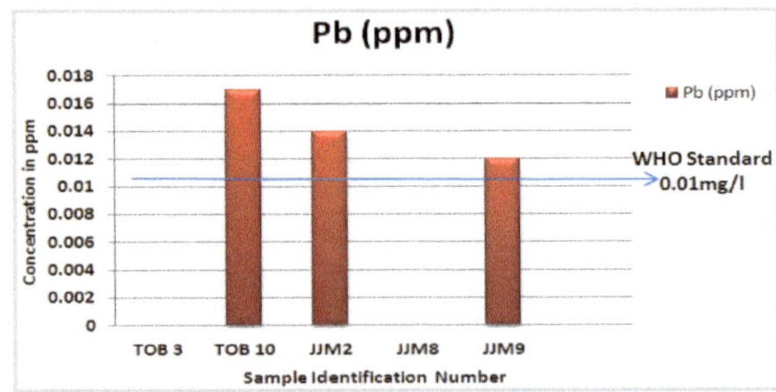

Fig 17. Comparing lead concentrations in water with WHO standard

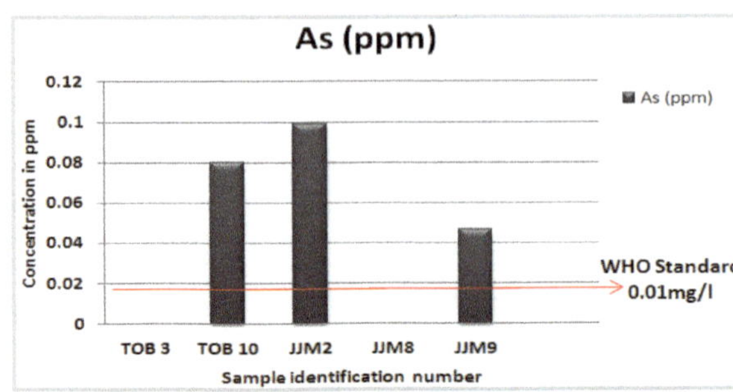

Fig 18. Comparing arsenic concentrations in water with WHO standard

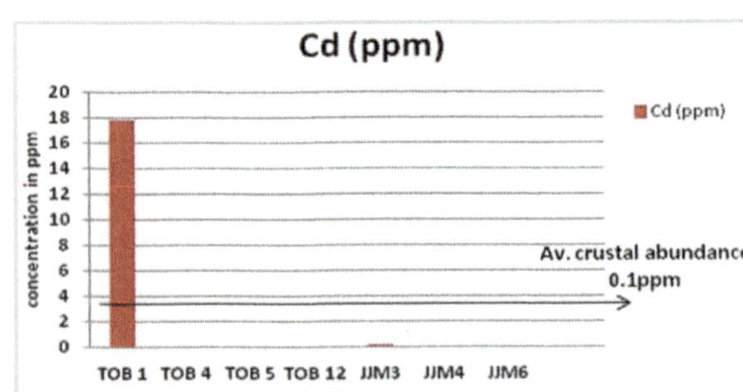

Fig 19. Comparing iron concentration in sediments with standard average crustal abundance

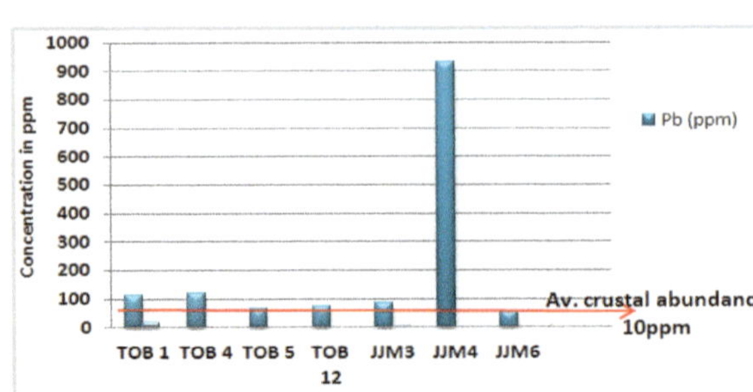

Fig 20. Comparing lead concentrations with standard average crustal abundance

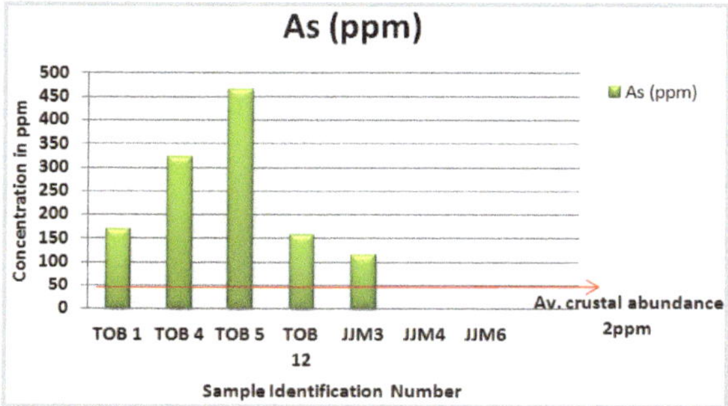

Fig 21. Comparing arsenic concentrations with standard average crustal abundance

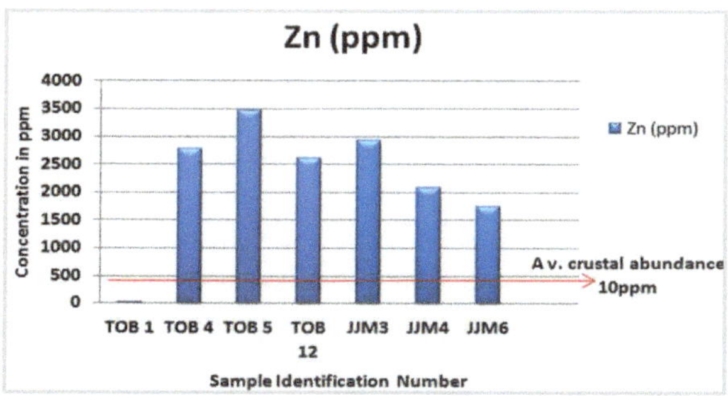

Fig 22. Comparing zinc concentrations with standard average crustal abundance

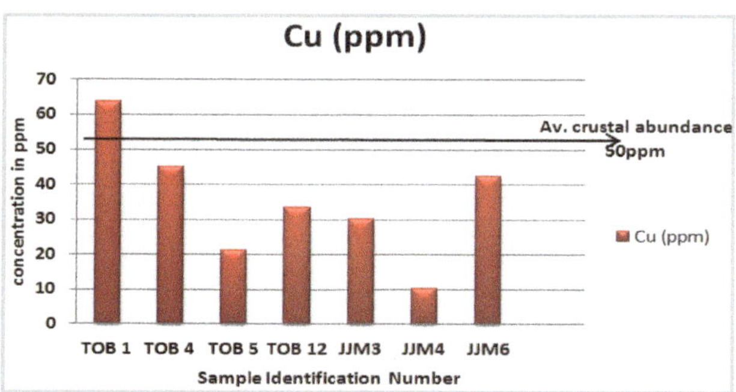

Fig 23. Comparing copper concentrations with standard crustal abundance

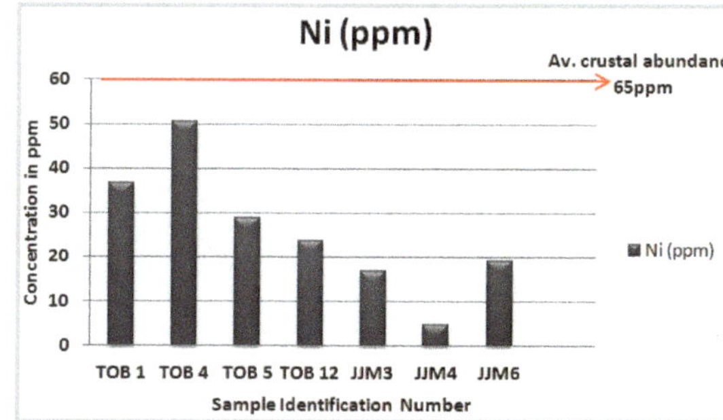

Fig 24. Comparing nickel concentrations with standard average crustal abundance

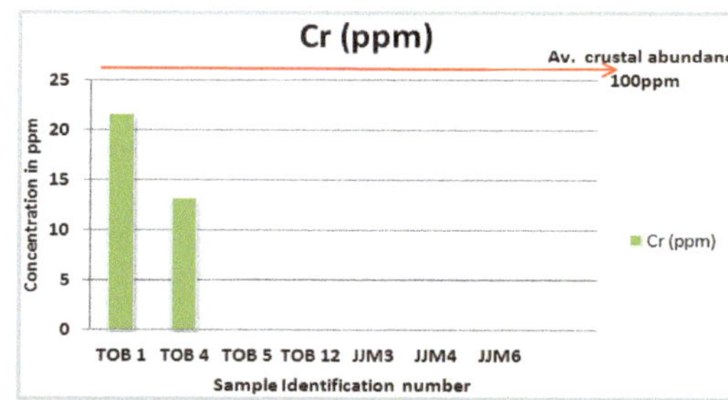

Fig 25. Comparing chromium concentrations with standard average crustal abundance

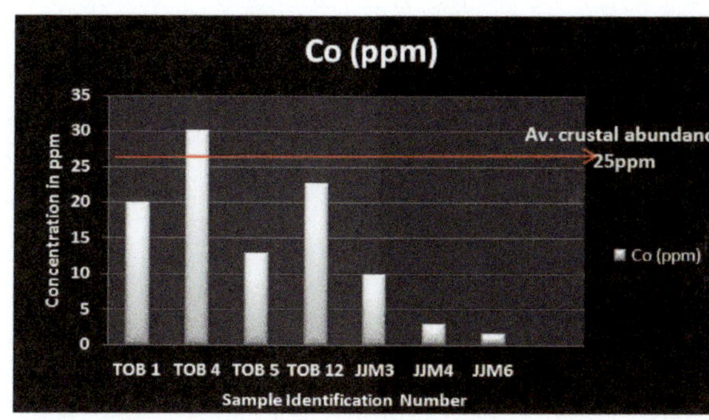

Fig 26. Comparing cobalt concentration in sediments with standard average crustal abundance

Table 3. *Chronic daily intake and health risks indexes of heavy metals through drinking water consumption*

Mean Concentration (microgram/ litre)		Co	Cr	Ni	Cu	Zn	As	Pb	Cd
		28	6.8	27.8	223	2.008	75.3	14.3	1
Chronic Daily Intake of metals (CDI)	Adults	0.7778	0.1889	0.7722	6.1944	0.5577	2.0833	0.3972	0.0278
	Children	0.8562	0.208	0.8502	6.9725	0.0614	2.3028	0.4373	0.0306
Health Risks Index	Adults	0.3889	1.20E-04	0.0386	0.1674	1.80E-03	0.0417	0.011	0.0556
	Children	0.4281	1.40E-04	0.0425	0.1884	2.00E-04	0.0461	0.0122	0.0612

N.B:

- Assuming that the average daily intake of water, D_W is 2 L/day for adult and 1 L/day for child (US EPA, 2011)
- Assuming that the average body weights, W_b to be 72 kg for adult and 32.7 kg for children (Muhammad et al., 2011; Khan et al., 2010; Jan et al., 2010).
- Oral toxicity reference dose (R_fD, μg/ (kg·day)) for Co, Cr, Ni, Cu, Zn, As, Pb, and Cd are 2, 1500, 20, 37, 300, 50, 36 and 0.5, respectively (Shah et al., 2012; Muhammad et al., 2010; US EPA, 2005, 2011).

7. Discussion

The general concentration of Cobalt in the soil ranges from 0.022 to 30.09ppm. Chromium concentration spans from undetectable limits to 21.64ppm while for nickel, the concentration ranges from between 0.02 to 50.67ppm. The concentration of copper ranges between 0.39 to 63.77ppm and zinc is between 0.04 and 3468ppm. Arsenic concentration falls below detection limit in some sample to

465.7ppm. The concentration of lead is between 0.012 and 9322ppm. Undetectable amount of cadmium is analyzed in the area up to a concentration of 17.74ppm. Its observed in the area that cobalt, chromium, zinc, arsenic, cadmium, nickel and copper have relatively low spatial distribution pattern in Agri and Bolya villages as well as its surroundings while high spatial distribution pattern of these elements are noticed in Jigawa ba da Goshi. However, lead is observed to have high spatial distribution pattern in almost all northern parts of the mapped area including Agri, Jigawa ba da Goshi and Bolya villages.

7.1. Water Samples

Generally, Chromium and Cobalt show high concentrations in all samples of the study area more than the WHO permissible limits [17] for water (0.001mg/l and 0.02mg/l respectively). Nickel concentrations in samples TOB 10 and JJM9 are lower than the WHO permissible limits [17] while in samples TOB 3, JJM 2 and 8 are higher WHO (0.01mg/l). Copper indicate 2 samples (TOB3 and JJM2) that have peaked above the WHO standard of 0.01mg/l. Zinc concentration in the area is generally higher than the WHO standard (0.2mg/l). Cadmium concentration in the study area is below the WHO standard (0.003mg/l) while lead concentration is shown to be high in samples TOB 10, JJM 2 and 9 more than WHO standard (0.01mg/l). Arsenic show Three (3) samples (TOB 10, JJM 2 and JJM9) with concentrations of above WHO permissible limits (0.01)[17].

7.2. Soils and Stream Sediments

The concentrations of cadmium in stream sediments of the study area is generally below the standard average crustal abundance (0.1ppm) except sample TOB 1 which contains 17.74ppm. Lead concentrations are above the standard average crustal abundance (10ppm) in the study area. Only two (2) samples have arsenic concentrations lower than the detection limit while the remaining samples are above the standard average crustal abundance (2ppm). The concentration of zinc in all the stream sediment samples is above the standard average crustal abundance (10ppm) except in sample TOB 1 which is 9.02ppm. Sample TOB 1 has high copper concentration in the area than the remaining samples. The entire concentrations of Nickel and Chromium are below the average standard crustal abundance (65ppm and 100ppm respectively). However, cobalt show that sample TOB 4 to have concentration higher than normal crustal abundance (25ppm).

7.3. Human Health Risks

The analyses show that the concentrations of heavy metals in water samples been the direct most consumed in the study area decreases in the order Cu>As>Co>Ni>Pb>Cr>Zn>Cd and all these toxic metals have concentrations greater than the WHO [17] admissible values for drinking water except cadmium. Health risk indices values obtained for these heavy metals either in children or adult are less than one indicating that it is safe and do not pose any human health risks. However, the intake of such heavy metals over a long period of time may increase the chance for serious human health problems.

8. Conclusion

This study has revealed that heavy metals such as cobalt, chromium, zinc, arsenic, cadmium, iron, nickel and copper have relatively low spatial distribution pattern in Agri and Bolya villages as well as its surroundings while high spatial distribution pattern of these elements are noticed in Jigawa ba da Goshi. However, lead is observed to have high spatial distribution pattern in almost all northern parts of the mapped area including Agri, Jigawa ba da Goshi and Bolya villages. Thus, the heavy metal concentrations in soils, sediments and water sources are considerably high relative to the recommended WHO values for water as well as the average crustal abundance. Whereas the calculated HRI for these heavy metals in potable water sources show that there is no immediate health risk for both adults and children, a long time intake may cause more serious health risks. On the positive side the spatial distribution of the heavy metals, especially Pb, As and Cu in the soils and sediments have great implications as to future mineral exploration targets.

References

[1] Salminen, R., Gregorauskienė, V., 2000. Considerations regarding the definition of a geochemical baseline of elements in the surficial materials in areas differing in basic geology. Applied Geochemistry, 15, 647-653.

[2] Johnson, C.C., Breward, N., Ander, E.L., Ault, L., 2005. G-BASE: baseline geochemical mapping of Great Britain and Northern Ireland. Geochemistry: Exploration, Environment Analysis, 5, 347-357.

[3] Lar, U.A., Ngozi-Chika, C. S. and Ashano, E. C., 2013. Human Exposure To Lead And Other Potentially Harmful Elements Associated With Galena Mining At New Zurak, Central Nigeria. Journal of African Earth Science 2(2).

[4] Madrid R, Biasioli M., Ajmone-Marsan F. 2008. Availiability and Bioaccessibility of metals in fine particles of some urban soils. Arch. Environ. Contam. Toxicol. 55,21.

[5] Nguyenh.I., Braun m., Szalokj I., Baeyens W., Van Grieken R., Leer Markers M. , 2008. Tracing the Metal Pollution History of the Tisza River Through the Analysis of a Sediment Depth Profile. Water Air Soil Pollut.doi: 10.1007/S11270-008-9898-2.

[6] Khan K., Yonglong L., Hizbullah K., Zakir S., Sardar K.L, Akbar A. K, Luo W., Tieyu W. 2013 :Health risks associated with heavy metals in the drinking water of Swat, northern Pakistan. Chinese Academy of Sciences, Beijing 100085, China. Nuclear Institute for Food and Agriculture, Tarnab, Peshawar, Pakistan. 21 May 2013.

[7] Jolly, Y.N, Islam A., and Akkat S. 2013. Transfer of metals from soil to vegetables and possible health risks assessment. SpringerPlus 2013, 2:385

[8] Prica M, Dalmacua B., Roncevic S., Kravtar D., Becelic M., 2008. A comparison of sediment quality results with acid volatile sulfide (AVS) and simultaneously extracted metals (SEM) ratio in Vojvodina (Serbia) sediments. Sci.Total Environ. 389, 235, 2008.

[9] Sanja M. S, Dragana S. D, Saczana S. T., 2011. Geochemical and Statistical Methods in the Evaluation of Trace Elements Contamination: an Application on Canal Sediments. Polish J. of Environ. Stud. Vol. 20, No. 1 187-199

[10] Shah M T, Ara J, Muhammad S, Khan S, Tariq S, 2012. Health risk assessment via surface water and sub-surface water consumption in the mafic and ultramafic terrain, Mohmand agency, northern Pakistan. Journal of Geochemical Exploration, 118: 60–67.

[11] Offodile, M. E. 1976. The Geology of the Middle Benue, Nigeria, Palecontological Institution of the University of Uppsala. Special Vol. 4.

[12] Benkhlil, J. M. 1986. Structure et Evolution Geodynamic du basin Inter Continemtale de la Benue. Nigeria. Thesis de Dotorat de-Etat University of Nice.

[13] Offodile, M. E. and Reyment, R. A. 1977. Stratigraphy of the Keana-Awe Area of the Middle Benue Region of Nigeria. Bull. Geol. Inst.. university of Uppsala, No. 7, Pp. 37-66.

[14] Rupert, L.H.; Neil, B.; Scott, D.Y.; Neil, M.J.C.; Andrew, M.T.; Ann, M.M. 2004. Assessing potential risk of heavy metal exposure from consumption of home-produced vegetables by urban populations. Environ. Health Persp., 112:215–21.

[15] Muhammad S, Shah M T, Khan S, 2010. Arsenic health risk assessment in drinking water and source apportionment using multivariate statistical techniques in Kohistan region,northern Pakistan. Food and Chemical Toxicology, 48(10): 2855–2864.

[16] Muhammad S, Shah M T, Khan S, 2011. Health risk assessment of heavy metals and their source apportionment in drinking water of Kohistan region, northern Pakistan. Microchemical Journal, 98(2): 334–343.

[17] WHO, 2008. Guidelines for Drinking-Water Quality 3rd edition, Vol.1, Recommendations. World Health Organisation, Geneva.

7

Ethnobiological Study of Svaneti Fungi and Lichens: History of Research, Diversity, Local Names and Traditional Use

I. Kupradze[1, *], A. Jorjadze[1], A. Arabidze[2], T. Beltadze[2], K. Batsatsashvili[2], N. Y. Paniagua Zambrana[3], R. W. Bussmann[4]

[1]Institute of Botany of IliaStateUniversity, Tbilisi, Georgia
[2]Institute of Ecology of Ilia State University, Tbilisi, Georgia
[3]Herbario Nacional de Bolivia, Universidad Mayor de San Andres, La Paz, Bolivia
[4]William L. Brown Center, Missouri Botanical Garden, St. Louis, USA

Email address:
Inga.Kupradze@gmail.com (I. Kupradze)

Abstract: Research about macro-fungi of Svaneti, a historical province of Georgia, started in the 1920th with a botanical expedition of R. Singer, under the auspices of the National Museum of Vienna. At present about 400 species of macro-fungi are known for Svaneti, 67 of which are edible (and used for food in Georgia), and 27 are poisonous. The history of Svaneti lichen research started with analysis of the material collected by H. Lojka and M. Dechy in 1884-1885 made by E. Vainio. As of now, 250 species of lichens are known for the region. In the present study ethnobiological interviews on Svaneti macro-fungi and lichens were conducted in 16 villages of Svaneti. Our analysis shows that the local population uses mainly 22 species of mushrooms for food, 10 of which are used most frequently: *Armillaria mellea* (Vahl) P. Kumm., *Agaricus arvensis* Schaeff., *A. campestris* L., *Cantharellus cibarius* Fr., *Lactarius piperatus* (L.) Pers., *L. deliciosus* (L.) Gray, *Macrolepiota excoriata* (Schaeff.) Wasser, *M. procera* (Scop.) Singer, *Ramaria flava* (Schaeff.) Quel. A comparison of the vernacular names given in the literature and mentioned in the interviews collected during the present study shows that the population of Svaneti must have had better knowledge of mushrooms in the past. This is evidenced by a much higher number of local names in literature: of about 50 vernacular names of particular species or groups of fungi found in the literature, only 12 are still used by the population. This paper presents various uses of macro-fungi in Svaneti as quoted in literature and based on interview data. Lichens, particularly, *Usnea* and *Bryoria* species, as well as *Ramalina thrausta* (Ach.) Nyl.are called "Pimpa" or "Pimpous" and "Chach". The only use of these species was by hunters as field-bedding to keep warm in the cold winters. In addition, we describe in the present paper how fungi and lichen uses were reflected in the local folklore.

Keywords: Macro-Fungi, Lichens, Ethnobiology, Georgia, Caucasus

1. Introduction

The first data on fungi of Svaneti, a historical province of Georgia, were published by R. Singer [1], who visited Svaneti in 1929 within the framework of a botanical expedition organized by the National Museum of Vienna. His paper provides a list of 95 species of macro-fungi found in Svaneti. Later the mycological investigation of the province was continued by I. Nakhutsrishvili (Agaricales s.l.) [2], V. Gulmagarashvili (Polyporales) [3], T. Anchabadze (Discomycetes) [4]. Based on current literature (mainly, [5]), and herbarium data, about 400 species of macro-fungi are known from Svaneti at present, of which 67 are edible (used for food in Georgia) and 27 are poisonous.

The history of Svaneti lichen research started with [6], listing 76 taxa, based on the analysis of the material collected by H. Lojka and M. Dechy in 1884-1885. Reference [7] mentions 26 taxa based on collections by S. Sommier and E. Levier. Reference [8] mentions a single species from material collected by F. Filarszky, reference [9] – two species, and reference [10] – 56 taxa identified from the collection of D.

Sosnovsky. Later research was continued by Tz. Inashvili in 1975-1977. At present 250 species of lichens are known from Svaneti.

No specific studies were, however conducted on the ethnobiology of the fungi (macro-fungi) and lichens of Svaneti and generally Georgia so far. However, vernacular names with descriptions of fungi and in some cases their use are present in a number of dictionaries [11-17].

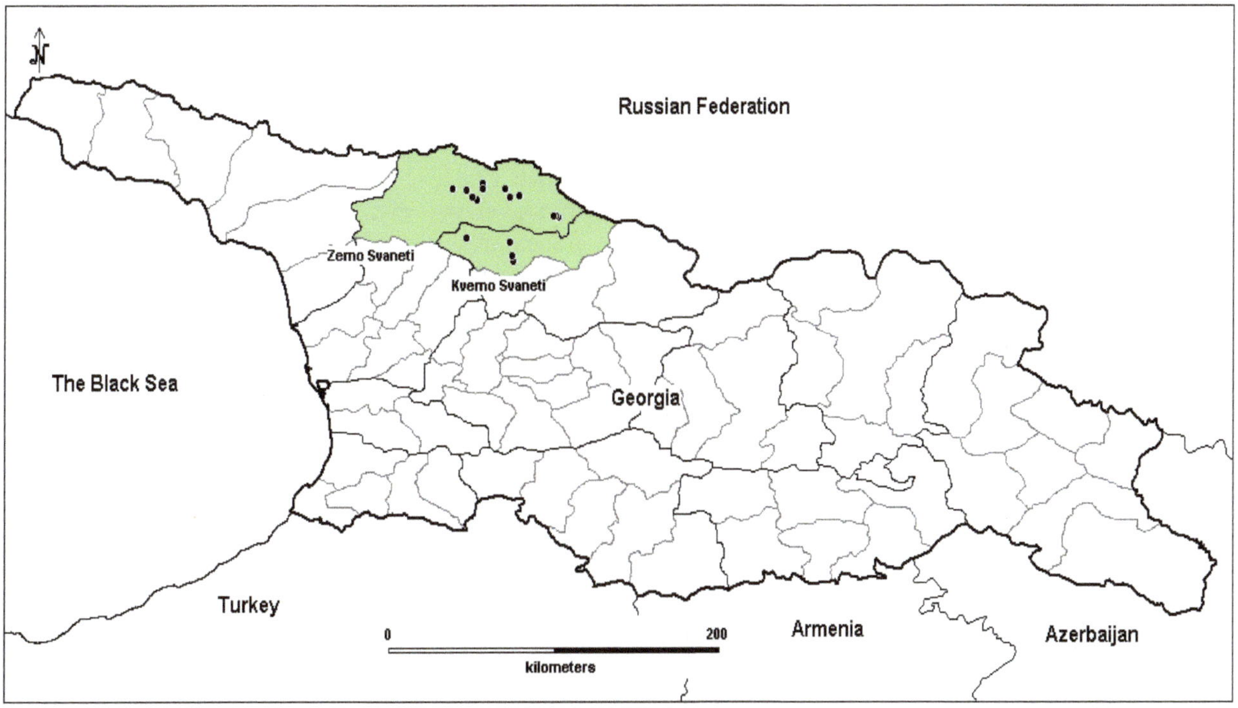

Figure 1. *Schematic map of the study area. The territory of Svaneti is colored and the villages where the interviews were collected are shown as black circles.*

2. Study Area

2.1. Location

Svaneti is a historical province of Georgia, located on the south-facing macro-slope of the western part of the Greater Caucasus. Svanetis Kavkasioni, a part of the Main Watershed Range of the Greater Caucasus, creates the natural northern border of Svaneti between Gvandra and Mt. Pasismta. From the west the region is bordered by the Kodori and Akibo ranges, from the south-west by the rivers Bordjali and Gandishis-Ghele, both tributaries of the river Enguri, as well as by the Bokunstha range. From the east Svaneti is bordered by the northern part of Lechkhumi range, and from the south by the southern part of the same Lechkhumi range, and the eastern part of Egrisi range. The Svaneti range divides the region into two depressions: Zemo (Upper) Svaneti and Kvemo (Lower) Svaneti creating a watershed between the Enguri and Tskhenistskali basins. The region has an altitudinal gradient from 800-4,500 m a.s.l. [18, 19].

2.2. Administrative Division

Administratively Zemo Svaneti belongs to Mest'ia municipality (with an area of 3,045 km^2), which is part of the Samegrelo-Zemo Svaneti Region (Mkhare) and includes Mest'ia, as its administrative center, and 134 villages of 16 communes: Becho, Ch'uberi, Etseri, Idliani, Ipari, Kala, Khaishi, Latali, Lakhamula, Lenjeri, Mulakhi, Nakra, Pari, Tskhumari, Ts'virmi, and Ushguli. Kvemo Svaneti is part of Lentekhi municipality (with an area of 1,344 km^2) and belongs to the Racha-Lechkhumi-Kvemo Svaneti Region (Mkhare). It includes the settlement Lentekhi as administrative center, and 59 villages of seven communes: Chikhareshi, Choluri, Kheledi, Khopuri, Rtskhmeluri, Tsana, and Zhakhunderi [20] (fig. 1).

Svaneti is a historical province of Georgia, located on the south-facing macro-slope of the western part of the Greater Caucasus. Svanetis Kavkasioni, a part of the Main Watershed Range of the Greater Caucasus, creates the natural northern border of Svaneti between Gvandra and Mt. Pasismta. From the west the region is bordered by the Kodori and Akibo ranges, from the south-west by the rivers Borjali and Gandishis-Ghele, both tributaries of the river Enguri, as well as by the Boq'unsta range. From the east Svaneti is bordered by the northern part of Lechkhumi range, and from the south by the southern part of the same Lechkhumi range, and the eastern part of Egrisi range. The Svaneti range divides the region into two depressions: Zemo (Upper) Svaneti and Kvemo (Lower) Svaneti creating a watershed between the Enguri and Tskhenists'q'ali basins. The region has an altitudinal gradient from 800-4,500 m a.s.l. [18, 19].

2.3. Climate

The whole western Greater Caucasus experiences the influence of the Black Sea and moist westerly winds. The mean temperature of the warmest months (July-August) in

Svaneti decreases from +22°C at relatively low altitudes to +7°C –1°C above 3,200 m a.s.l.; and the mean temperature of the coldest month (January) from +10°C to –30°C –35°C. The annual precipitation is 1,500-2,000 mm [21].

2.4. Vegetation Cover

The vegetation of the region includes montain forest, subalpine, alpine, subnival and nival zones and corresponds to the West Caucasian, i.e. Colchic, type of the vegetation vertical zonation [19, 22].

2.5. Population and Local Language

The region has a population of about 24,000 people; 99% of the population are Georgians [20]. Most inhabitants speak both Georgian and Svanetian. The Svanetian language belongs to the Kartvelian[1] group of the Iberian-Caucasian family of languages (there are several Kartvelian languages: Georgian, Zanian (Megrelian, Chanian) and Svanetian), and Svanetian is considered an old branch of the Proto-Kartvelian language. Nevertheless, it still has common basics and grammar structures with Georgian-Zanian languages; archaisms conserved in Svanetian are very important for linguistics. Svanetian does not have written language – historically the common colloquial language for all the Georgian tribes as well as formal language in the Georgian State(s) was Georgian – but rich Svanetian folklore provides sufficient material for linguistic research. The Svanetian language has four dialects: Balszemouri and Balskvemouri in Zemo Svaneti, and Lashkhuri and Lentekhuri in Kvemo Svaneti. In Kvemo Svaneti some researchers distinguish the third dialect, Choluruli, which is very close to Lashkhuri. Each of the listed dialects is only used in small groups of villages in Svaneti [23].

3. Methods

Research was conducted in July 2014. After obtaining prior informed consent for interviews, semi-structured interviews were conducted with participants in local villages.

Vernacular names provided by the respondents were analyzed in parallel to the names given in the dictionaries listed above.

Georgian names are transliterated according to a Romanization system for Georgian approved by US Board on Geographic Names (BGN)/Permanent Committee on Geographical Names (PCGN) 2009 Agreement.

Fungi and lichen species mentioned in the present paper are documented by collections stored at TBI; collection numbers of specimens from Svaneti are given in table 1 for each species. Species nomenclature follows [24].

4. Results and Discussion

Ethnobiological interviews were collected in 16 villages of Zemo Svaneti (Mest'ia, Ebut'i, Ieli, Labsq'aldi, Lakhamula, Mazeri, Tskhek'vani, Ts'virmi, T'viberi, Ushguli: Zhvibiani, Chvibiani and Murq'meli); and Kvemo Svaneti (Babili, Gvimbrala, Khopuri, Mananauri) (fig. 1).

4.1. Fungi

4.1.1. Mushrooms Used for Food

According to literature [2, 5, 25, 26] and herbarium data, 67 species of mushrooms use for food in Georgia occur in Svaneti (Table 1). Our analysis of interviews collected in Svaneti shows that the local population uses mainly 22 species of mushrooms for food, 10 of which are used most frequently: *Armillaria mellea* (Vahl) P. Kumm., *Agaricus arvensis* Schaeff., *A. campestris* L., *Cantharellus cibarius* Fr., *Lactarius piperatus* (L.) Pers., *L. deliciosus* (L.) Gray, *Macrolepiota excoriata* (Schaeff.) Wasser, *M. procera* (Scop.) Singer, *Ramaria flava* (Schaeff.) Quel. Morphologically similar species are often referred to by the same vernacular name, e.g. mushrooms morphologically close to *Boletus* s.l. are all called "Khortsa sok'o" or "Duma sok'o", *Agaricus* spp. as well as *Macrolepiota* spp. are called "Kama sok'o".

The local population interviewed recognized mushrooms to different degrees: 90-100% of the respondents recognized and used *Armillaria mellea* and *Lactarius piperatus*; 70-80% recognized and used *Agaricus arvensis, A. campestris, Cantharellus cibarius, Macrolepiota excoriata, M. procera, Ramaria flava, Lactarius deliciosus*; 30-60% – *Amanita caesarea* (Scop.) Pers., *Pleurotus cornucopiae* (Paulet) Rolland, *P. ostreatus* (Jacq.) P. Kumm., *Polyporus squamosus* (Huds.) Fr., *Boletus* s.l., and less than 25% – *Coprinopsis atramentaria* (Bull.) Redhead, Vilgalys & Moncalvo, *Bovista* spp., *Calvatia* spp., *Lycoperdon* spp., etc. Collectors paid major attention to the fruit-body shape, color and smell. They also considered brightly colored and / or pungent smelling mushrooms as well as unusually shaped ones as suspicious. Almost all mushroom species with white caps were considered "Kama sok'o", which is the major Georgian name for *Agaricus* species. This perception of local collectors often causes confusion of *Agaricus* with *Amanita* species, which leads to poisoning incidents with frequent lethal outcome. In some cases the death of whole families was mentioned by the respondents. Local custom responded to this problem by elaborating a kind of "safety test" for mushrooms when eating them and this custom is described in the local folklore: the mushroom dish was first eaten by the oldest member of a family together with a glass of vodka. If the mushroom appeared safe, the dish was then eaten by the other members of the family. This "test" unfortunately does nothing to avoid Amanita poisoning, which might manifest itself only days after ingesting the mushrooms. *Armillaria mellea* and *Lactarius piperatus* were considered the safest among the fungi. These species were both eaten freshly collected, and dried or salted (by an ordinary technique) for winter use. In Svaneti mushrooms are mainly roasted, stewed or used for "K'ubdari"[2] filling; soups are not prepared of mushrooms. Caps of some species, e.g. *Lactarius deliciosus* and *Macrolepiota* spp. were eaten raw by

[1] "Kartveli" means "the Georgian" in the Georgian language.

[2] "K'ubdari" is a Svanetian pie with beef filling.

shepherds.

Table 1. *Edible Fungi Occurring in Svaneti*

Scientific Name (Collection Nos.)	Major Georgian names according to the literature	Names mentioned in the interviews	Numberof respondents who mentioned the species	Season	Habitat; Substrate
Agaricus arvensis Schaeff.* (TBI5010029)	Mindvris kama	Kama	72	Jul-Oct	Meadows;Manure, soil
Agaricus campestris L.* (TBI5010030)	Mdelos kama	Kama	72	Jul-Oct	Grassy areas;Manure, soil
Agaricus silvaticus Schaeff. (TBI5010031)	Tq'is kama		0	Aug-Sep	Coniferous forest; Soil
Amanita caesarea (Scop.) Pers.* (TBI5009282)	Niq'vi	Neq'v	28	Jul-Oct	Primarily oak forest; Soil
Amanita fulva Fr. (TBI5009283)	Ts'ablisperi livliva		0	May-Oct	Forest; Soil
Amanita rubescens Pers. (TBI5009310)	Marts'q'vio		0	Jul-Oct	Forest; Soil
Amanita vaginata (Bull.) Lam. (TBI5009296)	Rukhi livliva		0	Jul-Oct	Forest; Soil
Armillaria mellea (Vahl) P.Kumm.* (TBI5009311)	Manch'k'vala	Manch'k'varai	85	May-Nov	Primarily beech forest; Trunks and stumps
Boletus edulis Bull.* (TBI5009295)	Datvis sok'o	Khortsa	33	Aug - Oct	Forest; Soil
Boletus erythropus Pers.* (TBI 5009316)	Ts'itelpekha mukhisdzira	–	33	Aug-Nov	Primarily oak forest; Soil
Bovista nigrescens Pers.* (TBI5009323)	Shavi pshuk'una	K'ok'oshol	9	Aug-Nov	Pastureland; Soil
Bovista plumbea Pers.* (TBI5010007, TBI5010008, TBI5009125, TBI5009126)	T'q'viisperi pshuk'una	K'ok'oshol	9	Aug-Nov	Pastureland; Soil
Calvatia candida (Rostk.) Hollós* (TBI5010002)	Tetri tavak'a	Chazhi pu	5	Aug-Nov	Pastureland; Soil
Calvatia gigantea (Batsch) Lloyd* (TBI5009322)	Giganturi pshuk'una	Chazhi pu	4	Jun-Sep	Subalpine meadow; Soil
Cantharellus cibarius Fr.* (TBI5009301)	Miklio	Miklio, Kat-nmiskonai	53	Jun -Nov	Primarily coniferous forest; Soil
Chlorophyllum rachodes (Vittad.) Vellinga (TBI5010040)	Rbilobts'itela ts'erosts'viva	–	0	Jul -Oct	Coniferous forest; Soil
Chroogomphus rutilus (Schaeff.) O.K.Mill. (TBI5010036)	Ts'iteli lorts'ovana	–	0	Jun -Nov	Pine forest; Soil
Clitocybe nebularis (Batsch) P. Kumm. (TBI5009309)	Mindvris sok'o	–	0	Aug-Nov	Forest edges; Soil
Clitopilus prunulus (Scop.) P. Kumm. (TBI5009288)	Alublisdzira	–	0	Jun-Sep	Primarily around Prunus spp.; Soil
Coprinellus micaceus (Bull.) Vilgalys, Hopple & Jacq. Johnson (TBI 5009337)	Mglis sok'o	–	0	May-Nov	Broad-leaved forest; Soil around stumps
Coprinopsis atramentaria (Bull.) Redhead, Vilgalys & Moncalvo* (TBI 5009351)	Silio	Kokodil	6	May-Nov	Grassland, deciduous forest; Soil, buried dead wood
Cortinarius violaceus (L.) Gray (TBI 5009361)	Lurjana	–	0	Jul -Oct	Forest; Soil
Craterellus cornucopioides (L.) Pers. (TBI 5009333)	Dzabra sok'o	–	0	Aug-Nov	Primarily beech forest; Soil
Fistulina hepatica (Schaeff.) With. (TBI 5009365)	Ghvidzla sok'o	–	0	Aug - Oct	Primarily oak forest; Dead wood
Flammulina velutipes (Curtis) Singer (TBI 5009344)	Zamtris sok'o	–	0	Sept - Dec	Primarily on Populus; Dead wood
Gomphidius glutinosus (Schaeff.) Fr. (TBI 5009197)	Nadzvnarebis lorts'ovana	–	0	Aug - Oct	Primarily spruce forest; Soil
Hericium coralloides (Scop.) Pers. (TBI 5009183)	Irmis bats'ara	–	0	Aug - Oct	Primarily beech forest; Dead wood
Hericium erinaceus (Bull.) Pers. (TBI5010032)	Gula sok'o	–	0	Aug - Oct	Primarily forest; Dead wood
Hydnum repandum Fr. (TBI 5009197)	Irmis tucha	–	0	Jul-Oct	Forest; Soil
Hygrophoropsis aurantiaca (Wulfen) Maire (TBI 5009169)	Tsrumiklio	–	0	Jul-Oct	Primarily coniferous forest; Soil, decaying wood

Scientific Name (Collection Nos.)	Major Georgian names according to the literature	Names mentioned in the interviews	Numberof respondents who mentioned the species	Season	Habitat; Substrate
Kuehneromyces mutabilis (Schaeff.) Singer & A.H.Sm. (TBI 5009184)	Zapkhulis manch'k'vala	–	0	Jul-Nov	Primarily beech forest; Living or dead wood
Laccaria laccata (Scop.) Cooke (TBI 5009156)	Martio	–	0	Jul-Nov	Forests; Soil
Lactarius deliciosus (L.) Gray* (TBI 5009175)	Mch'ada	Guiz, Guz, Ch'vadvai	63	Aug - Nov	Coniferous forest; Soil
Lactarius necator (Fr.) Karst (TBI 5009161)	Shavi arq'a	–	0	Jul-Nov	Primarily coniferous forest; Soil
Lactarius piperatus (L.) Pers.* (TBI 5009189)	Arq'a	Arq'ai	82	Jul-Nov	Primarily beech forest; Soil
Lactarius scrobiculatus (Scop.) Fr. (TBI 5009162)	Q'viteli pach'ich'a	–	0	Jul-Oct	Forest; Soil
Lactarius torminosus (Schaeff.) Gray (TBI 5009190)	Nabada sok'o	–	0	Jul-Nov	Primarily birch forest; Soil
Lactarius vellereus (Fr.) Fr. (TBI 5009185)	Savartskhela	–	0	Jul-Oct	Forest; Soil
Lactarius volemus (Fr.) Fr. (TBI 5009186)	Ch'ech'k'et'a	–	0	Jul-Oct	Primarily beech forest; Soil
Laetiporus sulphureus (Bull.) Murrill (TBI5009177)	Q'viteli abeda	–	0	May - Oct	Forest; Living or dead wood
Leccinum aurantiacum (Bull.) Gray. (TBI5009192)	Verkhvisdzira	–	0	Aug - Oct	Primarilypoplar stand; Soil
Leccinum scabrum (Bull.) Gray (TBI5009193)	Arq'isdzira	–	0	Aug - Oct	Primarily beech forest; Soil
Lycoperdon perlatum Pers.* (TBI5009067, TBI5009081)	Khorklebiani gudapshuk'a	Malatus	6	Jun-Nov	Forest, grassland; Soil
Lycoperdon pyriforme Schaeff.* (TBI5009090)	Mskhlisebri gudapshuk'a	Malatus	6	Jun-Nov	Various forests; Soil, decaying wood
Macrolepiota excoriata (Schaeff.) Wasser* (TBI5009304)	Tetri ts'erosts'viva	Kama, Irmispekha	72	Jun -Oct	Various forests, roadsides; Soil
Macrolepiota procera (Scop.) Singer* (TBI 5009305)	Ts'erosts'viva	Kama, Irmispekha	72	Jun -Oct	Various forests, roadsides, fields; Soil
Marasmius oreades (Bolton) Fr. (TBI5009287)	Jimla	–	0	May-Oct	Fields; Soil
Mycetinis alliaceus (Jacq.) Earle ex A.W.Wilson & Desjardin (TBI5010036)	Sanelebela	–	0	May-Oct	Various forest; Buried twigs and other debris
Neolentinus lepideus (Fr.) Redhead & Ginns (TBI5009435)	Kertslovani pirpitk'bila	–	0	May-Oct	Primarily coniferous forest; Stumps
Phallus impudicus L. (TBI5009416)	Kveq'nis gula	–	0	Aug - Oct	Primarily beech forest; Soil
Pleurotus cornucopiae (Paulet) Rolland* (TBI 5009421)	Machalo	Ts'iprash tk'ubul, Ts'ipi sok'	61	May-Oct	Primarily beech forest; Living or dead wood
Pleurotus ostreatus (Jacq.) P.Kumm.* (TBI 5009449)	Khetamkhali	Ts'iprash tk'ubul, Ts'ipi sok'	61	May-Oct	Primarily beech forest; Living or dead wood
Polyporus squamosus (Huds.) Fr.* (TBI5009436)	Dzerana	Dzero	23	May-Oct	Primarily beech forest; Living or dead wood
Psathyrella candolleana (Fr.) Maire (TBI 5009422)	Tetri tsrumanch'k'vala	–	0	May-Oct	Various forests; Living or dead wood
Ramaria flava (Schaeff.) Quél.* (TBI5010034)	Q'viteli sachechela	Irmis rka	43	Jun-Sep	Various forests; Soil
Russula adusta (Pers.) Fr. (TBI5009418)	Mura sok'o		0	Jun -Oct	Primarily coniferous forest; Soil
*Russula aeruginea*Lindbl. ex Fr. (TBI 5009447)	Mtredio	–	0	Jun -Oct	Primarily beech forest; Soil
Russula cyanoxantha (Schaeff.) Fr. (TBI 5009433)	Ghvinio	–	0	Jul-Sep	Primarily beech forest; Soil
Russula delica Fr. (TBI 5009426)	Tskhenis k'bila	–	0	Jun -Oct	Various forests; Soil
Russula emetica (Schaeff.) Pers. (TBI 5009414)	Bghavana	–	0	Jun -Oct	Primarily coniferous forest; Soil
Russula foetens Pers. (TBI 5009444)	Mq'rali khrashuna	–	0	Jun -Oct	Primarily birch forest; Soil
Russula nigricans Fr. (TBI	Moshavo khrashuna	–	0	Jun -Oct	Various forest; Soil

Scientific Name (Collection Nos.)	Major Georgian names according to the literature	Names mentioned in the interviews	Numberof respondents who mentioned the species	Season	Habitat; Substrate
5009430)					
Russula vesca Fr. (TBI 5009446)	Datuna	–	0	Jun -Oct	Deciduous forest; Soil
Russula virescens (Schaeff.) Fr. (TBI 5009439)	Khakhvilo	–	0	Jun -Oct	Deciduous forest; Soil
Sarcodon imbricatus (L.) P.Karst. (TBI5010035)	Irema sok'o	–	0	Aug - Oct	Primarily coniferous forest; Soil
Suillus luteus (L.) Roussel* (TBI 5009424)	Chveulebrivi duma sok'o	Khortsa soko	33	May - Oct	Primarily pine forest; Soil

* indicates the species mentioned in the interviews collected during the present study; Georgian names are given according to [25]; seasonality, habitat and substrate are given according to [25, 26].

4.1.2. Mushroom Names

Formerly the population of Svaneti must have had better knowledge of mushrooms, which is evidenced by literature data, in particular by a much higher number of local names in literature, in a number of cases accompanied by mushroom descriptions [11-17], than found in the current interviews (table 2). Reference [11] gives the highest number of vernacular names (30) with mushroom descriptions and in some cases descriptions of their use. The total number of vernacular names and their modifications in all the available dictionaries are about 50. The majority of these names is now lost in the population; the following 12 vernacular names are still remembered: "Arq'ai", "Ch'aduai", "Chazhi pu", "Dzero", "Guiz", "Guz", "K'ok'oshol", "Malatus", "Manch'k'varai", "Neq'v", "Ts'ipi sok' ", "Ts'iprash tk'ubul" (table 1). Three names: "Katmiskonai", "Irmispekha", "Irmis rka" mentioned by respondents were not found in any of the dictionaries considered. On the basis of mushroom descriptions provided by the above cited literature sources we managed to identify some of the species in question (table 2).

Table 2. *Svanetian Names and Descriptions of Macro-fungi Given in the Literature and Species / Species Groups Identified for the Vernacular Names Presented (Chol. – Choluruli dialect, Lshkh. – Lashkhuri dialect, Lnt. – Lentekhuri dialect, Bz.- Balszemouri dialect, Bk. – Balskvemouri dialect, Zs. – group of Zemo Svanetian dialects).*

Svanetian name	Latin name (identified by the authors)	Fungi / lichen description in the references cited
General names for fungi		
Kokod (Chol.)		Fungus [16]
Kokood		Fungus [16]
Sik', T'q'ubul		Fungus [12]
Sok'		Fungus [14]
Sok'il (Lshkh.)		Fungus [13]
T'q'ubul		Edible mushroom [27]
T'q'ubul (Bz., Bk., Lshkh.)		Fungus [14]
T'q'ubul (Chol.)		Edible mushroom [16]
Polypores		
Abed	*Fomes* sp.	Polypore [27]
Abed (Chol.)	*Fomes* sp.	Boiled down fungus[16]
Abed (Lshkh.)	*Fomes* sp.	A hard fungus, grows out from fallen trees; the fungus is boiled down in ash and kneaded to produce tinder [11]
Abed (Lshkh., Lnt.), Habed (Bz.), Hobed (Bk.)	*Fomes* sp.	Polypore [14]
Abedissok'o, Hobediak, Hobediakt'q'ubul		*Fomes fomentarius* [17]
Habed	*Fomes* sp.	Polypore[12]
Ha-bed, Ho-bed	*Fomes* sp.	Polypore [15]
Dzero	*Polyporus squamosus*	Eaten while fresh [11]
Gak'asok' (Lshkh.)	*Laetiporus sulphureus*	Walnut mushroom – yellow, non-edible [11]; walnut mushroom - yellowish, non-edible [13]
Gak'isok' (Chol.)	*Laetiporus sulphureus*	Walnut mushroom, useful for tinder production after boiling down; like Mamali soko, dry, used dry while treating "Dasuntkuli" (see explanation of "Dasuntkuli" in the text)

		with magic spells [16]; walnut mushroom [27]
Ts'iva	Not identified	Ts'iva is like Abedi, it gives tinder when boiled down, it easily flares up and does not die out [11]; Ts'iva is like Abedi and it transforms to tinder when boiled down [14]
Ts'iiva	Not identified	Tree fungus [16]
Puq'u (Bz.)	Not identified	Tree fungus used for tinder [14]
<u>Agarics</u>		
Arq'ai	*Lactarius piperatus*	White mushrooms with latex [16]
Arq'ai (Chol.)	*Lactarius piperatus*	White mushroom with white latex; poisonous when raw; its abundance betokened snowy winter [16]
Arq'ai (Lshkh.)	*Lactarius piperatus*	White, eaten roasted [11]
Arq'ai (Lshkh.), Arq'a (Lnt.)	*Lactarius piperatus*	White mushroom, Arq'a mushroom; Arq'a mushroom is eaten roasted [14]
Ch'aduai	*Lactarius deliciosus*	Red mushroom, eaten raw, roasted, boiled [11]
Chazhisok'	*Coprinus* sp.	Grows on horse manure, non-edible [11]
Ch'vadvai	*Lactarius deliciosus*	Edible mushroom, appears in summer and persists till autumn. It this mushroom predominated, people expected rich harvest [16]
Guiz(Zs.), Guz (Lshkh.)	*Lactarius deliciosus*	Mch'ada soko [14]
Kharik'bilai	*Russula delica*	Kharis k'bila - looks like Arq'a soko, is eaten roasted [11]
Manch'k'varai	*Armillaria mellea*	Manch'k'vala [16]
Manch'kvarai (Chol.)	*Armillaria mellea*	Manch'k'vala [16]
Manch'k'varai (Lshkh.)	*Armillaria mellea*	It is collected in spring, dried up and eaten in winter [11]; Manch'k'vala, collected in summer, dried up and eaten in winter in soup [13]
Neq'v	*Amanita caesarea*	Soft mushroom [11]
T'q'ubla	*Pleurotus ostreatus, P. cornucopiae*	Edible mushroom, grows on trees [11]
Ts'ipisok'	*Pleurotus ostreatus, P. cornucopiae*	Beech mushroom, edible [11]
Ts'iprasht'q'ubul		*Pleurotus ostreatus* [17]
Vijik'orgol (Lashkh.)	*Macrolepiota* sp.	Grows on ground, looks like an umbrella, edible [11]
Puffballs		
Chazhipu	*Bovista* sp., *Calvacia* sp.	Ts'khenis kuili - usual in riparian forest, rounded, small, becomes dusty when ripe. Grows in fields, looks like cheese, crashes like "suluguni"[3] when weighed upon; becomes dusty when ripe [11]
K'ok'oshol (Chol.)	*Bovista* sp.	Small rounded good-looking fungus [16]; Fresh fungus[27]
Malatus		*Lycoperdon* [17]
Malatus (Lshkh.)	*Lycoperdon* sp.	Meliis tsuili - grows on grass, white, later becomes black, edible while fresh [11]
Mavlatus (Chol.)	*Lycoperdon* sp.	Small white autumn fungi, after drying out it is full of black dust [16]; a fungus, which becomes dust as soon as it grows up [27]
<u>Not identified</u>		
Q'ivnul sok'		No other edible mushroom can be compared with it, good in March [11]
K'orich'ol (Lshkh.)		Blood red fungus, very small, good when roasted [11]
K'oroch'ol (Lshkh.)		White outside, blood red inside, very small, delicious when roasted [11]
Mavsoq'ian sok'		A fungus, which make one mad [11]
Matkhep sok' (Lshkh.)		Dapertkheba soko - spread over the ground, non-edible [11]
Piluai		A fungus, which make one mad [11]
Tsaira sok'		Elm fungus, black, edible, tasty [11]
Tsats'khvi sok'		Linden fungus, edible, tasty [11]
Ghumuri sok'		Spruce fungus, non-edible [11]
Ghuninichv		Khbostucha is good roasted; grows on pears and elsewhere [11]

[3]"Suluguni" is a kind of Georgian cheese.

Ghunitkhum		Khbos tava, edible [11]
Iebra sok' (Lshkh.)		Sweet cherry tree fungus, red, non-edible [11]
K'imeegh (Chol.)		White fungus, non-edible [16]
Kokodilar		A good mushroom of middle size, red, tasty when roasted [11]
Lichens		
Chach	*Usnea* spp., *Bryoria* spp., *Ramalina thrausta*	Beard of a man and a tree[27]
Pimpous	*Usnea* spp., *Bryoria* spp., *Ramalina thrausta*	Hair, tree moss [11]

Several species mentioned in the interviews were not identified on account of insufficiently detailed descriptions and absence of specimen material; the vernacular names of these species are: "Datvisq'ura" (5 interviews), "Ts'itla" (3), and "Dedopala" (2).

The collective names for mushrooms in various dialects of Svanetian language are: "Sok'", "T'q'ubul", "But'q'uli", "Kokod", "Sok'il", "Sik'" (Table 2). At present the population mainly uses the word "Sok'o" for mushrooms in general. Species specific Svanetian mushroom names are thus clearly being lost in everyday life.

4.1.3. Mushroom Oracles

In Svaneti (in particular, Lashkheti, a historical commune of Svaneti) mushroom observations were formerly used as an oracle: it was thought that abundant "Ch'vadvai" (*Lactarius deliciosus*) promised good harvest, and much "Arq'ai" (*Lactarius piperatus*) betokened snowy winters [27]. None of the respondents interviewed mentioned/had heard about such folklore.

4.1.4. Tinder

Formerly tinder was made of "Abeda soko" (*Fomes* sp.) – the name literally means "tinder mushroom". The population (mainly elders) still recalls how to prepare tinder. A birch polypore (*Piptoporus betulinus* (Bull.) P. Karst.) was thought to be the best material for tinder; it was well boiled with ash, well beaten, dried and ground, yielding tinder (B. Kakhberidze, Mest'ia, pers. comm., 2014). The younger population has already lost this knowledge.

4.1.5. Mushrooms as Medicine

The Svaneti population also used mushrooms for medicinal purposes. Almost 100% of the respondents collected already black (over-ripe) fruit-bodies of *Lycoperdon* and *Bovista* species from August to October, and kept them dry in glass pots during the whole year. The fungi were used to stop bleeding by pouring the contents of the fruit-bodies onto wounds; the powder was also used for treatment burns. The peel of a fruit-body was also put as a plaster on abscesses.

Another mushroom used for medicinal purpose is *Lactarius deliciosus*: a wine-like juice was produced by mixing this mushroom with sugar, and the juice was used for cough, and the treatment of diarrhea (D. Jinchradze, M. Kurasbediani and Sh. Kurasbediani, Babili, pers. comm., 2014). Several elders recalled that in their childhood their parents prepared a mushroom tincture, which was used by hunters to cure colds by rubbing the tincture onto body.

The medicinal use of fungi in Svaneti is also described in literature: "Patrik'i", "Patruki" was made of polypore and birch twigs; the twigs were sharpened like a pencil, the tip smeared with a mushroom based emollient and put into wound" [28]. "… „Mamali sok'o", was used dry while treating "Dasuntkuli" with magic spells" [16]. "Dasunuli" is the same as "Dasuntkuli" – when (a snake) touches something with its mouth and poisons it" [29]; Georgian traditional medicine knows a category of rash, whose appearance is explained by taking food products passed over or smelled by a mouse, snake, etc. [30]; "Dasunva" means "to smell" and "Dasuntkva" means "to breathe on" in Georgian) According to various descriptions provided by the literature sources (Table 2), the polypore used must be *Laetiporus sulphureus* (Bull.) Murrill.

4.2. Lichens

Lichens, particularly, *Usnea* and *Bryoria* species as well as *Ramalina thrausta* (Ach.) Nyl.are called "Pimpa" or "Pimpous" [11] and "Chach" [27] in Svaneti. Lichens were never used for food there. "Pimpa" or "Pimpous", literally means "a man with dense hair and beard" and is also described as "conifer bristle." These species were used by hunters as field-bedding to keep warmth in cold winters. The use of lichens is reflected in local folklore: one of the respondents told a story about a hunter who went to Mt.Zhibelakhvi, killed a Caucasian goat, and when he broiled its meat, a giant ogre came to the cave where the hunter stayed. The ogre asked for the broiled meat. The hunter gave him a piece but when the ogre asked for another one, the hunter thought that his unexpected guest would eat the whole goat and thought to dodge the ogre. He took the "Pimpous", wrapped himself in lichens, and set fire to them with a burning log. The ogre did the same but burnt himself, rushed out of the cave and fell into an abyss (R. Pirtskhelani, Mest'ia, pers. comm., 2014).

5. Conclusions

The following major conclusions can be drawn from the results of the present work: the local population of Svaneti currently uses 22 species of mushrooms for food, 10 of which are used most frequently. In the past the population of Svaneti must have had better knowledge of mushrooms, which is evidenced by literature data, in particular by the much higher number of local names. Fungi were originally not only used for food, but also for tinder production (polypores), for medicine (sticks for emollient application (polypores), medications for internal and external use (agarics, puffballs)). A limited number of lichens were used by hunters as field-bedding to keep warm in the cold winters. The uses of fungi and lichens are clearly reflected in the local folklore. Further investigations are needed to (a) clarify the identity of certain fungi indicated by respondents by vernacular names or usage descriptions / found in the literature without any descriptions or with insufficient descriptions to allow a identification to species or species group, (b) to compare fungi use with other regions of Georgia and the Caucasus in general. It is greatly recommended to publish local popular guidebooks on edible (67 recorded vs. 22 actually used by the local population), and poisonous (many respondents mentioned frequent cases of *Amanita* poisoning because of confusion with *Agaricus*), mushrooms occurring in the study region, to be disseminated among the population.

Acknowledgements

The Georgian authors also would like to thank the Shota Rustaveli Natonal Science Foundation, Georgia, for funding the field trip to Svaneti within the framework of the research project "Ethnobiological study of fungi and lichens of Georgia".

We would like to thank Ms. Chato Gujejiani who helped us in translation of Svanetian descriptions of a number of fungi and all the people of Svaneti, especially the residents of the villages of Zemo and Kvemo Svaneti, who kindly agreed to participate in the ethnobiological interviews, as well as those who helped the authors in organizing the expedition.

References

[1] R. Singer, "Pilze aus dem Kaucasus 2. Ein Beitrag zur Flora Swanetzens und einiger angrenzender Taler," Botanischer Centralblatt Beihefte, vol. 48(2), pp. 513-542, 1931.

[2] I. G. Nakhutsrishvili, Agarics of Georgia. Tbilisi: Metsniereba, 1975.

[3] V. Kh. Gulmagarashvili, New species of Aphillophorales for Georgia. Reports of the Academy of Sciences of the Georgian SSR, vol. 91(3), pp. 685-687, 1978.

[4] T. Anchabadze, Materials for study of Discomicetes of Svaneti, Bulletin of the State Museum of Georgia, vol. XX-A, pp. 111-127, 1960.

[5] I. G. Nakhutsrishvili, Flora of Spore-producing Plants of Georgia (A Conspectus). Tbilisi: Metsniereba, 1986.

[6] E. Vainio, Lichenes in Caucaso et in Peninsula Taurica annis 1884-1885 ab H. Lojka et M. Dechy collecti, Termeszetr. Fuz., vol. 22, pp. 269-343, 1899.

[7] A. Jatta, Lichenes in Sommier S. et Levier E. Enumeratio plantarum anno 1890 in Caucaso lectorum. Acta Horti Petropolitani, vol. XVI., 1900.

[8] M. Dechy, „Kaukasus". Reisen und Forschungen im kaukasischen Hochgebirge. Bd. III. Bearbeitung der gesammelten Materialien von F. Filarszky, E. Csiki, K. Papp, F. Schafarzi k und M. v. Dechy. Mit 36 Lichtdrucktafeln, Berlin: Dietrich Reimer (E. Vohsen), 1907.

[9] Degelius G., The Lichen genus *Collema* in Europe, Symb. Sot. Upsal., vol. 13, 1954, p. 2.

[10] O. Szatala, Lichenes in Peninsula Taurica et in Caucaso ab F. Komienski, D. Sosnowsky et E. Koenig collecti, Borbasia, vol. 4, pp. 70–96. 1944.

[11] A. Oniani, Collection of Svanetian Names of Trees and Shrubs (on Lashkuri Dialect), Materials on Japhetic linguistics, vol. VIII. Petrograd: Academy of Sciences Press, 1917.

[12] B. Nizharadze, Georgian-Svanetian-Russian Dictionary, Balszemouri dialect. Tbilisi: Universali, 2012.

[13] K. Dondua, Svanetian-Georgian-Russian dictionary (Lashkhuri Dialect). Tbilisi: Publishing house of Sulkhan-Saba Orbeliani Pedagogical University of Tbilisi, 2001.

[14] V. Topuria, Svanetian Dictionary. Tbilisi: Kartuli Ena, 2000.

[15] N. Ia. Marr, Extract from Svanetian-Russian Dictionary, Materials on Japhetic linguistics, vol. X. Petrograd: Russian State Academy Press, 1922.

[16] A. Liparteliani, Svanetian-Georgian Dictionary (Choluruli sub-dialect). Tbilisi, 1994.

[17] A. Makashvili, Botanical Dictionary. Tbilisi: Metsniereba, 1991.

[18] D. Ukleba, For physical-geographic characteristic of Zemo Svaneti, Manuscript of Candidate Dissertation, TSU, 1952, unpublished.

[19] R. Gagnidze and M. Davitadze, The Local Flora [Plant World of Georgia]. Batumi: GamomtsemlobaAch'ara, 2000.

[20] Major results of the first national general census of the population of Georgia (Population of villages of Georgia), vol. 2, T. Beridze, I. Archvadze and Ts. Eliadze, Eds. Tbilisi: State Department of Statistics of Georgia, 2003, pp. 143-146, 164-169,

[21] M. Kordzakhia and Sh. Javakhishvili, Climate of Georgia. Tbilisi: Ganatleba, 1971.

[22] N. Zazanashvili, R. Gagnidze and G. Nakhutsrishvili, Main types of vegetation zonation on the mountains of the Caucasus, Acta Phytogeographica Suecica, vol. 85, pp. 7-16, 2000.

[23] V. Topuria, Svanetian language (1965), Unpublished works on Svanetian language, in Works, vol. 2, V. Topuria. Tbilisi: Kartuli ena, 2002, pp. 150-219.

[24] P. M. Kirk, Species Fungorum (version Feb 2014), in Species 2000 & ITIS Catalogue of Life, 22nd December 2014, Y. Roskov, L. Abucay, T. Orrell, D. Nicolson, T. Kunze, A. Culham, N. Bailly, P. Kirk, T. Bourgoin, R.E. DeWalt, W. Decock and A. De Wever, Eds. Digital resource at www.catalogueoflife.-org/col. Naturalis, Leiden, the Netherlands: Species 2000, 2014.

[25] I. Nakhutsrishvili, Fungi of Georgia. Tbilisi: BunebaPrinti, 2007.

[26] K. Kanchaveli, I. Nakhutsrishvili and Tz. Inashvili, Key to Spore-producing Plants of Georgia, vol. 1. Tbilisi: Metsniereba, 1981.

[27] A. Liparteliani, Svanetian-Georgian Dictionary (Choluruli Sub-dialect). Tbilisi: Diogene, 2014.

[28] M. Burduli, Folk medicine in West Georgia (according to Svanetian Ethnographic Materials). Akhaltsikhe: Akhaltsikhe University, 2010.

[29] Explanatory Dictionary of the Georgian Language, A. Chikobava, Ed.-in-Chief. Tbilisi: Main Editorial Board of the Georgian Soviet Encyclopedia, 1986.

[30] Gurian Medical Book, J. Kharchilava, Comp., M. Asatiani Ed. Tbilisi: Pergamenti, 1993.

8

Algal Indication of Climatic Gradients

Sophia Barinova[1, *], Viktor Gabyshev[2], Mariyo Boboev[3], Lali Kukhaleishvili[4], Olena Bilous[5]

[1]Institute of Evolution, University of Haifa, Mount Carmel, Haifa, Israel
[2]Institute for Biological Problems of Cryolithozone SB RAS, Yakutsk, Russia
[3]Khatlon Scientific Center Academy of Sciences of the Republic of Tajikistan, Kulob-town, Tajikistan
[4]Tbilisi Botanical Garden and Institute of Botany, Tbilisi, Georgia
[5]Institute of Hydrobiology of NAS of Ukraine, Kiev, Ukraine

Email address:
barinova@research.haifa.ac.il (S. Barinova)

Abstract: Bio-indication is only possible because algal communities respond to environmental changes in a consistent way. In particular, the vectors of algal changes over latitudes and altitudes are well correlated with respective climatic gradients. The relationships between biotic and climatic changes are insufficiently studied so far, but regularities are analyzable provided the adequate geographic scale. Thus the impact of seasonality requires ecological study of the aquatic object as a whole or a considerable part of it. For the gradient analysis of altitudinal changes the sampling data must cover a region of diverse relief, whereas latitudinal gradients are traceable of sizeable parts of continents encompassing different climatic zones. In our studies, the impact of temperature changes is invariably significant when revealed with the help of bio-indication analysis. In the boreal realm, algal development most typically show three seasonal peaks correlated with the dynamics of ice cover, whereas two to single peak are distinguished down the latitudes. The latitudinal dynamics is fairly obvious in the quantitative relationships between diatoms, green, golden algae, and Infraspecies-Species variability. With climatic stress ascending to the north, the role of diatoms, but in the high Arctic regions drops significantly. The Infraspecific variability index increases from 1.09 up to 1.42. To the south, over the mountainous areas of Tajikistan, Georgia, Mediterranean and India, diatoms consistently decrease over the altitudinal range 200 – 2500 meters, while the other algal groups betray a less significant variation. Such regularities are more obvious with statistical correlation of climatic variables and freshwater algal diversity. Thus the distribution of phytoplankton species in the rivers of Yakutia and Chukotka over the gradient of DHI index and the duration of the ice free period is controlled by sun light intensity in the first place, but also reflects variation in the climatic stress resistance, the most prominent in the case of diatom algae.

Keywords: Freshwater Algae, Diversity, Ecology, Bio-Indication, Altitude, Latitude, Climate Change

1. Introduction

Climate is a complex dynamic system, and climate change is a cumulative effect of many interlinked natural processes. Sun luminosity is involved as well as the terrestrial factors related to the Earth rotation and revolution variables [1]. The long term periodicity is related to the rotation velocity effect on the mass distribution in the earth's crust and mantle [2] the hydrosphere effect producing a lower scale fluctuation [3], superimposed by the polar ice cap autocycles, with multiple feedbacks. The CO2 greenhouse effect among them falls under multiple regulation of hydrosphere, rock weathering, soil outgassing, and biomass production jointly damping fluctuations to a low amplitude cyclicity [2].

However, climatic variables can be considered relatively stable relative to the developmental time scale of algal communities. Factors such as latitude or altitude gradient, are not only linked to the main climatic characteristics, but also the key players in shaping factors of biota diversity as recognized in the current stage of biome evolution.

The existence and use of active systems of bio-indication confirms that algal communities respond to environmental change. The climatic conditions of the environment have a well-defined gradient when considered over a range of latitudes or altitudes of the habitats. Investigation of the relationship of freshwater algal communities with climatic gradient is still in the early stage [4,5], but it is already clear that community response has a certain regularity. The scale of the investigated region corresponds to the problem. Thus, to identify seasonality it can be the whole aquatic object or a portion thereof. For an altitudinal gradient it is the region comprising a steep topographic range as in the mountains. For the latitudinal gradient it has to be part a land mass of a considerable range of climatic gradient.

The aims of the current study are to reveal the response of high mountains and high latitude algal communities in Eurasia to the latitude and altitude-related climatic condition of its habitats. Thus, we try to implement diverse bio-indicational and statistical methods which represent some new approaches in freshwater algal diversity analysis of climate impact.

2. Study Area and Methods

Material for this study comes from the last twenty years of research in the rivers and lakes of Eurasia [6-17].

The main approach for algal diversity summarizing data comes from [18].

To identify the main distribution patterns of algae diversity over climatic gradients we used statistical methods that are those recommended by [19] for development of floristic and taxonomic studies, namely, the CANOCO Program [20] for Representational Difference Analysis (RDA) and Canonical Correspondence Analysis (CCA) [21,22] , GRAPHS program [23] for comparative floristic evidence, Statistica 7.0 Program for Stepwise regression analysis and Distance Weighted Least Squares calculation, and statistical techniques included in the PRIMER-v5 package [24] for the comparative inter-regional analysis of taxonomic diversity [6].

3. Results and Discussions

We have found that one of the most important climatic factors is temperature [25] and the aquatic algal communities react to changes in water temperature and sometimes it can only be seen through bio-indication analysis [9]. And, if in the boreal region of algal development has three peaks during free of ice-period [10], in the more southern areas it is about two [6], or southern up to one [11,12].

3.1. Latitudinal Distribution

So, in boreal regions seasonal fluctuations are regulating algal growth in respect to the level of nutrient supply. The fluctuations of the phytoplankton biomass show seasonal peaks in spring, autumn, and summer, associated with the main feature of boreal waterbodies – the complexity of their thermal structure [10]. At the same time, relationships between algal species diversity and water temperature can be revealed when we compare seasonal communities, in particular the summer ones which are the most rich and diverse. It can be seen in Fig. 1 for one of the large European rivers in Eurasian Boreal Realm, the Southern Bug River phytoplankton communities [12]. This approach lets us to avoid the impact of seasonality fluctuation to calculation result. In this case we choose summer phytoplankton communities for calculation of climatic parameters impact on the algal taxonomic structure and productivity.

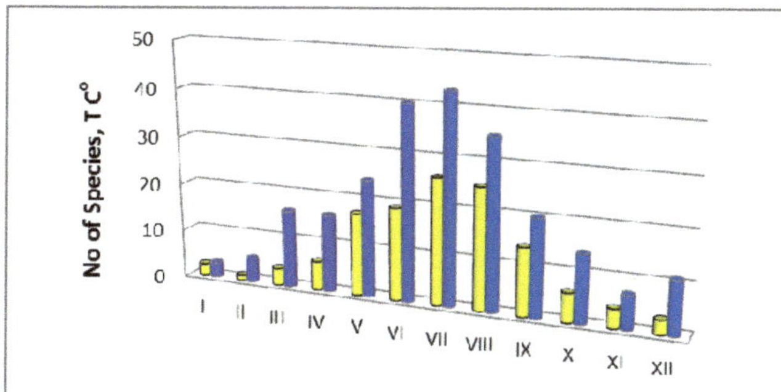

	I	II	III	IV	V	VI	VII	VIII	IX	X	XI	XII
T	2.5	1	3.5	6	17	19	26	25	14	6	4	3
No Species	3	5	16	16	24	40	43	35	20	14	7	11

Figure 1. Monthly water temperature and phytoplankton species richness in the Southern Bug River.

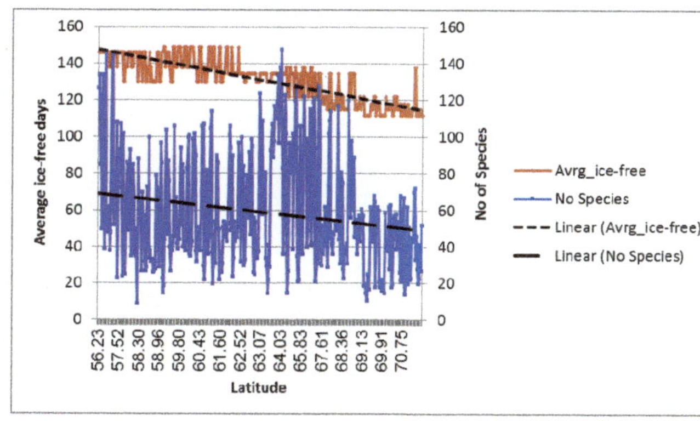

Figure 2. Distribution of ice-free days and species richness in rivers of Yakutia and Chukotka over latitude gradient.

As has been revealed, the latitudinal climate change from boreal region to the Arctic affects the distribution of diatoms, green and golden algae (Fig. 2) [13]. With the increase climatic stress the species richness sufficiently decreased with the number of ice-free days decreasing because the algal cells need photosynthetic radiation for it grows.

3.2. Altitudinal Distribution

Among the southern regions of Mediterranean, Tajikistan, Georgia, and India witness a decrease of diatoms with altitude in the interval from 200 to 2500 meters, with the other groups scarcely affected.

So, in the Eastern Mediterranean where altitude gradient changed from 440 m below sea level up to 2200 m above sea level (a.s.l.), the role of green algae and cyanobacteria is more significant than the diatoms (Fig. 3) [14].

Algal distribution in the Vakhsh River Basin shows the importance of altitude and associated climatic gradients (Fig. 4). Can be seen that the role of diatoms in aquatic habitats of the Pamir Mountains is significantly decreased with increasing of habitat altitude [15].

Similar distribution we revealed for the Caucasus Mountain algal diversity (Fig. 5) where diatoms also are a climate indicative group of algae [7,8].

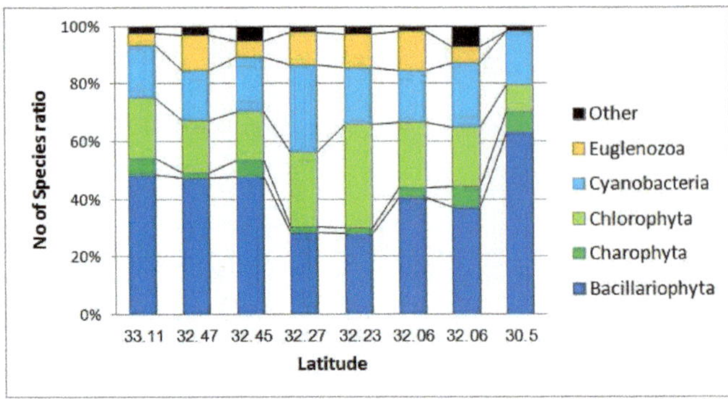

Figure 3. *Split Distribution of algal species richness in rivers of Eastern Mediterranean over latitude.*

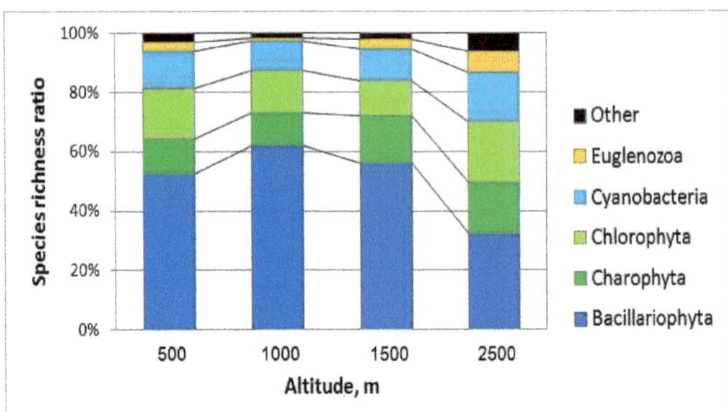

Figure 4. *Distribution of algal species richness in the Vakhsh River Basin habitats over latitude.*

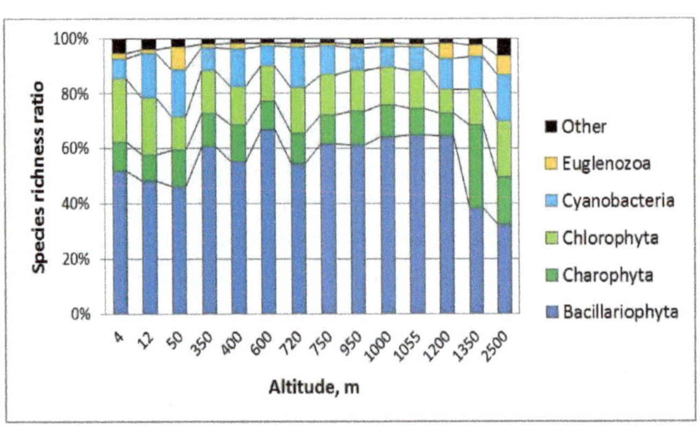

Figure 5. *Distribution of algal species richness in Caucasus Mountain aquatic habitats over latitude*

Climate influenced not only species richness in algal communities but also production variables such as abundance and biomass of riverine phytoplankton [13]. This let us to assume that sunlight intensity and environmental temperature is the major regulating factors for gamma-diversity distribution patterns.

Revealed relationship of algal diversity distribution over altitude of habitats seems like natural and therefore we can assume that the diatom algae can be marker taxa for the climatic influence on bio-diversity change.

3.3. Salinity - Global Climate Warming Relationships

The main influence on the Central Eurasia lakes comes from cyclic changes of moisture in the region, happening along with global warming. We revealed the statistically significant correlation between salinity of the Balkhash Lake [26], the forty eight northern Kazakhstan lakes [27], Ukrainian saline lakes [28], and water bodies of the Eastern Mediterranean [6] and the long-term dynamics of algal species richness which decreased with increasing of water salinity. Importance of salinity for productivity algal communities let us to assume that climate warming that provoke salinity increasing as a result of evaporation is especially high in arid and semi-arid regions of planet. Decreasing of species diversity and aquatic community productivity can decrease not only total biomass production but also increase CO_2/O_2 ratio in atmosphere when global temperature will increase.

3.4. Statistical Approaches

Statistical methods further clarify the relationship of climate change and diversity of algae. Thus, the distribution of phytoplankton species in the rivers of Yakutia and Chukotka on a gradient index DHI (Dynamic Habitat Index) [29] and the ice-free period shows the importance of sunlight intensity, which is understandable, but also inhibition of algae in the harsh conditions of the Arctic, can be traced more significant role in the communities of diatoms lower latitudes [13].

Statistical methods that we implemented to our material help us to reveal not only real data relation but also have prognostic properties. So, Fig. 6 shows the importance of DHI index and sunlight intensity for algal species richness.

Fig. 7 shows a decreasing role of diatom algae in the high-latitude Eurasian communities and correlated with the Geo-association index of latitudinal distribution combined with the direction of the river flow [13].

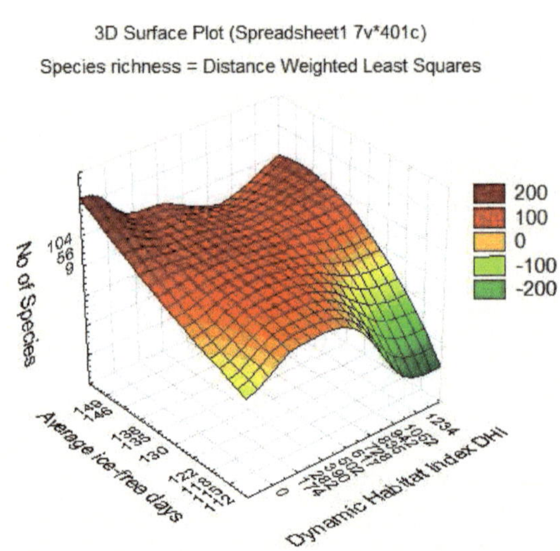

Figure 6. *Distribution pattern of phytoplankton species richness in phytoplankton of the large Eastern Siberia rivers over sunlight intensity (Average ice-free days) and landscape sensitivity (DHI).*

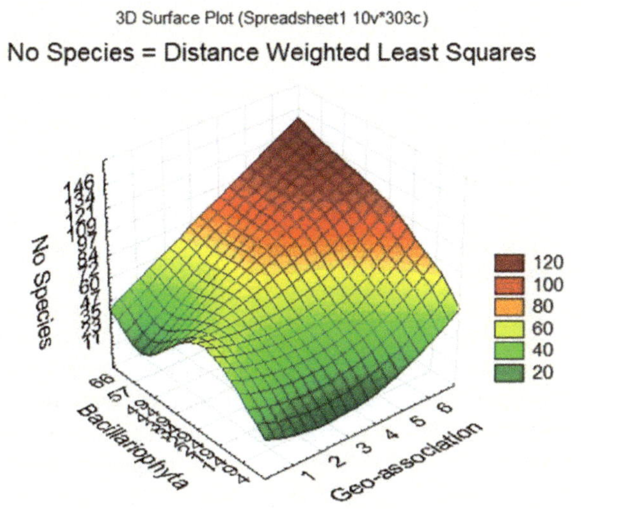

Figure 7. Distribution pattern of phytoplankton species richness in the large Eastern Siberia rivers over diatom numbers and Geo-association Index

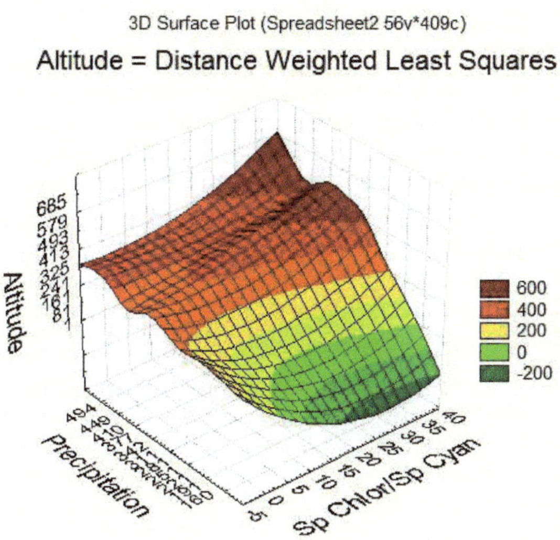

Figure 8. Altitude, precipitation and Chlorophyta/Cyanobacteria ratio relationships in the large Eastern Siberia rivers.

Despite the fact that the diatoms represent conspicuous response to climatic impact, indication of climate change is also possible with correlation of separate taxonomic groups. Fig. 8 show that green algae species richness increased with increasing of habitat altitude and altitude-related precipitation rate, while cyanobacteria decrease [13].

3.5. Infraspecific Variability

A notable feature of algal taxonomic structure in water bodies of southern regions of Eurasia is a large proportion of monomorphic species comparative with other algal floras of Europe, central - northern Asia and North America.

Such exceptionally low infraspecies variability might have been related to recent climatic instability and anthropogenic impacts that obliterated microhabitat differentiation and promoted highly tolerant monomorphic populations. In the both cases the Index Infraspecies/Species variability that the decrease in Hemisphere from north to south [30] can be related with climate changes and can be used as criteria of future warming. We added some new information from our recent research and represent it in Fig. 9.

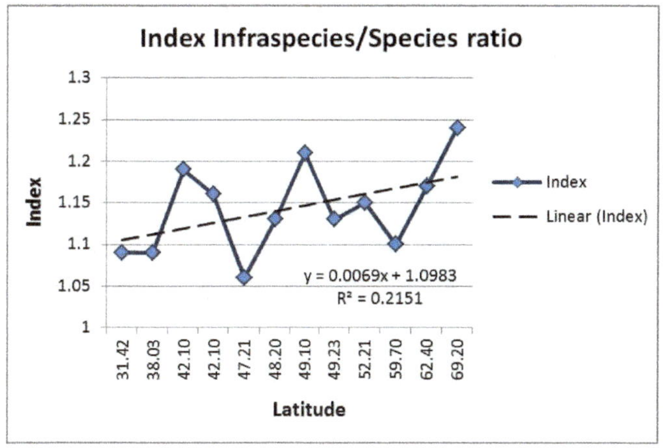

Figure 9. Index Infraspecies/Species variability over latitude of habitats in Eurasia.

Trend line shows sure rising of Infraspecies/Species Index due to the latitude of algal habitat. Remarkable that outstanding point such as 1.19 (Georgian Natural Reserves), 1.21 (Diatoms of Europe), 1.21 (Viliuy Reservoir, Yakutia) are summarized large sets of long-term scale information [30]. The Tajik Depression flora has even 1.43 Index [15], but it is not only large scale regional and time series, but also high mountain biome.

4. Conclusion

Algal bio-indication remains an efficient method for assessing not only water quality, but also climatic influence, involving the wide range of indicators and based on the ecological preferences of the species. Assessment of climate impacts on aquatic ecosystems, held the first time for gamma-diversity, shows the importance of bio-indication for the understanding of regional and global changes.

Recent studies show that the data on the diversity and ecology of algae can identify the climatic gradients influence, as well as reveals groups of aquatic organisms which are most clearly reflect their.

The impact of latitudinal and altitudinal temperature changes is invariably significant when revealed with the help of bio-indication analysis.

Infraspecies/Species Index is remarkable rising with latitude of algal habitat in the continent. The latitudinal climate change from boreal region to the Arctic affects the distribution of different groups of algae when the species richness sufficiently decreased with decreasing of number of ice-free days because the algal cells needs photosynthetic radiation for it grows.

The role of green algae and cyanobacteria is more significant than the diatoms with altitude of habitat rising.

Climate influenced not only species richness in algal communities, but also production variables such as abundance and biomass, where the diatom algae can be marker taxa for the climatic influence on bio-diversity change. This lets us to assume that sunlight intensity and environmental temperature is the major regulating factors for gamma-diversity distribution patterns.

We revealed the statistically significant correlation between salinity and the long-term dynamics of algal species richness which decreased with increasing of water salinity. Therefore, we assumed that the main influence on the Central Eurasia lakes comes from cyclic changes of moisture in the region, happening along with global warming.

Statistical methods that we implemented to our material helps us to reveal not only of the real data relation, but also having of prognostic properties showing the importance of sunlight intensity and altitude-related precipitation rate for algal species richness in purpose of revealing climatic trends.

Acknowledgements

We are thankful to the Israeli Ministry of Absorption that partly funded this research.

References

[1] Krassilov, V., Barinova, S. (2013). Sea level – geomagnetic polarity correlation as consequence of rotation geodynamics. Earth Science, 2(1): 1-8.

[2] Krassilov, V.A. (2003). Terrestrial Palaeoecology and Global Change. Pensoft: Sophia, 464 pp.

[3] Krassilov, V., Barinova, S., Rybnikov, S. (2014). Rotation Forcing of Tectonics and Climate. Earth Sciences, 3(3): 68-75. doi: 10.11648/j.earth.20140303.11

[4] Barinova, S. Nevo, E. (2012). Climatic and pollution impact on algal diversity of the freshwater ecosystems in Eurasia. In: Climate Change and Impacts. AcademyPublish.org, WY, USA: pp. 16-50.

[5] Medvedeva, L.A., Barinova, S., Semenchenko, A.A. (2012). Use of Algae for Monitoring Rivers in the Monsoon Climate Areas (Russian Part of Asian Pacific Region). International Journal of Environment and Resource, 1(1): 39-44.

[6] Barinova, S. (2011). Algal diversity dynamics, ecological assessment, and monitoring in the river ecosystems of the eastern Mediterranean. Nova Science Publishers, New York, USA, 363 pp.

[7] Barinova, S.S., Kukhaleishvili, L., Nevo, E., Janelidze, Z. (2011). Diversity and ecology of algae in the Algeti National Park as a part of the Georgian system of protected areas. Turk. J. Bot., 35 729-774.

[8] Barinova, S., Kukhaleishvili, L. (2014). Diversity and ecology of algae and cyanobacteria in the Aragvi River, Georgia. The Journal of Biodiversity. Photon, 113: 305-338.

[9] Klochenko, P., Shevchenko, T., Barinova, S., Tarashchuk, O. (2014). Assessment of the ecological state of the Kiev Reservoir by the bioindication method. Oceanological and Hydrobiological Studies, 43(3): 228-236.

[10] Barinova, S., Stenina, A. (2013). Diatom diversity and ecological variables in the Arctic lakes of the Kostyanoi Nos Cape (Nenetsky Natural Reserve, Russian North), Plant Biosystems, 147(2): 397-410. DOI:10.1080/11263504.2012.749956

[11] Barinova, S., Keshri J.P., Ghosh S., Sikdar J. (2012). The influence of the monsoon climate on phytoplankton in the Shibpukur pool of Shiva temple in Burdwan, West Bengal, India. Limnological Review, 2(2): 47-63. DOI 10.2478/v10194-011-0044-y

[12] Bilous, O., Barinova, S., Klochenko, P. (2013). Climatic influence on the phytoplankton communities of the upper reaches of the Southern Bug River (Ukraine). Transylvanian Review of Systematical and Ecological Research. The Wetland Diversity, 15(2): 61-86.

[13] Barinova, S., Gabyshev, V., Gabysheva O. (2014). Climate Impact on Freshwater Biodiversity: General Patterns in Extreme Environments of North-Eastern Siberia (Russia). British Journal of Environmental and Climate Change, 4(4): 387-407.

[14] Barinova, S., Krassilov, V.A. (2012). Algal Diversity and Bio-indication of Water Resources in Israel. International Journal of Environment and Resource, 1(2): 62-72.

[15] Barinova, S., Boboev, M., Hisoriev H. (2015). Climate impact on freshwater algal diversity of the South-Tajik Depression in a high mountainous extreme environment. Turk. J. Bot., 39(1), in press

[16] Barinova, S., Naiz Ali, Barkatullah, Sarim, F.M. (2013). Ecological Adaptation to Altitude of Algal Communities in the Swat Valley (Hindu Cush Mountains, Pakistan). Expert. Opin. Environ. Biol., 2(2): 1-15. doi:10.4172/2325-9655.1000104

[17] Barinova, S.S., Nevo, E. (2010). The Upper Jordan River algal communities are evidence of long-term climatic and anthropogenic impacts. J. Water Resource and Protection, 2: 507-526.

[18] Barinova, S. (2013). Multilevel approach in biodiversity analysis of freshwater algae. Expert Opinion on Environmental Biology, 2(2): 1-2. http://dx.doi.org/10.4172/2325-9655.1000e106

[19] Heywood, V. (2004). Modern approaches to floristics and their impact on the region of SW Asia. Turk. J. Bot., 28: 7-16.

[20] ter Braak, C.J.F., Šmilauer, P. (2002). CANOCO Reference Manual and CanoDraw for Windows User's Guide: Software for Canonical Community Ordination (version 4.5). Ithaca: Microcomputer Power Press.

[21] ter Braak, C.J.F. (1987). The analysis of vegetation-environment relationships by canonical correspondence analysis. Vegetatio 69: 69-77.

[22] ter Braak, C.J.F. (1990). Interpreting canonical correlation analysis through biplots of structural correlations and weights. Psychometrica 55: 519-531.

[23] Novakovsky, A.B. (2004). Abilities and base principles of program module "GRAPHS". Scientific reports of Komi Scientific Center, Ural Division of the Russian Academy of Sciences 27: 1-28. (In Russian).

[24] Clarke K.R., Gorley R.N. (2001). PRIMER v5: User Manual/ Tutorial. PRIMER-E, Plymouth.

[25] Barinova, S.S., Nevo, E. (2012). Climatic and pollution impact on algal diversity of the freshwater ecosystems in Eurasia. In: Climate Change and Impacts, AcademyPublish.org, Wyoming, U.S.A: pp. 16-56.

[26] Krupa, E., Slyvinskiy, G., Barinova, S. (2014). The effect of climatic factors on long-term dynamics of phytoplankton, zooplankton and macrozoobenthos of the Balkhash Lake (Kazakhstan, Central Asia). Advanced Studies in Biology, 6(3): 115-136. http://dx.doi.org/10.12988/asb.2014.4523

[27] Barinova, S.S., Bragina, T.M., Nevo, E. (2009). Algal species diversity of arid region lakes in Kazakhstan and Israel. Community Ecology, 10(1): 7-16.

[28] Klymiuk, V., Barinova, S., and Lyalyuk, N. (2014). Diversity and Ecology of Algal Communities from the Regional Landscape Park "Slavyansky Resort", Ukraine. Research and Reviews: Journal of Botanical Sciences, 3(2): 9-26.

[29] Coops, N.C., Wulder, M.A., Duro, D.C., Han, T., Berry, S. (2008). The development of a Canadian dynamic habitat index using multi-temporal satellite estimates of canopy light absorbance. Ecol Indicators, 8: 754–766. DOI: 10.1016/j.ecolind.2008.01.007.

[30] Barinova, S. (2011). The effect of altitude on distribution of freshwater algae in continental Israel. Current Topics in Plant Biology, 12: 89-95.

Life Characteristic of Cistanche Salsa (C.A. Mey.) G. Beck – Perspective Medicinal Plant of the South Peri–Balkhash Lake Saxaul Forests

N. G. Gemejiyeva, Zh. Zh. Karzhaubekova

Institute of Botany & Phytointroduction Ministry of Education and Science, Almaty, Kazakhstan

Email address:
ngemed58@mail.ru (N.G. Gemejiyeva), zhanna1322@mail.ru (Zh. Zh. Karzhaubekova)

Abstract: The present study was carried out within the framework of grant financing of scientific surveys per subject: Restoration potential of producing vegetation of solonchak cistanche (Cistanche salsa (C. A. Mey.) G. Beck) in the South Peri–Balkhash Lake area (2012–2014). The survey object is perspective medicinal plant encountered within the South Peri–Balkhash Lake area solonchak cistanche of the Orobanchaceae Vent. family, which in Kazakhstan is produced only at the territories of Almaty and Zhambyl regions and then it is exported to Korea and China as raw stock for production of a number of pharmacologically active compounds having wide spectrum of effect such as hypertension, sexual vigor, antioxidant activity, where at present time, the producing areas and reserves in these countries have diminished. The life characteristic of this species was studied for the purpose of balanced use and preservation of natural populations of cistanche as perspective source for production of plant-based preparations of tonic and antioxidant effect. As a result of surveys, the peculiarities of distribution and floristic composition of vegetation communities including solonchak cistanche located within the South Peri–Balkhash Lake area totaling at least 48 species of plants of 16 families were identified. As per results of field surveys, the inventory and monitoring of producing areas of cistanche were carried out before and after raw stock production. The reserves of raw stock materials were calculated and schematic maps on distribution of identified producing areas of cistanche within the surveyed area were compiled. The total hydroalcoholic extracts were received and preliminary phytochemical screening of cistanche vegetable stock collected in 2013–2014 in the area was provided. The stolons of studied plant were found to contain main groups of biologically active compounds such as flavonoids, coumarines, carbohydrates, amino acids, organic acids, alkaloids, triterpenoids, saponins. The structures of two substances were extracted and identified.

Keywords: Cistanche Salsa, Medicinal Plant, Kazakhstan

1. Introduction

Kazakhstan flora has great potential as a source of raw materials for production of contemporary and efficient phytopreparations. Among 6000 species of Kazakhstan flora nearly 1500 species are characterized by medicinal properties of which only 230 species are used in official medicine [1]. Among such poorly studied and unused in Kazakhstan official medicine species is wild-growing medicinal plant Cistanche salsa (C.A. Mey.) G. Beck of Orobanchaceae Vent. family, highly popular at foreign producers of medicinal products as a feedstock for production of a number of pharmacologically active compounds of wide range of effect such as tonic, sexual vigor and antioxidant [2].

In Kazakhstan this plant is harvested only at the territories of Almaty and Zhambyl regions and then it is exported to China, where to date the production areas and reserves have reduced [3]. Therefore, without scientific approach and balanced use of natural thickets, reserves of cistanche will be depleted in quick time.

The resource characteristic of this species was for the first time studied for the purpose of balanced use and preservation of natural populations of cistanche as perspective source for production of plant-based preparations of tonic and antioxidant effect being the foundation for development of scientifically-based algorithm of rational use of raw plant. The resource characteristic of this species includes description of ecologically-cenotic and biological features of species, dynamics of yielding capacity by years, data on raw stock in different phytocenosis, qualitative composition of raw materials, methods and duration of population restoration and methods of their protection [4].

2. Study Area & Methods

The goal of the study is analysis of life characteristics of perspective medicinal plant of Cistanche salsa (C. A. Mey.) G. Beck (fam. Orobanchaceae Vent.) which grows in saxaul forests the South Peri–Balkhash Lake area.

Cistanche order Cistanche Hoffmgg. et Link of the family Orobanchaceae Vent. includes nearly 20 species distributed in the Mediterranean, West and Central Asia countries. In «Kazakhstan Flora» [5], Cistanche Hoffmgg. is represented by three species: C. ambigua (Bunge) G. Beck, C. flava (C. A. Mey.) Korsh., C. salsa (C. A. Mey.) G. Beck), by last data [6, 7] there are five species among those which have already been mentioned above and C. mongolica G. Beck, and C. fissa (C. A. Mey.) G. Beck.

The most widely-spread is Cistanche salsa – perennial parasite plant with the height 10–40 cm, thick tufted stem (Fig. 1). The flower head has short cylindrical form of the height 5–25 cm and 5–8 cm wide and lush. The flowers are in spicate raceme, attached. The corolla is obtusely campanulate. Blossoms and fruits in April-May. The fruits represent egg-shaped capsules opening with two shells [5]. This species grows on solonetz and solonchaks, in desert steppes of almost all desert, small hummocky and piedmont Kazakhstan areas.

Cistanche practically has neither root system nor stems. The stolon (stem) is covered with chaffs. The plant vegetates on soil surface approximately two weeks. For this period the ripen seeds quickly break up and the plant dries out.

(a) (b) (c)

a – saxaul forest, b – Cistanche salsa in bloom stage, c – stolon

Figure 1. *Cistanche salsa (C. A. Mey.) G. Beck (family Orobanchaceae Vent.) in the South Peri–Balkhash area.*

Cistanche mainly parasites on the roots of the host plants Haloxylon Bunge order of Chenopodiaceae Vent. family, order Calligonum L. of Polygonaceae Juss. family, order Tamarix L. of Tamaricaceae Link family, thus attaching to the host roots and sucking out its nutritional substances. The relationship system formed between the parasitic plant and host plant is complicated which includes a number of adaptations from the side of the first plant and defense reactions from the second one.

It is very important for a parasite to ensure the contact with host plant beginning from seed germination. The seeds of many parasitic species do not germinate until they are found to be near the roots of host plant which exudes into the soil substances stimulating germination of seeds of a parasite and determining direction of its haustorium growth [8].

Survey methods include general resource, geobotanical and phytochemical investigations [9–12]. Resource survey was carried out by route–recognostic method [13] by using cartographic materials. The surveys were conducted in the south-east of Kazakhstan, in South Peri-Balkhash Lake area (Fig. 2) which located as consistent with botanic–geographic zoning within East–North Turan subprovince of North–Turan province of Iran–Turan subregion of Sakhara–Gobi desert area [14].

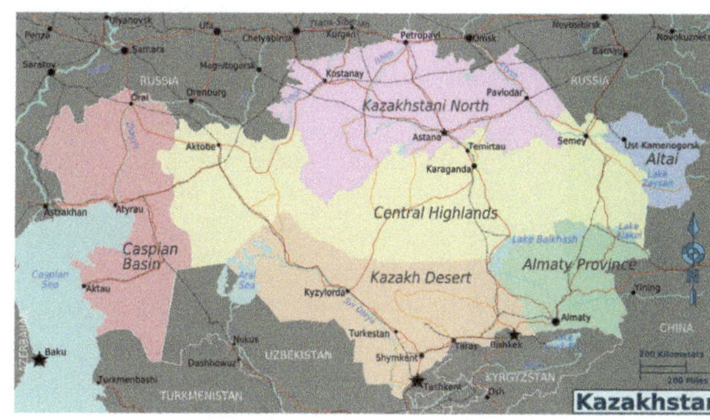

Figure 2. *Republic of Kazakhstan.*

The survey area characterized by continental climate, plain and hummocky-ridged relief. Soil cover formed by hydrogenic soils, solonchaks, takyrs, takyr–like and desert sandy soils. The vegetation cover characterized by wide development of black saxaul (Haloxylon aphyllum) and saltwort (Salsola orientalis) communities. Small sand areas occupied by saxaul–teresken–wormwood (Ass. Artemisia terrae-albae – Krascheninnikovia ceratoides – Haloxylon aphyllum) communities [15].

3. Results & Discussion

In autumn, 2012, route–recognostic survey was carried out for observation of natural plant population areas of cistanche from v. Bakanas to v. Kokzhide including outskirts of v. Kokzhide, Akkol, Akzhar, Ushzharma, Bereke and Bakanas. There were identified location areas of cistanche and pilot index plots for recording of cistanche intended for further monitoring of wild–growing populations in 2013–2014. Field surveys aimed at inventory of producing areas and monitoring of raw material base of cistanche were conducted before (April) and after (end of May–June) harvesting within Balkhash district of Almaty region.

As a result of surveys, the resource characteristic of natural populations was for the first time ever provided for Cistanche salsa in the South Peri–Balkhash Lake area within Balkhash district of Almaty region. It was established that the populations of cistanche growing in rhubarb–wormwood–saxaul (Bereke area), zhuzgun–saxaul (Ushzharma area) and wormwood–saxaul (Akzhar area) communities has high yielding capacity which are confined to takyr–like–solonetz and solonchak grey soils, and at some areas – to clayey, sandy soils.

Floristic composition of vegetation communities including C.salsa consists of at least 48 species of plants of 16 families having the biggest number of representatives of families: Chenopodiaceae (14), Asteraceae (6), Tamaricaceae (4), Poaceae (3), Limoniaceae (3) etc. Among associated cistanche species, 15 species were recorded: Allium schubertii, Arnebia decumbens, Cardaria repens, Erodium oxyrhynchum, Microcephala subglobosa, Nepeta micrantha, Nonea caspica, Scorzonera sericeo-lanata, Scorzonera purpurea, Strigosella africana, Hyoscyamus pusillus, Hypecoum parviflorum, Lepidium perfoliatum, Trigonella arcuata, Ziziphora tenuior. The vegetation communities including solonchak cistanche growing at pilot index plots of Akzhar, Ushzharma and Bereke were rich in floristic composition.

During field survey of natural populations of cistanche, the blooming specimens suitable for harvesting at the areas of 100 x 100 m^2 were identified and recorded prior to completion of large-scale harvesting of raw materials at the end of April 2013–2014, in addition, the density of raw stock, phytomass and size of model specimens of cistanche, yielding capacity of raw stock were identified at each index plot (Table 1).

Table 1. *Dynamics of yielding capacity of Cistanche salsa at key plots before harvesting (April, 2013–2014)*

Location of index plot and date of observation	Quantity of blooming cistanche specimens, units per 100x100 m^2	Density of raw stock reserve, spec./m^2	Yielding capacity of raw stock, kg/ha	
			fresh	dry
Bereke index plot (1). At 1.5 km westwards of turn to v. Bereke among saxaul forests. 30.04.2013	35	0.004	11.06±1.44	-
Bereke index plot (1). 29.04.2014	43	0.004	14.55±1.89	2.47±0.32
Kokzhide index plot (2). At 1 km from the turn of Bakanas-Kokzhide motor road to v. Kokzhide, by both sides of the road among saxaul forests. 1.05.2013	18	0.002	5.53±0.66	-
Kokzhide index plot (2). At 3 km from the turn of Bakanas-Kokzhide motor road to v. Kokzhide towards v. Akkol (9.4 km to Akkol) among saxaul forests not far from motor road. 30.04.2014	35	0.004	11.34±1.58	2.03±0.26
Akkol index plot (3). At 1 km from the turn of Kokzhide–Bakanas motor road to v. Akkol, by both sides of the road among saxaul forests. 1.05.2013	7	0.0007	1.94±0.25	-
Akkol index plot (3). At 1 km from gates to v. Akkol 30.04.2014	10 (left) 7 (right)	0.001 0.0007	2.83±0.37 1.98±0.27	0.40±0.06 0.28±0.04
Akzhar index plot (4). Turn to v. Akzhar, at 300 m to guidance sign to v. Akzhar 1.05.2013	35	0.0035	14.72±1.77	-
At 7–8 km southwards of v. Akzhar (4 a) (Tekturmys) on sand drifts among saxaul forests 1.05.2013	2	3	4	5
Akzhar index plot (4). At 800 m froм the turn to v. Akzhar 30.04.2014	70	0.007	25.76±3.35	-
Ushzharma index plot (5). At 1.6 km from the turn to v. Ushzharma, by both sides of the road. 1.05.2013	44 (91) (left) 21 (right)	0.0044 0.0021	14.37±2.01 6.86±0.96	2.49±0.32 1.19±0.17
Ushzharma index plot (5). 0.04.2014	28 42	0.0028 0.0042	11.04±1.21 14.72±1.91	-

According to our observations, in 2013, the number of specimens at pilot index plot of the size 100 x 100 m^2 ranged from seven to 70 thus averaging 49 specimens per hectare. However, the weight of raw stock of one specimen has significantly reduced ranging from 106.0 to 872.0 (at an average 276.6±30.42) g.

In 2014, the number of specimens at pilot index plots of the size 100 x 100 m^2 varied from seven to 73 thus averaging 47 specimens per hectare.

The weight of one raw plant specimen varied from 116.0 to 692.0 g (at an average 393.4±55.08), i.e. during 2013–2014, no significant changes per weight of raw plant specimens and density of reserves were detected.

As a result of field surveys, the inventory of producing areas of cistanche was done, the reserves of raw stock were calculated and location maps of cistanche producing areas recorded during 2013-2014 within the investigated territory of Balkhash district of Almaty region were compiled. The most productive areas were recorded in the outskirts of Akzhar village (within the territory of Kurtinskiy SI) and v. Ushzharma, Bereke, Kokzhide (within the territory of Bakanas SI).

The overall available inventory of raw stock of cistanche within the territory of Bakanas forestry during 2013–2014 amounted to 280–295 t at the area of 34835.0 ha respectively. Considering biological peculiarities of solonchak cistanche, the volume of possible annual production will not exceed 56-59 t of fresh raw stock (Table 2).

Table 2. *Information on results of resource survey of Cistanche salsa at the territory of Balkhash district of Almaty region*

Location of producing areas at the territory of Balkhash district forestries	Area occupied by Cistanche salsa, ha	Overall available inventory of raw stock, t	Volume of possible annual production of raw stock, t
2005 (as per results of works under economic agreement)			
Bakanas State Forestry Institution Bakanas Forestry (plots unknown)	41.500	111.2	22.2
Akkol Forestry (plots unknown)	7.300	38.1	7.6
Total	48.800	149.3	29.8
2013			
Bakanas State Forestry Institution Bakanas Forestry (plots 203, 204, 205, 213, 214, 215)	34.835	280.4	56.1
Akkol Forestry (plots 298, 299)	4.558	6.6	1.3
Kurtinskiy State Forestry Institution Akzhar Forestry (plots 1–35, 83)	6.333	99.9	20.0
Total	45.726	386.9	77.4
2014			
Bakanas State Forestry Institution Bakanas Forestry (plots 203, 204, 205, 213, 214, 215)	34.835	295.03	59.0
Akkol Forestry (plots 298, 299)	4.558	9.02	1.8
Kurtinskiy State Forestry Institution Akzhar Forestry (plots 1–35, 83)	6.333	91.01	18.2
Total	45.726	395.1	79.0

At the territory of Kurtinskiy SI of Akzhar forestry, the available inventory of raw stock of cistanche amounted to 91–99.9 t at the area of 6.333 ha respectively. Considering biological peculiarities of cistanche, the volume of possible annual production will not exceed 18–20 t of fresh raw stock. The archive data obtained in 2005 during implementation of works under economic agreement were used to compare them with the newly obtained data. Thus, in 2005, the weight of raw stock of one cistanche specimen reached up to 3 kg, while the average number of solonchak cistanche specimens per one hectare varied from seven to 23.

By the example of Akkol forestry, it may be noted that the area occupied by cistanche as well as available inventory and volume of raw stock harvesting have reduced from 2005 to 2013, although for 2005, the numbers and areas of plots at which cistanche reserves were recorded were unavailable. Unfortunately, it is impossible to compare by other data due to absence of specific and equal data.

The monitoring of producing areas of cistanche before and after harvesting of raw stock within the surveyed territory of Balkhash district of Almaty region showed that at index plots after harvesting of raw stock there were left "traces" in the form of deep hollows – pits, which need to be backfilled in order to prevent the roots of host plants such as Haloxylon Bunge, Calligonum L.,Tamarix L. from drying–out. Likewise, the remains of cut crowns of cistanche blossom clusters with immature seeds were recorded. As it turned out, local producers leave these cut blossom clusters purposely near the hollows in order that seeds would subsequently get into the soil once they are ripen and then attach to host plant thus indirectly contributing to preservation of seed regeneration.

Almost all detected prior to harvesting specimens were dug out, only a few specimens of cistanche were left intact at index plots due to later aftergrowth, i.e after large-scale collection of stolons.

At Ushzharma index plot, the aftergrowth of fresh cistanche specimens was recorded near the remaining dried-out last year specimens.

Continuation of monitoring surveys of the state of cistanche populations in 2014 allowed to establish that no significant changes were recorded per weight of raw stock and density of cistanche reserves during two vegetation seasons. The difference between obtained indices ranges within the error. However, this year, the factual harvesting areas were noticed to be shifted from the territory of selected index plots towards remote saxaul forests. In this connection, the local increase of yielding capacity of raw stock was recorded at certain index plots such as Bereke and Kokzhide.

The results of two-year observations (2013–2014) of the state of cistanche populations at index plots before and after harvesting of raw stock showed that density of raw stock at all index plots (Bereke, Kokzhide, Akzhar, Ushzharma) did not exceed 0.004 spec./m^2, except Akkol area where this index was the lowest ranging from 0.0007 to 0.001 spec./m^2.

The raw weight of specimens varied from 122.0 to 692.0 g (average 393.4±55.08) at Bereke index plot and from 178.0 to 550.0 g (average 326.6±45.71) at Akzhar index plot. The length of stolon without blossom cluster ranged from 13.0 to 45.0 cm, diameter of stolon varied from 2.5 to 6.5 cm. The length of blossom cluster reached up to 11–25 cm.

During two-year monitoring, the lowest yielding capacity of cistanche populations totaling 1.98±0.27 kg/ha was recorded in Akkol forestry. The highest yielding capacity of cistanche populations (14.55±1.89 kg/ha and 14.37±2.01) was recorded at Bereke and Akzhar index plots. The yielding capacity of cistanche populations at Ushzharma index plot was less totaling (8.52±1.10 and 9.25±1.20 kg/ha), as compared to 2013 it reduced at an average to 23-30% although the density of raw stock was the same as before. Evidently, this is due to the fact that vegetation is harvested without "resting" at one and the same areas.

The comparison of morphometric indices of model specimens of cistanche showed that populations of cistanche at Akkol plot had the least size and phytomass (224.3±29.16g). The biggest phytomass had specimens collected at Bereke plot (393.4±55.08 g), Akzhar (326.6±45.71 g) and Kokzhide (283.4±39.68 g). The average yielding capacity of cistanche at Bereke index plot in 2014 has increased for 30% thus totaling 14.55±1.89 kg/ha under the same density 0.004 spec./m^2.

At Kokzhide index plot, the productive populations of cistanche shifted at the distance of 3 km from last year harvesting areas as a result of which density of reserves and yielding capacity of raw stock (11.34±1.58 kg/ha) has increased almost twice as compared to 2013 data.

By yielding capacity, the most stable were populations of cistanche at Akzhar forestry whose maximal yielding capacity during 2013–2014 did not change much thus constituting 14.37±2.01 (14.72±1.77) kg/ha, although producing areas shifted inward saxaul forests for 800–1000 m. The difference between obtained data ranges within the error.

Overall, available local 2005 data per Bakanas and Akkol forestries obtained during works implemented under economic agreement are indicative of the fact that by now the areas occupied by cistanche at the territories of above–indicated forestries have reduced in a great measure and twice at certain areas. In order to determine the factual restoration potential of studied species, it is necessary to conduct multiyear monitoring of the state of natural populations of cistanche as well as dynamics of their reserves density and yielding capacity and morphometric indices.

For the purpose of preservation of natural populations of Cistanche salsa, it is recommended that next harvesting of raw stock at used producing areas should only be done after the land is allowed to rest for 4-5 years as new underground sprouts of cistanche are formed from the so-called root tubers –tuber–like formations at the place of contact with host plant during at least 3–4 years.

The herbal plant material was collected in natural populations for further production of extract from Cistanche salsa raw stock. The collected Cistanche salsa plant material at the territory of Bakhash district of Almaty region was dried–out, ground, weighed down and determined for good quality of studied plant raw material. The solvents for extraction of raw material were selected. Air–dry ground mass of plant material was treated with hydroalcoholic solvent three times. The extraction lasted three days under room temperature. The obtained total hydroalcoholic extract was filtered and then reduced to minimal volume and prepared for initial phytochemical screening.

Phytochemical survey of parasitic plants which are adapted to "host plants" is of big interest as many of these plants are used by local community for treatment of certain diseases but they are not included in the arsenal of scientific medicine due to their low chemical and biological study.

The first steps towards studying parasitic plants came out in the proceedings in [16, 17]. Reference [18] was described the early studies of the researchers of plant resources laboratory which shows the results of qualitative reactions for coumarines and saponins at parasitic plants exept Cistanche flava produced in the South Peri-Balkhash Lake area at the territory of Almaty region.

The preliminary phytochemical screening of Cistanche salsa samples collected during 2013–2014 in Almaty region was carried out. The fractional extraction of total water-alcohol extract was done by using ethyl acetate and n-butanol. The compounds β-sitosterol-β-D-glicoside was isolated from butanol extract. The β-sitosterol ($C_{29}H_{50}O$) was isolated from ethyl acetate extract [19]. These compounds were earlier extracted by [20] from Cistanche salsa. In addition, free monosaccharides, xylose, glucose and amino acids were found in the extracts, the study of amino-acid composition was started. The main groups of biologically active substances were established in stolons of studied objects such as flavonoids, coumarins, carbohydrates, amino acids, organic substances, alkaloids, triterpenoids, saponins which are consistent with reference materials [21].

4. Conclusion

Thus, the resource characteristic was provided for the first time ever for perspective medicinal plant Cistanche salsa grown in the South Peri–Balkhash Lake area, in addition, biological, phytocenotic as well as phytochemical peculiarities of this species and the most productive communities including Cistanche salsa were identified. The reserves and volume of possible harvesting of raw stock were calculated, distribution maps of detected producing areas of Cistanche salsa within the surveyed territory were compiled.

The recommendations were developed on optimal regime of harvesting and rational use of Cistanche salsa raw stock in the South Peri–Balkhash Lake area having all backgrounds for creation of domestic production of up-to-date dietary supplements, herbal preparations and efficient plant medicinal products based on renewable feedstock whose main exporter is Kazakhstan.

References

[1] L.M. Grudzinskaya, N.G. Gemejiyeva, N.V. Nelina, Zh.Zh. Karzhaubekova, The annotated list of Kazakhstan medicinal plants. Reference edition. Almaty, 2014, 228.

[2] S. Isabayev, Ecologic-biological peculiarities of cistanche dubious (Cistanche ambigua Hoffmgy et Link) of Moiynkum desert saxaul forests: autoabstract, PhD of Biology. Almaty, 2010, 22.

[3] K. Sarsenbayev, S. Isabayev, N. Kolosova, "New useful plant of Kazakhstan flora–dubious Cistanche," in Current ecological state of Peri-Aral Lake area, perspectives of solution of issues, Proceedings of International Scientific-Practical Conference. Kyzylorda, 2011, pp. 195–200.

[4] I.L. Krylova, Resource characteristic of medicinal plants as a scientific background of their rational use: autoabstract of Doctor of Biology. M., 1985, 50.

[5] Kazakhstan Flora. Almaty, 1965, vol. 8, pp.151-152.

[6] S.A. Abdullina, Checklist of vascular plants of Kazakhstan. Almaty, 1999, p. 123.

[7] M.S. Baitenov, Flora of Kazakhstan. Generic complex of flora. Almaty, 2001, vol. 2, p. 189.

[8] Guidebook on plant ecology. Reference source: http://www.kar-met.su (14.02.2014).

[9] Methodology for determination of medicinal plants' reserves. M., 1986, 50.

[10] A.A. Korchagin, "Species (floristic) composition of plant communities and study methods," in Field geobotany, vol.3, M.–L., 1964, pp. 39–60.

[11] V.M. Ponyatovskaya, "Recording of abundance and peculiarities of species location in natural plant communities," in Field geobotany, vol.3, M.–L., 1964, pp. 209–237.

[12] State Pharmacopoeia of the Republic of Kazakhstan. T. 1. Almaty, Publishing House "Silk Way", 2008, pp. 559–563.

[13] B.A. Bykov, Geobotany. Almaty, 1957, pp. 22–23.

[14] E.I. Rachkovskaya, I.N. Safronova, E.A. Volkova, The principles and the main units of regionalization. In book: Botanical geography of Kazakhstan and Middle Asia (desert region). Sankt Petersburg, 2003, pp. 379–382.

[15] O.M. Nasonova, "Feed characteristic of Balkhash vegetation of Almaty region," Proceedings of the AS KazSSR Institute of Botany. Almaty, vol. 11, 1966, pp. 3–26.

[16] I.G. Beilin, Parasitization and epiphytotiology. M., 1986, 351.

[17] I.G. Beilin, Flowering semiparasites and parasites. M., 1966, 118.

[18] O.U. Lushpa, F.M. Atalykova, "Chemical composition of some parasitic plants of Kazakhstan flora," KazSSR AS Bull., Biological edition, vol. 1, pp. 30–34, 1970.

[19] Zh.Zh. Karzhaubekova, N.G. Gemejiyeva, "Phytochemical investigations of Cistanche Hoffmgg. et Link," KazNU Bull., Biological edition, vol. 3/2 (59), pp. 288–290, 2013.

[20] H. Kiboyashi, H. Karasawa, T. Miyase and S. Fukushima, "Studies on the constituents of Cistanchis Herba II. Isolation and structures of New Iridoids cistanin and cistachlorine," Chem. Pharm. Bull., vol. 32, p. 1729–1734, 1984.

[21] E.N. Kapsalyamova, Development of composition and technology of medicinal form based on polyphenolic compounds of Cistanche salsa (C. A. Mey.) G. Beck: autoabstract of PhD of Pharmacy, Almaty, 2010, 24.

10

Peculiarities of Development of Phytoplankton as an Indicator of the Ecological State in Modern Hydrochemical Conditions of Masrik River

Hambaryan L. R.[1], Nalbandyan M. A.[2], Poghosyan A. M.[2]

[1]Department of Ecology and Nature Protection, Yerevan State University, Yerevan, Armenia
[2]Institute of Geological Sciences of the NAS Yerevan, Armenia

Email address:
lus-ham@yandex.ru (Hambaryan L. R.), marinen3@yahoo.com (Nalbandyan M. A.), ani-poghosyan@live.com (Poghosyan A. M.)

Abstract: This paper is related to the study of hydro chemical characteristics and the phytoplankton of the river Masrik. For planktonic study, algae were selected from 3 sampling sites: two from tributaries of the river and one from the river Masrik. Hydro chemical monitoring includes 9 sampling points. The study includes the spring of 2014 and the summer of 2013. According to the hydro chemical research of water, the pH ranged from 6.5-7.3. In 2014, the pH was lower compared with the spring of the previous year. Among tested anions, in some areas, relatively high level of sulfates were reached 80 -120 mg /L. Total hardness ranged from 0,6-5,6 mg-eqv /L, and the total mineralization - was ranged from 60 to 430 mg /L. Studies of heavy metals revealed a relatively high content of lead, nickel and chromium in water. Visual assessment revealed a significant organic pollution of Masrik river, which confirmed by high levels of BOD5 in water in 2014. In Masrik river mainly 3 groups of algae were detected: diatoms, green and blue-green. Dominant group was diatoms. Dominated types of algae were β-mezosaprobs (about 53%), which indicated the average level of organic pollution of the river. As a part of the green algae Spirogira sp. was found. The maximum number of phytoplankton in the investigated areas was 16,854,000 cells / L, and the biomass was 162g /m3, the minimum rates - respectively 464 000 cells /l in and 2.8 g/m3 for biomass. In water samples well developed whorls algae of the genus Chara were found, which presence indicates anthropogenic pollution. Thus, in a more acidic environment, the presence of high levels of organic pollution creates favorable conditions for the growth and development not only phytoplankton such as algae, but also for macrophytes, which indicates the reduction of ecological status of rivers.

Keywords: Ecological Monitoring, Water Quality, Saprobity, Phytoplankton, Hydrochemical Characteristics

1. Introduction

The main purpose of this work is an investigation of hydro chemical features, quantitative and qualitative compounds of phytoplankton in the river Masrik. This research is an example of interdisciplinary study of aquatic ecosystems, that are under the impact of numerous anthropogenic factors (agriculture, livestock, domestic and waste water), including mining (particularly gold) industry.

Study of water quality in the process of environmental monitoring involves the use of chemical, physical and biological parameters.

Rivers, especially in industrialized areas most exposed to pollutants such as heavy metals or hazardous organic pollutants. It is known that the water quality can be affected by chemicals such as pH, hardness, and oxygen. The changes in concentration can increase the toxicity of pollutants. Biomass of phytoplankton is considered as one of the most promising methods for assessing aquatic ecosystems in terms of the functioning of the primary autotrophic link. Data of phytoplankton in rivers indicate the presence and transformation of pollution, and can characterize the level of self-purification of aquatic ecosystem. It is also known that in the process of contamination of aquatic ecosystems, the prevailable species saprobity shows the presence of organic pollution of water [11].

It is very important that hydrobionts have integral response to changes in environmental conditions, fluctuation of water quality, thus, complex hydroecological studies represent special value and scientific interest, allowing to evaluate the interaction and mutual influence of hydrochemical and hydrobiological indicators.

Masrik River flows into Sevan Lake, which has a great importance for social-economic development not only for Armenia, but also the neighboring countries, as a promising source of fresh water [12, 13]. In this regard, the monitoring and research on the river water quality is important in terms of control of eutrophication processes and anthropogenic pollution of Lake Sevan ecosystems.

2. Study Area and Methods

The river Masrik flows through the territory of Ghegharkunik region. The river originates from the western slopes of the eastern part of the Sevan mountain ridge and flows into Sevan Lake. The Riverhead of Masrik river is at an altitude 2880 m. The length of the river is 45 km, catchment (watershed) area-685 km, the average slope in the upper reaches is 158%. The annual average flow rate of the river is 5.38 m^3/sec.

Hydrochemical and hydrobiological studies were included spring and summer of 2013 and 2014. Hydrochemical monitoring includes 9 sampling points (fig.1). In the samples, major ions, biogenic compounds and some heavy metals were measured.

For studying planktonic algae 3 points were selected: 2 from tributaries of Masrik river-Sotk (point 2) and Aziz (point 5) - and the last one from downstream of the river (point 9). As a result of development gold mining industry in Masrik river basin, especially in the basin of its tributary, the river waters characterized by an excessive intake of mineral suspended matter in them.

Photo 1. Photos of Sotk River (a), the main tributary of the River Masrik; River Masrik. (b); River Aziz (c), one of the left tributary of the River Masrik; the place on this tributary under the impact of the local man-made factor (d).

Mining waters characterized by high levels of electrical conductivity and salinity. These integrated indicators showed high level of mineralization in the water. Thus, the MPC of all ore elements, Cu, Zn and Ni was exceeded. Along with elements that were mentioned above, in mining waters the sulfate ion is dominated and exceed MPC 4-5 times.

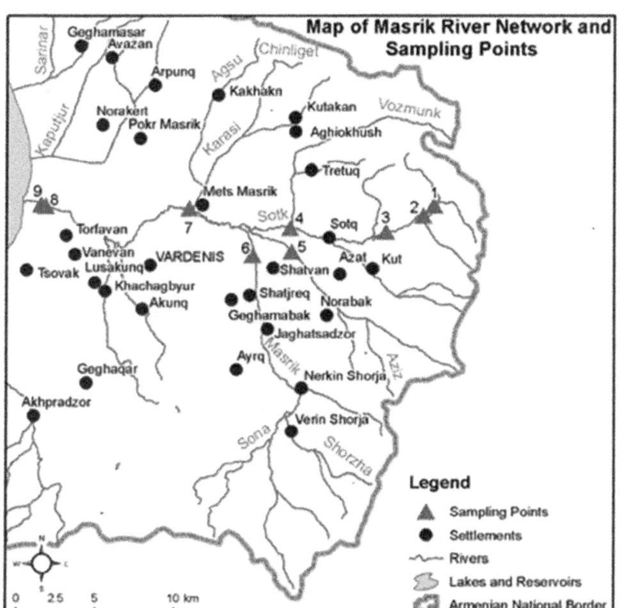

Sampling points description:

1. River Sotk-near mining factory
2. River Sotk-1.5km after Sotk mining factory
3. River Sotk-After Sotk mining factory, the first bridge
4. River Sotk- in the agricultural area
5. The left tributary ofMasrik river-Aziz - under the bridge
6. River Masrik near Shatvan village
7. River Masrik- near the Mets Masrik village
8. River Masrik-before the bridge of Ghille
9. River Masrik-after the bridge of Ghille

Fig. 1. The Map of Masrik River Network and Sampling Points

The MPC for HCO_3, NO_2, SO_4 and Mg was exceeded as well [1].

At the same time, it should be noted that the carbonate mode of geochemical landscape promotes natural neutralization of pollution of sulfate ion.

The quality of Masrik river waters is an integrated indicator of the ecological impact of mining industry on the environment. During an assessment of the ecological status of waters, it should be aware that the origins of the river forms within the biochemical province, where in different substrates is present natural higher contents of the individual components, which means high natural geochemical background, due to the influence of ophiolite rocks [9,10].

2.1. Sampling Methods

Before collecting of each sample, bottles were washed 3-4 times with testing water. The level of the bottled water samples met the requirements of samples selection-maximum. Collection, preservation and processing of algae were conducted according to standard methodology adopted in hydrobiology [3].

2.2. Chemical Analysis Methods

The table 1 shows the list of methods that were used for the analysis of water samples.

Different determinants were used for clarifying the species of algae [4, 5, 6, 7]. Quantitative analysis of phytoplankton samples produced in a Nageotte chamber (V=0.1 ml). The Nageotte chamber filled 7 times. Simultaneously with the determination of quantity, the measurement of cell volume of massive forms of phytoplankton was performed.

The Nageotte chamber filled 7 times. Simultaneously with the determination of quantity, the measurement of cell volume of massive forms of phytoplankton was performed.

The biomass was identified by individual volume of the cells for each type of phytoplankton in the sample. Relative weight of algae considered as 1.

The biomass of i species are calculated using following formula.

Table 1. *The methods for Determination of Different Elements and Compounds in Water*

N	Elements and Compounds	Methods of Determination	Sensitivity of Analysis Method	Links
1	Na^+	Flame-photometric analysis	0.1	ISO 9964-3
2	K^+	Flame-photometric analysis	0.3	ISO 9961-3
3	Ca^{2+}	Volumetric-trilonometric analysis		ISO 6058
4	Mg^{2+}	Volumetric-trilonometric analysis		ISO 7980
5	Cl^-	Volumetric method with silver nitrate		ISO 9297
6	SO_4^{2-}	Weights methods		ISO 9280
7	PO_4^{3-}	Colorimetric method with ammonium molybdate	0.01	ISO 6878
8	NO_3^-	Colorimetric method with sodium salicilate	0.2	ISO 7890-3
9	NO_2^-	Colorimetric method with Griess reagent	0.01	ÃÎÑÒ 4192-82
10	Ni^+	Colorimetric method with dimethylglixime	0.005	ISO 8288
11	Cr	Colorimetric method with diphenylcloride	0.005	ISO 11083
12	Cu^{2+}	Polarographic method	0.0005	Instrument manual
13	Pb^{2+}	Polarographic method	0.00005	Instrument manual

$$Bi = NiVi,$$

where

Bi-the biomass of i species, Ni-its quantity, Vi-average volume of the cell.

For assessment the level of contamination of Masrik river water saprobien system of Kolkwitz and Marson were used, particularly modified by Zelinka and Marvana. For calculation of saprobioty steps, the list of algae were used-indicators of saprobioty [8].

3. Results and Discussions

3.1. Hydrochemical Study of Masrik River Water

According to hydrochemical research of water, the pH ranged from 6.5-7.3. Compared with the spring of the previous year in 2014, pH was lower. Total mineralization varied within the limits of 60-430 mg/L, and water hardness averaged 0,6-5,6 mg-eqv/L. Moreover, it should be noted that the water of the river Sotk had higher salinity and hardness (3-5,6 mg-eqv/L), and according to the classification described as moderately hard. According to its ionic composition, the water of Sotk river can be classified as a hydrocarbonate- calcium-magnesium, and the water of Masrik river - hydrocarbonate-sulphate-calcium-magnesium-sodium.

In 2014, BOD5 in different parts of watercourses of the river amounted respectively - 10.8 (at upper- Sotk inflow), 4.2 (on average) and 4.5 (lower) parts of the river Masrik, indicating a relatively high level of organic contamination of the waters of the river.

Modern research and evaluation of surface water quality, that are presented in the article are based on national standards adopted in 2011 and developed with point of view of the basin management of the river watershed. According to the assessment of the content of major ions, nutrients and heavy metals in the water, by the norms of surface water quality and by taking into account geochemical peculiarities of the basins, the classification of the pollution level of the waters of Masrik river were performed [2].

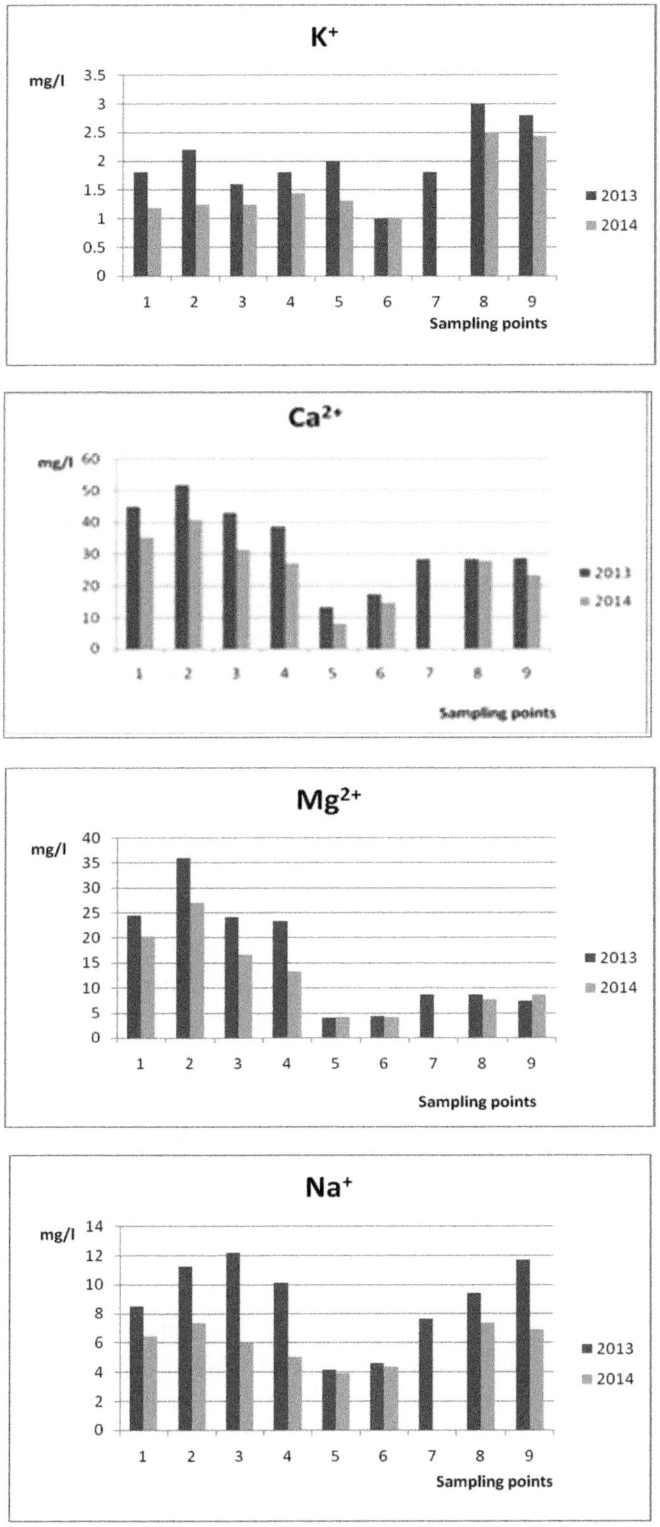

Fig. 2. The Level of Content of Cations in the Water of Masrik River in 2013 and 2014.

Assessment of concentrations and the pollution level by Na, Cl and SO_4 the river water belongs to II class, by K, Ca and Mg – to I class (fig. 2, 3). According to the evaluation of biogenic components NH_4 and $NO_2$the river water belongs to II class, by NO_3-to IV class, by PO_4-in upper (R.Sotk) and in the middle part of the river Masrik-to II class, but in the lowest part, in the river mouth - to the III class (fig.4). For heavy metals the following results were revealed – by Cu-I class, Ni and Pb-II class, Cr-III class (fig.5)

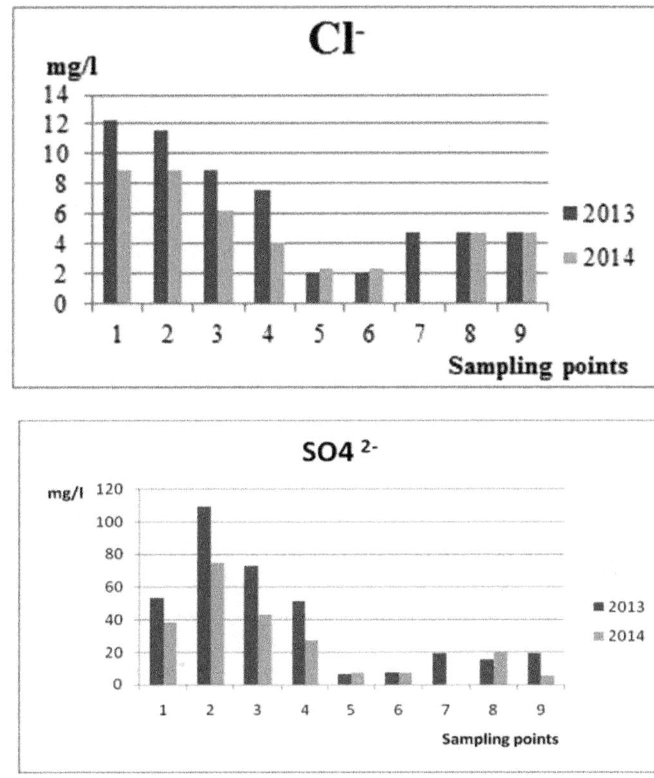

Fig. 3. The Level of Content of Anions in the Water of Masrik River in 2013 and 2014.

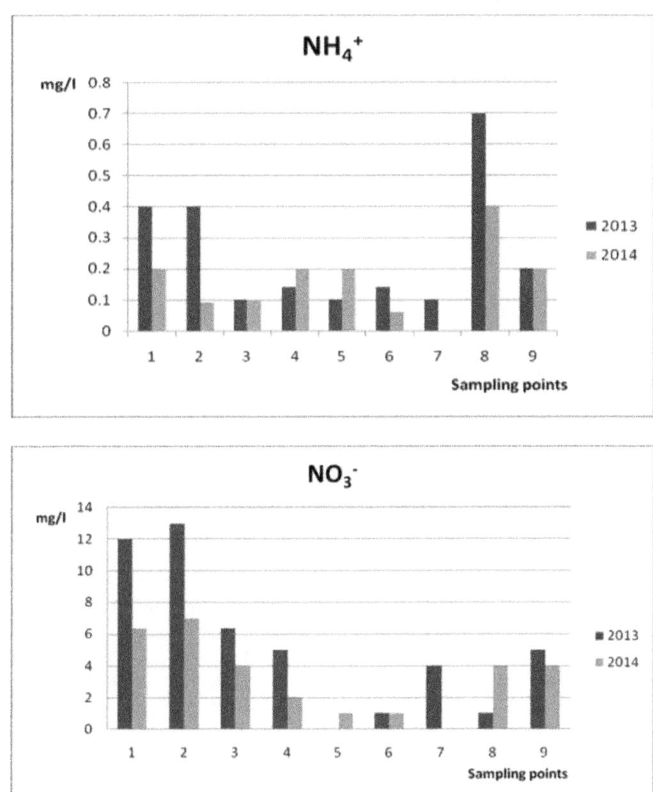

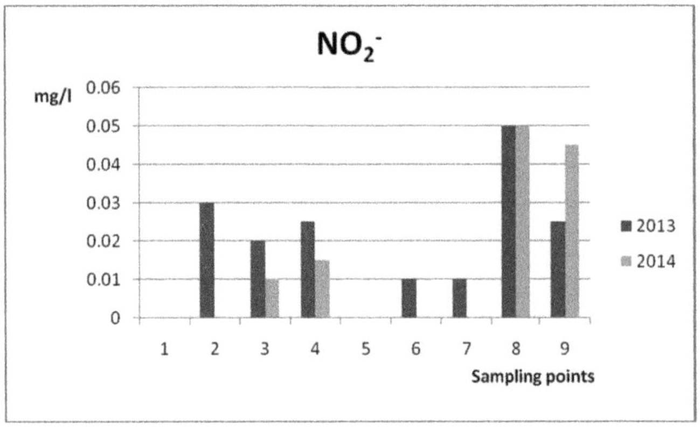

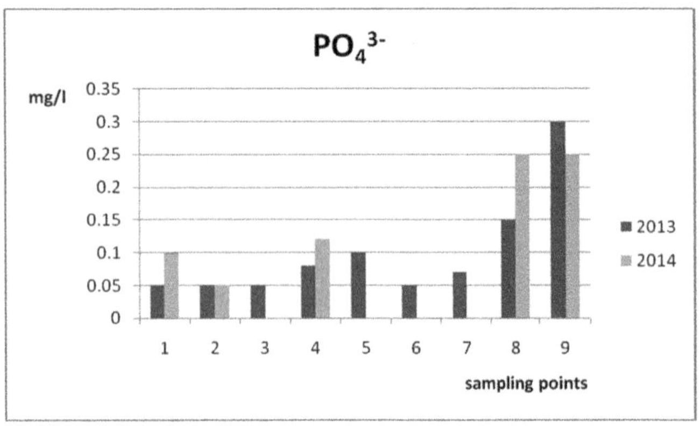

Fig. 4. *The Level of Content of Biogen Compounds in the Water of Masrik River in 2013 and 2014*

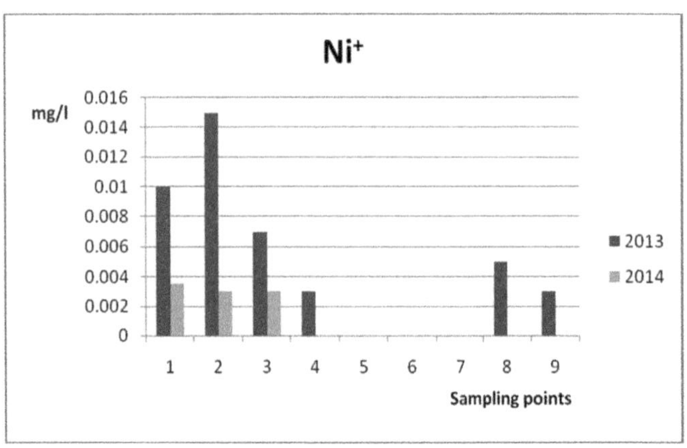

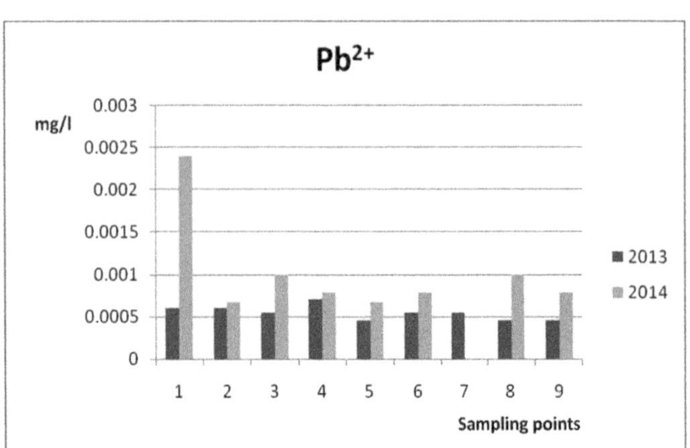

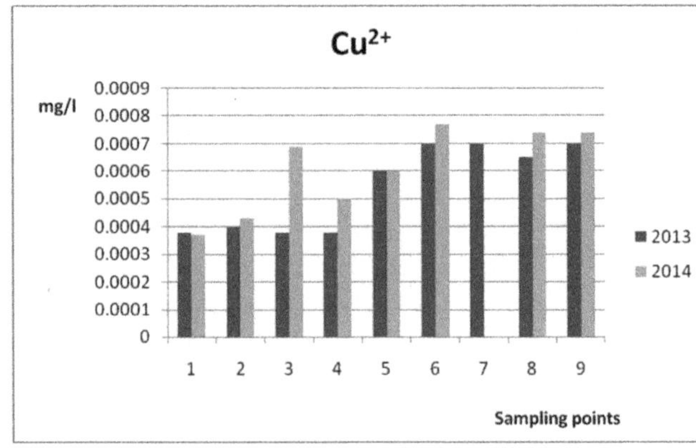

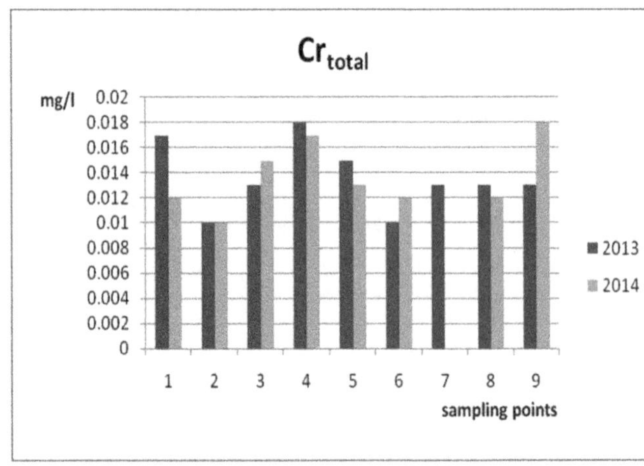

Fig. 5. *The Level of Content of Heavy Metals in the Water of Masrik River in 2013 and 2014.*

In the ores and rocks of Sotk ore deposit there are elements classified as a hazard of I and II class: respectively the group of Cd, Pb, As, Hg, Zn and the group of Cu, Co, Ni, Cr, Mo, Sb. Selected items in the mines are mainly presented as sulfide minerals which are not water resistant and are broken down in aerobic conditions forming higher acidic environment with a high mobility of releasing elements [9, 10]. Consideration of the content of heavy metals in ores shows the connection between the prevalence of this elements in ores and their high content in the waters of the river[9].

Qualitative and quantitative research on the composition of phytoplankton showed that the main communities were 3 groups of algae: diatoms, green and blue-green. Dominant group were diatoms. As a part of all groups of algae-dominated types were eutrophicators, β-mezosaprobs (about 53%), indicating about the average level of organic pollution of the river.

As a part of the green algae Spirogira sp. was found, which ecology is not inherent to the environment of reoplankton, as these species are widespread in the slowly flowing and stagnant waters, forming a mass of slime. The prevalence of this type indicates the presence of fecal contamination in the water.

3.2. Hydrobiological Study of the Water of Masrik River

The maximum number of phytoplankton in the investigated areas was 16,854,000 cells / l, and the biomass was 162g/m^3, the minimum rates - respectively 464 000 cells /l in and 2.8g/m^3 for biomass. High quantitative indicators of planktonic algae, also indicate the presence of favorable conditions for their mass development and active vegetation. In water samples were found well developed whorls algae of the genus *Chara*, which is not representative of the plankton and were not observed in the spring of last year, but their presence indicates technogenic pollution.

In a summer dominant group of phytoplankton were diatoms. They were 75% of the total phytoplankton and 79% of indicators of total biomass of phytoplankton of Masrikriver. (fig.6 (a,b)). Dominant diatom species of the genera were: *Stephanodiscus, Didimosphenia, Nitzhia, Synedra, Cocconeis, Navicula, Amphora, Diatoma, Ceratoneis, Cymbella, Fragilaria and Rhoicosphenia.* In summer, maximum rates of diatoms was registered in the estuary of the river Masrik amounting by the number of 468,000 cells /l, and by biomass in June, in the middle reaches, reaching 5.6 g/m^3. High rates of biomass caused by the development of the macrocellular Cocconeisplacentula form, which is a boreal species, developing at a pH above 7. This species is an indicator of polluted conditions and is characterizing beta mezosaprob level of contamination [14]

Blue-green algae in the studied period were in subdominant community, they were registered in all parts of the river. Quantitative indicators of blue-green algae were increased in August, the maximum indicators in numbering 164,000 cells/l and 0.33g/m^3, were found in the middle reaches of the river. The percentage of blue-green algae accounted for 19% of the size of the total population and 7% of the total biomass, respectively (Fig.6 (a,b)). They were mostly species of the genera *Aphanothece, Microcystis, Phormidium, Merismopedia.* These are mainly small-celled colonial toxic forms of algae.

Green algae were not observed in June. In August in the middle of the river were registered maximum values of green algae, which were amounted in the number of 132 000 cells/l and biomass 0.7g/m^3. By quantitative indicators green algae were accounted for 6% of the total population and 10% of the total biomass of the community. (Fig.6 (a,b)). Among the green algae, especially in filamentous forms, many types are occurring and developing in polluted conditions. Phytoplankton acts as an absorber of heavy metals and other chemical substances [15]. The dominant species were *Ankistrodesmusfalcatus, Dictiosphaerumpulchelum, Scenedesmusacuta, Oocystisparva, O. lacustris, Phacuslongicauta.*

From the literature it is known that some species of the genus *Scenedesmus* are highly toxic and it is one of the most promising species for cultivation in the laboratory. Species of *Scenedesmus*genus used for waste water treatment[16].

Ecological assessment of water quality of biomass of phytoplankton showed that the part of river in sampling point 9 appropriate to average clear water class.

Part of the river, particularly in sampling points 1-2 appropriate as a polluted water class, category of strongly polluted, and part of a river-sampling point 5 appropriate to the polluted water class, category extremely polluted.

It can be mentioned that the hydrochemical assessment of water corresponds to the environmental assessment of water quality of biomass of phytoplankton in points 9 and 2. In the point 5 the same water with hydro-chemical point of view refers to the less stained and observed maximum values of phytoplankton biomass associated with temporary anthropogenic impacts, which are formed the small artificial pound, that contributes an active vegetation of planktonic algae (Photo 1).

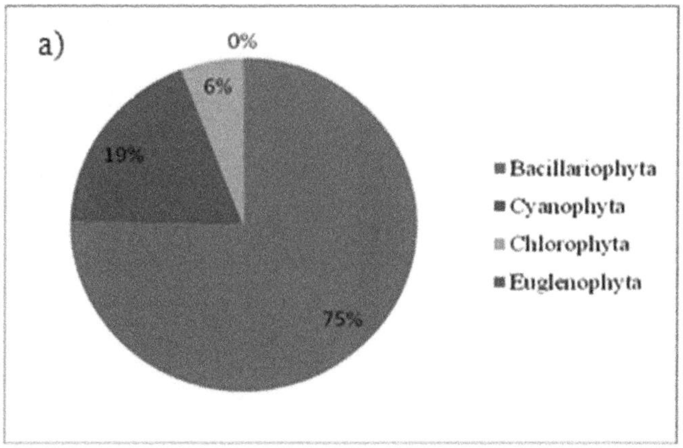

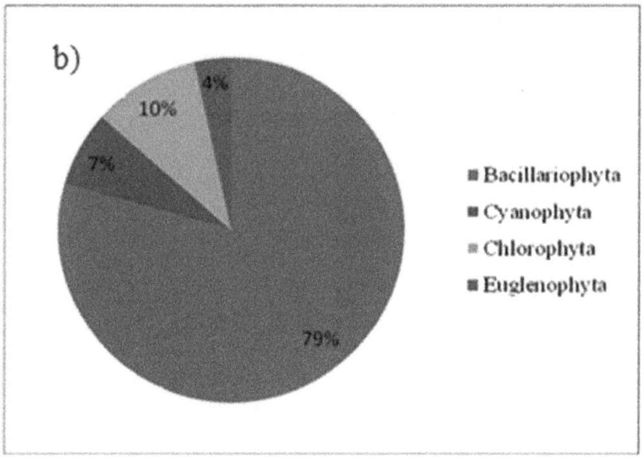

Fig. 6. *Percentage of major groups of phytoplankton Masrik River, in terms of the quantity (a) and biomass (b), summer 2014.*

This observation has a scientific interest in terms of prediction of environmental situation in the water in case of impact on the riverbed, which brings the changes of the water cycle and conditions.

4. Conclusion

Visual assessment of water samples revealed the presence of clusters of filaments of algae and peresistant inherent in municipal wastewater. These visual assessments revealed a significant organic pollution of river Masrik, which was confirmed by high levels of BOD5 in water.

According to the evaluation of hydrochemical characteristics, it can be concluded about the ongoing impact of Sotk gold deposit on the water quality of the river Masrik, that is expressed in a high level of sulfate ions in the water of the river Sotk (II class of pollution) and there is a stable high concentration of chromium along the entire length of the river to its confluence with Lake Sevan (III class of pollution). High concentrations of phosphate (III class) and nitrate (IV class) are the results of influence of domestic wastewater.

In the qualitative composition of phytoplankton, more than half (about 53%) are β-mesosaprobic species–indices of the level of pollution. The presence of blue-green and green algae in the river Masrik also indicates the averagelevel of ecosystem pollution.

In terms of phytoplankton biomass, water quality of the river Masrik in different parts of the river, was changed by class - from satisfactorily clean up until contaminated, by category -from slightly contaminated to heavily contaminated.

The comparison of water quality in 2013 and 2014 years allows to conclude that the presence of more acidic environment and a high level of organic pollution in 2014 created favorable conditions for the growth and development not only for phytoplankton blooms, but also formacrophytes, which indicates degradation of the ecological conditions of the river.

Acknowledgment

The authors wish to express their thanks to GIS specialist, senior researcher Arakelyan Alexander for technical support in design of this paper.

References

[1] Standards of maximum permissible concentrations that often can be found in natural waters of chemicals and substances of human origin. Approved by the Ministry of Health of RA and registered by the Ministry of Justice of RA, 2002 (In Russian).

[2] Norms for assurance of river water quality for each river basin, depends on peculiarities of the area. The government decision of RA. 27.01.2011(In Armenian)

[3] Guidelines for collecting and processing of materials in hydrobiological studies on freshwater bodies. Phytoplankton and its products. Leningrad, 1981. (In Russian)

[4] Kiselev I. A. The study of plankton of water bodies, 1950. (In Russian)

[5] Zabelina, M. M., Kiselev I. A., Proshkina-Lavrenko A. I., Sheshukova V. S., Identification Book of Freshwater Algae of the Soviet Union, Diatoms. SovetskayaNauka, Moskva. 1951 (In Russian)

[6] Hambaryan L., Shahasizyan I. Freshwater algae families determinant, Yerevan, 2014,YSU published, 61p.

[7] Makovinska Y. Manual for quantitive and qualitive monitoring of freshwater ecosystems phytoplankton community, 2013, 103 p. (In Russian)

[8] Standardized methods for studying water quality. Part III. Methods of biological water analysis. Appendix 2: Atlas of saprobic organisms. Moscow, Sekretariat SEV, 1977. 228 p. (In Russian)

[9] Evaluation of nowadays natural contamination of Sevan Lake basin from Sotq ore deposit. Determination of natural background on the land and in the lake basin. Institute of Geological Sciences NAS RA and scientific research company Georisk. Final report. Contract YE 013/08. Yerevan 2008, p. 48 (In Armenian)

[10] Assessment of the impact of the planned processing plant and mining tails on the environment in the basin of Lake Sevan. Institute of Geological Sciences of Armenia. Final Report. Contract GPMG 007/09. Yerevan, 2009, 55p.

[11] Hambaryan L. et al. Heavy metal pollution of the catchment basins of the Voghchi and Meghriget rivers (Armenia) and risks to the environment associated with water pollution. Academic journal of science. 2013, CD-ROM. ISSN: 2165 – 6282: 2(2): 255-268.

[12] Hovhannisyan R.H. Book "Lake Sevan, yesterday, today…..". – Yerevan: Press of NAS RA, 1994, 479 pp.

[13] Hambaryan L., Hovsepyan A., Khachikyan T. Influence of Lake Sevan catchment basin phytoplankton community structure on the same of the lake. Regional workshop on "Sustainable Management of Water Resources and Conservation of Mountain Lake Ecosystem of Asian Countries", Yerevan, Armenia, 25-29 June, 2014, pp.102-112

[14] Barinova S.S., Medvedev L.A., Anisimov O.V. Algae biodiversity indicators of environment. - Tel Aviv, 2006. - 498 p (In Russian)

[15] Nikanorov A.M. Scientific bases of water quality monitoring, S-Peterburg, 2005, pp.367-372. (In Russian)

[16] Oswald W.J. Microalgae and waste water treatment. In: Borowitzka M.A., Borowitzka L.J (eds) Microalgae biotechnology. Cambridge University Press, Cambridge, 1998, pp. 691-707

Evolution of African Human Trypanosomiasis in the Democratic Republic of Congo During the Year 2005

Achille Lutumba Suika[1, *], Zhang Liqiu[1], Anne Marie Nsaka Kabunda[2]

[1]College of Environmental Science and Engineering, Beijing Forestry University, Beijing, China
[2]College of international law, Zhongnan University of Economics and Law, Wuhan, China

Email address:
lutumba.achille@yahoo.fr (A. L. Suika)
[*]Corresponding author

Abstract: The illness has caused most devastating and lethal epidemics at the beginning of the colonial era in the Democratic Republic of Congo. Between 1885 and 1920, parasites have decimated the population of several infected areas. The second major outbreak started in 1920 and reached its peak in the 1930s and 1940s. At this time, a very impressive means for a policy based on active research, patient care and sanitation of the community living environment. According to the national fight against (sleeping sickness), African human trypanosomiasis, in the eve of independence, the rate of African human trypanosomiasis had a very low level (0.01%). After independence, due to the relaxation of the shares, the destruction of health facilities and the care of the sick, it was observed that the disease has gradually increased in almost all old houses. After independence, due to relaxation of the actions, the destruction of health facilities, as well as the care of the sick, it was observed that the disease gradually increased in almost the all old houses. Currently, trypanosomiasis remains a major health problem in the Democratic Republic of Congo. The government, assisted by United Nations Agencies and International NGOs, has, since 1968, a specialized service to combat the disease. This is the national program for the fight against African human trypanosomiasis that rules the national policy in this area.

Keywords: Glossina, Trypanosomes, African Human Trypanosomiasis

1. Introduction

African human trypanosomiasis (sleeping sickness) is becoming a major endemic to Africa, protozoan, and seems to be in all area of life in our country. It is not only a rural disease, but soon as urban conurbations due to several factors that contribute to persistent movements of populations [1-2].

It is important to consider the fight against sleeping sickness as a specific problem, the tragic in human terms, if it is not solved. The World Health Organization (WHO) said that 55 million people living in rural areas are exposed to Tse Tse Glossina, flies responsible for the disease [3].

This terrible disease is certainly with the slave trade, and other factors, the main cause of the exodus from Africa. At the end of the 19th century and in the beginning of the 20th, lethal epidemics of African human trypanosomiasis (sleeping sickness) broke out in many major outbreaks in Africa and particularly in the Democratic Republic of Congo [4]. Health problems that were allowed during colonization and the beginning of independence in our country have become widespread in rural and urban areas because of the destruction of basic infrastructure and the abandonment of reorganization measure.

The control of trypanosomiasis is one of the priorities of the African continent, which is part of the research agenda established by the World Health Organization, focusing on the major tropical diseases. In our country, the program is implemented by the national program for the fight against African human trypanosomiasis.

The problem is that, in the light of the information gathered by the national program for the fight against African human trypanosomiasis in 2005 can confirm the thesis of great writers such as GENTILINI and GOLVAN that African human trypanosomiasis (sleeping sickness) which is becoming an endemic not only to Africa, but also in the Democratic Republic of Congo, its spread has no borders, and goes beyond urban areas.

African human trypanosomiasis (sleeping sickness) is a group of parasitic protozoonoses trypanosome brucei group due to the transmitted by the bite of the community (Tse Tse flies Glossina), characterized clinically by the occurrence of meningoencephalitis in spontaneous development is always fatal. It occurs in parts of Africa and the black station is strictly located roughly between the twelfth and fifteenth degree north latitude south latitude [3].

This parasitosis formerly known as sleeping sickness, is called the African human trypanosomiasis, which was adopted by the World Health Organization (WHO). It presents a threat to more than 60 million people, of which 10 million are under permanent surveillance.

It is classified with malaria, schistosomiasis, onchocerciasis, and other major diseases caused the most disabling and making worst African life [5].

The elements of the existence of African human trypanosomiasis in each of these areas of health are ecological, geological, botanical, and socio-economic, those who promote, on the one hand, the contact between hosts and vectors (the community) (man and domestic animals in particular) and, on the other hand, they set up the disease without medical intervention.

1.1. Biology

The trypanosomes are flagellated protozoa sanguineness. They live in the blood, where they are not within the longitudinal cell and multiply by fission [3]. Training is between 18 and 30 days after the infecting meal and digestion and can in turn transmit the parasite and left all his life.The trypanososme is transmitted during blood feeding of the tsetse fly injects saliva with thousands of trypanososmes youth. those are extra celleulaires, vievent in the liquid inside and migrate to different organs. Three forms are distinguished: long, squat and intermediate [4-14].

The extra cellular protozoan flagellates in the fusiform 40μ m up to long and thin (host) or when they are short and stocky 15μm (christidia) [1]. The liquid in their hosts, or the blood, lymph node, the cerebrospinal fluid or saliva (glossine) median and economic nucleus flagellum form an undulating membrane surrounding the cell body.

There are two forms of trypanosomes in Africa: *Trypanosoma brucei gambiense* African human trypanosomiasis (sleeping sickness) in charge of the west and central Africa and *Trypanosomiasis brucei rhodesiense* which occurs in eastern and southern africa [6].

Two species are recognized as human pathogens, namely:

* *Trypanosoma cruzy*, head of the American trypanosomiasis or Chagas disease is in south America, and the vector is a bug.

* *Trypanosoma brucei*, located in regions where the tsetse fly. With only three species provided by the community, it is of *Trypanosoma brucei gambiense* and *Trypanosoma brucei rdodesiense, Trypanosoma brucei brucei*, the first two are responsible for african trypanosomiasis and while the latter is the animal trypanosomiasis [7-8].

The trypanosomes are first in the body and in the community as from epimastigote and midgut digestive tract and is found in the salivary glands of the bug into a form trypanostigote.

1.2. Localization and Classification

These flies have acquired the historic name of tsetse fly from the name of MATEBELE tribe of south Africa spread by the explorer Livingston, Glossine used by entomologists due to the shape of the trunk with a tongue when they are at rest.

Areas, or there is the risk of being bitten by a tsetse fly are generally as follows: the secular trees and other places of camping in the countryside, the area around the drill into bodies of water, the vegetation near places baignade and fetching water on the banks of rivers and streams, savanna residents, forests or cemeteries planted with trees, fishing and vegetation surrounding villages [4, 13-14].

There are several species living in tropical Africa. The community are strictly in the hydrophilic species that live in the forests and rivers which form the palpalis group glossina fuscipes tachynoides mainly linkage, and give the trypanosoma brucei gambiense in almost any part of the west and central Africa, characterized by a general trend with slow, sometimes from the emitted but the pig and human in the forests for the last two.

And in the dry savanna species that prefer to form the group moristans as glossina moristans, glossina pallidipes and glossina swynnertoni, they convey the trypanosoma brucei rhodesiense in east Africa by differentiating the previous epidemiology, its history and its evolution and rapid, acute and fatal. [9].

1.3. Diagnostic

Chagas disease can be diagnosed by laboratory tests on blood and c.s.f. (cerebrospinal fluid) is [1].

*A the phase of blood lymphato

The elements of suspicion: the numbering format packet indicates anemia and lymphocytosis.

The diagnosis of certainty: the indirect immuno fluorescence was positive, but requires high rates because of frequent cross reactions with the malaria parasite in blood by thin or other enrichment technology.

* A phase of brain

Abnormalities of the spinal cephalo series: the c.s.f. (cerebrospinal fluid) meaning of the acronym? is clear but high strung, containing 20 500 cells / mm^2 phagocytoses.

1.4. Treatment

The treatment depends on the type of infection and the stage of the infection. Several drugs, anti-protozoan flagellates are available [7, 8, 9-10].

To deal with the early stages of human trypanosomiasis *Trypanosoma brucei gambiense*, lomidine pentamidine and suramin sodium. it is effective only in the lymphatic blood stage infection. Its can cause discomfort with sweating, abdominal pain; it is administered by intramuscular route every other day to see every day for six weeks. Suramin sodium is active in phase lymphatic blood, it possesses a certain nephrotoxicity imposing a search proteinuria before each injection, it becomes infected intravenously.

For the treatment of the advanced stages of the disease when the nervous system is achieved in the treatment of *Trypanosoma brucei gambiense* and *Trypanosoma rhodensiense* admiring the melarsoprol arsobal is administered strictly intravenously, any leakage of liquid causes a pressure ulcer patients must be to young for 12 hours intolerance manifests to the third and fourth injection hyperthemia, nervous and digestive disorders.

For the treatment of human trypanosomiasis in later stages of *Trypanosoma brucei gambiense*: Eflornithrine. This product is no longer available today. It is administered by infusion at 4 doses during 7to 14 days.

1.5. Prophylaxis

It is the implementation of strict control measures, the glossine and the trypanosomiasis.

The control of the glossines in four levels:

Large scale use of insecticides; the creation of the barrier by clearing forests and scrub in order to protect the villagers and the improved genetic control regions, and regions with sterile male healthy and trapping by the use of biconiques CHARLES LAVAISSIERE traps which are impregnated with insecticides.

The fight against the trypanosomes:

Diagnosis and care of patients, to protect subjects and control the reservoirs of the virus, animals, and humans.

The prophylaxis was circulated and not the scourge out of central Africa [11-12]. Experience has shown that the failure in the application of the preventive measures was to ransom back the disease.

Finally, this control is carried out at international level with mobile teams, specialized screen patients' clinical and parasitological and initiate and control the outcome of therapeutic treatments.

2. Environment, Materials and Methods

2.1. Environment

The epidemiological data of African human trypanosomiasis (sleeping sickness) available in the services of the national program for the fight against African human trypanosomiasis were collected through the health areas in all the provinces of the vast territory of the Democratic Republic of Congo, which has an area of 2345000 km^2 with a population of 11 provinces, and a density of 60 million. 21 inhabitants / km^2, 12 600 000 population and 2 462 831 population covered by the national program for the fight against African human trypanosomiasis.

2.2. Materiel

The realization of a need for material investigations consisted mainly of the annual reports of the national program for the fight against African human trypanosomiasis and the ministry of health in 2005. In these documents, we find all the information on the population of the provinces with: age, sex, the level of social and the development of human trypanosomiasis in the Democratic Republic of Congo, Africa [11, 13-14].

2.3. Methods

To collect the data necessary for the completion of our work, we have used the following methods:

a) The documentation allowed us to gather all the information of annual reports and new technologies to fight against African human trypanosomiasis.

b) The samples and the collection of data necessary for the achievement of the tables and figures.

c) Epidemiological surveillance data to calculate the rate of new cases per year, and the rate of infestation in the provinces, in order to better understand the evaluation item by item and understand the scientific, socio-economic, educational and epidemic of endemic on the various according to the segments of the population statistics.

In order to make better use of the data, the formulae:

$$\mathrm{T.I} = \frac{N.C * 100}{P.T.Ex}$$

$$\mathrm{P.C} = \frac{P.E * 100}{P.T.Ex}$$

3. Results

After investigation and observation, we present the results in the form of tables, graphs and a discusion for better understanding charts and graphs, here the legend of some abbreviations used:

T.I: Rate of infection

P.C: Population covered

N.C: New cas

N.E: Level of endemicity

Z.E: Epidemic area

P.E: Exposed population

Z.S.: Health area

Z.S.T.H.A: Area of health of human african trypanosomiasis (sleeping sickness)

U.M: Mobile unit

P.T.R: Total population census

P.T.Ex: Total population is discussed

Table 1. *State of the Human African Trypanosomiasis in Democratic Republic of the Congo in 2005.*

ZE	PE	ZS	ZS THA	UM	PTR	PTEx	PC	NC	T.I	NE
BAS CONGO	1700000	27	11	5	189275	173639	10,21	279	O,16	AVERAGE
BANDUNDU	3800000	38	31	13	1013455	819240	21,56	4621	0,56	SEVERE
MANIEMA ET KATANGA	1200000	81	16	4	174221	163352	13,61	667	0,41	AVERAGE
KASAI	2000000	58	35	15	286365	290114	14,51	2720	0,94	SEVERE
EQUATEUR	3000000	33	21	2	1019894	935926	31,20	1320	0,14	light
KINSHASA	6000000	22	8	1	47859	75578	12,60	331	0,44	AVERAGE
PROVINCE ORIENTALE	300000	47	3	3	64618	66822	22,27	431	0,64	SEVERE
TOTAL GENERAL	12600000	307	125	47	2795687	2524671	20,04	10369	0,41	AVERAGE

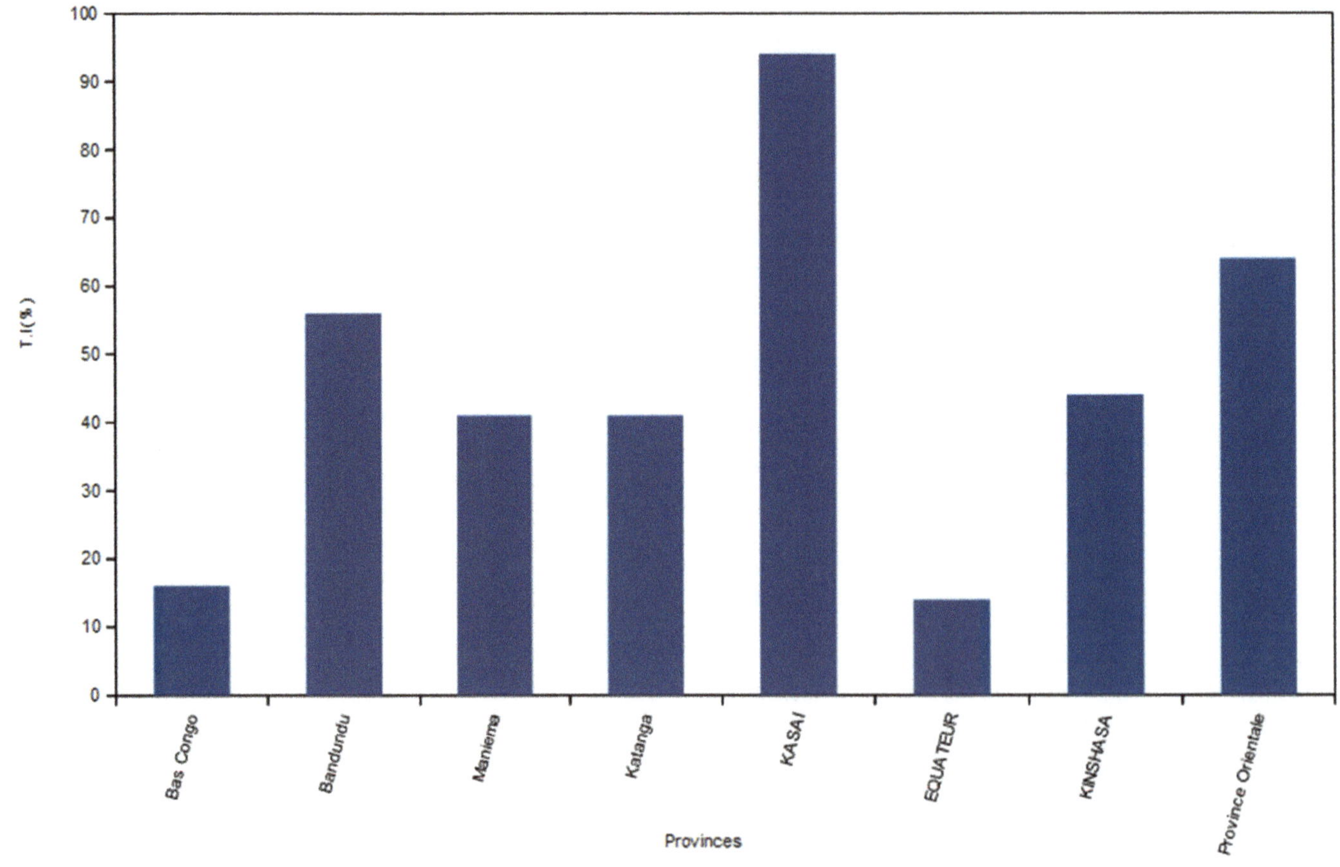

Fig. 1. *State of the Human African Trypanosomiasis in Democratic Republic of the Congo in 2005.*

Table 2. *Evolution of Human African Trypanosomiasis in 2005, depending on the circumstances and the stage of the disease.*

Z E	Active testing				Passive testing				TOTAL NC
	first stage	second stage	unknown	TOTAL	first stage	second stage	unknown	TOT.	
BAS CONGO	43	27	0	70	20	116	0	136	206
BANDUNDU	1848	622	68	2538	1689	591	8	2288	4826
MANIEMA	167	26	3	196	59	28	0	87	283
KATANGA	16	29	2	47	55	6	0	61	108
KASAI ORIENTAL	407	417	50	874	1191	77	6	1274	2148
KASAI OCCIDENTAL	167	77	7	251	254	108	5	367	618
EQUATEUR	79	187	3	269	560	65	5	630	899
KINSHASA	22	20	10	52	191	13	8	212	264
PROVINCE ORIENTALE	440	185	0	625	130	123	1	254	879
TOTAL GENERAL	3189	1590	143	4922	4149	1127	33	5309	10231
%	64 ,8	32,3	2,9	48,1	78,1	21,2	0,7	51,89	

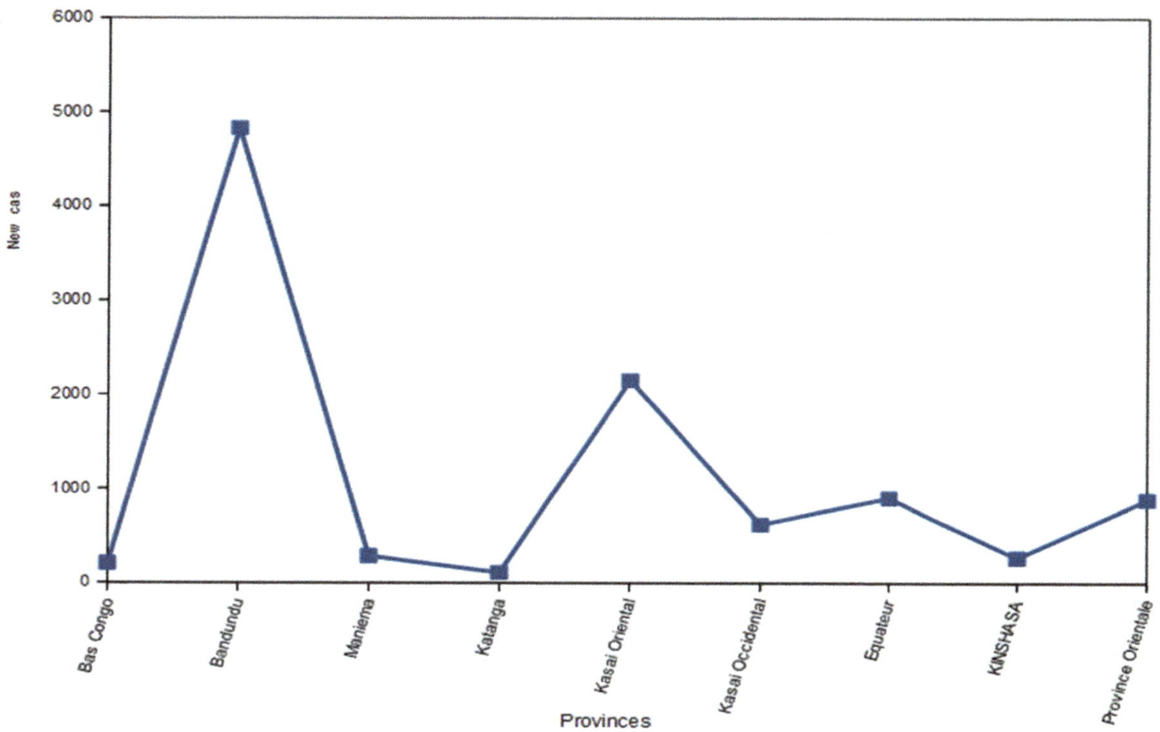

Fig. 2. Evolution of Human African Trypanosomiasis in 2005, depending on the circumstances and the stage of the disease.

Reading this chart, we observed that, during the year 2005, 10 231 new cases of African human trypanosomiasis were reported. In all endemic areas, 2 462 831 were examined. The average of the total population that is exposed is 9, 55% of the population. The overall infection rate was 0.42%.

The coverage of the population has reached about 20%. As a result, the number of people already infected at the moment is probably about 5 times higher, and in the absence of effective measures on a large enough scale, the transmission will increase significantly in the near future.

Bandundu is head of displays with 47% of the new cases in the whole country, Kasai oriental and eastern province 21% and 9%.

If the incidence of this disease is still insignificant, it should be emphasized that all patients do not show up in the screening centers either because they are unaware or shame or diversion by traditional healers and pastors that prophesy miracles cures.

Reading this chart, we observed that the 10 231 patients African human trypanosomiasis were reported during the year 2005, the agreement is 7338 cases (71, 72% in the first stage, lymphatic blood), 2717cases in the second stage (26.55% meningo encephalopathy), 176 cases (1,72%) about the diagnosis of stage n has not been made because the refusal of the lumbar puncture (no repeat, the pregnancy of the patient.

*Active detection is 48, 11 4922% thereof:

Table 3. *Active testing.*

First stade	Second stade	inconnu
3189	1590	143
64,8%	32,3%	2,9%

*Passive testing is 51,89 5309% thereof:

Table 4. *Passive testing.*

first stade	Second stade	inconnu
4149	1590	33
78,1%	21,2%	0,7%

These results confirm the presumptions of the program against human African trypnaosomiasis on Advanced glossina, year by year to the center of life and therefore the emphasis means of health education of the population especially those living common peripherals and extension of the means of struggle.

4. Discussion

Throughout the Democratic Republic of Congo, the new endemic areas, we found that 12 600 000 people are exposed (23% of the country's population), 2 806 751 were from among those identified in 2005 and 2462831 were examined.

The population of 10 231 new cases representing 0, 42 of the national average. The results can be compared to those presented by the national program for the fight against African human trypanosomiasis until 2005, these services have been rising in all the endemic areas of work have been taken as a working model for the field of national program for the fight against African human trypanosomiasis [9, 10-11]. Also was identified by the national institute for scientific research of France and the institute of tropical medicine in Belgium on African human trypanosomiasis in west Africa, particularly in Burkina Faso and east Africa in Uganda is confirmed by the fact that the African human trypanosomiasis is evolving in the means of implementation the traffic and the screen [12].

The vegetation and climate of our country to create the

conditions for the proliferation of skin, as well as all provinces with the exception of the altitude are threatened by this disease, which is not dealt with in life, breaking the social and economic consequences that we know of.

5. Conclusion

The investigation on the problems of African human trypanosomiasis in the Democratic Republic of Congo was assigned as the first test of this condition through the different provinces, as long as the level of infection, and the new cases. Analysis of variance showed that there was an upward trend of the disease and the publications of the national program for the fight against African human trypanosomiasis supports [12].

Due to the vegetation and climate; the Democratic Republic of Congo has not only the conditions for the development of the agro pastoral sector but also the proliferation of large homes of the tsetse fly. and trypanosomes.

With the exception of a few areas across the country, with altitudes of fly vectors of trypanosomes causing skin, very serious damage to the health of the actors in the agricultural sector, but also their families, livestock and livestock in some provinces in trypano resistant animals.

The new cases detected are the effort of the hero in the shadows, and the extent of work if all households were under control.

The results of the national program for the fight against African human trypanosomiasis should address not only the government, but also the agencies involved in funding to help complete the work of its policy is to eradicate the African human trypanosomiasis, both at the level of the vehicle, the causative agent for the treatment and allow the population to return to normal social and economic life.

Acknowledgements

All those who in one way or another, we have made a significant contribution, but their names are not mentioned in these lines here, i should like to express our deep gratitude. It is a pleasure to think China Scholarship Council (CSC) and Beijing Forestry University for them support regarding this study.

References

[1] GENTILINI, M. Medecine tropicale, ed. Flammarion, Paris (1989) 1-839.

[2] GOLVAN, J. J. Element de parasitology medicale, ed. flammarion, Paris (1983) 1-571.

[3] ANONYME conseil scientifique internantional pour la recherché et la lutte contre les trypanosomiases: 24eme reunion de l''OUA tenue à Maputo (1995) 1-3.

[4] MULUMBA, M. Elements de protozoologie,ed. biometrix kinshasa RDCongo (2003) 1-35.

[5] WERY, M. Trypanosomaise africaine, maladie du sommeil, encyclopedie medicale chirurgicale, ed.elsevier SAS paris (2000) 1-20.

[6] LAVEISSIERE C, GREBAUT, P et al. Les glossines vectrice de la maladie du sommeil, ACEAC, yaounde /cameroun (2000) 1-240.

[7] WERY, M. Protozoologie medicale, ed. de boeck et larvier S A les trypanosomiases africaines (1995) 1-110.

[8] EUZEBY Les parasitoses humaines d'origines animales ed. flammarion, paris (1984) 1-324.

[9] CLAUDE Manuel de lutte contre la maladie du sommeil en Afrique centrale et occidentale, vol. (2000) 1-48.

[10] DURIEZ, R. Encyclopaedia universalis, trypanosomiase zygophycées corpus 23, SA (1982) 1-919.

[11] LUTUMBA, S. Etat de la THA dans la zone de santé de Mont ngafula de juillet en decembre 2004,CRPA /UPN, kinshasa /RDCongo (2004) 1-38.

[12] DIANGO, KASHONDO et al. Guide pratique de zoologie fascicule1 les protozoaires, CRPA /UPN, kinshasa /RDCongo (1985) 1-28.

[13] KALASA, A. Dispersion des mouches tse tse, glossina fuscipes quanzensis à kinshasa, CRPA /UPN, kinshasa /RDCongo (2000) 1-43.

[14] KALUNGA, F. Taux d'infestation trypanosomienne des glossines, glossina fuscipa quanzensis, cas des foyers de Matadi mayo et kimwenza à Mont ngafula, Mémoire, UPN/ kinshasa /RDCongo (2003) 1-49.

12

Study and approbation of *ex situ* conservation methods for preservation of the biodiversity of wild relatives of cultivated plants of Kazakhstan

Gulnara Sitpayeva, Tansara Murzatayeva, Saniya Inerbayeva, Karina Makhmudova

Institute of Botany and Phytointroduction, Almaty, Kazakhstan

Email address:
m.tansara@mail.ru (T. Murzatayeva)

Abstract: Researchers of the laboratory of seed growing and plant protection of RSE "Institute of Botany and Phytointroduction" CS MOS RK are in the process of studying the techniques of the leading seed banks of the world, of their assessment and approbation using seed samples of wild relatives of the cultivated plants of Kazakhstan at their placing for storage. At the Institute of Botany and Phytointroduction of the Ministry of Education and Science of the Republic of Kazakhstan within the scientific-technical program "Botanical variety of wild relatives of cultivated plants of Kazakhstan as a source of enrichment and preservation of the agrobiodiversity gene pool for realization of the Food programme" for 2013-2015 the work on creation of seed bank of wild relatives of cultivated plants (WRCP) of Kazakhstan is being carried out. Since 2013 the employees of the Institute carry out collecting seed material of WRCP. In the present work the results of positive germinating test and planting the seeds of caper grassy (*Capparis herbacea* Willd.) collected in Kazakhstan are presented.

Keywords: *Ex situ*-Preservation, Germinating Test, *Capparis herbacea* Willd., Seed Bank, Kazakhstan

1. Introduction

In this research work the methods of storing seeds of wild plants in Kazakhstan on the example of the organization and operation of the seed bank of the wild relatives of cultivated plants (WRCP) were studied. Man always engaged in farming and cultivation of plants suitable for food, saved seeds to plant them next year. Modern seed banks pursue the same goal – to keep the genetic diversity of cultivated and wild plants, to ensure the stability of our food supply, ensure the restoration of crops after global catastrophes and provide an important source of material for scientists. Seed samples can be used to develop more productive crops and species that are resistant to pathogens and climate changes [1]. In the world there are more than 1,750 gene banks [2]. Some of them maintain local plants; others act on the global level. Consultative Group on International Agricultural Research, founded in 1971, includes 47 member countries and 11 gene banks, saving 650,000 samples. Genetic diversity is very important for all organisms. If all of them were same, then one pathogen or cataclysm could destroy everything at once. Genetically diverse populations possess more chances to have individuals who can survive and pass on their useful traits to the next generations.

The aim of this work is studying and approving methods of *ex situ* preservation of seed samples of the wild plants of Kazakhstan.

Research tasks: - to find appropriate conservation methods of seed of the wild plants of Kazakhstan; - to test the available techniques in the conditions of the seed bank of the Institute of Botany and Phytointroduction; - to carry out statistical analysis of linear dimensions and weight of the seeds of *Capparis herbacea*; - to carry out the germination of seeds of *Capparis herbacea* Willd.; - to identify ways of further use of the data obtained from the study.

In general, seed banks holding wild species follow the instructions of the International genebank for collecting, processing, storage and use of seeds. But during last 10-20 years the problem and lack of knowledge had formed and it led to more focused researches of ways of seeds of various species. For example, now more eco-geographical understanding of the behavior of seeds during conservation exists and more knowledge about the relative longevity of orthodox seeds is available, and that is why we can determine in advance what seeds should be stored using cryo-technology. Research on the development of seeds determined when it was better to collect the seeds which were more resistant to drying and long-term storage; also the best understanding of how the development of seeds can vary between species was formed; it became possible to use them for recovery of species or reintroduction or for evaluating the other ways of their use.

For *ex situ* conservation of plant diversity in general seed banks hold about 4.6 million accessions, 64 food and forage crops covered by the multilateral system of beneficial use "International Treaty Plant Genetic Resources for Food and Agriculture" [3]. Seed banks create seed germplasm for crop study and other international researches. Seed germinating is the first step in the search of genes to improve quality, productivity and to overcome biotic and abiotic stresses. Seed banking had already been used for *ex situ* conservation of wild species that were not the relatives of cultivated plants. Until then, it was adopted the Global Strategy for Plant Conservation (GSPC / GSPC) in 2002, thousands of seed samples of wild species were placed into the bank for long-term storage due to the efforts of the Millennium Seed Bank Partnership [4]. One of the revised objectives of the GSPC is that at least 75% of the species of seeds in the world should be included in the *ex situ* collection, and only 20% of them are suitable for rehabilitation and restoration program "2020".

Seed banks usually store seed corresponding to the genebank standards [5, 6]. There are no specific standards for seed storage of wild plant species, and most of the information obtained through the study of crops, with the exception of some species due to the fact that the behavior of the samples is different from the expected. While most of the rules are suitable for similar cultivated and wild species, as well, there are differences between them. Expenses for long-term storage and medium storage varies, for long term storage a large amount of material is required (the temperature must constantly remain low), while the medium-term storage, the base material is distributed over several years.

It is expected that the collection of wild species in seed banks will play an even greater role in the rehabilitation and reconstruction of species in their natural habitats [7, 8].

The reason why we have chosen caper grassy as the object of our study is the need to introduce into the culture of the alternative sources of polyunsaturated fatty acids of vegetable origin.

It is well known that people consuming a large number of seafood containing fatty acids from the family Ω-3, less often get sick with metabolic diseases of the civilization, less often suffer from atherosclerosis, coronary heart disease, a cancer of a mammary gland and a thick gut, thrombosis of vessels and asthma [9]. In the conditions of sharply continental climate seafood is not available in large volumes. The alternative sources of polyunsaturated fatty acids are known, it is a series of oil-yielding plants. Attempts to fill up the range of vegetable oils at the expense of cultivars of foreign breeding and their inclusion into the local cultivars of oil-bearing crops, did not give the expected result. In this connection creation of a collection of the domestic wild vegetable species resistant to environmental stresses, which are the alternative sources of the polyunsaturated fatty acids and their further use in breeding programs is urgent.

It is known, that sufficient receipt of polyunsaturated fatty acids (omega-3 and omega-6) is necessary for normal realization of the following processes in a human organism: development and maintenance of the brain function; realization of visual process; the response of the immune system; participation in the synthesis of hormones. The European department of safety of foodstuff (EFSA) has confirmed the fact of improvement of the health state on the background of adequate consumption of the polyunsaturated omega-3 fatty acids with food [10, 11]. Further, we will be interested in the fatty acids necessary for maintenance of health, not developed by the human organism – unreplaceable or essential fatty acids (EFA). In the EFA composition 5 polyunsaturated fatty acids (PUFA), linoleic, linolenic, arachidonic, eicosapentaenoic and docosahexaenoic, are distinguished. Their amount in the human organism depends on that, how many fats and oils a human person eats. Fatty acids are the basic building blocks not only in the fats contained in tissues of the person, but also taking place in foodstuff. They are the important energy source for any organism [12].

Evans and Barr in 1928 showed the growth delay and fertility decrease of rats whose diet contained no fats but contained vitamins A and D. Further, it was shown that this syndrome of insufficiency can be treated by means of adding into the food linoleic, linolenic and arachidonic acids. Other symptoms of the given syndrome are squamous dermatitis, necrosis and defeat of uric system; the syndrome usually does not result in a lethal outcome. EFA are contained in enough amounts in vegetable oils and in small amount – in animal tissues [13]. People whose diet did not contain EGAs were also affected by squamous dermatitis; in their case infringements of the lipids' transport were noticed. At usual feeding menu adult people had no symptoms of lack of essential fatty acids. Suckling children receiving artificial feeding with the insignificant content of fats were also affected by squamous dermatitis, which was easily treated with the linoleic acid preparation. The disorders connected to the EFA-lack was also observed in patients whose vital activity long time was supported only due to the intravenous feed almost devoid the fatty acids. For prevention of such disorders it is necessary the share of essential fatty acids would be (on caloric content) not less than 1-2 % from the general requirement for calories [13].

The countries having access to sea riches for producing the food additives containing PUFAs can use sea fish, but in mankind disposal there are also riches of flora. Researches devoted to use of the vegetable material as a source of the PUFA [14-13] are also carried out; in plenty the PUFAs omega-3 and omega-6 are contained in linen, soya and nut oils [17-19]. These acids also preset in the other vegetable oils, seeds of sunflower, in the peanut, almonds, avocado and soybeans.

The researches of the chemical compound and the characteristic of buds, fruit, seeds and oil of capers [20, 21] showed that they could be the source of the PUFAs. Absence of toxicity and presence of the expressed antiulcer activity of the capers' fruit oil [22] were confirmed.

Caper grassy or caper spiny (*Capparis herbacea* Willd. / *Capparis spinosa* L.) is a perennial shrub that throws its creeping branches whose length is 150 cm. Its leaves are light green and covered with wax, entire, elliptic and petiolate. Capers blossom and fructify since May till August and even October. Flowers are solitary, rare beauty, reveal in the morning and close at noon. Out of the white tinged with pink flower long purple anther stalks stretch out. Fruits are oval-shaped, fleshy and green. They contain a lot of nodular brown seeds [23-25].

They grow on clay and rubbly soils often on solonetzic soils of deserts, the slopes of the lower belts of mountains and their loops, waste grounds and in crop fields.

Growth area of caper grassy: Crimea, Caucasus, Central Asia, Southern Europe, Balkans, Asia Minor, Iran, Afghanistan, Western China.

Distribution of *Capparis herbacea* Willd. in the territory of Kazakhstan: Prikaspy, Turgaysky region, Northern and Southern Ust-Urt, Betpakdala, Moyun-kum, Pribalkhashye, Turkestan, the Dzhungar, Ile and Kyrgyz Alatau, Karatau, Western Tien Shan.

Economic value: Flower buds are edible and known under the name of "capers". Their fruits ("cucumbers") are also edible. They can be used both in fresh, and in a marinated condition. Oil of seeds is suitable in food. Capparis herbacea is a good melliferous herb. Their nutrition value in 100 grams: proteins – 0.8 g; fats – 0.1 g; carbohydrates – 2.5g; ashes – 0.1 g; water – 85 g. Capers are low-calorie and dietary product containing 14 kcal per 100 g of fresh product. Calorie of the canned capers is 23 kcal per 100 g of product. They are recommended to use by everyone including humans suffering from obesity [23-25].

2. Research area & Methods

The work was carried out under the conditions of the seed bank in the laboratory of seed growing and plant protection of the Institute of the Botany and Phytointroduction. For the germination test the seeds of *Capparis herbacea* Willd. (further *Capparis herbacea*) collected in the Turkestan region of South Kazakhstan region were selected. In the present work, we used methods applied in seed banks developed by N.I. Vavilov Scientific-research institute of plant growing [8] (Russia, Genebank Standards) and by the ENSCONET [1] (EU): 1) Cleaning and morphological description of seeds; 2) Basic germination test; 3) Documentation maintenance.

The following procedure was used for germination of *Capparis herbacea*: Seeds germinate very badly; they differ by a stretched (for several years) germination. Before sowing the seeds were incubated for 10 days in the refrigerator at +2°C. Then, the seeds first were dipped for 30 seconds into boiling water (100°C) and then for 20 min they were being cooled in the cold (about 0°C) water.

Then the seeds were left for 1 day in a wet cotton pouch at room temperature, and then 20 seeds in Petri dishes on wet filter paper at a temperature of 20-25°C. After the seedlings reached 0.5-0.7 cm in length, they were put into the soil in pots for germination. At the bottom of the pots claydite was stacked. The composition of the soil for germination: finished soil mix for houseplants and river sand in the ratio 1: 1. The seedlings of *Capparis herbacea* were mounted to a depth of 3 cm. After seedling emergence the capers were under the observation every day. At present two plants are noted to have already tillered. By the occurrence of the growing season these plants are planned to be placed either into the larger vegetation pots or to be planted at the experimental plot of the IBP.

Thus we have planned the introduction of *Capparis herbacea* in conditions of IPB for its further introduction and use in biochemical researches and breeding purposes.

3. Results & Discussion

All seed lots were thoroughly cleaned off all related parts. In some cases, samples were laid in the seed bank in the form in which they were received. In this case, it was documented. Seeds of wild species were cleaned manually using the cleaning devices (sieves and rubber floats). Before each new sample the sieves and graters were cleaned. It is recommended that all cleaning operations must be carried out under the hood to filter dust or to make cleaning in the standard protecting masks. During cleaning, the seeds were checked for damages using an electron microscope and a digital magnifier. Vegetable wastes that were accumulated during the seed cleaning process were destroyed by controlled incineration. Seeds in already treated batches were checked visually and by touch to determine the "empty" seeds.

In each sample the mass of seeds of the WRCP was determined. Depending on the size of the seed there was measured the weight of 10, 20, 50, 100, 500, and 1000 seeds. In each sample the linear dimensions of the seeds (length and width) were determined. After the determination of the mass and linear dimensions of the seeds we began to describe the seed samples intended for banking. To do this, we purified some number of seeds from the pericarp or scales. Then seeds were placed on the object micrometer and photographed with a digital magnifying glass. Thereafter, the color and shape of the seeds were described. Having detected the samples damaged by pests or diseases affected, we gave them to a specialist in plant protection. And then we placed them in quarantine.

The results of the preliminary germination test of seeds of *Capparis herbacea* can be found in Table 1. Thus, from 523 seeds set on germination there were 128 seeds germinated that was 25.5% of the total amount. 15 of 128 seedlings sprouted in the pots that were 2.9% of the total number of seeds set for germination.

Below in Figure 1 the photo of seedlings of *Capparis herbacea* is presented.

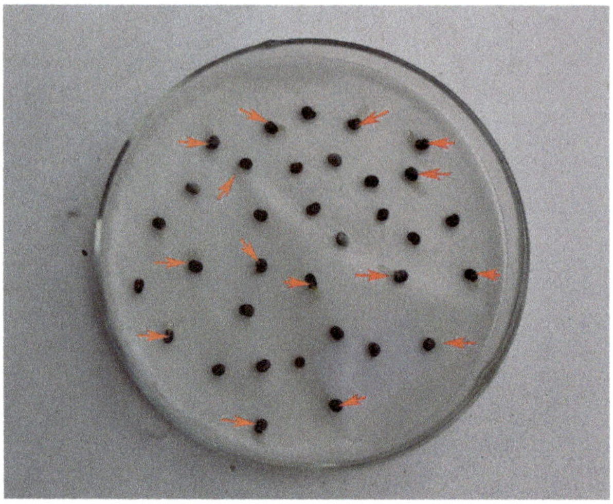

Figure 1. *Seeds of Capparis herbacea: Germination Test, Seedlings, 28.08.14*

In Figure 2 you can see the first plantlet of *Capparis herbacea* in the vegetation pot.

Figure 2. *First plantlet of Capparis herbacea in the vegetation pot, 09.09.14*

In figure 3 you can see adult plants of Capparis herbacea in the vegetation pots.

Figure 3. *Capparis herbacea plants in the vegetation pot, 23.02.15*

In Table 2 the description of the seed samples of *Capparis herbacea* is presented.

4. Conclusion

Researches and obtained data led to the following conclusions:

1. Study of literature let us concluded:
    - Creation of seed banks of wild plants is national importance for a separate state and international significance as a whole, as it allows us to preserve the genetic diversity of flora;
    - Scientists around the world, being in close cooperation with each other, developed requirements for genetic banks of cultivated plants, these requirements have further formed the basis of seed banks of wild species of plants;
    - For the successful placing a sample into the seed bank it must first be cleaned, dried (15%) tested for moisture content, and tightly packed;
    - Samples laid down for storage into the seed bank were divided into two collections: active (short-term storage) and a basic (long term storage). But in both cases it is better to select a long-term storage conditions (-20 ° C);
    - In a month after storage under long-term storage conditions the seed samples must be subjected to the basic germinating test (2 replications of 100 seeds);
    - It is always keeping documented records that are entered into the database. The exchange of data between the seed banks is compulsory;
    - Each 5-10 years the seed samples of the basic collection being in long-term storage should be subjected to regeneration test;
    - The samples may be sent to the applicant (seed banks or other interested parties).
2. As a result of working with samples of the seed bank of the RSE "IBP" the seed samples of *Capparis herbacea* were described (linear dimensions of the seeds, shape, color, weight of 1000 seeds, the presence/absence of pests and diseases).
3. In the laboratory conditions the positive germination test of seeds of *Capparis herbacea* was carried out.
4. In the laboratory conditions the first plantlets of *Capparis herbacea* were grown in the vegetation pots.
5. To start the introduction of *Capparis herbacea* its planting out at the experimentation plot of the IBP was planned.

Table 1. *Number of seeds of Capparis herbacea Willd.: Germination Test*

Name of the sample	Total number of seeds (pieces)	Number of the Germinated Seeds (pieces)	Number of the Sprouted Plants in the Pots (pieces)	
			with the Cotyledonous Leaves	with the Real Leaves
Capparis herbacea Willd.	523	128	4	11

Table 2. *Description of Seed Samples of Capparis herbacea Willd., Collected in 2013*

Name and description of the sample	Photo of the sample
1. *Capparis herbacea* Willd. Floristic region – (13B). 18.07.2013. Mangistau area, Beineou region, Beineou vicinity, Ustyurt, Kuiken gorge. N 45°12′724′′, E 54°57′763′′. 220 m of true altitude. Fruit is a juicy berry; fruitcase is oblong inversely egg-shaped. Kidney-shaped seeds. Dorsal side is convex. The lower end slightly depressed and slightly twisted. Seed surface smooth. The color is dark brown or brown. M_{1000} = 6.29 g; length 3.32±0.14 mm; width 2.27±0.05 mm.	
2. *Capparis herbacea* Willd. – Floristic region – (13B). 9.07.2013. Mangistau area, Beineou region, gorge, 22nd km from Manaty-ata. N 44°08′194′′, E 53°16′196′′. 221 m of true altitude. Fruit is a juicy berry; fruitcase is oblong inversely egg-shaped. Kidney-shaped seeds. Dorsal side is convex. The lower end is slightly depressed and slightly twisted. Seed surface is smooth. The color is from dark-brown till light brown M_{1000} = 2.3 g; length 2.67±0.09 mm; width 2.3±0,11 mm.	
3. *Capparis herbacea* Willd. Floristic region – (13 B). 18.07.2013. Mangistau area, Beineou region, South Ustyurt, the 30th km to the North from the holy place Beket ata. N 43°46′599′′, E 54°05′706′′. 320 m of true altitude. Fruit is a juicy berry; fruitcase is oblong inversely egg-shaped. Kidney-shaped seeds. Dorsal side is convex. The lower end is slightly depressed and slightly twisted. Seed surface is smooth. The color is brown. M_{1000} = 24 g; length 2.75±0.05 mm; width 2.07±0.08 mm.	
4. *Capparis herbacea* Willd. Floristic region – (21). 28.06.2013. South Kazakhstan area, Turkestan region, Near Karatau plain (south macro slope), at access road to Turkestan from the west side. Highland plain. Ephemer wermuth community with capers. N 43°25′25.5′′, E 068°00′29.1′′. 232 m of true altitude. Fruit is a juicy berry; fruitcase is oblong inversely egg-shaped. Kidney-shaped seeds. Dorsal side is convex. The lower end is slightly depressed and slightly twisted. Seed surface is smooth. The color is from dark- till light brown. M_{1000} = 4.13 g; length 3.06±0.06 mm; width 2.8±0.07 mm	
5. *Capparis herbacea* Willd. Floristic region – (27). 2.08.2013. Zhambyl area, Merken region. West part of the Kyrgyz Alatau ridge, dry gorge to the East from Taraz. N 42°53′463′′, E 71°35′970′′. 1032 m of true altitude. Fruit is a juicy berry; fruitcase is oblong inversely egg-shaped. Kidney-shaped seeds. Dorsal side is convex. The lower end is slightly depressed and slightly twisted. Seed surface is smooth. The color is from dark-brown till dark fulvous. M_{1000} = 6.6 g; length 3.58±0.11 mm; width 2.65±0.08 mm.	

Acknowledgements

This work was made possible due to the state funding of the Republic of Kazakhstan in the frame of the scientific-technical program No. 0199 "Botanical variety of wild relatives of cultivated plants of Kazakhstan as a source of enrichment and preservation of the agrobiodiversity gene pool for realization of the Food programme", 2013-2015.

References

[1] ENSCONET (2009). *ENSCONET.* Curation Protocols & Recommendations. www.ensconet.eu/download

[2] Fiona R.H., Probert R.J. (2013). Advances in seed conservation of wild plant species: a review of recent research. *Conservation physiology*. Vol. 1. pp. 1-11. http://conphys.oxfordjournals.org

[3] ISTA (2013). International Rules for Seed Testing. *International Seed Testing Association*, Bassersdor.

[4] Millennium Seed Bank Project website www.kew.org/msbp

[5] FAO/IPGRI (1994). Genebank Standards. Food and Agriculture Organization of the United Nations, International Plant Genetic Resources Institute, Rome.

[6] FAO (2013). Draft genebank standards for plant genetic resources for food and agriculture. http://www.fao.org/agriculture/crops/core-themes/theme/seeds-pgr/conservation/gbs/en/ (accessed April 2013).

[7] Merritt DJ, Dixon KW (2011). Restoration seed banks – a matter of scale. *Science* 332: pp. 424–425.

[8] Plilipenko G.I. Methods of the *ex situ* conservation of the genetic plant resources (2013). *Presentation* http://www.myshared.ru/slide/392931/

[9] http://www.norwesol.ru/most-important-omega-3#dpa

[10] http://www.nutri-facts.org/rus/nezamenimye-zhirnye-kisloty/essential-fatty-acids/kratkii-obzor

[11] Dolecek TA, Granditis G. Dietary polyunsaturated fatty acids and mortality in the Multiple Risk Factor Intervention Trial (MRFIT) // *World Rev Nutr Diet.* 1991. Vol.66. pp. 205-216.

[12] http://club.uliabell.e-gloryon.com/0907721872

[13] http://humbio.ru/humbio/biochem/001a55c7.htm

[14] Kamysheva I.M. Development of technologies of complex processing the amaranth seed for food goals: Thesis of the candidate of sciences (Technique). Saint-Petersburg: All-Russia scientific-research institute of fats, 2000. – 20 pp. *(published in Russian)*

[15] Tsekhanovskiy S.N., Morozova Z.F., and Sergiyevskaya I.V. Application in clinical practice the "Vegetable oil "Forma" (made of amaranth seeds) // New and unconventional plants and perspectives of their use: 5th International symposium. – M., 2003. Vol 1. pp. 98-100. *(published in Russian)*.

[16] http://life-spb.ru/others/amaranth-oil-clinical-research.htm

[17] Tyutyunnikov B.N. Chemistry of lipids – M.: Kolos, 1992. – 448 pp. *(published in Russian)*

[18] Bezzubov L.P. Chemistry of lipids. M.: Food industry. 1975. – 280 pp. *(published in Russian)*

[19] Scherbakov V.G. Chemistry and biochemistry of treatment the oil seeds – M.: Food industry. 1977. 180 pp. *(published in Russian)*.

[20] http://www.minclinic.ru/drugs/lekarstvennie_svoystva_pishevih_rasteniy/kapersy.html

[21] Akgül A.and Ôzcan M. Some compositional characteristics of capers (*Capparis spp.*) seed and oil // GrasasyAceites. – 1999. – Vol.50., Fase.1. – pp. 49-52.

[22] Rakhimov I.F. Biochemical and pharmacological features of sea-buckthorn and caper oils in Tajikistan. Dissertation of doctor of sciences (Medicine): 14.00.25. Dushanbe: Chemistry, 2006. 254 pp. *(published in Russian)*

[23] http://innature.kz/articles.php?article_id=1651

[24] http://edaplus.info/produce/caper.html

[25] http://www.plantarium.ru/page/view/item/8076.html

Territorial and Branch Assessment of Water Management Balance of Georgia Based on Differential Estimation and Modeling of Economic Impact on Water Resources

Nino Kezevadze

Directorate, ltd “Georgian Water and Power”, Tbilisi, Georgia

Email address:
nkezevadze@georgianwater.com

Abstract: Political and social-economic coups in the 1990s of the last century in Georgia, as well as in countries of the former USSR and East Europe has left its mark on the water management problems of those countries. Thus, for alternative assessment of probable water economy balance first priority is to select a representative period in dynamics for utilization of water resources. For Georgia such period is 1980s of the last century, as the most active period in historic economic development. The indicated period is characterized by comparatively reliable water management statistics, with stable socio-economic background and maximum anthropogenic load on water resources. For the very period branch and territorial structure of water management balance have been studied. Based on that alternative forecast have been assessed for the first quarter of the current century,according to the planned scenario during the above mentioned period. The indicated studies have been carried out based on differential assessment and mathematical modeling of economic load on water resources. Water management models have been developed for certain categories of water consumers using economic, demographic and ecological criteria.

Keywords: Water Resources, Water Management Balance, Mathematical Modeling

1. Introduction

Political cataclysms occurring in the country in the 1990s of the last century predetermined recessionary drastic changes in the social and economic sphere. Correspondingly,a new reality has had influence on water economy topics. The unpredicted development scenario of demographic and business processes hasmade topical to study spatial and time dynamics of water management based not only on statistical data, but also on mathematic modeling using generalized parameters and criteria which are determined by indirect method.

Assessment of spatial and time dynamics requires conformity of quality of territorial differentiation with the accuracy of determination of water utilities, components and level of detail on the one hand, and time aspect, i.e. adequate selection of the study period, on the other hand.

Branch and territorial assessment of water economy balance of Georgia may be carried out taking into account only landscape and business activity of the area, which determines the locality of territorial distribution of the elements of water and water economy balance. From this point of view, the first and the most important stage of territorial differentiation is to divide the territory of Georgia into two natural and economic regions – West and East regions. The present article deals with branch and territorial topics of water management balance.

Taking into consideration the above complexities for assessment of spatial and time dynamics of water economy balance, especially problematic is to select a representative period, which at more adequately reflects the possibilities of economic development of the country from the point of view of hydro recourse potential. Such representative period for this article is the 1980s of the last century as the most active period in the economic development of the country with comparatively reliable water economy statistics, stable social and economic background and maximum anthropogenic impact on water resources.

2. Overview of Water Resources and Water Management Balances of Georgia

Water budget analysis (hydrologic inventory) is carried out based on comparison of certain economic branches, i.e. water used by main water consumers with the existing resource potential. On this basis first stage of the water balance calculation is to establish the resource potential of the area.

The resource potential of rivers in Georgia comprises transit and local flows. Local river flows, established and adjusted on the basis of water resource balance assessment, excluding transit flows of River Mtkvari, Chorokhi and North Slope of the Caucasus makes up 57, 9 km3, and steady flow –19, 6 km3 (Steady flow's components are underground flows, flows regulated by water reservoirs and total useful groundwater storage). In addition to West Georgia appropriate indicators are 43, 3 and 12, 3 km3, and for East Georgia – 14, 6 and 7, 2 km3.

According to world standards and taking into account demographic data, Georgia belongs to average river resource serviced area. This indicator (Specific water provision) for complete flow equals to 11 thousand m3/year and for underground flow – 3, 5 thousand m3/year per capita. While the world average indicator is 11 and 4 thousand m3/year correspondingly. Naturally unequal geographical distribution of water resources of Georgia between west and east regions is reflected on the surface as well as an underground specific water provision. Surface and underground specific water provision in West Georgia equals to 19 and 5 thousand m3 correspondingly, and for East Georgia – 5 and 2 thousand m3/year per capita. In addition, this inequality is deepened by the unequal location of the population according to height. In particular, 88 % of the population of Georgia is concentrated in the lowland. For this reason lowland of East Georgia is less serviced by river resources – 0, 45 thousand m3/year, and mountain regions of West Georgia are characterized by high water provision – 317 thousand m3 /year per capita.

The unequal distribution of the population, according to altitude, in its turn, has functional relations with location of economic branches which indicates to the fact that share ofindustrial production produced in the lowland of Georgia in comparison with average geographical indicator in the 80is of the last century was 90 %. This fact naturally specifies deepening inequality of anthropogenic impact and correspondingly widening of the water resource deficit in the lowland area of Georgia.

Over a century hydrology and its allied sciences pay attention to issues of anthropogenic effects on water resources. Initially changes of water regime as a result of different economic activities have been studied. Later scientific and technological revolution has made the issue of qualitative transformation of natural waters topical. Lately hydrologic range of problems has become a priority. It is related to modern climatic and possible anthropogenic changes. Scientists of State Institute of Hydrology, Water Problem Institute and Geography Institute under Academy of Sciences of former USSR and its republics. Methods for calculation of hydrological changes and transformations have been developed, alternative predictive assessments of possible states of water resources have been carried out all over the world for certain regions and countries [1,2].

Indicated methods and predictive assessments for Georgia, as a mountainous country, or an area with high-altitude, length regularities length have been carried out based on basic research of altitudinal regularities of water resources and water balance elements, which has been studied for decades at the Vakhushti Bagrationi Institute of Geography of Academy of Sciences of Georgia [3]. As for water management balance and its predictive assessments, they are subject to the principle of altitudinal regularities and they require interpretation according to altitudinal belts and natural zones.

It should be noted that for the entire world, as well as for majority of Non-CIS States, predictive assessments of water management balance have been justified. As for post-socialist space, including Georgia, the transformation of water resources and water balance from the beginning of the 1990s have been developing under unpredictable scenario. Obviously it is the result of those political, social and economic changes which involved former Soviet Union and Eastern Union countries. Against the background of the world economic growth, economic downfall emerged in those countries in the 1990s, which in its turn,had an impact on the quantity and quality of water resources. Over the period of modern social and economic crisis, at the expense of sharp decline in industrial and domestic water consumption, total water consumption, volume of recycled and reused water, as well as volumes of waste waters and non-recoverable loss, have decreased. To a certain extent, it has improved quantitative and qualitative state of water resources. This tendency has been kept to this day. Though it is evident that it is temporal because of economic depression and under depression the state will not be able to preserve economic stability for a long period. Prerequisite for overcoming economic and social crisis in the country is industrial expansion, development of agriculture, tourism and communal services. Certainly those processes will thoroughly change components of water economy balance for the worse. This is the reason why, in the process of economic use of water resources against the background of topical economic load,it is necessary to substantively change management principles, activate protective measures in terms of quantity and quality.

It should also be noted that the economic crisis in Georgia has been followed by bringing to a stop certain metering and control systems. Under shortage of statistical data for the given period, it is possible to establish expenses on water economy balance elements, first of all, on water intake, used water and sewerage, based on theoretical and empirical methods. Though using similar methods for observation and metering period is frequently inevitable, as metering of water economy elements has always been characterized by poor accuracy.

3. Mathematical Model of Water Management Balance

For the purpose of differential assessment and mathematical modeling of economic load on water resources the author of this article has developed and submitted empirical formula and algorithms, based on which water management balance should be calculated in the case of statistical data unreliability as well as during predictive research.

Main formulae of water management balance may be represented in the following way:

$$\sum_{i=1}^{n} Q_{i,b} = \sum_{i=1}^{n}(Q_{i,n} - Q_{i,o} + Q_{i,m}) \quad (1)$$

where $Q_{i,b}$ is the volume of water taken at a certain time by i-consumer (industry, population, agriculture, etc.), $Q_{i,n}$ – volume of used water, $Q_{i,o}$ –recycled water consumption and $Q_{i,m}$–volume of water lost on transportation.

Volume of used water:

$$\sum_{i=1}^{n} Q_{i,n} = \sum_{i=1}^{n}\left(Q_{i,g} - Q_{i,c} + Q_{i,e} + Q_{i,o}\right) \quad (2)$$

where $Q_{i,g}$ is taken back loss, $Q_{i,c}$– volume of waste water and $Q_{i,e}$–volume of water returned to the riverbed.

With an account of (2) final definition will be as follows:

$$\sum_{i=1}^{n} Q_{i,b} = \sum_{i=1}^{n}(Q_{i,g} - Q_{i,c} + Q_{i,e} + Q_{i,m}) \quad (3)$$

As it was mentioned above, drawing up water management balance requires availability of large data base about resources and functions of the basin, quantitative, as well as qualitative, which, in most cases, leading bodies of water utility system are not able to provide. Under the conditions of deficit of materials and unreliable assessment of the current water economy balance processes may be carried out only by rough estimate based on economic, water and ecological standards. Main algorithms for calculating water economy elements may be represented in thefollowing way:

Water consumption volume in domestic sector in the course of time (t)

$$Q_1(t) = K_1\left[\sum_{j'=1}^{R} q_{j'}(t)Z_{j'}(t) \sum_{j''=1}^{R} q_{j''}(t)Z_{j''}(t)\right] \quad (4)$$

where $q_{j'}$ and $q_{j''}$– are water use rates for city and village population (l/day per capita), and j′ and j″ – settlement category, $Z_{j'}$ and $Z_{j''}$ – corresponding number of the population of the settlement categories, and K_1–coefficient for regulating measurements.

In industrial – business sector

$$Q_2(t) = K_2 \sum_{j'''=1}^{R} q_{j'''}(t)V_{j'''}(t) \quad (5)$$

where $q_{j'''}$is water consumption per production unit (specific water-retaining of the product), and $V_{j'''}$ –volume of the corresponding industrial production.

In agriculture, particularly in livestock sector

$$Q_3(t) = K_1\left[\sum_{j''''=1}^{R} q_{j''''}(t)Z_{j''''}(t)\right] \quad (6)$$

where $Z_{j''''}$ and $q_{j''''}$ –correspondingly numbered ofj''''type and water consumption norm.

Volume of water necessary for j type culture in irrigation farming during vegetation period $(\mathbf{t'})$ may be expressed in an algorithm:

$$Q_4(t) = \sum_{j=1}^{R} N_j(t)F_j(t) ** \sum_{j=1}^{R}\begin{Bmatrix} 0, & when W(t') \geq E_0(t') \\ & or \\ E(t') - W(t'), & when W(t') < E_0(t') \end{Bmatrix}(t')F_j(t) \quad (7)$$

where $N_j$is irrigation rate, F_j–culture acreage, E_0–evaporation during vegetation period, and W – total moisture of the area (infiltration)for the same period. It characteristics for rehydration resources which constitute evaporation and ground water inflow.

Consumptive water use during a certain (**t**) period will be:

$$Q(t) = \sum_{i=1}^{n} Q_i(t) = \sum_{i=1}^{n} \sum_{j=1}^{R} K_i q_{ij}(t)Y_{ij}(t) \quad (8)$$

where q_{ij} is specific flow intensity per unit index, $Y_{ij}$is the volume of the corresponding index.

A comparison of total volume of water consumption calculated by proposed an empirical formula and algorithms to the existing resources is an availability index (water resources) of the area.

As for quality state of water in natural objects, it is defined by pollution and self-purification of the reservoir (basin). Pollution depends on point and diffuse sources, and self-purification of the reservoir (basin), on the intake object, the number of its resources and primary quality. For assessment of self-purification, first of all necessary volume of water for watering, we propose the following image:

$$Q_p = \max Q_{j(\epsilon 1;R),p} = \max\left\{\sum_{i=1}^{n}\left[\frac{P_{ij,c}/Q_{i,c} + K_{j,a}}{K'_j Q_{i,c}}\right]\right\}, \; by j \epsilon 1; R \quad (9)$$

where$P_{ij,c}$is i-total concentration of waste waters discharged in the water object against j component, $K_{j,a}$ –primary concentration of the intake object against j component, K'_j is the maximum permissible concentration of j component.

In figure (9) reciprocal dilution for water consumption i sector Δij is expressed by the following formulae:

$$\Delta ij = \max_{j(\epsilon 1;R)}\left[(K_{ij,c} + K_{j,a})/K'_j\right] \quad (10)$$

If the intake object is capable to satisfy the conditions:

$$Q_a \geq Q_s + Q_p = Q_s + \left[\max \sum_{i=1}^{R} \Delta ij \; (\epsilon 1; n)\right]Q_{i,c} \quad (11)$$

It may be considered that saturation quality corresponds to self-purification of the basin and thus qualitative load is allowed for the specific water object.

In equation (11) $Q_a$is water flow in the intake object, Q_s–sanitary flow of the water object, i.e. minimum flow, which meets sanitary standards of the river bed.

It should also be noted that water quality criteria are fixed indexes only in the concrete situation. As usual, they must have the determinate flexibility (from the point of view of one sign deviation – to reduction). In particular, if for short-term task maximum permissible concentration is acceptable K'_j, for long-term tasks the following condition must become deeper:

$$K''_j < K'_j, \text{i.e.,} \quad \sum_{j=1}^{R} K''_j/K'_j \le R \qquad (12)$$

where K''_j is concentration of natural water against component j.

4. Variational Water Management Balance of Georgia

Based on differential assessment and modelling of economic load on water resources, alternative water management balance has been developed in Georgia.

As at present Georgia actually faces the necessity of economic revival, particularly production economy, it is of special interest water management balance of Georgia during the most active period of economic development. As it was mentioned above in the dynamics of usage of water potential of Georgia for assessment of water management elements, up to now, the most presentable was the situation in the 1980s of the last century. It was a period of maximum economic activity, in which water economy statistics were comparatively reliable, social-economic background stable and anthropogenic impact on water resources was highest possible.

As a result of the survey, it was established that at the end of the 1980s, total water intake from water bodies was 4860 mln.m^3, which made up 10 % of river runoff and 28 % of stable flow. At the same time, 30 % water intake (1500 mln.m^3) falls to the share of transportation losses. Approximately 40% (1570 mln.m^3) water usage was accounted for non-returnable losses. Annual volume of waste waters was 1720 mln.m^3, and 19800 mln.m^3 of pure river water became unfit, i.e. 34 % of total river runoff in Georgia and almost total stable flow (fig.1).

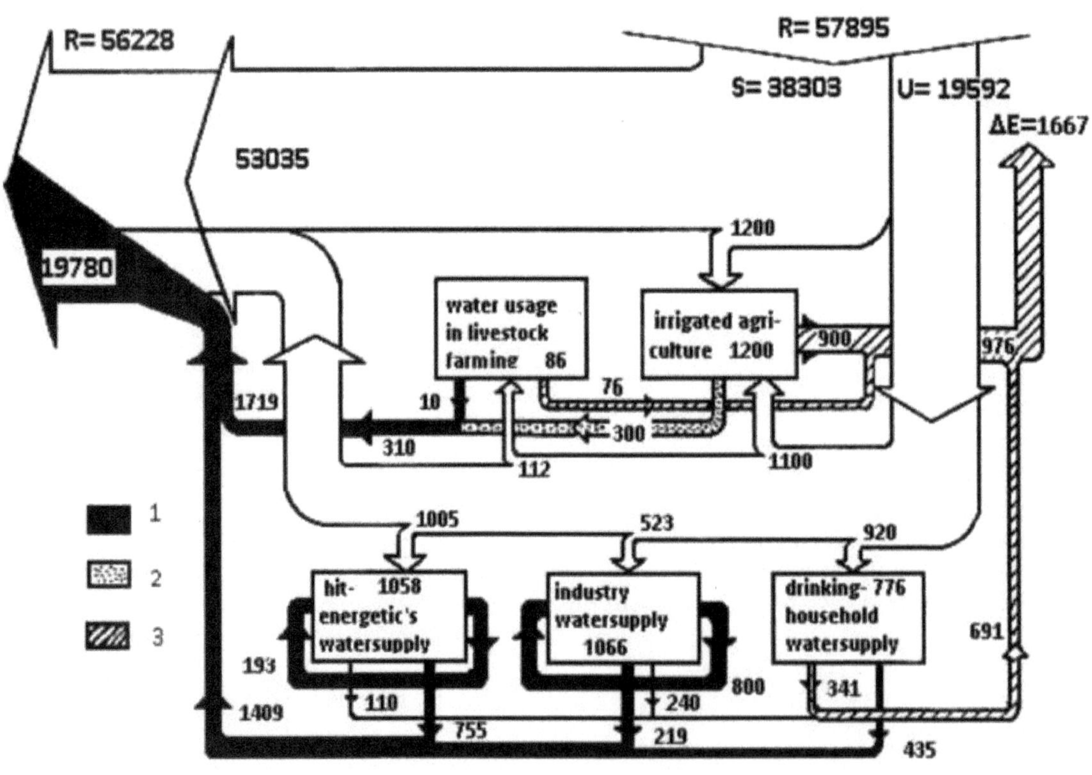

R – Total resources; U – Stable resources; S – Non-stable resources;E – Non-returnable loss.
1 – Polluted waters; 2 – Returned waters; 3 – Non-returned waters.

Fig. 1. *Table of Water Resource Usage at the end of the 1980s (mil.m^3).*

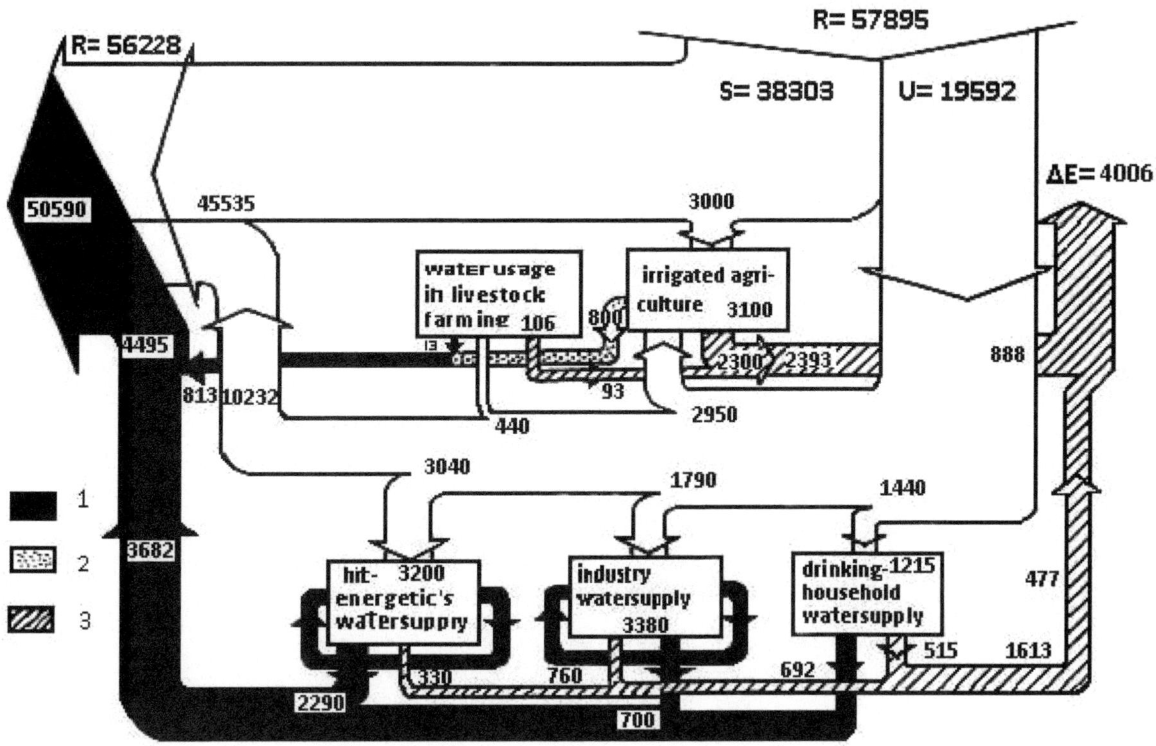

R – Total resources; U – Stable resources; S – Non-stable resources;E – Non-returnable loss.
1 – Polluted waters; 2 – Returned waters; 3 – Non-returned waters.

Fig. 2. *Table of Long-term Usage of Water Resources of Georgia (Option 1of the forecast) (mil.m^3).*

The given data indicate to the critical situation of water resources of Georgia from the point of view of quality. The cause was in the low performance of cyclic/circulating water supply (19 % of water usage), as well as treated waste waters (27 % of total waste waters, from which only 4 % were subject to biological treatment).

Proceeding from the principles of indicated water management, water industry faced heavy prospects. Without substantial changes of principles and standards of usage of water resources, in compliance with the state plan of development of national economy, by the first quarter of the 21st century, 12400 mln.m^3 water would be necessary for water supply, which is 20 % of river resources of Georgia. Non-returned loss would reach 4000 mln.m^3, and volume of waste waters - 4500 mln.m^3, for dilution of which, even in case of artificial treatment, it would be necessary 50000 mln.m^3 water (95 % of river resources of Georgia) (Fig.2).

Thus we can conclude that in the nearest future water resources of Georgia would threaten not quantitative but mostly qualitative deficit.

As it has been repeatedly mentioned, water and waterindustry balances drawn up for entire Georgia do not adequately reflect the true picture of territorial distributionof water resources and branches of the economy. For this purpose a special survey has been carried out at the major territorial differentiation - for West and East Georgia separately.

From the point of view of water management balance tension East Georgia situation is of special interest.

As a result of the water management balance assessment of East Georgia it was established that the volume of water necessary for dilution of waste waters during the reporting period quantitatively is the same category as East Georgia river runoff resources (Fig.3).

If we take into consideration expected growth of the population, as well as further development of water use branches, according to state plans of national economy (Master plan for Development of National Economy of Georgian SSR in 1985-1988),in the case of maintaining modern principles of economic management, the total amount of used water from the beginning of 21 century would increase by 2,5 times. Correspondingly volume ofwaste waters would reach 3,2 km^3. Even if the total volume of waste waters underwent high quality biological treatment, for dilution it would be necessary 22 km^3 flow, which exceed river resources of East Georgia by 1,5.

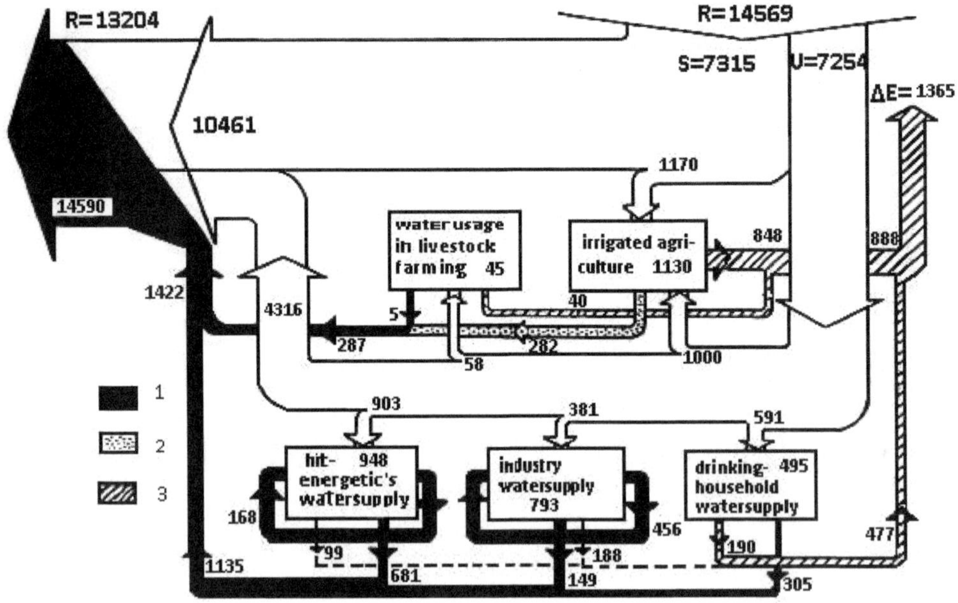

R – Total resources; U – Stable resources; S – Non-stable resources;E – Non-returnable loss.
1 – Polluted waters; 2 – Returned waters; 3 – Non-returned waters.

Fig. 3. *Table of Use of Water Resources of East Georgia in the 1980s of the Last Century (mil.m^3).*

Thus, in the 1980s in East Georgia was facing depleting water resources quantitatively, as well as qualitatively which required to pursue an immediate strategic course of rationalization of water resources.In particular isolation ofeconomic water recycling from natural one, i.e. introduction of preventive principle in the mechanism of water resource protection – fighting causes and not results. Some practical measures connected with this principle are:Transition toclosed (cycle) water supply in industrial enterprises, maximum sewage sanitation, industrial and municipal wastewaters and their use as far as possible for technical purposes in those enterprises not requiring high quality water.At the same time, use of domestic sewage after propertreatment for irrigation of natural grasslands and technical cultures in some regions. It must also be introduced principle of universal saving of water industry at the expense of reduction of water retaining in the product, through improvement of scientifically grounded irrigation standards and methodology, in irrigation farming.

Taking into account the above mentioned principles, in case of rationalization of the water industry for expected changes in population and the same scale of economy development long-term table may be represented in the following way (Fig.4):

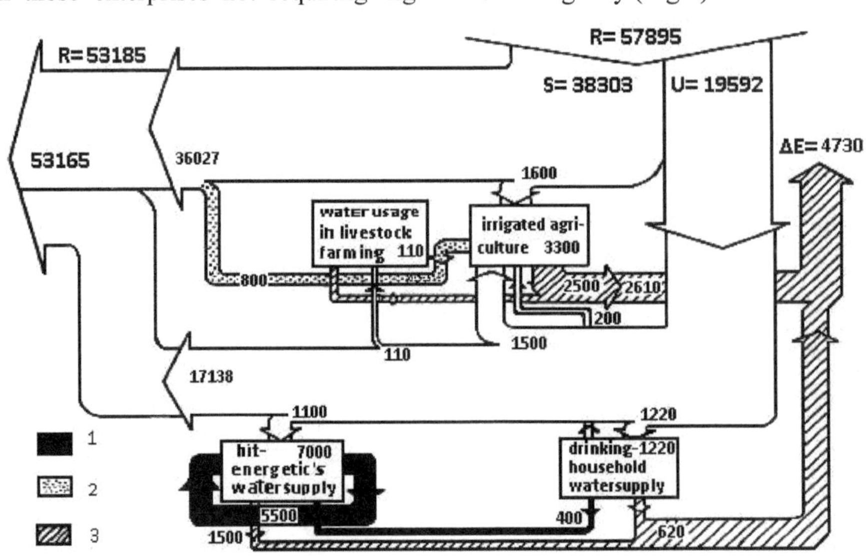

R – Total resources; U – Stable resources; S – Non-stable resources; E – Non-returnable loss.
1 – Polluted waters; 2 – Returned waters; 3 – Non-returned waters.

Fig. 4. *Table of Long-term Use of Water Resources of Georgia (Option II of the forecast) (mil.m^3).*

According to the given predicted option of water industry balance of Georgia non-returned loss compared to the actual one would grow by 3 times and reach 4800 mln.m^3, but it would occur against the background of stopping waste waters in the water body. The main effect of rationalization of water economy primarily lies in protection of water resources from qualitative exhaustion which is presented as a rationalized option of the forecast. It should also be noted that this option is somehow idealized, but it can play environmental quality benchmark, which must be strictly seek.

5. Conclusion

The territory of Georgia is characterized by an uneven distribution of water resources in the West and East regions which causes uneven supply of the population with resources of full flow and interflow in those parts. On that basis if during the representative period selected by us for entire Georgia, waste waters polluted 1/3 of the total resources of Georgia, water volume in East Georgia, necessary for wastewater treatment, has already been proportionate to theexisting resources.

Moreover, according to the scenario of long-termeconomic development of the first quarter of the currentcentury with the existing principles of water economy in the period under study for dilution of effluent waters, it would be necessary 95% of the water resources of Georgia and in East Georgia volume of water necessary for sewage sanitation would exceed the existing local resources by 1,5 times.

Taking into consideration the above mentioned in the period under study, the country has already been facing the necessity of urgent and complete reorientation of the principles of water economy. As a result of implementation of the recommended scheme of its rationalization and optimization which implied increase of quantitative impact on water recourses compared to the modern one at the expanse of increase of irretrievable water consumption, however it would be possible to avert the threat of qualitative exhaustion of water resources at the expanse of the maximum possible stopping of pollution of rivers with waste waters.

References

[1] M.Lvovich. World water resources and its future. M., 1974, 448 p. (Rus.)

[2] M.Lvovich. Water and Life. M., 1986, 254p.(Rus.)

[3] Water balance of the Caucasus and its geographical features (consistency with natural laws).Tbilisi,1991, 142p.(Rus.)

[4] N.Koronkevich, I.Zaitseva. Change in specific water consumption in the Russian Federation for the last decades. Water resources, 1999, volume I, 90-95.(Rus.)

[5] I. Bitz. Trend analyses, forecasting and long-term planning of water consumption in river basin.Scientific basis for rational utilization, conservation and management of water resources. Part I, M., 1983,78-85(Rus.)

[6] J.Alcamo, P.Döll, Th.Henrichs, F.Kaspar, B.Lehner,Th.Rösch, S.Siebert. Global estimates of water withdrawals and availability under current and future 'business-as-usual' conditions. Hydrological Sciences Journal, 2003, volume 48, No. 3, pp. 339–48.

[7] F.Molle, D.Vallée. Managing competition for water and the pressure on ecosystems. WWAP (World Water Assessment Programme). United Nations World Water Development Report 3: Water in a Changing World. Paris/London, UNESCO Publishing/Earthscan, 2009,150–159.

[8] Guidelines on Strategic Planning and Management of Water Resources. ST/ESCAP/2346. New York: United Nations, 2004, 94p.

[9] Dinar, A. and Saleth, R. M. Can water institutions be cured? A water institutions health index. Water Science and Technology: Water Supply, Vol. 5, No. 6, 2005,17–40.

[10] Pres, A. 2008. Capacity building: a possible approach to improved water resources management. Water Resources Development, Vol. 24, No. 1,123-134.

[11] GWP-TAC (Global Water Partnership – Technical Advisory Committee). Integrated Water Resources Management. TAC Background Papers 4. Stockholm, GWPTAC, 2000.http://www.gwpforum.org/gwp/library/TACNO4.pdf

14

Assessment of potentially toxic metals in soil and sediments of the Keana Brinefield in the Middle Benue Trough, Northcentral Nigeria

Sallau Adamu[1], Mangs Ayuba[2], Adamu Murtala[2], Lar Alexander Uriah[2]

[1]The Carter Centre, Tudun Wada, P.O. Box 7772 Jos, Plateau State Nigeria
[2]Department of Geology, University of Jos, Jos, Nigeria

Email address:
adamusallau@yahoo.com (S. Adamu), mangsdanmangu@gmail.com (M. Ayuba), murtalasulea@gmail.com (A. Murtala), ualexanderlar@yahoo.co.uk (L. Uriah)

Abstract: This research was conducted in the Keana area of the middle Benue trough endowed with abundant mineral deposits (brine/salt, lead-zinc sulfide ores, barite and limestone) and has a long history of mining activity. Small scale mining activity in the area is currently increasing in intensity causing immeasurable damage to the environment and to the surrounding communities. Potentially toxic metal releases and remobilization from the mining sites into the surrounding arable soils and stream sediments poses serious human health risks since the surrounding communities live and grow their food from such environment. This research assesses the accumulation and the extent of contamination of Potentially Toxic Metals (PTM) in the vicinity of the mining environment using standard pollution indices such as geo-accumulation index (I_{geo}), Enrichment Factor (EF), Contamination Factor (CF) and Pollution Load Index (PLI). A total of 40 samples were analyzed for As, Ba, Co, Cr, Cu, Mo, Ni, Pb, Sc, Sr and Zn, using ICP-OES geochemical technique. Analysis of geochemical data shows that the soil and stream sediments in the area are extremely polluted by Zn, Cr and Ni, having I_{geo} values of 4.9, 3.8 and 2.9 respectively. Igeo levels of Pb, Cr and Ba indicate moderate to no pollution. Zn and As with EF of 128.60 and 87.66 respectively, are severely enriched in the environment, while Pb, Ni and Cr with EF between 6 and 8, are moderately enriched. The calculated CF and PLI values for Cu, Sr and Sc in the soil and sediments remain very low, indicating these media are not contaminated by these elements. A pollution load index of < 1 as seen in most locations for those elements, indicate no pollution. The continued exposure of the populations living in this area particularly those in the mining locations who depend on the soil (for subsistence farming) and water from ponds, wells and boreholes (for drinking, cooking and other domestic uses), would result in various health risks and concerns, if not checked or abated. Lead, for example is known to be toxic even at very low exposure, with chronic lethal effects. This study recommends an immediate plan for analysis of the quality of drinking water and some staple crops grown in the area to determine the levels of these noxious metals and uptake by plants, to be followed by a comprehensive mitigation or remediation plan.

Keywords: Potentially Toxic Metals, Contamination, Human Health, Soils, Sediments, Benue Trough

1. Introduction

The Keana brinefield is endowed with widespread occurrences of lead-zinc, baryte, limestone, as well as the brines, in a geology dominated by continental and marine sediments represented by sandstone, shale, mudstone and limestone, and is associated with mineralized hydrothermal veins that are a consequence of the tectonic rifting that led to the emplacement of the Benue Trough. The area is characterized by a proliferation of artisanal mining activities as a result of the mineral deposits. This area is located between Longitudes 8° 45′ 00″E and 8° 15′ 25″E and Latitudes 8° 05′ 00″ N and 8° 15′ 00″ N in the southern quadrant of the Lafia topo degree sheet 231SE.

This area is an important location, known for the production of NaCl salt, used for diverse domestic and industrial purposes since the early 15th century. Also, mining activities for the production and processing of baryte ($BaSO_4$) and lead-zinc (Pb-Zn sulphides) have also become highly ubiquitous. The mining of these mineral deposits has become the main economic and revenue earning activities of the people of the area, besides subsistence farming. Salt and brine are used for human consumption, while mine wastes, tailings, abandoned mine ponds, etc, practically releases potentially toxic trace elements into agricultural and non-agricultural soils, sediments, drinking and irrigation water, and may be absorbed by humans, plants and subsequently into the food chain, thereby affecting the environment, human and animal health.

In recent times, a lot of attention, including a global consensus, has been drawn, to the impact of human activities such as mining, on the environment, not minding the socio-economic benefits of the people and country where such mineral deposits are exploited. Some studies [14], [5] , [7] have demonstrated the negative impacts of heavy metals such as Pb, As, Cd and so on, on the environment and human health resulting from exposure through food, water and inhalation of these toxic metals. These and other studies [18], [6] have also revealed that the geographical distribution and occurrence of certain diseases could be correlated to the presence of trace elements in the geologic environment.

The justification for selecting the Keana area for this study is owing to the many years of domestic consumption of edible salt and other uses, coupled with the occurrence, mining and processing of the listed mineral deposits versus the impact of these on the environment and human health. Furthermore, the research has become necessary in view of the dearth of information (data) on trace element levels in soil, sediments and water, as the people continue to depend on these basic life support media for their survival. This paper therefore seeks to assess the distribution and contamination levels of elements such as Pb, As, Zn, Cd, Cr, Se, I, and likely impact.

1.1. The Study Area

The Benue Trough is geologically divided into the upper, middle and lower depositional sub basins, stretching from the Gombe area in the north to the Abakaliki area in the south, towards the Niger Delta. The Keana Brinefield is located in the middle and is part of a long narrow stretch of sedimentary basin, extending from the Gulf of Guinea upwards to the northeast. Stratigraphically, the area is made up of five rock units which are essentially continental and marine sediments. These include the Asu River group of sediments (sandstone-shale series) which are the oldest, the Awe formation (sandstone-shale-siltstone), the cretaceous Keana formation (sandstones), EzeAku formation (sandstone-shale-mudstone) and Awgu formation as the youngest, is entirely shale. The major structural feature in the area is the Keana anticlinorium with the Brinefield and environs located along the western limb of the anticline.

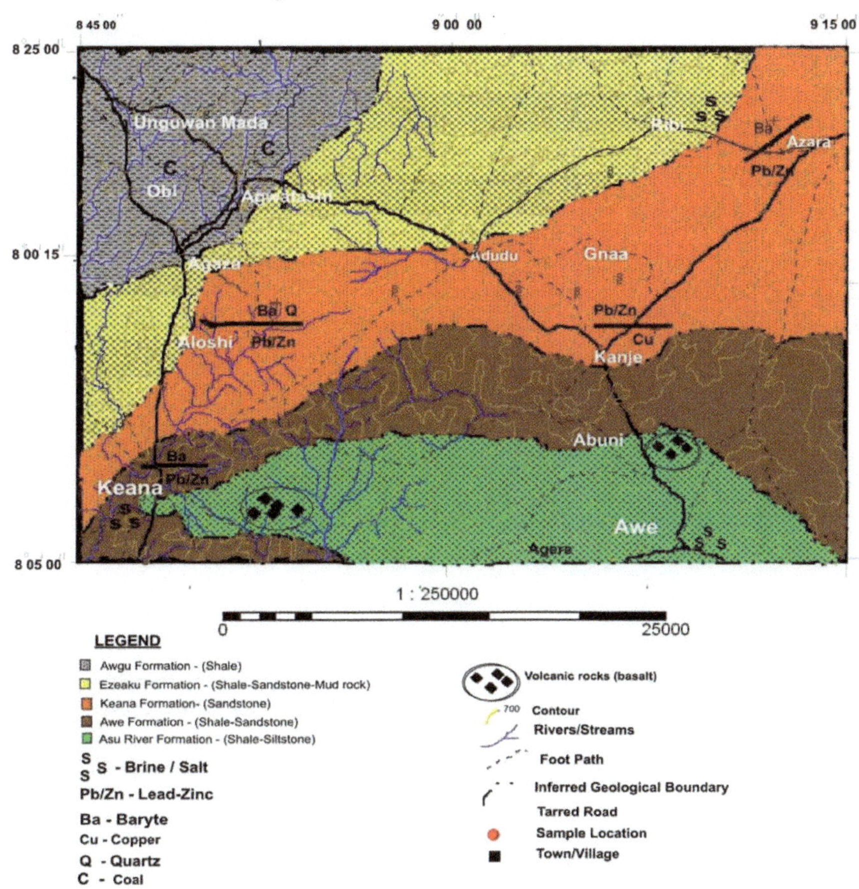

Fig 1. Geological Map of the Keana Brinefield

The study area is endowed with enormous occurrences of mineral deposits such as brine/salt, lead zinc, baryte and limestone. The mineral deposits are found around the axial part of the Keana anticline. Salt is mined in many locations around Keana; while baryte is mined in Aloshi, Bature, Chiata and Keana (Fig. 1). The brines are hosted by the Awe formation which is transitional between the Asu River and the Keana formations, with the brines oozing out along the east and western limbs of the Keana anticline in valleys and depressions undercutting the Awe formation. The baryte and lead-zinc minerals are found in veins and lodes, and occur with quartz and the carbonates of Mg, Fe and Ca (siderite and ankerite) with most pronounced ridges found in the area marking the mineral veins.

The mineralization (baryte, lead-zinc) is believed to be associated with mineralized hydrothermal veins that are a consequence of the tectonic rifting that led to the emplacement of the Benue Trough, connected with the stress that developed following the continental separation of Africa and the South American plates [12], [13] [1].

2. Materials and Methods

2.1. Sampling and Sample Preparations

Two major media namely soil and sediments were sampled with a total of 40 samples collected (33 soil, 7 stream sediments. Samples were collected from mining areas, cultivated and uncultivated farmlands. Sample locations are shown in Fig. 2.

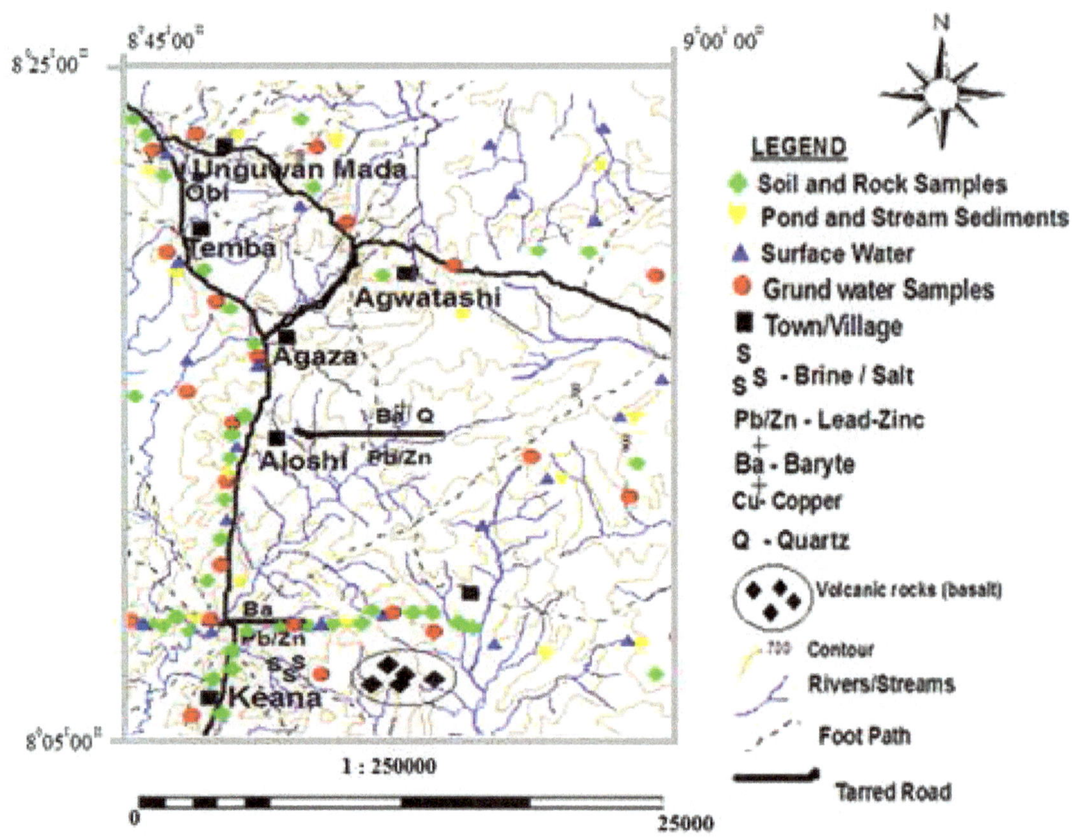

Fig 2. Map of the Keana Brinefield Showing Sample Locations

The soil and stream sediment samples were collected in special clean, zipped and sterilized sample bags, well labeled according to the location, date of collection for easy referencing as well as adequate precautionary measures were taken during the collection, packing, transportation and storage. The soil samples were collected in the N-E and S-W directions but the stream sediment samples were randomly distributed and so was the sampling pattern. In the geochemical laboratory, the soil and stream sediment samples were pulverized and made to pass through a 0.067mm mesh size, after which a 100 mg of the now powdered soil sample was weighed into a teflon crucible and then dissolved in aqua regia after 6 hours of heating the solution to dryness on a hot plate (250°C). The crucible containing the sample is then allowed to cool. Two (2) ml of 2 M HCl is added and then topped with de-ionized water to about ¾ full. It is re-heated on the hot plate and again allowed to cool. The content is diluted to 100 ml and filtered using size 42, 125 mm diameter, ashless filter paper into a flat bottom flask ready to be run on the analytical equipment. The instrument was calibrated prior to the introduction of sample by measuring in-house standards and blank solutions, the results turn-out have an accuracy range of ± 2–5 % depending on the number of standards and concentrations used.

All the samples were analyzed using the Inductively Coupled Plasma Optical Emission Spectrometry (ICP-OES) at the Geology Department of the University of Jos. This technique was chosen because of its numerous advantages in multi-media tolerance, multi element and low detection limits, its precision and accuracy, among others. A total of 26 elements were analyzed and include Ca, As, Cd, Co, Cr, Cu, Fe, Mg, Mn, Mo, Ni, Pb, I, Sb, Se, Sr, Ti, V, Ba, K, Na, La, Zn, P, S and SC, with only half of these discussed.

2.2. Data Analysis

The assessment of the levels of enrichment or contamination of the trace elements in soil and sediments was carried out using standard pollution measurement indices such as the index of geo-accumulation (I_{geo}), enrichment factor (EF), contamination factor (Cf) and the pollution load index (PLI), as well as the use of Surfer^R software, to determine the spatial distribution of the elements.

2.3. Calculation of Geo-Accumulation Index (Igeo)

The geo-accumulation index is generally used to determine the anthropogenic contamination in sediments as introduced by [10], [11] and corroborated by prominent works like [3]; [8]; [9] and [19]. This index allows us to evaluate the contamination levels by comparing present concentrations with background levels. The I_{geo} is expressed using the following Muller equation:

$$I_{geo} = \log_2 \frac{Cn}{1.5Bn}$$

Where C_n is the measured concentration of the given heavy metal examined in the soil or sediment, B_n is the geochemical background value of the element, 1.5 is incorporated in the relationship to account for possible variation in background data (the background matrix correction factor) owing to lithogenic effects, according to [8].

The geo-accumulation index, consist of seven grades (0 to 6) based on the increasing numerical value of the index and ranges from unpolluted to extremely polluted. The standard Igeo values are presented below (Table 1).

Table 1. *Classification of Index of Geo-accumulation (Igeo).*

I_{geo} Value	Grade	Classification
≤0	0	Unpolluted
0-1	1	Unpolluted to moderately polluted
1-2	2	Moderately polluted
2-3	3	Moderately polluted to strongly polluted
3-4	4	Strongly polluted
4-5	5	Strongly polluted to extremely polluted
>6	6	Extremely polluted

2.4. Calculation of Enrichment Factor (EF)

Enrichment Factor is used to evaluate the level of soil contamination and seeks to know the possible natural or anthropogenic input and impact in soils and sediments. In this assessment, aluminum (Al) was used as the immobile element to differentiate between natural and anthropogenic components and is associated with crustal rocks. Al is also the normalizing element assumed not to be consequentially enriched owing to local contamination, [15]. Baseline values were adopted from [24].

EF of a heavy metal in sediments can be calculated using the following equation;

$$EF = \frac{\left[\frac{M}{A1}\right] \text{Sample}}{\left[\frac{M}{A1}\right] \text{Background (crust)}}$$

Where EF is the enrichment factor, M_{sample} and $M_{background}$ are the concentrations of the investigated elements in the sample and crustal materials (background) while Al_{sample} and $Al_{background}$ are the concentrations of Al in the sample and crustal abundance respectively.

According to [16], EF values lower than and around 1.0 indicates that the element in the sediment originated predominantly from the crustal/background material and /or weathering process. EF values greater than 1.0 displays anthropogenic origin of the element.

According to [19];

EF =< 3 indicates minor or minimal enrichment,
EF = 3-5 indicates moderate enrichment,
EF = 5-10 indicates moderately severe enrichment,
EF = 10-25 indicates severe enrichment,
EF = 25-50 indicates very severe enrichment,
EF > 50 indicates extremely severe enrichment.

As the EF values increases, the contribution of the anthropogenic origins also increases [15].

2.5. Calculation of Contamination Factor (CF) and Pollution Load Index (PLI)

Pollution Load Index (PLI) is obtained as a product of the measured contamination factors of the different trace elements, while CF itself is the quotient obtained by dividing the concentration of the elements. Generally, the PLI is calculated by obtaining the n-root from the n-Cfs measured.

The pollution load index and contamination factor are expressed, as developed by [17] by the following equation:

$CF = C_{metal} / C_{background\ value}$

$PLI = {}^{n}\sqrt{(CF1 \times CF2 \times CF3 \ldots \times CFn}$

Where;

CF = contamination factor

n = number of metals

C_{metal} = metal concentration in polluted sediments

$C_{background}$ = background value of the metal

According to the equation, PLI value of > 1 is polluted whereas PLI value of < 1 indicates no pollution.

Contamination factor and level of contamination advanced initially by [10] and modified by [11] and several other workers and universally used, is shown below (Table 2);

Table 2. *Classification of Contamination factor and level of contamination*

Contamination Factor	Level of Contamination
Cf < 1	Low contamination
1 ≤ Cf < 3	Moderate contamination
3 ≥ Cf < 6	Considerable contamination
Cf > 6	Very high contamination

3. Results and Discussion

The concentrations of trace elements in soil and stream sediments collected and analyzed from the study area are presented in Tables 3 and 4 respectively. Tables 5, 6 and 7 represent the Igeo, EF and CF/ PLI of soil while Tables 8, 9 and 8 show the calculated Igeo, EF and CF/PLI of stream sediments in the area. A summary of the interpretation of the pollution indices (Igeo, EF, CF/PLI) of the two sample media are presented in Tables 5, 6 and 7.

From the results, zinc (Zn), barium (Ba), chromium (Cr), nickel (Ni) and iodine (I) appear to be more concentrated in both stream sediments (>1800ppm) and in the soil samples (>1000ppm) than the other trace elements. The mean concentrations of Zn, Ba, Cr and Ni in the stream sediments (in ppm) of 7427, 2253, 2174 and 1176 respectively, are higher than concentrations in soil.

Unlike Zn which display very high concentrations in virtually all locations, iodine generally occur below detection limit in 70 percent of soil samples and has maintained this trend in sediments, in most of the locations. The very high values of iodine recorded in some 30% of the samples, occur in locations around cultivated lands, connecting that, perhaps fertilizers containing iodine have been used, and or other anthropogenic sources.

The mean concentrations of strontium-Sr (150.87 ppm), cobalt-Co (129.88 ppm) lead-Pb (105.38 ppm) and copper-Cu (50.18 ppm) in stream sediments are higher than their levels in the soil, having (in ppm) 85.34, 22.41. 50.24, 27.27, for Sr, Co, Pb and Cu respectively. Arsenic (As) has maintained a narrow range of concentrations (33.95 ppm - 43.54 ppm) in both soil and sediments samples while molybdenum (Mo) remain the least concentrated in both sample media.

The mean values of Zn, Ba, Cr and Ni are higher than reported global trace elements concentrations but consistent with those of mineralized areas. This study area has widespread sulphides (lead-zinc and baryte) mineralization thus making the concentrations of elements like Zn and Ba very high and uncommon. However, the occurrence of molybdenum, for example, which ranges between 1.61 ppm to 3.11ppm in both soil and sediments, resemble its abundance in crustal or parent rocks, as does Sr.

Generally, the natural content of most elements in soils stems from parent rocks while high concentrations are related to mineralization. Abundance of these elements in sediments, especially in a sedimentary terrain such as the Keana area, is a function of clay fraction, soil organic matter (SOM) as well as other physico-chemical condition such as Eh and pH, and thus argillaceous sediments contain more (like Pb) than sands, sandstones and limestone. In mining areas, elements such as Pb may be dispersed due to the erosion and chemical weathering of tailings. The severity of these processes depends on chemical characteristics, and the minerals present in the tailings [2].

Table 3. *Trace Elements Concentrations (in ppm) in Soil of the Keana Area*

Sample Location #	I	Mo	Zn	As	Pb	Co	Cr	Cu	Ba	Ni	Sr
2	555.90	0.00	1410.00	0.00	73.01	111.40	89.18	71.09	60.23	53.73	32.09
5	0.00	0.00	144.00	0.00	23.34	6.54	111.20	16.49	218.70	13.72	18.31
6	1200.00	0.00	0.00	0.00	129.60	104.00	9879	80.97	138.30	7322.00	255.90
8	0.00	0.00	851.80	49.99	26.31	4.51	40.74	14.26	146.00	7.38	18.27
9	0.00	0.00	1854.00	0.00	61.68	9.53	167.00	40.19	124.90	21.90	201.80
11	2412.00	0.00	1255.00	16.54	46.73	11.29	84.34	20.22	956.30	18.62	28.96
12	0.00	0.00	990.40	0.00	14.70	31.26	74.54	25.62	227.20	53.84	109.20
14	0.00	0.00	1998.00	71.20	66.92	10.39	233.90	35.16	36.87	155.60	237.60
15	0.00	6.54	1617.00	57.01	28.00	2.92	15.42	9.09	99.31	7.97	75.38
17	879.50	0.00	644.20	14.29	33.02	10.44	42.74	19.07	313.10	13.33	22.95
19	0.00	0.00	2590.00	75.99	63.10	8.01	126.80	20.47	44.72	15.76	229.20
20	144.80	2.21	1088.00	79.00	27.10	4.18	26.57	5.53	41.96	7.01	21.69
21	0.00	0.00	479.80	14.28	17.47	1.34	33.29	6.91	500.50	6.06	17.82
22	0.00	0.00	2732.00	213.10	121.40	6.41	50.03	37.16	64.68	14.37	213.60
23	5752.00	9.66	1592.00	123.80	36.50	3.88	23.11	10.95	2270	6.66	24.06
24	2531.00	0.00	763.80	4.70	37.49	11.04	55.54	18.65	61.83	21.95	28.77
26	0.00	1.48	1271.00	101.20	25.04	4.15	24.80	7.65	58.30	4.99	73.45
28	0.00	1.50	755.10	15.44	20.83	7.10	28.83	10.78	71.70	7.47	23.71
30	2256.00	7.85	1471.00	66.60	20.59	2.89	21.86	5.64	104.70	13.78	21.47
33	0.00	110	2281.00	101.20	64.95	4.34	45.32	31.41	76.57	12.43	270.20
34	0.00	0.00	699.20	54.29	23.49	3.16	24.84	9.20	42.19	7.56	20.07
35	2849.00	7.08	1093.00	94.19	18.57	3.75	25.48	5.03	81.39	5.04	21.60
36	0.00	0.00	623.90	30.14	23.82	2.10	28.31	6.19	101.40	6.87	20.34
38	0.00	0.00	0.00	78.42	60.50	12.10	853.50	26.25	78.51	633.20	195.90
41	6562.00	5.40	1203.00	54.27	24.61	4.84	18.84	6.97	136.10	7.45	23.30
43	0.00	0.00	436.60	13.99	27.97	10.25	58.74	13.75	60.45	11.60	23.80
44	0.00	0.00	782.60	85.49	25.25	5.42	49.08	8.55	212.20	10.94	21.78

Sample Location #	I	Mo	Zn	As	Pb	Co	Cr	Cu	Ba	Ni	Sr
45	5747.00	0.00	2934.00	0.00	3.46	16.41	127.60	72.05	59.57	107.80	39.07
47	2150.00	0.00	670.10	5.91	25.14	6.06	50.31	11.08	60.10	10.34	20.02
50	2350.00	0.00	1066.00	15.94	28.98	5.93	94.68	15.17	435.10	15.96	24.12
51	0.00	0.00	0.00	0.00	224.70	164.70	282.40	85.17	1369	47.11	22.73
52	0.00	0.00	3958.00	0.00	147.70	136.10	196.10	125	155.70	112.90	422.10
53	2112.00	0.00	971.10	0.00	86.09	12.99	238.00	42.13	252.10	28.84	36.97
MC	1136.40	1.61	1218.96	43.54	50.24	22.41	400.67	27.70	262.44	266.19	85.34

MC = Mean Concentration

Table 4. *Trace Elements Concentrations (in ppm) in Stream Sediment of Keana Area*

Sample Location #	I	Mo	Zn	As	Pb	Co	Cr	Cu	Ba	Ni	Sr
3	1018.00	0.00	46110	0.00	192.40	220.40	14150	86.18	509.80	12170	78.13
18	0.00	10.07	1789.00	53.29	65.57	3.95	32.40	25.59	92.86	9.57	216.70
24	1509.00	0.00	152.90	0.00	68.36	16.82	291.20	41.76	7621.00	17.17	147.40
27	0.00	0.00	1850.00	48.55	83.63	9.09	52.46	39.65	248.40	24.18	252.50
42	2394.00	0.00	676.40	0.00	90.11	195.90	90.33	44.02	1942.00	83.60	41.23
49	0.00	0.00	671.30	0.00	232.10	463.00	398.00	153.50	5250.00	111.80	270.90
94	7697.00	11.73	745.80	135.80	5.48	0.00	207.60	23.54	113.30	21.92	49.21
MC	1802.57	3.11	7427.91	33.95	105.38	129.88	2174.57	59.18	2253.91	1776.89	150.87

MC = Mean Concentration

Table 5. *Index of Geo-accumulation of Trace Elements in Soils of the Keana Brinefield*

Sample Location #	I	Mo	Zn	As	Pb	Co	Cr	Cu	Ba	Ni	Sr	Sc
2	3.24	0.00	0.47	0.00	0.19	0.19	0.53	0.48	0.12	0.15	0.02	0.03
5	100.33	0.00	0.00	0.00	1.04	3.02	47.33	2.35	0.08	81.63	0.35	0.00
6	0.00	0.00	2.76	2.13	0.21	0.13	0.20	0.41	0.08	0.08	0.02	0.02
8	0.00	0.00	6.00	0.00	0.50	0.28	0.80	1.17	0.07	0.24	0.28	0.00
9	201.67	0.00	4.06	0.71	0.38	0.33	0.40	0.59	0.53	0.21	0.04	0.03
11	0.00	0.00	3.21	0.00	0.12	0.91	0.36	0.75	0.13	0.60	0.15	0.14
12	0.00	0.00	6.47	3.04	0.54	0.30	1.12	1.02	0.02	1.73	0.32	0.00
14	0.00	0.73	5.23	2.43	0.22	0.08	0.07	0.26	0.06	0.09	0.10	0.01
15	73.54	0.00	2.08	0.61	0.27	0.30	0.20	0.55	0.17	0.15	0.03	0.04
17	0.00	0.00	8.38	3.24	0.51	0.23	0.61	0.60	0.02	0.18	0.31	0.00
19	12.11	0.25	3.52	3.37	0.22	0.12	0.13	0.16	0.02	0.08	0.03	0.01
20	0.00	0.00	1.55	0.61	0.14	0.04	0.16	0.20	0.28	0.07	0.02	0.01
21	0.00	0.00	8.84	9.10	0.97	0.19	0.24	1.08	0.04	0.16	0.29	0.00
22	480.93	1.08	5.15	5.29	0.29	0.11	0.11	0.32	1.26	0.07	0.03	0.02
23	211.62	0.00	2.47	0.20	0.30	0.32	0.27	0.54	0.03	0.24	0.04	0.07
24	0.00	0.16	4.11	4.32	0.20	0.12	0.12	0.22	0.03	0.06	0.10	0.01
26	0.00	0.17	2.44	0.66	0.17	0.21	0.14	0.31	0.04	0.08	0.03	0.04
28	188.63	0.88	4.76	2.84	0.17	0.08	0.10	0.16	0.06	0.15	0.03	0.01
30	0.00	1.27	7.38	4.32	0.52	0.13	0.22	0.91	0.04	0.14	0.37	0.00
33	0.00	0.00	2.26	2.32	0.19	0.09	0.12	0.27	0.02	0.08	0.03	0.01
34	238.21	0.79	3.54	4.02	0.15	0.11	0.12	0.15	0.05	0.06	0.03	0.01
35	0.00	0.00	2.02	1.29	0.19	0.06	0.14	0.18	0.06	0.08	0.03	0.01
36	0.00	0.00	0.00	3.35	0.49	0.35	4.09	0.76	0.04	7.06	0.27	0.00
38	548.66	0.60	3.89	2.32	0.20	0.14	0.09	0.20	0.08	0.08	0.03	0.02
41	0.00	0.00	1.41	0.60	0.22	0.30	0.28	0.40	0.03	0.13	0.03	0.04
43	0.00	0.00	2.53	3.65	0.20	0.16	0.24	0.25	0.12	0.12	0.03	0.02
44	480.51	0.00	9.50	0.00	0.03	0.48	0.61	2.10	0.03	1.20	0.05	0.26
45	179.76	0.00	2.17	0.25	0.20	0.18	0.24	0.32	0.03	0.12	0.03	0.04
47	196.49	0.00	3.45	0.68	0.23	0.17	0.45	0.44	0.24	0.18	0.03	0.04
50	0.00	0.00	0.00	0.00	1.80	4.79	1.35	2.48	0.76	0.53	0.03	0.12
51	0.00	0.00	12.81	0.00	1.19	3.96	0.94	3.65	0.09	1.26	0.58	0.00
52	176.59	0.00	3.14	0.00	0.69	0.38	1.14	1.23	0.14	0.32	0.05	0.07
53	0.00	0.00	0.00	0.00	0.00	0.00	0.00	0.00	0.00	0.00	0.00	0.00

Table 6. Contamination Factor (Cf) and Pollution Load Index (PLI) of Trace Elements in Soil of the Keana Brinefield

Sample Location #	I	Mo	Zn	As	Pb	Co	Cr	Cu	Ba	Ni	Sr	Sc	PLI
2	1.35	0.00	0.01	0.00	0.01	0.03	0.01	0.03	0.00	0.01	0.00	0.00	1
5	41.81	0.00	0.00	0.00	0.04	0.44	1.13	0.17	0.00	4.53	0.00	0.00	0
6	0.00	0.00	0.04	0.45	0.01	0.02	0.00	0.03	0.00	0.00	0.00	0.00	1
8	0.00	0.00	0.10	0.00	0.02	0.04	0.02	0.08	0.00	0.01	0.00	0.00	0
9	84.03	0.00	0.07	0.15	0.02	0.05	0.01	0.04	0.00	0.01	0.00	0.00	1
11	0.00	0.00	0.05	0.00	0.00	0.13	0.01	0.05	0.00	0.03	0.00	0.02	1
12	0.00	0.00	0.10	0.65	0.02	0.04	0.03	0.07	0.00	0.10	0.00	0.00	1
14	0.00	0.40	0.08	0.52	0.01	0.01	0.00	0.02	0.00	0.00	0.00	0.00	1
15	30.64	0.00	0.03	0.13	0.01	0.04	0.00	0.04	0.00	0.01	0.00	0.00	1
17	0.00	0.00	0.14	0.69	0.02	0.03	0.01	0.04	0.00	0.01	0.00	0.00	1
19	5.04	0.14	0.06	0.72	0.01	0.02	0.00	0.01	0.00	0.00	0.00	0.00	1
20	0.00	0.00	0.03	0.13	0.01	0.01	0.00	0.01	0.00	0.00	0.00	0.00	1 4
21	0.00	0.00	0.14	1.94	0.04	0.03	0.01	0.08	0.00	0.01	0.00	0.00	0
22	200.39	0.60	0.08	1.12	0.01	0.02	0.00	0.02	0.00	0.00	0.00	0.00	1
23	88.17	0.00	0.04	0.04	0.01	0.05	0.01	0.04	0.00	0.01	0.00	0.01	1
24	0.00	0.09	0.07	0.92	0.01	0.02	0.00	0.02	0.00	0.00	0.00	0.00	1
26	0.00	0.09	0.04	0.14	0.01	0.03	0.00	0.02	0.00	0.00	0.00	0.00	1 4
28	78.59	0.49	0.08	0.60	0.01	0.01	0.00	0.01	0.00	0.01	0.00	0.00	0
30	0.00	0.71	0.12	0.92	0.02	0.02	0.01	0.07	0.00	0.01	0.00	0.00	1
33	0.00	0.00	0.04	0.49	0.01	0.01	0.00	0.02	0.00	0.00	0.00	0.00	1
34	99.25	0.44	0.06	0.86	0.01	0.02	0.00	0.01	0.00	0.00	0.00	0.00	0
35	0.00	0.00	0.03	0.27	0.01	0.01	0.00	0.01	0.00	0.00	0.00	0.00	1
36	0.00	0.00	0.00	0.71	0.02	0.05	0.10	0.05	0.00	0.39	0.00	0.00	0
38	228.61	0.33	0.06	0.49	0.01	0.02	0.00	0.01	0.00	0.00	0.00	0.00	1
41	0.00	0.00	0.02	0.13	0.01	0.04	0.01	0.03	0.00	0.01	0.00	0.00	1
43	0.00	0.00	0.04	0.78	0.01	0.02	0.01	0.02	0.00	0.01	0.00	0.00	0
44	200.21	0.00	0.15	0.00	0.00	0.07	0.01	0.15	0.00	0.07	0.00	0.03	0
45	74.90	0.00	0.03	0.05	0.01	0.03	0.01	0.02	0.00	0.01	0.00	0.00	1
47	81.87	0.00	0.06	0.14	0.01	0.02	0.01	0.03	0.00	0.01	0.00	0.00	1
50	0.00	0.00	0.00	0.00	0.07	0.69	0.03	0.18	0.00	0.03	0.00	0.01	1
51	0.00	0.00	0.21	0.00	0.05	0.57	0.02	0.26	0.00	0.07	0.00	0.00	1
52	73.58	0.00	0.05	0.00	0.03	0.05	0.03	0.09	0.00	0.02	0.00	0.01	1
53	0.00	0.00	0.00	0.00	0.00	0.00	0.00	0.00	0.00	0.00	0.00	0.00	1

Table 7. *Enrichment Factor (EF) of Trace Elements in Soil of the Keana Area*

Sample Location #	I	Mo	Zn	As	Pb	Co	Cr	Cu	Ba	Ni	Sr
2	687.64	0.00	67.52	0.00	8.67	47.93	6.30	15.07	0.49	8.86	5.29
5	0.00	0.00	14.06	0.00	5.65	5.74	16.03	7.13	3.66	4.62	6.16
6	0.00	0.00	0.00	0.00	0.00	0.00	0.00	0.00	0.00	0.00	0.00
8	0.00	0.00	162.94	126.15	12.48	7.74	11.50	12.08	4.78	4.86	12.04
9	0.00	0.00	0.00	0.00	0.00	0.00	0.00	0.00	0.00	0.00	0.00
11	6671.91	0.00	134.38	23.36	12.41	10.86	13.33	9.59	17.54	6.87	10.68
12	0.00	0.00	22.37	0.00	0.82	6.34	2.48	2.56	0.88	4.19	8.49
14	0.00	0.00	0.00	0.00	0.00	0.00	0.00	0.00	0.00	0.00	0.00
15	0.00	116.20	834.23	387.99	35.82	13.52	11.74	20.77	8.78	14.16	133.95
17	1778.92	0.00	50.44	14.76	6.41	7.34	4.94	6.61	4.20	3.59	6.19
19	0.00	0.00	0.00	0.00	0.00	0.00	0.00	0.00	0.00	0.00	0.00
20	844.13	17.16	245.52	235.17	15.17	8.47	8.85	5.53	1.62	5.45	16.86
21	0.00	0.00	193.47	75.96	17.47	4.86	19.82	12.34	34.56	8.42	24.75
22	0.00	0.00	0.00	0.00	0.00	0.00	0.00	0.00	0.00	0.00	0.00
23	28967.35	64.86	310.35	318.36	17.65	6.79	6.65	9.45	75.82	4.47	16.16
24	3135.69	0.00	36.63	2.97	4.46	4.76	3.93	3.96	0.51	3.63	4.75
26	0.00	19.70	491.79	516.55	24.03	14.43	14.17	13.11	3.86	6.65	97.89
28	0.00	4.70	68.67	18.52	4.70	5.80	3.87	4.34	1.12	2.34	7.43
30	14333.87	66.54	361.79	216.08	12.56	6.40	7.94	6.15	4.41	11.67	18.19
33	0.00	0.00	0.00	0.00	0.00	0.00	0.00	0.00	0.00	0.00	0.00
34	0.00	0.00	155.72	159.50	12.97	6.32	8.17	9.07	1.61	5.80	15.40
35	20806.81	68.96	309.00	351.26	13.02	9.53	10.63	6.30	3.94	4.91	21.03
36	0.00	0.00	157.23	100.20	14.89	4.75	10.53	6.91	4.38	5.97	17.66
38	0.00	0.00	0.00	0.00	0.00	0.00	0.00	0.00	0.00	0.00	0.00
41	36176.30	39.66	256.73	152.78	13.02	9.28	5.94	6.59	4.97	5.48	17.13
43	0.00	0.00	33.40	14.12	5.31	7.05	6.63	4.66	0.79	3.06	6.27
44	0.00	0.00	107.84	155.39	8.63	6.72	9.98	5.21	5.01	5.19	10.34
45	1881.60	0.00	37.18	0.00	0.11	1.87	2.39	4.04	0.13	4.71	1.71
47	5083.88	0.00	61.34	7.14	5.71	4.98	6.80	4.49	0.94	3.26	6.31
50	4787.48	0.00	84.07	16.58	5.67	4.20	11.02	5.30	5.88	4.34	6.55
51	0.00	0.00	0.00	0.00	15.41	40.92	11.53	10.43	6.48	4.49	2.17
52	0.00	0.00	0.00	0.00	0.00	0.00	0.00	0.00	0.00	0.00	0.00
53	2653.13	0.00	47.22	0.00	10.38	5.68	17.08	9.07	2.10	4.83	6.19

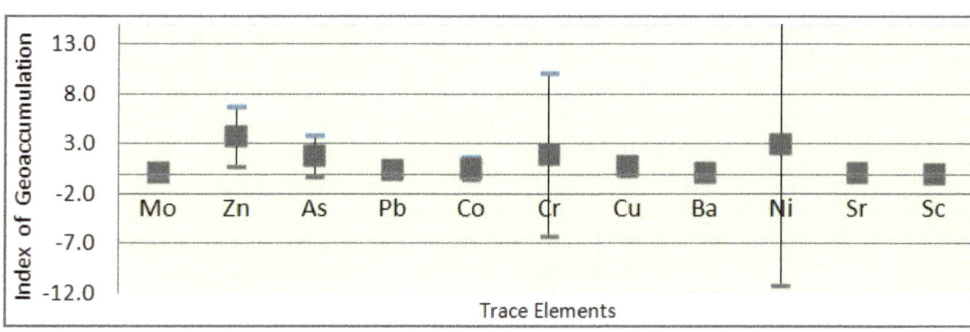

Fig 3. *Igeo of trace elements in soil of the Keana Brinefield*

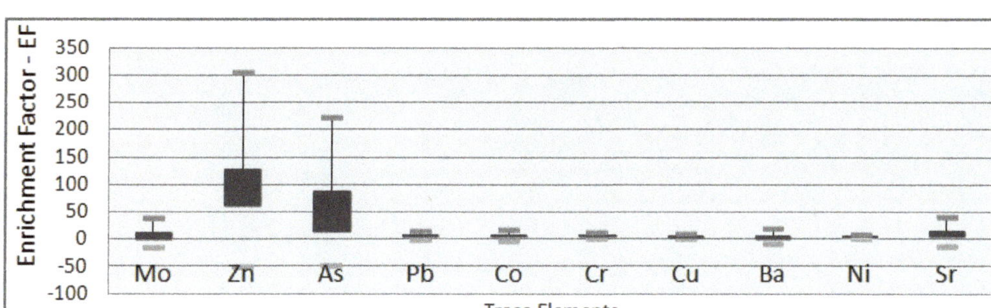

Fig 4. *EF of trace elements in soil of the Keana Brinefield*

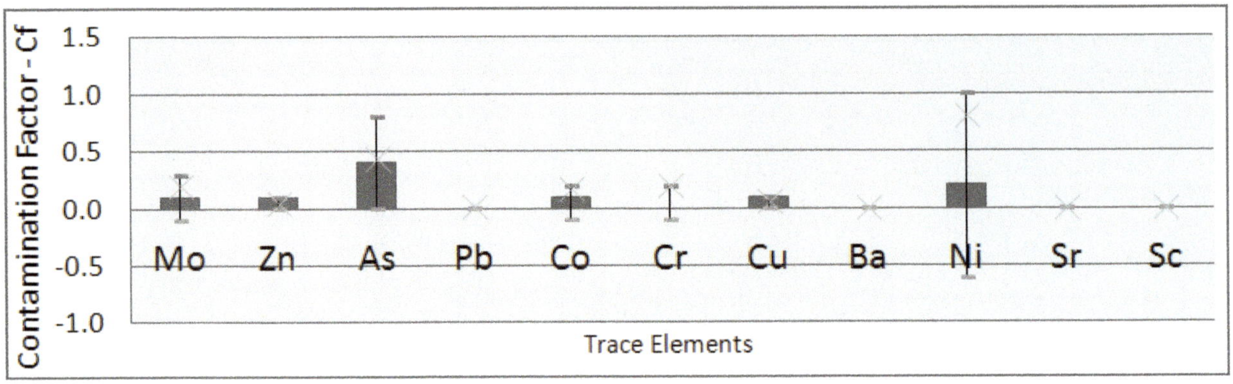

Fig 5. *Cf of trace elements in soil of the Keana Brinefield*

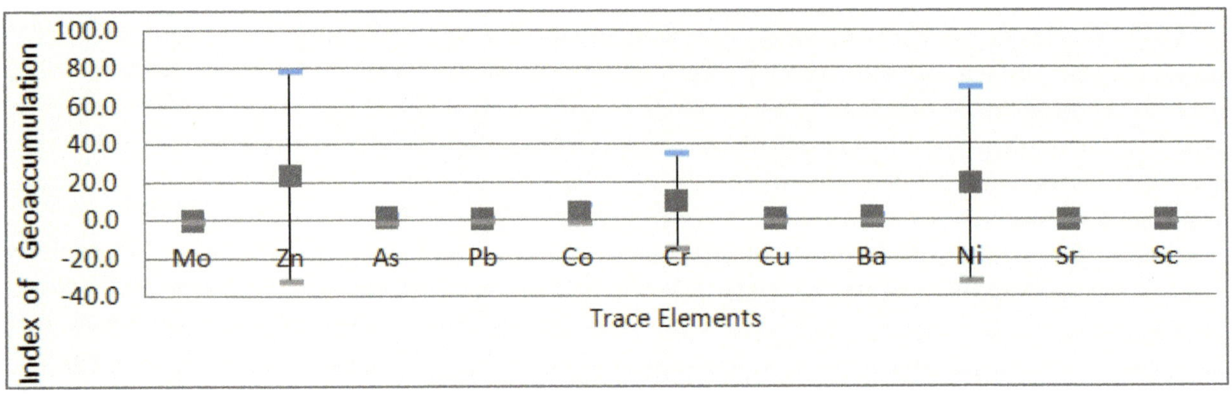

Fig 6. *Igeo of trace elements in stream sediments of the Keana brinefield*

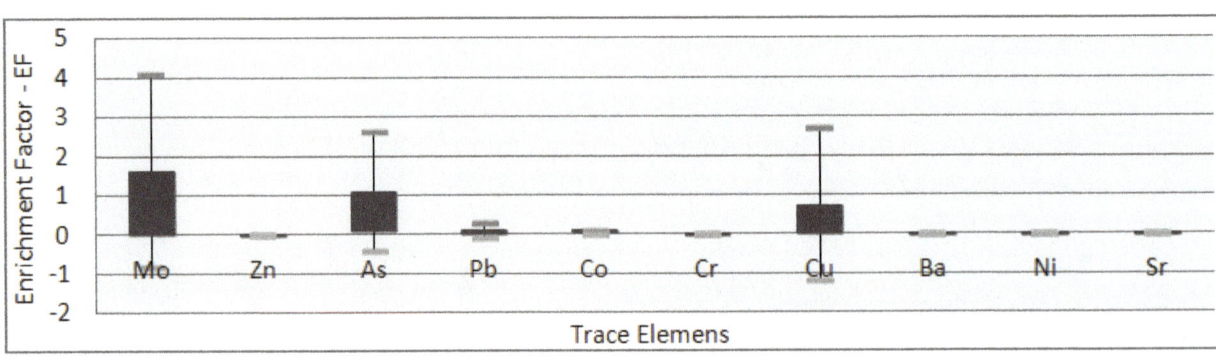

Fig 7. *EF of Trace Elements in stream sediments of the Keana brinefield*

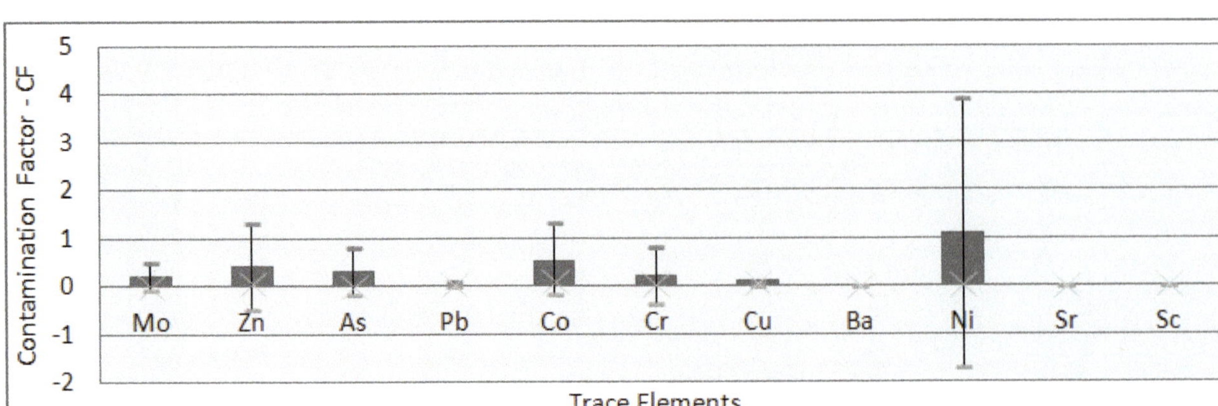

Fig 8. *CF of trace elements in stream sediments of the Keana brinefield*

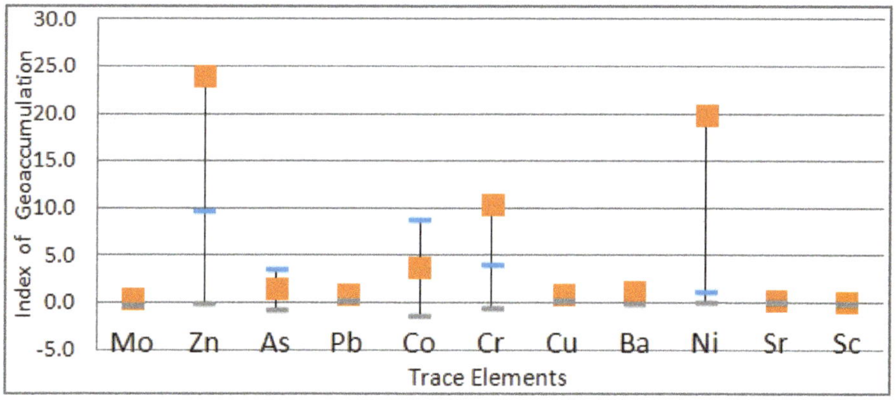

Fig 9. *Index of Geoaccumulation of elements in Stream Sediments of the Keana Brinefield*

Table 8. *Summary of I_{geo} Values of Trace Elements in the Keana Soil*

Elements	I_{geo} Value	Class	Remarks
Mo	0.18	1	Unpolluted to moderately polluted
Zn	3.81	4	Strongly polluted
As	1.86	2	Moderately polluted
Pb	0.39	1	Unpolluted to moderately polluted
Co	0.55	1	Unpolluted to moderately polluted
Cr	1.91	2	Moderately polluted
Cu	0.74	1	Unpolluted to moderately polluted
Ba	0.14	1	Unpolluted to moderately polluted
Ni	2.95	3	Moderately to strongly polluted
Sr	0.12	1	Unpolluted to moderately polluted
Sc	0.03	1	Unpolluted to moderately polluted

Table 9. *Summary of I_{geo} Values of Trace Elements in stream sediments of the Keana Brinefield*

Elements	I_{geo} Value	Class	Remarks
Mo	0.35	1	Unpolluted to moderately polluted
Zn	4.85	4	Strongly polluted
As	1.45	2	Moderately polluted
Pb	0.85	1	Unpolluted to moderately polluted
Co	3.78	1	Unpolluted to moderately polluted
Cr	1.70	2	Moderately polluted
Cu	0.85	1	Unpolluted to moderately polluted
Ba	0.33	1	Unpolluted to moderately polluted
Ni	0.62	3	Moderately to strongly polluted
Sr	0.21	1	Unpolluted to moderately polluted
Sc	0.03	1	Unpolluted to moderately polluted

Table 10. *Summary of the EF Classification of Trace Elements in Soil of the Keana Area.*

Element	EF Value	Classification
I	3872.99	Extremely severe enrichment
Mo	12.05	Moderate – severe enrichment
Zn	128.60	Extremely severe enrichment
As	87.66	Extremely severe enrichment
Pb	8.59	Moderate – severe enrichment
Co	7.65	Moderate – severe enrichment
Cr	7.04	Moderate – severe enrichment
Cu	6.08	Moderate – severe enrichment
Ba	6.01	Moderate – severe enrichment
Ni	4.30	Moderate enrichment
Sr	14.53	Severe enrichment

3.1. Using the Pollution Indices for Assessment

Contamination measurement indicators used for the assessment of soil and sediments in the study area include geo-accumulation index, enrichment factor, contamination factor and pollution load index. Average concentration of background values for world soil and sediments used in the various equations are taken from [21]. Calculated values of the pollution indices are presented in Tables 5 to 9.

Geo-accumulation index (Igeo):

Using the Muller scale for Igeo, zinc (Zn) is the most enriched heavy metal in soil of the study area, having I_{geo} value of 4.9 and 3.81 (Table 8, Figs. 3 and 6) in soil and sediments respectively, and an Igeo class of 4 and 5, indicating that the soil of the Keana area is strongly to extremely polluted, by Zn. Cobalt in the stream sediments has Igeo of 3.8, similar to the Igeo exhibited by Zn and indicates the sediment is strongly polluted by Co, which conversely has a lower Igeo of 0.55 in soil indicating the soil is unpolluted. Arsenic, in both soil and sediment maintains an Igeo of 1.86 and 1.42 respectively, indicating that the soil is moderately polluted by As. This is similar to the Igeo values displayed by Cr, having 1.91 and 1.73 respectively for soil and stream sediments. Another element in the moderate to strong pollution level in the soil is Ni with Igeo value of 2.95, while Mo, Pb, Cu, Ba, Sr and Sc all have Igeo values of less than one (1), in the two sample media indicating the soil and sediments of the Keana area are unpolluted by these elements (Fig.9).

Enrichment Factor (EF):

Trace elements assessment using EF in the Keana area shows that soil of the area is extremely severely enriched by Zn and As, having EF values of 128.60 and 87.66 respectively. However, in stream sediments, the two elements have very low EF values of 0.01 and 1.11 respectively, indicating minor or minimal enrichment. Molybdenum has an EF value 12.05 in soil indicating moderate-severe enrichment but shows minimal level in stream sediments with EF of 1.63. Other elements such as Pb, Co, Cu, Ba, Ni and Sr have EF values between 6 and 8 indicating moderate to severe enrichment in the soil, while Ni is of moderate enrichment with an EF of 4.30. These elements however have EF less than one (1) in stream sediments, indicating minimal enrichment levels.

Contamination factor (Cf) and pollution load index (PLI):

Both Cf and PLI values of elements in soil and stream sediments of the Keana area are less than one (Figures 5 and 8). This indicates that the there is no contamination by these elements in the two media

In summary, the assessment of the degree of accumulation of heavy metals such as Pb, Zn, As, etc, in soil and stream sediments of the Keana area, calculated using index of geo-accumulation (I_{geo}), enrichment factor (EF), contamination factor (Cf) and pollution load index (PLI), shows that soil and stream sediments in the area are highly polluted by Zn, Cr, Ni and As going by geo-accumulation index. Zn and As are also similarly severely enriched in the two media when EF is measured. In all indices measured, Mo, Co, Cu, Ba, Sr and Sc levels in soil and sediments remain less than one, indicating the media are not contaminated by these elements. A pollution load index of < 1 as seen in most locations (Table 6) indicates no pollution.

The ubiquitous mining activities characterized by indiscriminate dumping of mine wastes, tailings and other foreign materials, coupled with the intense weathering in the area (which predisposes the minerals hosting the trace elements), have contributed to the re-mobilization and re-distribution of the heavy metals into the surrounding soil and sediments. The main source of these trace elements is therefore anthropogenic.

The distribution of these elements by both natural and anthropogenic processes can result in deficiency or toxicity of these trace elements (such as Pb, As, Cr, Se) in the environment and can be detrimental to plants and animals. Pb, for example is known to be toxic even at low exposure levels and has acute and chronic effects on humans (neurological, renal gastrointestinal and reproductive effects), especially children and pregnant women who are very vulnerable in nature, particularly in the tropics. Deficiencies in water or soil of iodine, selenium, and zinc, in the same vein, are attributable to iodine deficiency disorders -IDDs (such as goiter), human immune-deficiency syndrome and slow or stunted growth rate respectively [6], [25].

4. Conclusion

The high concentration of trace elements such as Zn, Cr, Ni, As and others in soil and sediments, relate to their properties as transition elements, are easily oxidized to insoluble complex anions which are readily absorbed and co-precipitated in the sediments with Fe^{3+}, Al^{3+} and Si^{4+} [23]. In addition, these elements must have also remobilized from the surrounding basaltic rocks rich in these elements, coupled with remobilization from the widespread sulphides mineralization characterizing the study area.

The use of pollution indices in this research, on the basis of geo-accumulation index and enrichment factor has shown that the soil and sediments in the area are extremely contaminated with Zn, Cr, Ni and As. Pb and Ba are moderately contaminated while the contamination levels of elements such as Mo, Cu, Co, Sr and Sc appear to be insignificant. Contamination factor and pollution load index analysis both indicate no pollution of these elements in the study area.

The continued exposure of the populations living in this study area particularly those in the mining locations who depend on the soil (for subsistence farming) and water from ponds, wells and boreholes (for drinking, cooking and other domestic uses), would result in various health risks and concerns, if not checked or abated. This study would recommend an immediate plan for analysis of the quality of drinking water and some staple crops grown in the area to determine the levels of these noxious metals and uptake by plants, to be followed by a comprehensive mitigation or remediation plan. The mining regulatory body in the country should move to control the proliferation of mining and processing of the minerals as well as the uncontrollable disposal of toxic mine tailings and dumping of other wastes.

Acknowledgements

The authors would like to thank Nasarawa State Government and the University of Jos for their financial support and to the anonymous reviewers for their input.

References

[1] Benkhelil, J., Dainelli, P., Ponsard, J.F., Popoff, M. and Saugy, L. (1988). The Benue Trough: Wrench fault related Basin on the border of the Equatorial Atlantic. In. Manspeizer, W. (Ed.) Triassic to Jurassic Rifting and the Opening of the Atlantic Ocean. Amsterdam. *Elsevier Publishing Co.*

[2] Da Silva, E.F., Zhang, C., Pinto, L.S., Patinha, C., & Reis, P. (2004). Hazard assessment of arsenic and lead in soils of Castromil Gold mining area, Portugal. *Applied Geochemistry,* 19, 887-898.

[3] Forstner U., Ahlif, W., & Calmano, W. (1993). Sediment Quality Objectives and Criteria Development in Germany. *Water Science Tehcnology,* 28, 307-316

[4] Hakanson, L. (1980). An ecological risk index for aquatic pollution control and sedimentological approaches. *Water Research Journal,* 14, 975-1001

[5] Lar, U.A., & Sallau, A.K. (2005). Trace element geochemistry of the Keana brinesfield, middle Benue trough, Nigeria. *Journal of Environmental Geochemistry and Health,* 27(4), 331-339.

[6] Lar, U.A., & Tejan, A.B. (2008). Highlights of some environmental problems of geo-medical significance in Nigeria. *Journal of Environmental Geochemistry and Health,* 30, 383-389.

[7] Lar, U.A. (2013). Trace elements and health: An environmental risk in Nigeria. *Earth Sciences,* 2(3), 66-72, London: Science Publishing Group.

[8] Loska, K., Cehula, J., Pelczar, J., Weichula, D., & Kwapuyinski, J. (1997). Use of enrichment and conatmination factors together with geo-accummulation indexes to evaluate Cd, Cu and Ni in the Rybnic water reservoir in Poland. *Water, Air and Soil Pollution,* 93, 347-365.

[9] Lokeshewari, H., & Chandrappa, G.T. (2006). Heavy metals content in water hyacinth and sediment of Lalbagh Tanks, Bangalore, *Indian Journal of Environmental Sciences Engineering,* 48, 183-188

[10] Muller, G. (1969). Index of geo-accumulation in sediments of the Rhine river. *Geological Journal,* 2(3), 108-118.

[11] Muller, G. (1979). Heavy metals in the sediment of the Rhine river. *Geological Journal,* 79, 778-78

[12] Offodile, M.E. (1976b): A hydrogeochemical interpretation of the middle Benue and Abakaliki brinefields. *Journal of Mineralogy and Geology*, 13, 2-15.

[13] Olade, M.A. (1976). On the genesis of lead-zinc deposits in Nigeria's Benue rift (Aulacogen): A reinterpretation. *Journal Mining Geology,* 13(2), 20-27

[14] Ogola, J.S. (2003). Impact of gold mining on the environment and human health: A case study in the Migori gold belt in Kenya, *Environmental Geochemistry and Health,* 24, 141-158

[15] Sutherland, R.A. (2000). Bed sediments associated trace elements in an urban stream, Oahu, Hawaii. *Envrinmental Geology,* 39, 611-627.

[16] Szefer, P., Pempkowiak, J., Skwarzec, B., & Bojanowiski, R. (1996). Distribution and co-associations of selected metals in seals of the Antarctic. *Environmental Pollution,* 83, 341-349.

[17] Tomlinson, D. C., Wilson, J. G., Harris, C. R., & Jeffrey, D. W. (1980). Problems in the assessment of heavy metal levels in estuaries and the formation of a pollution index. *Helgoland Mar. Resources* 33, 566-575.

[18] Davies, T.C. (2003). Some environmental problems of geomedical relevance in East and Southern Africa. In: geology and Health (pp.139-144). Elsevier. Oxford University Press

[19] Chen,Y.W., Belzile,N. and Gunn, J.M. (2001) Antagonistic effects of selenium on mercury assimilation by fish population at Sunbery metal smelter. Limnol Oceanology 46; 1814-1818

[20] Hakanson, L. (1980). An ecological risk index for aquatic pollution control. A sedimentological approach. Water Res., 14: 975-1001.

[21] Kabata-Pendias, A. and Pendias, H. (1999) Trace elements in soils and plants, 3rd ed., CRC Presss, Boca Raton, FL

[22] Zhang, K.L., (1998) Keshan disease. In : Geng, G.Y. (Ed) Epidemilogy Vol.3, 2nd Edition, People Publishing House, Beijing, pp378-397.

[23] Lar, U. A., Walsh, A., Ashano,C., Ogezi, A.E. and Nixon, P. (2000). Mobilization of major and trace elements (including REE) in weathered basaltic rocks of the Jos Plateau. African Journal of Natural Sciences Vol.3 pp40-56.

[24] Kabata-Pendias, A., & Sadurski, W. (2004). Trace Elements and Compounds in Soil. In Merian, E., Anke. M., Ihnat, M. & Stoepppler, M. (Eds.). Elements and Their Compounds in the Environment, 2nd Ed. (pp 79–99), Weinheim: Willey

[25] Sallau, A.K. (2014). Environmental assessment of trace elements distribution in the Awe-Keana Brinefield, middle Benue trough Nigeria. An unpublished PhD thesis in the Department of Geology, University of Jos, Nigeria (In Press)

15

Has Ever Brahmaea Ledereri Rogenhofer, 1873 Inhabited the Colchis Refugium

Arnold Gegechkori[1], Eter Didmanidze[2]

[1]Department of Biodiversity, Faculty of Exact and Natural Sciences, Iv.Javakhishvili Tbilisi State University, Tbilisi, Georgia
[2]Georgian National Museum, Tbilisi, Georgia

Email address:
arngegechkori@yahoo.com (A. Gegechkori), lepidopterology@mail.ru (E. Didmanidze)

Abstract: The biogeographical background. There are two refugial centers in Transcaucasia: Colchis (along the Black Sea coast) and Hyrcan (Hyrcania) (along the Caspian Sea shore). Elements from Europe and Asia and their biogeographical subdivisions compose flora and fauna, which combined with the diverse geomorphology (Laurasian, Gondvanian), the presence of the impressive mountain ridge of the Great Caucasus, its insularity during 100 million years (from Cretaceous time), have resulted in a high taxonomic (species, genera) richness and endemism. The both refugiais characterized by a rather uniform thermical indexes, considerably the former one. The both shelters consist of an extraordinary abundance and diversity of thermophilic woody flora (trees, shrubs and vines) belonging to Arcto-Tertiary Geoflora. Except species typical for nemoral biota, there are extremely rare species of tropical-subtropical origin, which became extinct in South Caucasus during the Tertiary (at the close of the Pliocene) climate deterioration. In response to cooling climate and physiographic changes, some rare genera became restricted to both refugia of Transcaucasia. Most of them are now confined chiefly within the Hyrcanianrefugium (species of *Albizia, Gleditsia, Panthera, Hystrix*). Among the invertebrate fauna of Tertiary origin should be considerably mentioned species of remarkable nocturnal Brahmin moth – *Brahmaea* Walker. The interdisciplinary study of the orthodox view regarding the vicariant event between two species of moths as thoughthey are indigenous of Colchis (*B. ledereri*) and Talysh (*B. christophi*) does not actually exist. The reason circumstances to express such opinion remain important due for a variety of reasons which are given below.

Keywords: Colchis, Hyrcan, Refugia, Biogeography, *Brahmaea ledereri*

1. Introduction

In the moderate climate belt of the North Hemisphere is represented broad-leaved deciduous, summer-green forest zone (biome). In biogeographical literature, it is determined as a Temperate Broad-Leaved Deciduous Forests (TBDF) [1, 2]. By geobotanists this thermophilic dendroflora are proposed as the Arcto-Tertiary Geoflora [1, 2]. Sometimes TBDF complex mentioned as a Nemoral Flora [3] within these communities participate some warm climate loving summer-green arboreal plants (sweet chestnut, beeches, oaks, zelkova, etc.), conifers (yew tree, tsuga, sequoia, etc.) [4]. After the Pliocene climate deterioration, representatives of the TBDF grow preserving in local places, restricted in refugial areas [5]. Two the largest and richest with the Arcto-Tertiarry relicts' regions in the western Eurasia are situated in the south Caucasus and adjacent some countries [6, 8]. One of it, Hyrcanian province (subprovince) includes the Caspian Sea coast area and northern slopes of the Alborz mountains range in North Iran. Its smaller northwestern site, Talysh, represented in the southeast Transcaucasia (Republic of Azerbaijan) [7]. The second large harbor is Colchis. This region, the main part of anEuxinian province (subprovince) covers eastern and southeastern part of the Black Sea (coast area and adjacent slopes of the Greater and the Lesser Caucasus). It includes main part of West Georgia, Krasnodar district of Russia and north-eastern Turkey [6]. All of these regions and districts are present the Caucasus Ecoregion (480.000 sq.km.) [6].

Physical-geographically, the two regions extend from $35^0$to 45^0 latitude [6].

Among the Caucasus biomes, the floristic and faunistic diversity and infrequency (relict and endemic plant and animal species) of forests of mentioned two refugia are one of the most striking feature of the Ecoregion [8,9]. The central place of the most humid area is occupied by south Colchis with West Georgia's province, Adjara ($41^0$21'and $41^0$53' N), refugium in refugium, with considerably humid and warm temperate (oceanic) climate. January mean temperatures are above 0^0C, and maximum precipitation is known from the Mt Mtirala ('Crying Mountain' – from Georgian), namely 4520 mm, which is record data for the Caucasus Isthmus [9], and the Former Soviet Union (FSU).

Figure 1. *The typical site of the Mtirala gorge (Adjara).(All photos belong to A. Gegechkori).*

Except the narrow sea coastal area, it is typical mountainous district of Georgia, which occupied Adjaro-Imeretian (Meskheti Range) mountain system of the Lesser Caucasus [6]. Adjara with adjacent another province of West Georgia – Guria and neighboring to the northwestern part of Turkey is richest in endemic and relict species. Orographically it is represented as part of Western Lesser Caucasus (WLC), which is largely known as a Major Forest Refugium. WLC belongs to the Eastern Euxinian or Colchianphytogeographic province [10].

If phytocenoses of Talysh is favorable with its archaic representatives in the canopy (species of *Parrotia, Quercus, Acer*) [7-8], the Colchis shelter is impressive with its understory (especially species of *Rhododendron*) [5].

Derivative forms of relict fauna are an inseparable component of Arcto-Tertiary woody communities of both refugia, especially to Talysh. Mentioned should be species of following genera: *Brahmaea* Walker, *Isophya* Wattenwyl, *Parandra* Latreille, etc. [11].

2. Results and Discussion

Endemic and relict species of any refugia are of exceptional biological and biogeographical interests. Refugial centers sheltering such species are frequently referred to as a biodiversity hotspot. They have a prominent role in conservation biology [12].

The most vivid example of relict and at the same time endemic species are manifested by genus *Brahmaea* (Brahma – the creator, major god of Indian religion) [11,28]. Brahmin moths, family Brahmaeidae, of 28 species are found in the tropical latitudes of Ethiopian and Oriental zoogeographical Regions, and extending north into the warmer area of temperate localities of Palearctic.

The family takes its taxonomic place in superfamily Bombycoidea and consists itself by two subfamilies: Dactyloceratinae and Brahmaeinae [13]. Adults of Brahmin moth species are large: wingspan varies from 50 to 180 mm. Species occur in tropical-subtropical biomes have circular 'eye' markings on their wings ('Owl Moths'). This complex of moths regarding to independent genus *Brahmophthalma* Mell by some entomologists [14, 15]. The other ones are from temperate climate biomes, without these markings - to *Brahmaea* (sensustricto) [16]. The Brahmid moths have most complicated wing patterns of any Lepidoptera [16]. Leaf feeders, larvae are recorded in Asclepiadaceae and Oleaceae [17,18,19].

Brahmaidae at present include four recognized genera: *Brahmaea* Walker (type genus of the family), genus represented in Oriental and Palearctic Regions (8 species); *Acanthobrahmaea*Sauter, a monotypic European genus; *Calliprogonos* Mell, a monotypic Chinese genus; *Dactylocerus* Mell, an Afrotropical genus of about eight species; by some authors within mentioned family is considered also *Lemonia* Hübner in Palearctic with about twenty species [20,21,22]. The range of *Brahmaea* spp. in the temperate climate area is complex, involving several refugia. Present-day as a minimum 5-6 species of this genus have become an endemics of local regions of the Palearctic: *B. tancrei* Austaut (Russian Far East, Korea), *B. certhia* Fabricius (Sino-Korean, but from southern China most likely more), *B. litserra* Hao, Zhang & Yang(northern China), *B. japonica* (Butler) (from Japan), *B.* (*Acanthobrahmaea*) *europaea* Hartig (from Italy), *B. ledereri* Rogenhofer (Turkey and adjacent small area of Syria). Hyrcanian *B. christophi* Staudinger by some authors treated as a subspecies of *B. ledereri* [23, 24], subspecies – *B. ledereri zaba*de Freina occurs in extreme easternward of Anatolia (Turkish Kurdistan region). It shows intermediate features between latter two species [25].

Despite mentioned data, the Brahmaeidae seem to have been little studied and literature sources are very scanty [16]. There are apparently 15 species and 4 genera in China alone [16]. Sometimes information about Brahmid moths similar to genera including in Brahmaeidae, are quite controversial. For example, *B. ledereri* and *B. christophi* are mentioned in Chinese complex of these moths[14]; then, according to another source, *B. ledereri* in Chinaamong the other species of this genus feeds on *Ligustrumsinense* [26].

Therefore, "Eurasiatic species of Brahmaeidae present polymorphic populations with an uneven geographic distribution. The disputed taxonomy of this group is also due to the occurrence of populations that show morphological characteristics intermediate between related species" [20].

Among relict and endemic species of South Caucasus

refugia species of genus *Brahmaea* have emerged as of great interest to biogeographers due disjunct distribution area [7,11] but are there within two refugia – Colchis and Hyrcanvicariant event of Brahmid moths?

It should be stressed that among the relict fauna of Colchis, stenopaleoendemic *B. ledereri*is very problematic species for this region till today. A controversial information of its occurrence in perhumidhabitats of Colchis lowland and foothills arises chiefly from very old, questionable and with a large probability incorrect records. It seems to be doubtful the presence of this species along the eastern and southern Black Sea coast supposedly as a result of wrongly given locality in scientific literature [27]. Lepidopterologist Dr. E. Didmanidze in her's monograph 'The butterflies of Georgia' [28] wrote: 'In collections of the museum (Georgian National Museum – A.G.) only one specimen of *B. ledereri* is preserved, label of which is '1875' (Romanoff's collection)'. In another unpublished paper Didmanidze and Goginashvili [30] underline: 'The first summarized information on Brahmin moths of South Caucasus is summarized by N. Romanoff (1885) [65] in which is indicated *Brahmaealedereri* from "Batumi province".

The next important information includes the catalogue by Staudinger and Rebel (1901) [29]. In this work as a first site ofBrahmin ledereri's moth finding, followed to Rogenhofer, who described *B. ledereri,* is named 'Taur.os' (with following sites in German language: 'Gebirgsgebeitvom Bulghar-daghbis Gjaur-daghmit den OrtenHadjin, Zeitun, Marasch and Akbes (Eibes)' [30].

Figure 2. *Brahmaealedereri from collections of the Georgian National Museum.*

Latter existence of a Brahmin moth in Adjara (Batumi) is indicated by famous Russian lepidopterologists Ryabov [31] in his valuable 'Butterflies of the Caucasus'. This view seems to be 'turning point', i.e. a series of articles based on misunderstandings, as though the presence of 'relict moth' in Western Lesser Caucasus (Adjara). This viewpoint was largely motivated to the former Soviet Union's entomologists have been manifested 'everybody and together': tropical-subtropical fauna including *B. ledereri* of West Georgia as 'Colchis' ones. In this context, to show typical erroneous scholarly sources, should be quote Didmanidze [28]: 'It is very rare species of Tertiary Period and endemic of Colchic (mixed broad-leaved forests –A. G.), included in the 'Red Book of the USSR'; or 'on the territory of the Caucasusoccur only two species of this scanty tropical genus, *B. christophi*is one of them'; then 'on the territory of Georgia lived Colchicbrahmaea – *B. ledereri* Rogenhofer 1873, but after the beginning of XX century, in spite of regular inspection of typical area, this species was not found anymore'. This point of view that this moth could be considered extinct, i.e. Regionally Extinct (RE) in West Georgia is shared by almost all the entomologists of the Caucasus and the former USSR [31,38-44].

Despite of mentioned viewpoint, Lederer's moth as a live species of Colchis biota is still indicated by some authors [7,11,28,32-37,64].

A complex of reasons has led co-author of the present article [27] to look carefully into the relict distribution of this archaic species upon in humid ecosystems of the Black Sea coastal area.

Colchis as a region is situated in the temperate climate zone supporting temperate and humid biome with broadleaf and mixed forests. It has an extremely impoverishment species of humid-tolerant tropical-subtropical and Mediterranean, sub-arid type of subtropical flora and fauna within this (Colchis) marginal parts of their range (f. e. species of *Hymenophyllum*, *Arbutus*, *Erica*, etc.) [41-44]. However, viewed over the entire Quaternary, if any taxa survived glacial-interglacial cycles in Transcaucasia, with a great probability it should be found in a contemporary refugial locations. One of such stable shelter is Colchis forests composed of thermophilic broad-leaf deciduous and coniferous dendroflora. The latter was not endangered through changes in timber composition [5, 9, 10,45, 46].

It is doubtful whether species like *B. ledereri* has become extinct across all Colchis, still less in southern Colchis, i.e. outpost region for many relictual plant and animal species. For example, relict moth *Phassusschamyl*, the salamander of Miocene time *Mertensiellacaucasica*, another 'dying relict' (plant) *Epigaeagaultheroides*, and others [28,45-47].

B. ledereri unlikely to be extincted because of in South Colchis still exists required to this moth large areas of hygro-thermophilous, close to primeval broadleaf forests with complex habitats.

This is especially true for some gorges of Adjara district, which are really impressive with lush flora, sheltering a number of outstanding ancient living organisms [44-46, 48]. Some of most important from this viewpoint sites are represented as a Natural Parks (Kolkheti National Park, Mtirala National Park) or reserves (Kintrishi State Nature Reserve, Kobuleti State Reserve). Some reserves in Adjara were founded in the 1950s and 1960s, i.e. after 50-60 years later of as though extincted data of *B. ledereri.*

It should be stressed that Mtirala NP (and neighboring other sites) with its Arcto-Tertiary deciduous forests develops as in the extremely moist gorges is more 'tropical site' than those of Talysh (Azerbaijan), which shelter another relict species - *B. christophi* [8, 11, 41].

The same requiring habitats of Brahmid moths are largely represented along the neighbouring Turkish Caucasus (part of WLC), in the Pontic Mountains (Pontids) and Lazistan, which runs as a narrow stripe along the same southern coast of the Black Sea. It should be stated that Lederer's moth is not recorded in Lazistan till today [49, 50].

Despite Brahmin moth's official conservation status as Regionally Extinct (RE), sightings are still continuous, though none has been conclusively proven. It should be mentioned two recent attempts of positively decision of this problem. For example, a joint expedition carried out by Zaguliaev (St.-Petersburg), Didmanidze (Tbilisi), Jobaba (Kobuleti, i.e. Adjara). During 3 years (1983-1985) was collected a rich fauna of butterflies including nocturnal moth species [30]. At the same time pictures (photos) of the moth were allotting to residents of different gorges of Adjara to recognize the imago, thus being identities unmistakably to this species of Brahmid moth; Jobava himself was ready to take pictures of life-cycle stages of *B. ledereri*.

Another attempt to collect this species of moth was put into practice in 2010-2011 (in Adjara) and 2012 (in adjacent to Adjara Turkish part of Colchis refugium, Çoruh River Valley). One of the main goals of expeditions was to collect caterpillars of moth on, as though food plant, *Osmanthus decorus* in both states. The material was collecting by light trapping (PRK-2 lamps) at night, netting by day, and looking for caterpillars in daytime. In four different elevations of the Kintrishi River Gorge outstanding by its relict species of vegetation light traps were used including constantly operative Sakharov's insectrapping. The last trip was granted by the Rustaveli National Science Foundation. All these attempts to collect *B. ledereri* had no results [30].

It is difficult to judge whatever during last century microorganism diseases (fungal and others) through epidemiology can cause severe, demonstrable impact against vulnerable relict invertebrate animal species of Colchis, including Lederer's moth.

As regards to predators, species of genus *Brahmaea*, similar to most tropical butterflies, to escape predation, adult insects will hide through mimicry (cryptic coloration: on the example of *B. Ledereri* the ground color is dark-brown, ornamented with many parallels, finally wary-curly lines of white-rose color). Important cryptic selective advantage provides theirs slow flying (during daytime adults reston tree trunks). Cryptic anatomy and coloration is exemplified by large caterpillars as well [28].

This fact that ecological parameter of some temperate deciduous forests is attractive for exotic insects is confirmed by colonizing of Colchic 'rainforest' with saturnid moth. During 20th century western Georgia was enriched with tropical-subtropical Ailanthus silkmoth (*Samiacynthia*). It was introduced in Georgia for sericulture development. The moth did not domesticate as the silkworm unlike *S. cynthiaricini*in India (it is the only completely domesticated silkworm other than *Bombyxmori*). Indigenous populations of *S. cynthia*occur in China and Korea. In some countries of all continents, like Caucasus, it escaped from cultivation and naturalized [28]. It 'adapted to the natural conditions, has become a part of the Caucasian peacockbutterflies fauna and today is distributed in the total subtropical belt of the Georgian Black Sea coastal line' [28].

Figure 3. *The stand of the Osmanthusdecorus in Adjaristskali valley (Adjara).*

According to Ryabov [31], Borodin et al. (Eds.) [33] and Didmanidze [28], Didmanidze and Petrov [40], the caterpillars of *B. ledereri*feed on *Osmanthus* (*Phillyrea*) *decorus* and as it seems on other arboreal plants of Oleaceae, within the community usually found *Castanea sativa* in over forest and *Alnusbarbata*and *Corylus*sp. in second stratum of forests. The stand of *O. decorus* usually occurs in subcanopy in Adjara and Guriadistricts (Georgia) and in the adjacent Turkish Caucasus [48].

This record requires confirmation because of, as it was mentioned, there are no professional entomologists and naturalists who recorded feeding larva, i.e. insect (moth) host plant (*O. decorus*) associations in West Georgia and Lazitan.

According to Mandjavidze [48], *O.* (*Phillyrea*) *decorus*, is evergreen shrub or small tree. It is one of the most moist-tolerant species of its genus. *O. decorus*is restricted in Adjara to foothills and low mountain's slopes. Species typical within the group of noble hardwood include Oriental beech, sweet chestnut, European hornbeam, Oriental spruce, common ash tree (*Fraxinus excelsior*), Caucasian persimmon; understory is represented by *Laurocerasus officinalis, Rhododendron ponticum, Staphyllea pinnata, S. colchica, Ilex colchica, Vaccinium arctostaphylos*, etc. In marginal parts of forests it forms stands with Colchic boxwood. Therefore, in this community of deciduous forest there are some species of fam. Oleaceae (*Fraxinus, Osmanthus*), i.e. host plants of the moth.

If we take into consideration this fact that in typical Colchic woody community were organized numerous field excursions from Georgian National Museum and Zoological Institute RAS (St.-Petersburg) supplied by night moth trapping equipment, unlike to Hyrcanian forests, caterpillar-plant interactions research, as it was underlined, gave no positive results.

Finally, an additional confusion regarding the presence ofLederer's Brahmin in West Georgia should be originated from recent historical toponyms of the Colchis (Kolkheti) region. For example, 'Pontus' being an ancient name for the region around the Black Sea, and by extension, for the sea itself (en.wikipedia.org./wiki/Toponimy). Probably, these circumstances motivated entomologists of early 20th to incorrect determine the toponime of *B. ledereri* finding site of the moth if it exists indeed. The same situation occurs regarding to principality of Kutaisi. In this principality during the Tsarist Russian Empire (19th) besides Imereti district of West Georgia was included Guria adjacent to Adjara. Hence, when in the past time for *Osmanthusdecorus*was mentioned that this shrub grows in principality of Kutaisi it should be understood not Imereti region (Kutaisi is capital of this district) but Guria just from latter region was described *Phillyreamedwedewi*(i.e. *Osmanthusdecorus*) by Sredinski [48]. The same view seems to be based on misunderstanding for lepidopterologists. For example, they are emphasized as first collecting sites of discussed moth Batumi city (capital of Adjara – A.G.) vicinity [30,40], or moth's distribution in 'Adjara and Lazistan' [40], and believe that 'species is distributed in western Georgia, Colchis lowland. It is described from the Black Sea coast area (Lazistan) and inhabits in the same habitats, Colchic forests' [31], it is 'endemic and relict species of Pontic Province' [40], 'endemic of Colchis' [32, 33], 'endemic of Colchis and adjacent territory of the Black Sea coast of Turkey' [34], 'Colchis *B. ledereri*and Talyshian*B. christophi*'[35], 'Adjarian*B. ledereri*' [37, 64], etc.

Based on entomologists' data, now Lederer's moth is included in all regional (Caucasus, former USSR) and international (IUCN) Red Data Books and Red Lists [30, 42].

Today, *B. ledereri* is included in all 'Environmental Impact Assessment' types of issues dedicated to Colchis region of Georgia entitled 'Animals, included in the Red Data List of Georgia (2006), which can be found within the work area' (f.e., National Report on the State of the Environment of Georgia (2007-2009), 'Follow-up of Recommendation #120 (2006) on a European strategy for the Conservation of Invertebrates. Report by the Goverments. Strasbourg, 3 November 2011).

Figure 4. *Brahmaeachristophi – new emerged from pupa imago (photo by M. Soufi, 2013).*

The most vivid example of relict and at the same time endemic species are manifested by *Brahmaeachristophi*, one of the most beautiful moths in the invertebrate fauna of the extreme south-east Caucasus and Northern Iran [27-28, 30-44, 51-55].

B. christophi much resemblesTurkish vicarious species – *B. ledereri*, but is slightly larger (the wing-span reaches 110-130 mm (maximum – 160mm) vs*B.* christophi (95-107 mm) [28]. The latter species cover the whole Hyrcanian area, including Iranian part of the southern Caspian basin (Gilian, Mazandaran, Gorgan). The co-author of present work collected some specimens of this species with lepideptorologists R. Effendi and E. Didmanidze during several excursions in 1970s. At present four specimens of the moth are deposited in the collections of Georgian National Museum (Azerbaijan, Astara: 1910; Talysh, Hyrcanian relict forest, 14.04. 1974 (leg. Gegechkori), 17.04.1974 (leg. Didmanidze) [30].

Figures 5-6. *The forest composed by Parrotiapersica in overstory; Brahmaeachristophi caught in 1974 (Talysh).*

The insect prefers the most damp and shady forests of gorges at the Talysh mountain from lowland up to 1,000m a.s.l, i.e. the environment of its inhabitationwillmore or less resemble that of the subtropical-warm and humid climate of ancient geological epochs, when the waters of the Sarmatian Sea washed the coastal zone of the Caucasus Peninsula. The single glance at the genus *Brahmaea* is enough to became sure of how rich and lush thermophilous woody communities covered the Caucasus at the boundary of the Paleogene and Neogene Periods. Similar to Iran, Christoph's moth caterpillars feed on the leaves of native ash-tree (*Fraxinus excelsior coriariifolia*) [35]. Belonging to the relicts of immemorial time (about 30 mya), *B. christophi* close to extinction (Critically Endangered, CE) [37, 53].

A few publications refer to studies of the peacockbutterflies and considerably family Brahmaeidae of Northern Iran [54-56]. As a relict species of Miocene time *B. christophi*is now present in only a few populations in the Hyrcanian region (mentioned sources and pers. Observ. – A.G.). Present information was collected from local naturalists, from collections of museums of local National Parks during excursion of Gegechkori in Caspian forests (2012, X).

In Northern Iran this species occurs both in mesophilous Hyrcanian forests and more xeric variant of the same forests nearby to open woodlands and scrubs in Golestan Province northern of Golestan National Park (GNP).

One of the most xeric location of distribution *B. christophi* occurs near Totli Tamak village. This village and its surroundings are located northwestern part of GNP (300m away of NP border, northwestern Iran (620m a.s.l.) (Soufi, pers. comm.), annual precipitation ranging from 220 to 280 mm (Khorozyan, pers. comm.). Precipitation increases to 400mm at the slopes of the Bylikooh Mountain area, which is adjacents to mentioned village (Soufi, pers. observ.).

The local semi-lush forest, where the new emerging Christof's moth was recorded (900m a.s.l., mentioned mountain area), consists of leading species of noble hardwood of Hyrcanian region. Major forest-building species include: *Quercu scastaneifolia, Parro tiapersica, Carpinus betulus, Fraxinus excelsior* ssp. *coriariifolia*; *Diospiros lotus*, *Ficus hyrcana* are represented as subcanopy trees; *Populus caspica, Alnus glutinosa, Salix* spp. are confined in riverine; *Crataegu spentagina, C. monogyna, Mespilus germanica, Prunus divaricata, Danae racemosa* frequently occur as understory in these communities. The ground vegetation is manifested by the presence of *Primula heterochroma, Viola alba* (Soufi, pers. comm., pers. observ.).

In adjacent semi-dry habitat mentioned communities are largely outcompeted by more xerophillous communities of *Zelkova carpinifolia* and *Pyrus boissieriana.* In the more xeric variants of the site is phytocenoses, where the leading species are shrubs like *Paliurus spina-christi* and *Rhamnus pallasii* (pers. observ.).

The flight of the moth is observed in April, at night. In daytime adults (the imago) were observed resting with wings spread out on a tree trunk of the dominant forest tree, chestnutleaf oak.

Figure 7. *The trunk of the magic oak (Quercuscastaneifolia) near TotliTamak village (about 900m a.s.l.).*

Adult of *B. christophi*was observed after emerging from the pupa in the spring, which already indicate the species can permanently bread in not far of rural landscapes situated in the borderland near GNP. This site has now been rather intensively utilized for agriculture and logging. Thus, at present it occurs regularly in this locality.

The species is endangered by changes in forestry management in non-preserved sites of the moth's range (Soufi, pers. comm.).

In collections of the museum of GNP is exhibited one specimen of *B. christophi.* It was taken in 1980 by ranger also from transition zone of NP. TangeGol similar to TotliTamak village, it is characterized by mosaic semi-lush and semi-dried landscapes.

Figure 8. *Brahmaeachristophi (upper specimen) in collection of the Museum of Golestan National Park.*

German and Austrian scientists (Gutleb and Wieser) have discovered Cristoph's moth in TangeGol (2001) as well (Soufi, pers. comm.).

From mentioned peculiarities can be concluded:

1. Unlike all other species of the genus *Brahmaea, B. christophi* in the Iranian part of the Hyrcanian subprovince is ecologically most plastic species. It inhabits both Gilan-Mazandaran humid 'rainforest' and semi-dry, northeastern part of Iran (northern of GNP); it occurs on the one hand in undisturbed non-urban sites of GNP and some other reserves, and on the other hand in area represented in surrounded by modern homogenous rural land; GNP provide a meeting area for the Caspian Sea coastal and foothills taxa with semi-arid mountain taxa of living organisms, e.g. in the case of the transition zone between the Euro-Siberian (with Euxino-Hyrcanian province) and Irano-Turanian taxa.

2. Outside conservation area, occupying sub-xerothermic sites on the border of National Park and severely reduced numbers in typical habitats population of Christoph's moth are highly endangered.

Currently, as it was already mentioned, Talysh (Azerbaijan) and adjacent Caspian Sea coastal area of Iran represent relict ecosystems of summer-green warm temperate forests i.e. Major Forests Refugium, sheltering many rare and endangered species including stenopaleo endemics of plants and animals. As unique biome including unique forest composition and floristic and faunistic richness worldwide, it is urgent goal of environmentalists to synergize all activities in order to conserve native biodiversity like, for example, Hyrcan and Golestan National Parks. Therefore recognizing of having great value to strict conservation and management policies, it is urgent need to establish Iran-Azerbaijan cross- and transboundary cooperation for Hyrcan highly complex biodiversity conservation [57].

Figure 9. *The pristine forest of Hyrcanianrefugium, Mazandaran province.*

On the other hand, review of literature sources and private findings of Freina [23,25] in southwestern, eastern and southeastern Turkey have make the further light on distribution pattern of *B. ledereri.* His research on taxonomy [23], chorology and biology [25] of the moth showed following: firstly, the range of moth includes part of Western Asia, mainly Asia Minor (Turkey) and adjacent small part of Syria. Therefore, this range does not encompass the Northern Anatolian Mountains, i.e. Pontic Mountains and their northern, far more humid or perhumid slopes that face to the Black Sea (annually rainfall in Lazistan ranges from 2000-3000 mm per year), and then, Western Transcaucasia (south Colchis); secondly, the moth's range in Asia Minor is characterized by disjunct distribution (see map): a) population of Central Taurus Mts, i.e. mountain complex dividing the Mediterranean coastal region of southern Turkey (prov. Antalya) from the central Anatolian plateau; b) population of the upper reaches of the Euphrates River (Eastern Anatolia); c) population of surrounding the city of Elazig (Eastern Anatolia); d) population of Hakkari Province (the southeastern corner of Anatolia) and adjacent Syria. The largest range of discussed moth is represented in Central Taurus Mts, where *B. ledereri* was collected together with other moths, *Saturnia pyri, Nola ronkayorum*, etc. [58]; thirdly, ecological requirement of *B. ledereri* in mentioned sites of Turkey is quite different from those of represented in hygro-mesophilous relict mixed forests of Colchis. In Asia Minor this species prefers chiefly deciduous forests limited by riparian gallery forest represented in riversides of gorges and valleys of local mountains. Dominant vegetation composed of woody species, such as *Fraxinus excelsior, Platanu* ssp., *Eleagnus angustifolia*, secondary wild grapevine (*Vitisvinifera*), fig tree (*Ficuscarica*), hazel-nut (*Corylus*sp.), etc. As a result of human activities surrounding landscapes in some regions support only sparsely treed forests consist of overstory deciduous and evergreen species of oak (*Quercus infectoria, Q. libani, Q. persica*). Understory represented by species of *Crataegus, Prunus, Lonicera, Sorbus, Juniperes* and others. Leaf feeding caterpillars are found only on ash tree (*F. excelsior*).

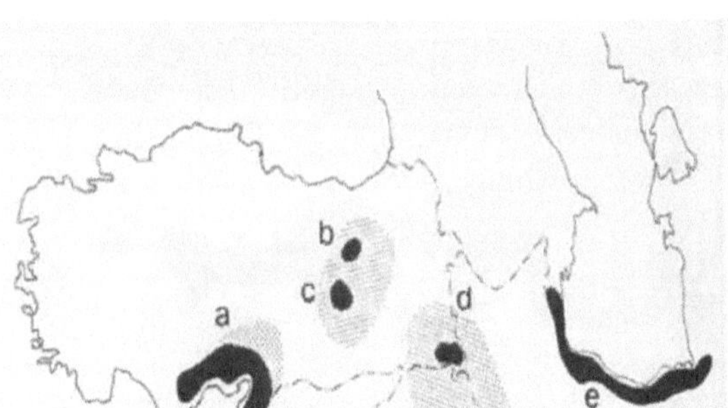

Figure 10. *The distribution sites of BrahmaealedereriRogenhofer, 1873 in southern Caspian basin area and Anatolia plateau: a) Brahamea ledereri ledereri Freina de, Taurus Mts population; b) the same subspecies population in the upper reaches of Euphrates River (site Tanyeri); c) the same population in Elazig; d) B. ledererizabaFreina de, population from Hakkari; e) B. ledererichristophi (Stgr.) from Alborz and TalyshMts area [25].*

It seems that in studied area Lederer's moth occurs from 1,000 to 2,000m a.s.l. The lower amplitude of altitudinal distribution of the moth found in the central Taurus (1,000-1,600m), the highest elevation moth reaches occurs about 1,800m (upper reaches of the Euphrates River), and in Hakkari province the altitudinal distribution areas varies from 1,200 to 1,650m.

Riparian forests, which are surrounded by xeric shrublands and patches of sparsely forested sites, dramatically reduced by human's activity from the historical past. Thus, this is a wide spectrum of semi-arid habitats with clearly expressed sub-Mediterranean climate (Taurus Mts) location (a) and continental climate locations (b,c), or sharply expressed continental climate (very hot and dry summer and cold winter) location (d) sites of Turkey (see fig. 1).

Vicariant event, in which part of a once widely distributed population of Brahmid moth becomes isolated in a spacially localized patches in Turkey, should be explained by intense folding of the Anatolian plateau. This tectonic event that caused this local landmass to emerge and largely change its morphology is the eastward displacement of the Anatolian microplate, driven from the northward displacement of the Arabian plate [59], considerably after rising Anatolian diagonal, which caused the separation of East and West Anatolian biota [60].

Current distribution area of Lederer's moth is considered with suggestion of Tarkhnishvili et al. [10]. From these authors, viewpoint during Last Glacial Maximum climates suitable for forest vegetation existed in six regions of western Asia: Colchis, western Anatolia, western Taurus, and upper reaches of the Tigris River, Levant and the southern Caspian basin. Currently, among named six regions the most biodiverse area of the forests is represented only in Colchis and Hyrcania but other four regions are almost treeless, although, as it seems, deforestation within the Fertile Crescent region, i.e. cradle of agriculture caused by human activity [61]. Now only in some gorges of Anatolia plateau survive the derivatives of former mesic forests just in these microrefugial shelters are found some forest biome dependent invertebrate fauna including *B. ledereri*. Thus, current range oftwo sister species of *Brahmaea* (*B. ledereri* and *B. christophi*) have shaped an area from southern Caspian basin to the headwaters of the Euphrates and on the other hand to the Mediterranean basin (Central Taurus).

According to Freina [25], the possible range of *B. ledereri* is not investigated in adjacent to eastern Anatolia, northern Iraq and Zagros Mountain range (Iran).

Did the mesic forests actually occur during LGM in the Zagros Mts, which currently have covered by quite different biomes in result of the xerophilization of the whole region? Neontological data support the presence of isolated refugia in the central Zagros Mts, i.e. in the Irano-Turanian part of Iran. In this site of range at an altitude of 1,730 m a.s.l., annual precipitation about 600-700mm and with very cold winter, but much of winter precipitation is snow, which cover the area until the end of March. In this area was discovered disjunct stand, remnant population of *Pterocaryapterocarpa* (*P. fraxinifolia*) [62].

At first glance this extraordinary discovery confirms that in Late Tertiary some gorges of Zagros Mts were not extensively influenced by continental climate and were covered by thermophillousArcto-Tertiary communities of plants resemblance of Euxino-Hyrcanian forests. Hence, these mountainous systems as a natural bridge between Alborz and Taurus Mts have provided to be effective area to harbor some relict representative of the Tertiary biota including with a high probability *B. ledereri*. In other words Zagros Mts might have played a role of transitional area to connect the Hyrcanian refugee center of the Caucasus Ecoregion with those of cryptic glacial refugia of Asia Minor and other regions of the Near East.

On the base of the present (real) and imaginative (speculative) distribution of two species of the genus *Brahmaea* should be proposed the following range of the moth: in a southern Caspian Sea area (northern Iran) on the one hand, and Anatolia plateau on the other hand. Within large refugium two taxa are concentrated from AlborzMts (in general Hyrcania), then throughout with larger probability of the occurrence in a dry segment of the Zagros Mts, and via extreme eastern part of the Anatolia plateau to the Taurus Mts (upper reaches of the Euphrates River), almost adjacent to its western part of the Mediterranean Sea.

The present-day distribution of two taxa of *Brahmaea* reflects a history of past wide distribution of tropical-subtropical and Arcto-Tertiary biotas, which today is largely replaced by modern, increasingly impoverished Hyrcanian nemoral biota and Anatolian Turkey plateau's semi-deserts with gorges sheltering some remain of former TBDF forests.

As we see this disjunct range of two species of the genus *Brahmaea*is distinguished by various climatic characteristics, the orographic heterogeneity, the geological and edaphic conditions as well as human impact considerably in Fertile Crescent area.

Among above indicated six regions, where the scientists hypothesize the presence of forest refugia there, the largest and most forested, major refugia is Colchis.

Therefore, in westernmost stretches of the Transcaucasia, in Colchis as well as in Hyrcania, an independent population of *Brahmaea* should be survived. This region with northern slopes of Pontic Mts undoubtedly satisfy the relict moth's ecological requirements. But today the eastern and south-eastern Black Sea coastal area should be considered as the region outside the confirmed refugium, where Brahmaea moth truly survived. On the other hand, without paleontological data there is no reason to postulate that some bottleneck event took place on the territory of Asia Minor.

Thus, recognizing this assumption as though *B. ledereri* lived in Colchis appears to be erroneous, I [27] hypothesize an extreme version of scenario to formation of the range of both paleoendemic taxa of Brahmin moths in Asia Minor and southern Caspian basin (Hyrcania). Under the most likely simulation regarding to origin of the common ancestor of *ledereri-christophi*e volutionary lineage should be occurring within four or five forested or suggested mesic forest vegetation regions of Western Asia, most of which are represented today as cryptic refugia. Among six LGM refugial regions *B. ledereri* should be excluded from Colchis refugium. At the same time we once more emphasized that there are no abiotic factors, which made it an endangered and extinct species.

Floristic and faunistic studies of Indo-Malayan and other tropical origin species of Hyrcanianrefugium (species of *Albizia, Gleditsia, Panthera* (including *P. tigris, P. pardus*), *Hytrix, Brahmaea*, etc.), support a viewpoint of the above modelling simulation.

There are some other explanations, which should be used as an additional suggestions for my [27] personal view. For example, geographical (latitudinal) and biogeographical (hystorical) reasons. In general, for all mentioned thermophilous tropical-subtropical species from northern margin of the tropical-subtopical latitudes (i.e. Hyrcanian part of Iran and Azerbaijan Republic) Colchis temperate climate latitudes with its abiotic factors, including, except temperature and precipitation, also photoperiodism, are insuperable ecological and geographical (latitudinal) barrier, preventing further northward distribution of *Brahmaea*'sautoecology organisms. It is worthy to remember, that the Hyrcanian refugium is located 4^0c5^0 southward thanthat of the Colchis refugium.

Unlike Colchis, for a long time of historical past, south and south-eastern Asia, as a part of Oriental zoogeographic region, was an important contributor to the Hyrcanian part of northern Iran provided northern territories with thermophilous tropical species [7, 63].

This straight biological connection between two remote regions was shifted during the current (Quaternary) series of large-scale environmental changes (glacial events) causing desertification of some regions of Near East including central and eastern parts of the Iranian plateau. This dryland area converted into bottleneck to decrease the potential to expansion of thermo- and hygrophilous species in some marginal areas of higher latitudes of subtropic and subtropic-temperate border areas.

Thus, few species of the ancient genus *Brahmaea* (some species of which are found in tropics and subtropics) are scattered now within such remote but from climatic viewpoint more or less similar (except part of Golestan province of Iran and Anatolian Plateau) places of Northern Hemisphere.

On the example of allopatrically dispersed species united in genus *Brahmaea*, could be provided valuable information for our understanding of global climate changes patterns, its influence of living organisms eliminations and surviving processes, the role of harbours – refugia – during major Ice Ages in preservation some archaic thermophilous plant and animal taxa of Tertiary Period [27].

3. Conclusions

Comprehensive analysis of vicarious event between two species of Brahmid moths – *Brahmaea ledereri*and *B. christophi* as though occurring in two refugia of Transcaucasia, Colchis and Talysh, opened new views and permit provide another conclusions. Here, on the base of research the ranges of two species of peacockbutterflies and their bioecology, carrying out mainly by some entomologists [23,34-35] we pointed out another relation between Euxino-Hyrcanian and Anatolian biotas. There are actually allopatric ranges of two relict species of moths, but the disjunction has occurred not between relict moths of Brahmeidae of Colchis and Hyrcanian refugia, as it was accepted as an unconditionally dogma among Soviet Union's entomologists. Split the common range of genus *Brahmaea*, as it seems, actually took place in geological past (Pliocene), although not between Transcaucasian two famous refugia – Colchis and Talysh but between Hyrcanianrefugium on the one hand, and Anatolian cripticrefugial areas, on the other hand [27].

The latter region is represented at more lower (35^0-38^0) latitudes unlike the Colchis occurring at higher latitudes (40^0-45^0). This geographical phenomenon, as it seems, became major distributional limit ecological factor, preventing penetration complex of subtropical (Hyrcanian/Anatolian) organisms including *B. ledereri* within Colchis.

Despite its prehumid and warm climate of Colchis Triangle, one of the main abiotic factors - seasonal (insufficient) cycles of light (photoperiodism) became unsuitable for *B. ledereri*and another organisms of its autoecology, whereas at lower latitudes light is available for the same complex of organisms during year-round.

Acknowledgements

Research partially supported by CEPF (2005-2006) (research project 'Biodiversity Hotspots – Caucasus') [42]. We thank MrRastaghi the director of the Golestan National Park and its staff; the great help of PhD student Mahmoud Soufi and his family are appreciated for support of my trip to Golestan province, Dr. Eslami from the University Natural Science and Agricultural (city of Qaemshahr) for providing the research activity of MazandaranHyrcanian forested areas (Iran). The same support from director of Mtirala National Park (Adjara), D. Khomeriki; head of zoological expeditions in Turkey(2005-2007), Prof. R. Zhordania. We are much indebted to LaliGegechkori (Tbilisi) for linguistic improvements.

References

[1] A. Engler, „Versuch einer Entwicklungsgeschichte insbesondereder Pflanzenwelt seit der Tertiärperiode," II.Leipzig: Wilhelm Engelmann. 1882, 14,p. 386.

[2] D. I. Axelrod, "Origin of deciduous and evergreen habits in temperate forests," *Evolution*,1966,20, 1,pp. 1-15.

[3] H. Walter, „Die Vegetation der Erde," Vol. I. *Progress*, Moscow, 1968,p. 422. (translation in Russian).

[4] A. N. Kryshtofovich, "The main characteristic feature of development of the Tertiary flora of Asia," *Izv. Glavn. Botan. Sada RSFSR, 29,* 1930, 3-4, pp. 391-401. (in Russian).

[5] A. G. Dolukhanov, "Forest vegetation of Georgia," *Universal*, Tbilisi, 2010,p. 532 (in Russian).

[6] L. Williams, N. Zazanashvili,G..Sanadiradze, A. Kandaurov, (Eds.) "An Ecoregional ConservationPlan for the Caucasus," WWF, KFW, BMZ, CEPF, MacArthur Foundation. Tbilisi, Printed by Signar Ltd, 2006,p. 222.

[7] A. M. Gegechkori, "The results of biogeographical study of Arcto-Tertiary refugia (Colchis and Talysh) of the southern Caucasus," *Annals of Agrarian Science,* 2011a,9,1:16-33pp.

[8] A. A. Grossheim, "Flora of Talysh," *Izd. Narkomzem Azerb.* SSR, Tiflis, 1926,p. 273.

[9] K. Browicz, "Chorology of the Euxinian and Hyrcanian element in the woody flora of Asia*,"Plant Systematics and Evolution",* 1989, 162, pp. 305-314.

[10] D. Tarkhnishvili, A. Gavashelishvili, L. Mumladze, "Palaeoclimatic models help to understand current distribution of Caucasian forest species," *Biol. J. Linn. Soc:* 2012, 105, 1, pp. 231- 248.

[11] A. M. Gegechkori, "Biogeographical characteristic of the Tertiarian organisms of the Talysh part (South Caucasus) of the Hyrcanian refugium," *Annals of Agrarian Science,* 2012b, 10, 1,pp. 8-16.

[12] N. Myers, R. A. Mittermeier, C.G. Mittermeier, G. A. B. da Fonseca, J. Kent, "Biodiversity Hotspots for conservation priorities,*" Nature,"* 2000, 403, pp.853-858.

[13] J. Minet, "The Bombicoidea: Phylogeny and higher classification (Lepidoptera: Glossata), "*Entomologica scandinavica,"* 1994, 25, 1, pp. 63-88.

[14] H. F. Chu, W. Ling-Yao, "The Brahmaeidae of China (Lepidoptera),*"Acta Entomologica Sinica,* 1997, 20,1, pp. 83-86.

[15] R. Spicciarelli, „La psichedel Frassino: European owl moth. Consiglio Regionale della Basilicata," Finiguerra Arti Grafiche-Lavello (pz), 2004,pp. 67-72.

[16] A. Marson, Rearing & breeding, Brahmaea hearseyi in the UK.huffmantaxidermy.com/.../amarsonrearing.htm. 2004.

[17] P. C. Rougeot, "Les Bombycoides (Lepidoptera-Bombycoidea) de l'Europeet du BassinMēditerranēen, Tome I: Lemoniidae, Bombycidae, Brahmaeidae, Attacidae, Endromididae. Fauna de l'Europe et du BassinMēditerranēen, " Masson et Cie, Paris, 1971, 5, p. 159.

[18] W. A. Nässig, Treadaway, C. G. "The Brahmaeidae (Lepidoptera) of the Philippines," *Nachrichten des Entomologischen Verein Apollo Suppl.*, 1998,17, pp.425-440.

[19] J. B. Heppner, "Brahmin moths (Lepidoptera: Brahmaeidae)," *Encyclopedoa of Entomology. Springer,* 2008. pp. 561-562.

[20] J. W. Brown, "Records of Hypaurotis crysalus (Edwards) (Lycaenidae) from Western Mexico," *Journal of Research on the Lepidoptera,* 1988(89),27,2,pp. 135-143.

[21] W. A. Nässig, I. W. B. Nye, "Brahmaea Walker, 1855 (Insecta, Lepidoptera): proposal confirmation of Bombyx certhia Fabricius, 1973 as the type species," *Bulletion of Zoological Nomenclature,* 1991, 48, 2,pp. 137-139.

[22] A. W. Nässig, S. Naumann," Saturniidae and Brahmaeidae of the Palearctic Region: how many species are there?" *Entomologica Romanica,*2010,15, pp. 17-20.

[23] De. J. J. Freina, "Neuere kenntnisseüber die Biologie und Systematik von *Brahmae ledereri* Rogenhofer 1873. Und deren verbreitung in Kleinasien (Lepidoptera, Brahmaeidae), "*Mitt. münch. ent. Ges.,* 1985, 74,pp. 77-90.

[24] L. Lehmann, R. Zahiri, "Results of a lepidopterological expedition to North andNorthwest Iran in summer 2007 with new record for Iran (Lepidoptera) (plates 19-22)," *Esperiana Band*,2011, 16,pp. 135-165.

[25] De. J. J. Freina, "Eineneue population von *Brahmaea ledereri* Rogenhofer 1873 (*Brahmaea ledereri zaba*ssp.n. in Turkisch-Kurdistan, sowienachweis der konspezifitäz von *Brahmaea ledereri* Rogenhofer, 1873, und *Brahmaea christophi* Staudinger, 1879 (Lepidoptera, Brahmaeidae)," *Entomofauna*, 1982,3, 9,pp. 129-139.

[26] *Ligustrumsinense*. Chinese privet. In: wiki.bugwood.org/uploads/Ligustrum.pdf.

[27] A. M. Gegechkrori, "Has ever *Brahmaea ledereri* Rogenh. Inhabited the Colchis refugium unlike the Hyrcan refugium?" conference.ens-2015.tsu.ge./en/lecture/view/262. 2015.

[28] E. A. Didmanidze, "The butterflies of Georgia," *Studia 99*, Tbilisi, 2005,p. 85.

[29] O. Staudinger, H. Rebel, "Catalog der Lepidopteren des paläarktischen Faunengebieter,"Berlin, 1901, p. 411.

[30] E. Didmanidze, N. Goginashvili, "On the existence of Lederer's Brahma (Lepidoptera, Brahmaeidae) in the biomes of Georgia", Report (Rustaveli National Science Fundation Grant, #1-8-38), 2012, pp. 1-9 (in Georgian).

[31] M. A. Ryabov, "Butterflies (Lepidoptera) of the Caucasus," In: *Zhivotni Mir Kavkaza* [Animal World of the Caucasus]. *Izd. AN SSSR*, M.-L. 1958, 5, pp. 351-375 (in Russian).

[32] A. A. Mirzoyan, I. D. Batiashvili, "Rare insects," *Lesnaia Promishlennost*, Moscow, 1982. p. 165. (in Russian).

[33] A. M. Borodin, A. G. Bannikov, V. E. Sokolov, [Eds.] "Red Data Book of the USSR,"Moscow. *Lesnaya Promishlennost* 1984, Vol. 1, pp. 279-280. (in Russian) .

[34] R. E. Effendi, "To study of peacock butterfly fauna of Transcaucasia (Lepidoptera, Saturnidae. Brahmaeidae)," *Materiali Sessii Zakavkazia* [Materials of Session of Transcaucasia], Tbilisi, 1968,pp.536-538 (in Russian).

[35] R. E. Effendi, "Some stages of formation of the Lepidoptera fauna of Transcaucasia", *Zool. Zhurnal*, 1971, 1,5, pp.696-703 (in Russian).

[36] R. E. Effendi, "Some relict and endemic species of Macrolepidoptera of Talysh" *Materiali Sessii Zakavkazia* Po Zaschite Rastenii [Materials of Session of Transcaucasia about Plant Protection], Erevan, 1967, pp. 545-547 (in Russian).

[37] S.V. Aliev, R. E. Effendi, Sh.Y. Abdulaeva, "Order Lepidoptera of butterflies (Saturniidae, pp. 247-248; Brahmaeidae, pp. 248-249),"In: Animal World of Azerbaijan, Vol.II. Invertebrates, PhylumArthropoda (Musaev, M.A. et al., Eds). *Elm*, Baku (in Russian), 1996.

[38] E. Didmanidze, N. Ninua, "Some groups of animals from Strict Reserves of subtropic zone of Transcaucasica," Tbilisi, *Metsniereba,*1999, pp. 1-45 (in Russian).

[39] E. Didmanidze, I. Skhirtladze, N. Ninua, L. Gurgenidze, V. Chkhikvadze, "On some endemic, rare and disappearing species of the Georgian fauna: new approaches to problem solution," Tbilisi, *Khatula,*2002. p. 65. (in Russian with English summary).

[40] E. Didmanidze, V. Petrov, "Review of fauna of peakockbutterflies (Lepidoptera: Saturniidae, Brahmaeidae) of Southern Caucasus from collection of Georgian National Museum," *Proceedings Natural and Prehistoric Section. Georgian National Museum*, 2011, 3, pp. 166-181 (in Georgian, with English summary).

[41] A. M. Gegechkori, "Relicts and endemics and Georgia's biodiversity at the background of the Caucasus orogenesis," In: *Biological and Landscapes Biodiversity of Georgia*. WWF Georgia, Tbilisi, 2000, pp. 83-96 (in English and Georgian).

[42] A. M Gegechkori, "Caucasus – treasure of the Nature,"CEPF, WWF, *Meridiani,*2007, pp. 168. (in English and Georgian).

[43] A. M. Gegechkori, "Biogeography, chapter 1 (Biomes of the World),"*Tbilisi StateUniversity Publ. House*, 2008, p. 527 (in Georgian, with English summary and contents).

[44] A. M. Gegechkori, "The results of biogeographical study of Arcto-Tertiary refugia (Colchis and Talysh) of the southern Caucasus," *Annals ofAgrarian Science*, 2011a, 9,1, pp.16-33.

[45] Z. Kikvidze, M. Oshawa, "Adjara. East Mediterranean refuge of Tertiary vegetation," In: Oshawa M. et al. — Eds. *Chiba Univ.,*1999, pp. 297-315.

[46] T. Denk, N Frotzler,. N. Davitashvili, "Vegetational patterns and distribution of relict taxa in humid temperate forests and wetlands of Georgia (Transcaucasia),"*Biol. J.Linn. Soc.* 2001, 72 ,2, pp. 287-333.

[47] D. Tarkhnishvili, U. Kaya, A. Gavashelishvili, I. Serbinova, "Ecological divergence between two evolutionary lineages of the Caucasian salamander: evidence from GIS Analysis," *Herpetological journal*, 2008, 18, pp. 155-163.

[48] D. V. Mandjavidze, "Relict forests of Adjara," *Metsniereba, Tbilisi* (in Russian), 1982, p. 237.

[49] T. Gokturk, F. Bucak,T. Artvinli, "Mammalian fauna of Artvin, "*African Journal of Agricultural Research*, 2011, 6 (6), pp. 1418-1424 .

[50] E. Karaçetin, H. J. Welch, "Red Book of butterflies in Turkey," *Doga Koruma Merkezi.* Ankara, Turkey. 2011, pp. 1-125.

[51] A. M. Gegechkori, "The two refugia of the southern Caucasus, Colchis and Talysh: a comperative biogeographical analysis," *International Conference of IGU*, Santiago, Chile (acreditacion. fisa.cl/ugi/…/ GegechkoriArnold_ 20112324VO3ZVH_f.doc.),2012a.

[52] A. M. Gegechkori, H. Joosten, (Eds.) "The biomes of the Caucasus," *Draft for the IMCG Field Symposium*, Georgia/Armenia, Sept. 2009. Greifswald, 2009,p. 169.

[53] Red Data Book of Azerbaijan – Wikipedia. https://ru.wikipedia.org/.../Список_Насекомых_занесенных_в_Красную. . .

[54] H. Mohammadian, "Biological diversity of Lepidoptera of Iran (Geographic distribution of 2200 species)," *Shabpareh publications*,2006, p. 386.

[55] L. Lehmann, R. Zahiri, "Results of a lepidopterological expedition to North and Northwest Iran in summer 2007 with new record for Iran (Lepidoptera) (plates 19-22),"*Esperiana Band*,2011, 16,pp. 135-165.

[56] List of moths of Iran - Wikipedia, the free encyclopedia 2014.en.wikipedia.org/wiki/List_of_moths_of_Iran.

[57] M. Sharifi, "An overview of ecological potential and the outstanding universal value offorests resources of I. R. Iran with respect to climate change,"Regional Workshop 'Forests, Rangelands and Climate Change in the Near East Region' 20-22 September, 2011, Cairo, Egypt, 2011, pp. 1-10.

[58] S. Beshkov, "*Nola ronkayorum* sp. n., a new species from Bulgaria and Turkey (Lepidoptera: Nolidae)," *Phegea,* 2006, 34(1), pp. 17-32.

[59] P. Lymberakis, N. Poulakakis, "Three continents claiming an archipelago: the evolution of Aegean's herpetofaunal diversity," *Diversity*,2010,2, pp. 233-255.

[60] P. H. Davis, "Distribution patterns in Anatolia with particular reference in endemism," In: Davis, P. H., Harper, P. C., Hedge, I. C. (Eds.). Plant life of Southwest Asia. *Royal Bot. Gard.,* Edinburgh, 1971, pp. 15-27.

[61] D. Zohary, M. Hoph, E. Weiss, "Domestication of plants in the Old World," *Oxford UniversityPress*, 2012, pp. 1-280.

[62] H. Akhani, M. Saliman, "An extant disjunct stand of *Pterocarya fraxinifolia* (Juglandaceae) in the central Zagros Mountains," W. Iran. *Willdenowia*, 2003, 33, pp. 113-120.

[63] N. K. Vereshchagin, "Mammals of the Caucasus,".In: *Zhivotni Mir SSSR*[Animal World of the USSR], 1959.V 5. *Izdatelstvo AN SSSR*,M.-L. (in Russian), p. 704.

[64] Lederer'sbrahmin, sin: http://ru.wikipedia.org/wiki/Брамея-Ледерера

[65] N. M. Romanoff, "Les Lepidopteres de la Transcaucasie,“ In: Memuares sur les Lepidopteres‘. St.-Petersbourg, 1885. v.2, pp. 3-118.

16

Quantitative Assessment of Permissible Loads on Georgia's Soil

David Kereselidze[1], Lia Matchavariani[2, *], Vazha Trapaidze[1], Lamzira Lagidze[3], David Svanadze[4], Giorgi Bregvadze[1], Zaal Gulashvili[1]

[1]Department of Hydrometeorology, Faculty of Exact & Natural Sciences, Ivane Javakhishvili Tbilisi State University, Tbilisi, Georgia
[2]Department of Soil Geography, Faculty of Exact & Natural Sciences, Ivane Javakhishvili Tbilisi State University, Tbilisi, Georgia
[3]Department of Nature Use Management, Faculty of Exact & Natural Sciences, Ivane Javakhishvili Tbilisi State University, Tbilisi, Georgia
[4]Department of Geography, Faculty of Exact & Natural Sciences, Ivane Javakhishvili Tbilisi State University, Tbilisi, Georgia

Email address:
david.kereselidze@tsu.ge (D. Kereselidze), lia.matchavariani@tsu.ge (L. Matchavariani), v.trapaidze@yahoo.com (V. Trapaidze), lagidze_l@rocketmail.com (L. Lagidze), datosvani@posta.ge (D. Svanadze), gbregvadze76@gmail.com (G. Bregvadze)

Abstract: Environment, including different types of soils, is permanently influenced by human activity, natural and technogenic disasters, etc. These influences weaken stability of ecosystems, landscapes, soils, especially in the case when these influences exceed the value of the maximum permissible standard. Every natural effect causes weakening of soil stability. That's why it is important to know the values of the lads, which cause failure of proper functioning of soils. At present averaged values of observation data are considered as maximum permissible norms of different loads on soils. In the meantime, actual indices of soil loads are stipulated by many variable factors. Existing standards of permissible loads on soil don't meet requirements, which provide proper functioning. Additional demands, related through time to operating conditions and probabilistic-statistical characterizing of factors causing processes of soil degradation in time, must be raised upon actual standards. Absence of scientifically substantiated permissible load norms is a reason of soil erosion that as the final result leads to land depopulation and loss of ethno-cultural traditions. That's why in the given work take into account factors, which cause regulatory values of parameters of soil loads. Time of possible loss, fertility to the marginal degradation was calculated for all main types of Georgia's soils.

Keywords: Loss Fertility, Soils, Calculation, Georgia

1. Introduction

Observations on natural environment show that in the era of modern industrial development the nature experiences various kinds of the undesirable effects, and as a result deterioration of its conditions takes place, it degrades. Disturbances and certain changes of the natural environment are the elements of united global processes. That's why both common physical laws and individual regularities caused by specific indicators of separate elements are clearly reflected in geographic system and its elements. Temporal and space variability of indicators best of all determines the condition of the Environment and threat of hazard risks.

Accurate quantitative assessment of environmental condition is almost impossible, that's why the application of approximation methods is acceptable, to which end is necessary to use the current mathematical possibilities and select the most appropriate model, which makes possible a problem solving using a valid approximation.

The environment consists of a great number of elements, which are accompanied also by a great many processes; it is desirable, but virtually impossible to take into account all these elements, that's why it is important to select the basic elements, which play the leading role in the environment's functioning.

In our case a soil has been selected as the basic element of the environment, since the existence and future of mankind are directly related to it.

The work objective is a quantitative assessment of condition of fertile soils of Georgia and elaboration of approximation (probable) methods for calculation of risks of hazardous undesirable situations.

The necessity of these methods is driven by soil preservation with minimal losses that in its turn will promote harmonic coexistence of mankind with environment and radical improvement of the current situation.

2. Research Method

Soil, as well as natural environment consists of a great number of elements and processes, which accompany these elements. Consequently, disordered motion permanently occurs in the soil, which in this regard is similar to Brownian motion, which is usually studied using stochastic kinematic equations.

The process of soil functioning can be described by Kiyoshi Ito's stochastic differential equation [13,14]:

$$dy(t) = m_0(t)dt + \delta(t)dx(t) \tag{1}$$

where $y(t)$ is soil depth, $m_0(t)$ and $\delta(t)$ are average rate and mean square deviation of soil depth variability, $x(t)$ is a random value of Gaussian process.

Equation (1) can be rewritten as follows:

$$dy = mdt + \delta dx(t) \tag{2}$$

Equation (2), which describes variation of soil depth leads us to diffusive distribution of soil during its permanent functioning, which under certain conditions is represented by the solution of Fokker-Planck-Kolmogorov equation:

$$\frac{\partial P}{\partial t} + m\frac{\partial P}{\partial y} - \frac{\delta^2}{2}\frac{\partial^2 P}{\partial y^2} = 0 \tag{3}$$

Fokker-Planck-Kolmogorov equation according to classification of partial differential equations is an equation of parabolic form. For its solution, it is necessary to know initial and boundary conditions of a given function, and namely boundary conditions determine the form of a solution of equation (3).

When implementation of process (2) is by monotonous nature, boundary and initial conditions are of the following form:

$$P(y,t) = \left|y = -\infty = P(y,t)\right|_{y=\infty}^{0} \tag{4}$$

$$P(y,t) = \Big|_{y=0=\delta(y)} \tag{5}$$

where $\delta_{(y)}$ is delta-function.

Solution of equation (3) with conditions (4) and (5) is known as fundamental and they are of following form:

$$p = \frac{1}{\delta\sqrt{2\pi t}}\exp\left[+\frac{(y-\mu t)^2}{2\delta^2 t}\right] \tag{6}$$

Here parameter μ, which represents a comparison of the admissible limit value of soil depth with an average rate of soil depth variability.

The density of a transient diffusive kind is precisely described by differential equations. Since soil degradation is caused by the continuous action of external factors, variability of random processes will be of monotonous nature. Proceeding from this fact we can suppose that soil degradation process is of the monotonous nature with an average rate (m) and mean square deviation (δ) of persistent variables.

Trouble-free operation of soil during degradation $p(t)$ is the same as a probability of failure of the undesirable condition, or soil reliability during t time [7, 13, 14].

$$P(t) = \Phi\left(\frac{t-\mu}{\alpha\mu\sqrt{t}}\right) \tag{7}$$

From the standpoint of theory of reliability a functional possibility of happening of undesirable situation can be characterized by areas of limit states, beyond of which, as it is considered, a soil have a failure. Generally a violation of conditions of soil functioning is considered as soil failure. Mathematical expectation $m(t)$ and dispersion δ_t^2 of failure of undesirable condition are calculated using the following formulas:

$$M_t = \mu\left(1 + \frac{\mu\alpha^2}{2}\right) \tag{8}$$

$$\delta_t^2 = \mu^3\alpha^2\left(1 + \frac{5\mu\alpha^2}{4}\right) \tag{9}$$

where: $\mu = \frac{I_r}{m}$ and $\alpha = \frac{\delta}{\sqrt{I_r m}}$.

When using normed random variable Z [8], the expression (7) will be as follows:

$$\frac{\mu - t}{\alpha\mu\sqrt{t}} = Z \tag{10}$$

In this case Z is a static reserve of resistivity $Z = R - P$, which causes the probability of trouble-free operation of soil.

Time, in the course of which a soil reaches the boundary (limit) value I_r of degradation, can be determined by squaring of equation (10) and afterwards by solution of receiving quadratic equation.

$$t^2 - \left(2 + Z^2V^2\right)\mu + \mu^2 = 0 \tag{11}$$

where $V = \frac{\delta^2}{m^2}$;

$$t = \mu\left[\frac{2+Z^2V^2 \pm \sqrt{\left(4+Z^2V^2\right)^2 \cdot V^2Z^2}}{2}\right] \quad (12)$$

By means of formula (12) it is possible to determine the degradation rate for the soil.

Similarly, we know that $\mu = \frac{I_r}{m}$, so we can determine Z_v:

$$I_r = \frac{2tm}{(2+V^2Z^2)-\sqrt{V^2Z^2(2+V^2Z^2)}} \quad (13)$$

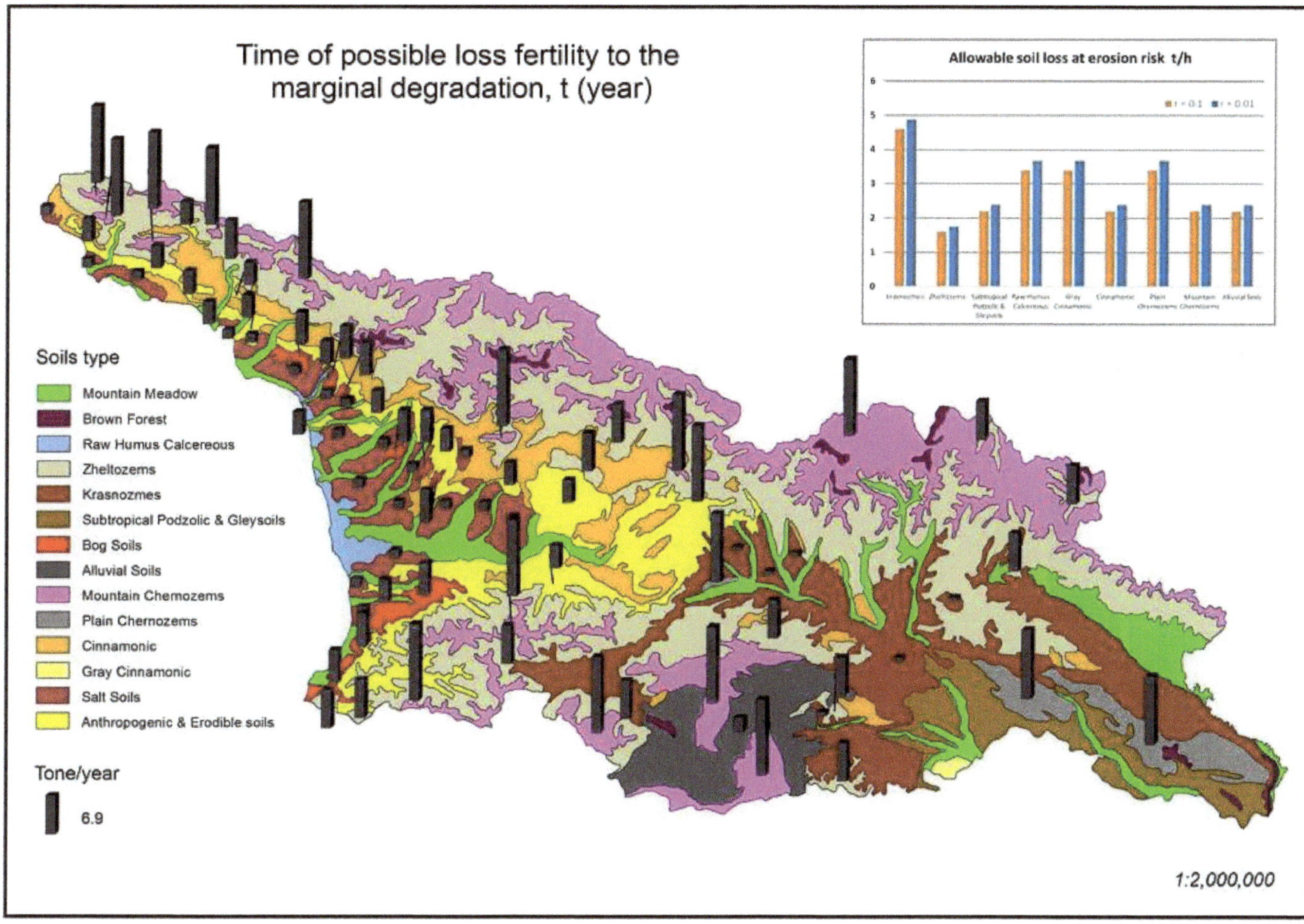

Figure 1. *Time of possible loss of soil fertility to the marginal degradation, t (year)*

3. Results

As we mentioned above, time, during which soil reaches boundary values of degradation, is calculated according to the formula (12). Humus percentage was selected as a determining parameter of soil degradation. The average rate of its variability was determined on the basis of following consideration that deep-earth irreversible soil transformations vary from 2-3 minutes from 8-10 minutes, i.e. 5 years in average [3,4], the mean square deviation of average rate of variation of humus percentage δ is calculated by means of observation data processing, if they are not available, then using approximation methods of mathematical statistics [15]. Approximate mean square deviation is simple and quickly determined with the use of following formula (14):

$$\delta = \frac{X_{\max} - X_{\min}}{K} \quad (14)$$

where, $X_{\max}$, $X_{\min}$ in our case are maximum and minimum values of humus percentage in the soil; K – coefficient, which is depended on the number n of observed data.

For the first time this method was offered by Tippet [15], table 1.

Determination of permissible rates of soil degradation makes possible the soil protection from catastrophic washouts, which have tended to grow. At the same time by their means it is possible to formulate criteria of the decision making, with the use of which soil protection and recovery measures will be planned.

Researchers have been carried out at the various types of Georgian soils in order to determine permissible rates of soil degradation. Studies gave us the following results: if the duration of soil existence is 1000 years, while humic layer depth is 20 cm, then maximum permissible soil losses within the risk limits $0,01 < r < 0,1$ are calculated by following formula [5, 10, 11, 12].

$$H_0 = \frac{H_0 - \delta_{H_0} H_0 - H_D}{t} \qquad (15)$$

where H_0 is the initial soil resource; H_D is a maximum permissible rate of soil degradation; δ_{H_0} is the mean square deviation of initial soil resources. Obtained data are given in the table, and on the basis of these data the map for permissible soil losses in Georgia is compiled (figure 1).

The time during which an each soil reaches limit values of degradation, was determined according to a formula (12) and using above mentioned approximation methods of mathematical statistics. These values are given in table 2.

Table 1. *Calculation of the K coefficient, which is depending on the number n of observed data [15]*

n	0	1	2	3	4	5	6	7	8	9
0	-	-	1.12	1.69	2.05	2.32	2.53	2.7	2.84	2.97
10	3.07	3.17	3.25	3.33	3.40	3.47	3.53	3.58	3.64	3.69
20	3.73	3.77	3.81	3.65	3.69	3.93	3.96	3.99	4.02	4.06

Table 2. *Limit values of soil degradation*

	Soil type	Soil depth, cm	Profile humification, cm	Humus content, %	Absolute altitude z, dm	Time before complete degradation, year
1	Mountain-meadow soils	Small 50-80	average & small, 10-20	5-10 15-20	1800-2800 West Georgia 2000-3500 East Georgia	13.8
2	Mountain-forest-meadow soils	Small 50-80	average & small, 10-20	4-8	1800-2600	3.6
3	Mountain-meadow-chernozem -like	small or average 70-80	deep, 20-30	8-12	above 1800 (2000)	5.5
4	Brown forest soils	Average 50-100	average, 15-20	5-10	800-2000 West Georgia 900-2000 East Georgia	6.9
5	Black, brown forest soil	Small 60-90	too deep, 40-50	4-9	1100-1600	6.2
6	Raw Humic Calcereous soils	Average 100-150	clearly expressed, 15-20	2-5	400 (600) – 1600 (1800)	1.6
7	Mountain Chernozems	Average or small	deep, 30-35	5-8	1200-1900	2.6
8	Plain Chernozems	deep	deep, 40-50	5-15	60-650	12.1
9	Cinnamonic soils	average or small	average & deep, 20-35	2.5-3.5	500 (700) – 900 (1300)	0.74
10	Meadow cinnamonic soils	deep	deep, 20-30	2-3	500 - 900	0.60
11	Gray-cinnamonic soils	Deep 150-200	deep, 20-30	2-3	650 - 700	0.60
12	Meadow Gray	Deep 150-200	average & small, 10-15	1-2	135 - 200	0.42
13	Salt soils	Deep 150-200	small, 10-15	1-1.5	135 - 700	5,9
14	Krasnozems	deep	deep, 20-40	3-10	300 - 500	17,9
15	Zheltozems (Yellow Soils)	average and deep	average & small, 10-15	5-9	500 - 700	4.1
16	Subtropical Podzolic soils	average and small	average, 15-20	2-5	30 - 200	1.6
17	Podzolic-Gley soils	average and small	average, 15-20	3-5	up to 100	1.3
18	Bog soils	Deep 150	average & deep, 15-20	10-15	up to 50	
19	Alluvial soils	average and deep, 100-125	deep, 15-30	1.5-3	different natural zones	

4. Conclusions

Offered method, based on the theory of reliability makes possible the forecasting of soil degradation with the use of a minimal amount of initial data and doesn't require complicated calculations, that makes it easy-to-use under actual operation conditions.

The elaborated approach of forecasting of degradation processes gives us an opportunity to optimize the parameters of different soils for tolerant value of risk of origination of the unfavorable state of soil as a result of degradation.

Given the data of observations on changes of soil properties under the influence of anthropogenic and other factors, the offered method will make possible the forecasting of the degradation process with approximation sufficient for practice.

The method offered for forecasting of risk of origination of the unfavorable state of soils is sufficient for reasonable conclusion on the expediency of soil use for different purposes.

Growing rates of soil degradation under different influences require the fast formation of new areas of scientific knowledge on soil degradation, which, hopefully, will become a one of the basic fields of soil science.

The authors are aware that for the offered method the following fact is important. Since this method has no restrictions it can be used for the solution of other tasks such as: assessment of ecological sustainability of soils having global biospheric importance, sustainability of soil systems and for other goals.

References

[1] A.G. Bondarev, I.B. Kuznetsov, Problems of degradation of physical properties of soils in Russia and ways of their solution, Eurasian Soil Science, №9, 1999, 1126-1131.

[2] G.B. Dobrovolsky, U.D. Nikitin, Soil function in biosphere and ecosystems, M. Nauka, 1990, 260.

[3] B.A. Zimovets, N.B. Khitrov, G.N. Kochetkova, N.P. Chizhikova, Assessment of degradation of irrigated soils, Eurasian Soil Science, №9, 1998, 1119-1126.

[4] B.A.Zimovets, Ecology and melioration of soils of dry steppe zone, V.V. Dokuchaev Institute of Soils, Moscow, 1991, 249.

[5] D.N. Kereselidze, L.G. Machavariani, B.B. Kalandadze, V.Z. Trapaidze, Permissible rates of Georgian soil losses during erosions, Eurasian Soil Science, №4, 2013, 438-446

[6] M.S. Kuznetsov, G.P. Glazunov, Erosion and soil protection, Moscow, MSU Publishing House, 1991, 334.

[7] A.N. Kolmogorov, On analytical method in the theory of probability, Successes of mathematical sciences, Issue 5, 1938.

[8] I. Likesh, I. Lyacha, Basic tables of mathematical statistics, Finansy i Statistika (Finances and Statistics), Moscow, 1985, 356.

[9] Ts.E. Mirtskhulava, Quantitative assessment of maximum permissible landscape loads, Bulletins of RAS, Geographic Series, Moscow, №3, 2001, 68-74.

[10] Ts.E. Mirtskhulava, Soil degradation and ways of forecasting of undesirable conditions during irrigation, Eurasian Soil Science, № 12, 2001, 1503-1510.

[11] Ts.E. Mirtskhulava, Hazards and risks in some water and other systems, Tbilisi, Nauka (Science), 2003, 540.

[12] Ts.E. Mirtskhulava, Environmental disturbance (forecasting of disturbance risk), Measures on its reduction, Institute of Hydroecology of GNAS, Tbilisi, 1993, 438.

[13] L.L. Sveshnikov, Applied methods of theory of random functions, M. Nauka, 1968, 512.

[14] S.B. Pogrebinsky, V.P. Strelnikov, Designing and reliability of multiprocessor computers, M. Radio i svyaz (Radio and communication), 1998, 168.

[15] N.A. Tolokontsev, Calculation of mean square deviation, Procceedings of the third meeting on application of mathematical methods in biology, Publishing house of Leningrad University, Leningrad, 1961, 83- 93

[16] R.P.C. Morgan, Soil Erosion and Conservation, Blackwell Publishing, USA, 2005, 304.

Important Plant Areas of the Peri-North Tien Shan Subprovince as Perspective Protection Areas of Phytobiodiversity

G. M. Kudabayeva, P. V. Vesselova, M. P. Danilov, B. M. Sultanova

Institute of Botany &Phytointroduction Ministry of Education and Science, Almaty, Kazakhstan

Email address:
kgm_anita@mail.ru (G. M. Kudabayeva), pol_ves@mail.ru (P. V. Vesselova), michaelpetrovich@mail.ru (M. P. Danilov), sultanovab@mail.ru (B. M. Sultanova)

Abstract: This article describes the identification results of Important Plant Areas within the Peri-Tien Shan botanic-geographic sub-province. It was shown that identification of Important Plant Areas for Kazakhstan territory is the crucial tool for conservation of botanic diversity as well as rare, endemic and relic species. The results of the survey and their compliance with the criteria were shown by the example of three among 32 detected important areas. The botanic value of areas and their phytocoenotic diversity were demonstrated. The rare, endemic and relic species of each area were recorded.

Keywords: Important Plant Areas, Peri-North Tien Shan Botanic-Geographic Sub-Province, Floristic, Phytocoenotic Diversity

1. Introduction

Considering the significance of biodiversity conservation, Kazakhstan has signed and ratified global ecological conventions [1,2]. With the aim to comply with the convention obligations, Kazakhstan has developed the National strategy and action plan aimed at conservation and balanced use of biodiversity [3].

In 2002, the Global strategy of plant conservation developed at the 6th Conference of Convention Parties in the Hague was accepted. It was recommended to develop the regional and national strategies of plant conservative, which would consider the specifics of regions.

In Kazakhstan the environment is under active, pressing which affects the number of natural components of biocoenosis or even leads to extinction of certain species. Considering that it is a continuing process, the development of biodiversity conservation measures and first of all, rare flora specimens are of particular actuality.

Kazakhstani species requiring protection (rare, endemic and relic species) are provided in the Red Data Book of Kazakhstan [4]. If their number in the first edition was 308, then in the List of rare and endangered species of fauna and flora their number amounted to more than 387 species. This points to the fact that it is necessary to conduct additional measures aimed at detection of location areas of rare species as well as their plant communities for further possibility of conservation.

The issue of necessity to detect the most Important Plant Areas for the purpose of their further conservation was proclaimed in 1995 at the first pan-European plant conservation conference of Planta Europa in Hyeres (France). Consequently, the criteria for identification of Important Plant Areas (IPA) were developed and the terminology was defined. By 2007, the lists of IPA were prepared for the most countries of Central, East and Southern Europe. In CIS countries this work was done in Belarus (2007), partly for Ukraine and Armenia [7], Uzbekistan.

For the purpose of identification of the most valuable areas, the methods implemented within the framework of the project "Important Plant Areas of Central and East Europe" (2002-2004) were developed. In Russia, the strategy of important areas at the regional level was developed within the framework of the project "Strategy of plant conservation of Altay-Sayan Ecoregion" as well as in separate regions [8, 9].

The purposeful study for identification of Important Plant

Areas within Kazakhstan were not conducted. However, the experience obtained during the preparation of the Regional State Cadastre of plants of certain regions of Kazakhstan (East Kazakhstan, Aktobe, South Kazakhstan, Mangistau, Zhambyl and Kzylorda) and regional Red Data Books is available.

In addition, the detailed up-to-date layout of botanic-geographic zoning [10] was compiled for desert areas of Kazakhstan which in fact can be used further as the basis for creation of IPA network, whereas methodologically it will allow identifying IPA with account of not only floristic diversity but also regularities of composition of vegetation cover.

Overall, the important areas can consequently be involved in the evaluation process of SPNA network and detection of their gaps. However, it should be noted that IPA is not a specific kind of protected area, but is the most important tool for efficiency checking of already existing SPNA within Kazakhstan.

2. Study Area & Methods

Preliminarily conducted works have determined the necessity to develop the project «Important Plant Areas of Kazakhstan - monitoring basis of vegetation status (by the example of the Peri-North Tien Shan botanic-geographic subprovince)» aimed at possibility to detect areas characterized by high botanic value.

The selection of piedmont plains of Tien Shan as a model area was stipulated by a number of factors. First of all, it should be noted that these areas represent ecotone (transitional) cross-pollinated strip whose floristic composition combines the elements of desert plains and mountain ecosystems. In addition, piedmont plains of subprovince are densely populated and are intensely used in agriculture (plough land, pasture) and industry. As a result of the combined negative impact of man-made factors on natural ecosystems and first of all, on their vegetation component it was recorded that the composition and structure of vegetation cover have simplified across all areas. Namely, at these areas, it is necessary to detect and conserve natural areas having a high degree of biodiversity. The important argument for the study of piedmont areas consists in unique species composition having a high degree of endemism as well as availability of relics and threatened species.

Given that regional ecosystem of the Peri-North Tien Shan subprovince is intensely impacted, this area is covered insufficiently with the network of Specially Protected Natural Areas (SPNA) while in hard-to-reach mountains this network in the form of nature reserves and national parks is dense enough. Our work aimed at the study of IPA may fill in the gap in regard to environmental protection. Certain areas are perspective for establishment of new SPNA.

Survey targets are Important Plant Areas of piedmont plains included in the composition of the Peri-North Tien Shan botanic-geographic subprovince within piedmont plains of the North Tien Shan. Important plant area means natural or semi-natural area with high botanic diversity and (or) area which maintains the unique community of rare, threatened species and (or) endemic species of plants and (or) plant community of high botanic value [6].

Survey methods: common floristic, geobotanic, cartographic.

During the identification of Important Plant Areas, the developed criteria were used:

A – availability of threatened species within the area:

A(i) — plant species, under the threat of global extinction;

A(ii) — plant species under the threat of extinction in Europe;

A(iii) — endemic endangered species listed in the National Red Data Book and not included in A(i) or A(ii);

A(iv) — endangered subendemic narrowly-areal species not included in A(i) or A(ii) and spread in adjacent countries;

A(v) — rare, endangered and protected species.

B - overall high species richness of flora or richness in plant species having any particular importance;

C – availability of endangered habitats in the area.

3. Results &Discussion

Overall, 32 IPAs were identified within the Peri-North Tien Shan botanic-geographic subprovince of the total area 13321 ha. The criteria developed by EUNIS and adapted for Altay-Sayan Ecoregion (Russia) were used during the identification of Important Plant Area. The area of identified plots in each case is unique taking into account specific conditions.

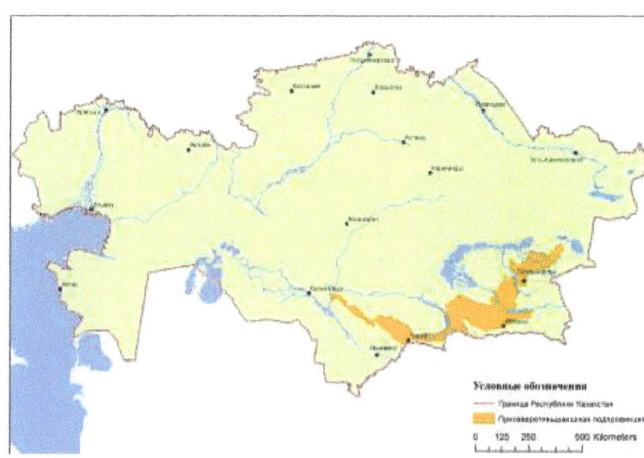

Figure 1. *Layout of Peri-North Tien Shan botanic-geographic subprovince*

The area of the Peri-North Tien Shan botanic-geogrpahic subprovince occupies piedmont plains of Tien Shan (piedmonts of Zailiyskiy Alatau, Chu-Iliyskiy mountains, Dzhungarian Alatau, Karatau).

Among the most important identified botanic areas are Tyrnakty mountain area (Chu-Iliyskiy mountains), Azat (Zailiyskiy Alatau), Samsy (Zailiyskiy Alatau piedmonts).

3.1. Important Plant Area – Tyrnakty (Chu-Iliyskiy mountains)

Location: Almaty region, Zhambyl district. Coordinates: N 43°52'13,4" E 075°21'53,5". Absolute elevation 1070-1092 m. Area – 371 ha.

Proposed Tyrnakty IPA refers to the piedmont plain of Chu-Iliyskiy mountains. Site elevation ranges within 1070-1092 m above MSL. Terrain relief is gently-rugged plain with intersteeply-sloping depressions intersected by the branches of temporary water channels. Water regime is automorphous and semi-terrestrial characterized by seasonal atmospheric and additional surface moistening of the area.

According to botanic-geographic zoning [10], vegetation cover of Tyrnakty IPA is referred to Peri-North Tien Shan piedmont subprovince, Dzhungarian-North Tien Shan province, Iran-Turan subregion, Sakhara-Gobi desert area. V.N. Khramtzov [11] has referred the Chu-Iliyskiy mountains to botanic-geographic type of the Peri-North Tien Shan semi-shrubs (*Artemisia heptapotamica, A. sublessingiana*) deserts with grass plants (*Stipa richterana, S. sareptana*). The intrinsic components of these phytocoenoses are ephermers (*Alyssum turkestanicum, Taeniatherum crinitum, Anisantha tectorum, Eremopyrum orientale, Trigonella arcuata*) and ephemeroids (*Poa bulbosa, Gagea olgae, Eremurus inderiense, Tulipa kolpakovskiana*). Floristic diversity of Tyrnakty IPA is represented by 52 species of vascular plants.

In the allocated site *Niedzwedzkia semiretschenskia* B. Fedtsch., which is the only representative of the family *Bignoniaceae* Juss. in natural flora of Kazakhstan, was noted. *Niedzwedzkia semiretschenskia* is the rarest endemic, highly decorative species of monotype order growing only in the South Kazakhstan in the east part of Chu-Iliyskiy mountains.

Figure 2. *General view of Tyrnakty IPA*

Figure 3. *Flowering specimens of Niedzwedzkia semiretschenskia B. Fedtsch.*

Figure 4. *Fruit-bearing specimens of Niedzwedzkia semiretschenskia* B. Fedtsch.

The species is included in the Red Data Book [4] of Kazakhstan. As to its taxonomic status, there are different opinions. According to Grierson understanding, this species is referred to *Incarvillea* Juss. order – *Incarvillea semiretschenskia* (B. Fedtsch.) Grierson [12]. However, in the survey work of V.V. Fisyun [13], based on comparative analysis of fruit structure, endosperm development, pollen morphology, availability of sticky glands, it can be concluded with certainty that this species is regarded as self-consistent order *Niedzwedzkia,* which is of high priority and complies with original description of B. Fedchenko. Uniqueness of morphological features as reproductive organs (fruit – six-winged ligneous capsule) as well as a life form (plant with multiple stems lignifying at the bottom), geographic hiatus with other family representatives (Himalaya and Pamir-Altay) are indicative of long-term duration of establishment period of this relic taxon. According to evidence of Kamelin R.V. [14] «there is no such possibility to date precisely its age, but there is no doubt that its age is significant, but not younger than Miocene age and it is most likely older». In addition, he thinks that "mesophilic and thermophilic character of the flora of their arising and long-term evolution in-situ in conditions of low competence and increasing xerophilization" contributed significantly to the establishment of this species. At present, this species has legal conservation status embodied in the List of rare and threatened species of flora and fauna [5], in the materials of the Republican Red Data Book [4]. However, the areal of this species does not fall within the protected area (although it is located nearby) of historic-cultural and nature reserve-museum "Tanbalytas" (included in the list of UNESCO World Heritage site, whereas the main function is to conserve archeological monument and petroglyphs [15]. The species grows at sloping, stony-rubbly hillsides, sloping tops located within the strip of intensified pasturing thus causing real threat to conservation of relic species. The data of I.I. Kokoreva et al. [16] are indicative of significant reduction of floristic and phytocoenotic diversity of communities with *Niedzwedzkia semiretschenskia* due to increase of anthropogenic impact.

In addition, two species listed in the Red Data Book were recorded within proposed IPA: *Juno kuschakewitsczii, Juno caerulea.* Likewise, three species of tulips were encountered within Tyrnakty IPA [4, 15, 16, 17, 18,] listed in the Red Data Book of Kazakhstan. Among them are Albert tulip (*Tulipa alberti*) - rare, endemic species, Greig tulip (*Tulipa greigii*) - rare, endemic species with contracting areal and Regel tulip (*Tulipa regelii*) - rare, threatened, endemic species.

Concentration of rare, relic and endemic species included in the National Red Data Book allows to identify this area as an Important Plant Area. It is necessary to carry out the protective measures as the loss of such unique botanic objects is irreplaceable.

3.2. Important Plant Area - Azat (Piedmont and North Part of High Foothills of Zailiyskiy Alatau Northeastwards of Talgar Town)

Location: Almaty region, Talgar district. Coordinates: N 43°20'33.4" and E 077°16'36.4". Absolute elevation – 1044 m. Area – 112 ha.

IPA is located at the transition boundary North Tien Shan and Peri-North Tien Shan botanic-geographic subprovinces. Elevation point ranges within 1100-900 m above the sea level. The landscape is a deeply cut valley of the temporary waterway of the eastern slope of the high foothills, that adjoins to the alluvial loss into the piedmont plain.

Figure 5. *General view of Azat IPA*

Vegetation cover is quite diverse [19]. At the bottom of ravines and at the lower parts of north slopes, the brushwood of high shrubs with parts of wild fruit forests are spread. The structure of tree and shrubbery vegetation includes high shrubs and low trees such as *Berberis sphaerocarpa, Crataegus songarica, Malus sieversii, Frangula alnus.* The lower tier is occupied by *Lonicera altmannii, Rosa alberti, Spiraea hypericifolia, Euonymus semenovii, Rubus idaeus.* The grass tier is represented by *Aegopodium tadshikorum, Agrimonia asiatica, Artemisia dracunculus, Brachypodium sylvaticum, Dictamnus angustifolius, Eremurus robustus, Euphorbia lamprocarpa, Festuca gigantea, Geum urbanum, Impatiens parviflora, Leonurus turkestanicus, Ligularia heterophylla, Origanum vulgare, Poa nemoralis* etc. *Iris sogdiana* is often encountered at the bottom of ravines. Species abundance of tree and shrubbery communities often exceeds 50 species per 100 m^2.

The semi-dominant dry meadows and meadow steppes are common on north slopes and forest fringes of tree and shrubbery vegetation, including *Brachypodium pinnatum, Bromopsis inermis, Dactylis glomerata, Elymus tianschanigenus, Phleum phleoides, Poa angustifolia.* Such species are intrinsic among the representatives of wild grasses as *Achillea millefolium, Betonica foliosa, Bunium setaceum, Hypericum perforatum, Inula macrophylla, Paeonia anomala, Potentilla impolita, Vicia tenuifolia etc.*

Shrub and meadow-mesophilic communities are surrounded by ephemeroid-grassy-sagebrush communities at loessial slopes of ridgy low-hill terrains. Along with the representatives of desertified and dry steppes (*Artemisia sublessingiana, Poa bulbosa, Festuca valesiaca, Kochia prostrata, Stipa sareptana*), including associated ephemeroids (*Crocus alatavicus, Gentiana olivieri, Holosteum umbellatum, Geranium transversale, Iridodictyum kolpakowskianum* and etc.), large ephemeroids and wild grasses can be encountered in these communities which are intrinsic to semi-savannas of southernmost regions of Central Asia such as *Crambe kotschyana, Elytrigia trichophora, Eremurus tianschanicus, Scabiosa songarica* and etc. Xerophilic shrubs like *Cerasus tianschanica, Ephedra equisetina, Rosa platyacantha* can be encountered by south slopes.

The botanic value of the area consists in big phytocoenotic diversity. Among vegetation communities there dominates phytocoenosis with high level of floristic richness. Flora includes endemic species of Zailiyskiy Alatau included in the National Red Data Book [4]: *Atraphaxis muschketovii, Erysimum croceum, Tulipa ostrowskiana.* Tertiary relicts of broad-leaved forests are present in tree and shrubbery vegetation which are common with South Siberia taiga [10] – *Brachypodium sylvaticum, Festuca gigantea, Geranium robertianum.*

At Azat IPA, 109 species of higher vascular plants were recorded among which there are 16 species of trees and shrubs. The representatives of forests, shrubs, meadows, deserts and Middle Asia savanna-growing species are present in ecological spectrum.

As consistent with the criterion *A(iii)*, ten threatened species listed in the Red Data Book of Kazakhstan were recorded within IPA [4]: *Armeniaca vulgaris, Atraphaxis muschketovii, Malus sieversii, Crocus alatavicus, Erysimum croceum, Gymnospermium altaicum, Iridodictyum kolpakowskianum, Tulipa kolpakowskiana, Tulipa ostrowskiana.*

Figure 6. *Atraphaxis muschketovii*

Figure 7. *Malus sieversii*

The high level of floristic richness with significant participation of rare and endemic species (criteria B) is intrinsic to IPA. Flora include a wide range of botanic-geographic elements. A big diversity of multispecies vegetation communities was recorded at loesses and boulder-cobble deposits of high and low piedmonts within the studied area (criteria C).

The area is frequently used for pasturing. The main part of vegetation cover is assessed as background. However, the ecosystem undergoes a significant transformation. The alien to local flora species like *Ulmus pumila* and *Ulmus laevis* were encountered among brushwood of trees and shrubs. The brushwood of quarantine weeds *Pseudosophora alopecuroides* occupy large areas.

The state of rare and threatened species is satisfactory. The degree of anthropogenic impact varies from weak to strong.

Unfortunately, the piedmonts of Almaty region are protected improperly by the network of Specially Protected Natural Areas (SPNA) - reserves and SNNP. SPNA occupies the main areas in middle and high altitudes. There is an acute need to protect vegetation cover of piedmonts by including them in SPNA or by the creation of a specific botanic nature reserve that may include such areas as Azat IPA.

3.3. Important Plant Area - the Samsy River Valley (Piedmonts of Zailiyskiy Alatau)

Location: Almaty region, Zhambyl district. Coordinates: N 43°15'12.1" and E 076°06'28.3". Absolute elevation – 888 m. Area – 74 ha.

Samsy IPA is located within piedmonts of the west part of Zailiyskiy Alatau. Elevation points range within 1000-800 m above the sea level.

Terrain relief is low mountain, hummocky-steeply-sloping, dissected by the channels of temporary water streams and complicated by ridges with outcrops of conglomerates of boulder-cobble deposits. The landscape of hummocky piedmonts smoothly runs into sloping piedmont plain.

The ephemeroid-sagebrush vegetation communities such as *Artemisia sublessingiana, Carex pachystylis, Poa bulbosa, Festuca valesiaca, Kochia prostrata, Stipa sareptana* intrinsic to piedmonts of Zailiyskiy Alatau are prevalent at the area. At the beginning of vegetation period and in the first half of summer, the numerous ephemers and ephemeroids such as *Anemone gortschakowii, Ceratocephala testiculata, Crocus alatavicus, Eremurus altaicus, Gagea ova, Gentiana olivieri, Holosteum umbellatum, Geranium transversale, Iridodictyum kolpakowskianum* and etc. play a significant role in vegetation structure. The sparse petrophyte grouping including representatives of *Juno* order is encountered at the outcrops of wind-blown conglomerates. Along temporary water courses and river banks there are brushwood of shrubs and meadows. High altitude-zonal location contributes to a big diversity of ephemers and ephemeroids in vegetation cover.

Floristic diversity at Samsy IPA includes 48 species of higher vascular plants whose majority consists of ephemers and ephemeroids adapted to Mediterranean regime of atmospheric precipitations with their maximum in spring.

Phytocoenotic structure and flora composition of IPA is in large measure determined by the composition of underlying mountain rocks - boulder-cobble conglomerates cemented by carbonate mortar which can be accounted for encountering of *Juno almaatensis*, species which was earlier regarded as extinct.

Juno almaatensis was first described byacademician N.V. Pavlov. However, the available herbarium specimens did not retain the color of the crown - one of the main systematic feature of taxon. A.I. Vvedenskiy was doubtful of this species as to its pertaining to individual order, considering that it may represent the diversity of another junona species. However, during a number of years, the specimens of this species were not encountered at the place where they first had been described. The taxon was regarded as lost in the area of their natural habitat. In 2009, A.A. Ivaschenko [20] found the specimens having the intrinsic green color of the crown near v. Samsy. These specimens were found at geographically another places, although ecological conditions of growth were similar to typical. During field surveys aimed at identification of IPA, not only the specimens *Juno almaatensis* were again collected in the outskirts of v. Samsy but also the area of taxon growth was expanded.

When analyzing species of Junona order of North Tien Shan, I.I. Kokoreva [16] has traced a number of morphological consequently changing transition forms having a certain similarity with the description of *Juno almaatensis*. It is necessary to continue studying this taxon in order to identify its systematic status.

Figure 8. *General view of Samsy IPA*

Figure 9. *Juno almaatensis*

One of substantiations for IPA identification was growth of other rare flora species. Five threatened endemic species listed in the National Red Data Books were recorded within the territory of proposed Samsy IPA (criteria *A(iii)* [4]: *Crocus alatavicus*, *Iridodictyum kolpakowskianum*, *Juno almaatensis*, *Juno kuschakewiczii*, *Tulipa kolpakowskiana*. It should be pointed out to exclusive participation of ephemers and ephemeroids (criteria B) in flora richness. The outcrops of conglomerates and their wind-blown rock slides and alluvial deposits with groupings of petrophytes are of certain interest.

The territory is often used for pasturing. The main part of vegetation cover of the territory is assessed as background with spots of minor degree disturbance. Among the types of anthropogenic impact are pasturing, roads, steppe fires. The degree of anthropogenic impact is transitional ranging from minor to moderate.

On the basis of identified IPA, the botanic nature reserve should be arranged in order to protect ephemers and ephemeroids, particularly for narrowly distributed endemic species *Juno almaatensis*.

4. Conclusion

Considering the vastness of Kazakhstan territory as well as necessity to carry out the inventory works for localization of botanic diversity, the work implemented for identification of Important Plant Areas within Peri-North Tien Shan botanic-geographic subprovince is the initial phase for creation of IPA network. In the following, the database of Important Plant Areas representing reserve fund of the Republic gene pool will be created.

Identification of Important Plant Areas within Kazakhstan territory is the crucial tool for conservation of botanic diversity as well as rare, endemic and relic species. Selection of piedmont areas (taking into account peculiarities of floristic composition and man-made impact in the region) as an object for IPA identification is topical for the possibility to make decisions on their conservation.

Currently, 32 IPAs identified within Peri-North Tien Shan botanic-geographic subprovince are not included in the system of Specially Protected Natural Areas (SPNA) of Kazakhstan. However, they are characterized by a high degree of species diversity and species, both listed in the Red Data Book of Kazakhstan and endemic species grow in these areas.

Proposed areas (Tyrnakty, Azat, Samsy) are characterized by the richness of species composition. The flora composition includes unique relic species as well as rare, narrowly-endemic species. For the purpose of conservation of unique objects, part of surveyed areas were recommended to be put under protection status.

The results of works aimed at identification of Important Plant Areas within Kazakhstan territory were approved at 7th Conference Planta Europa «Plants for people, people for plants» (Crete, 2014).

References

[1] Convention on Biological Diversity (in Russian) (1995). Switzerland, 34 p.

[2] UN Convention to Combat Desertification (1994). Paris, 66 p.

[3] National Strategy and Action Plan on conservation and balanced use of biodiversity (1999). Almaty, 336 p.

[4] KazSSR Red Data Book (1981). Alma-Ata, V. 2, 263 p.

[5] Resolution of the RoK Government. On approval of the List of rare and threatened species of flora and fauna: approved as of 31st October, 2006, No.1034.

[6] Sh. Anderson, Identification of Important Plant Areas. Guidelines on selection of IPAs in Europe and development basis of these rules for other world regions (2003). M., 39 p.

[7] I.I. Artemov, A.Yu. Korolyuk, N.N. Laschinskiy, I.E. Smelyanskiy, Identification criteria for Important Plant Areas in Altay-Sayan ecoregion (2007). Novosibirsk: Siberian Ecological center, 106 p.

[8] Important Plant Areas of Kemerovo region (2009). Kemerovo, 112 p.

[9] I.A. Boriskin, O.D. Chernova, E.A. Bondarevich, Identification of Important Plant Areas of south-east of Zabaikalskiykrai (2011). // Proceedings of 2nd International Internet Conference. Kazan, pp. 22-27.

[10] Botanic geography of Kazakhstan and Central Asia (within desert area) (2003). St.Petersburg, 423 p.

[11] V.N. Khramtzov, Piedmont deserts // Botanic geography of Kazakhstan and Central Asia (within desert area) (2003). St.Petersburg, pp. 157-166.

[12] Rare and threatened species of USSR Flora requiring protection (1981). L.: Science, 262 p.

[13] V.V. Fisyun, Materials on systematics and ecology of *Niedzwedzkiasemirechenskaya* // Botanic herbaria materials of AS KazSSR Institute of Botany (1982). Alma-Ata, No.12, pp. 49-57.

[14] R.V. Kamelin, On generic endemism of Central Asia Flora (1965) // Botanic magazine, V. 50, No. 12, p. 1708.

[15] O.V. Marynich, S.A. Nigmatova, Valuable botanic objects of historic-cultural and nature reserve-museum "Tamgaly" (2005) // Botanic surveys in Kazakh Altay. Ridder, pp. 145-148.

[16] I.I. Kokoreva, I.G. Otradnykh, I.A. Syedina, V.V. Lysenko, Rare plant species of North Tien Shan (2013). Almaty, 208 p.

[17] Kazakhstan Flora (1956-1966). Alma-Ata, V. 1-9.

[18] A.A. Ivaschenko, Kazakhstan tulips (2005). Almaty, 192 p.

[19] P.V. Veselova, G.M. Kudabayeva, S.K. Mukhtubayeva, Peculiarities of floristic composition of communities of piedmont Peri-North Tien Shan deserts (2012)// Vegetable world and its protection. Almaty, pp.31-34.

[20] A.A. Ivaschenko, O.V. Belyalov, New finding of Junonaalmatinskaya (*Juno almatensis*Pavl.) (2009)// Terra magazine, No. 1(9), pp. 58-63/

Brownian Motion and Coagulation Process

Djamel Ghernaout[1, 2, *], Abdulaziz Ibraheem Al-Ghonamy[2], Ahmed Boucherit[1], Badiaa Ghernaout[3], Mohamed Wahib Naceur[1], Noureddine Ait Messaoudene[2], Mohamed Aichouni[2], Ammar Abdallah Mahjoubi[2], Noureddine Ali Elboughdiri[2]

[1]Department of Chemical Engineering, University of Blida, Blida, Algeria

[2]Binladin Research Chair on Quality and Productivity Improvement in the Construction Industry, College of Engineering, University of Hail, Ha'il, Saudi Arabia

[3]Department of Mechanical Engineering, University of Laghouat, Laghouat, Algeria

Email address:

djamel_andalus@hotmail.com

Abstract: This review discusses the Brownian motion and coagulation/flocculation (C/F) in water/wastewater treatment. In water/wastewater treatment processes, pertinent questions relating to Brownian motion and C/F are often asked. Some of these questions are: Brownian motion and molecular agitation are favourable or not to separation processes? As high salinity (seawater) decreases disorder, increasing surface water salinity would be a convenient water treatment process or not? The processes of C/F are used to remove dissolved substances and colloids from water in order to assure efficient settling.

Keywords: Colloid, Brownian Motion, Coagulation, Flocculation, Water/Wastewater Treatment

1. Introduction

In water treatment engineering, some questions are asked: Brownian motion and molecular agitation are favourable or not to separation processes? After coagulation/flocculation (C/F) and settling, two separate phases (limpid water + mud) are present, disorder is minimised or not? Freezing decreases disorder or not? Freezing treats or not water? High salinity (seawater) decreases disorder or not? Increasing surface water salinity would be a convenient water treatment process or not? In other words, this review tries to find links between Brownian motion and coagulation.

The processes of C/F are used to remove dissolved substances and colloids from water in order to assure efficient settling [1,2].

2. Colloids

Colloidal particles are defined as aggregates of atoms and/or molecules; their density is near to water density (~1) and their diameters are small enough that gravity is not able to settle them. Consequently, they remain in suspension [3]; this is why they are called stable dispersions [4]. Their stability originates from the reciprocal repulsion between colloids [5]. However, their stability may be disturbed by applying some chemical products. Coagulation is defined as the key unit process where such reactants are injected in order to destabilise the colloids repulsion, thus pushing them to form bonds together [6]. This chemical process is usually encountered before the unit operation of flocculation [7]. The colloids are frequently responsible of the turbidity and sometimes of the colour that make water undrinkable; consequently, these fine particles should be completely eliminated from water [8].

This section reviews some prerequisite topics, which are indispensable in the comprehension of C/F process, including behaviour and stability of colloids and Zeta potential [8,9].

2.1. Colloid Behaviour

In its great fraction, the suspended solids in surface waters are constituted with materials, such as silica, with density of 2.65 [10]. As their sizes are ranged from 0.1 to 2 mm, they may be easily removed from water by settling. On the other hand, when their diameter is less than 10^{-5} mm (10 μm), they require sometimes a year to settle only 1 mm. Moreover, when a particle is in this size range, we may assist to the appearance of the turbidity and/or colour of water, rendering the water unpotable. The elimination of colloids by decantation is always applicable in the case if their settling velocity is relatively rapid, i.e., some meters per hour. For these reasons, C/F process acts by disturbing the reciprocal repulsions between colloids pushing them to form bonds together and constitute agglomerates for efficient decantation [6]. Colloids may be classified in the diameter interval of nm to µm. Colloids are considered as agglomerates of some hundreds of atoms and/or molecules, even if a distinguished molecule, e.g., in the case of proteins, is big enough to be classified as a colloidal particle [4,8].

A colloidal suspension is constituted with two phases: (1) the *dispersed phase*, or the *solute*, and (2) the *dispersion medium*, or the *solvent* [4]. Each phase can possess all the three states of matter, i.e., solid, liquid, and gas. As an illustration, we may have the dispersion medium as a liquid, and the dispersed phase as a solid. This suspension is known as a *liquid sol*. Sometimes, the dispersion medium is a gas and the dispersed phase is a solid. This system is known a *gaseous sol*, and the best illustrations are dust and smoke. Table 1 presents the various types of colloidal constitutions. In the C/F process, in the case of water/wastewater, attention will be accorded to the solid being dissociated in water [6,11]. Sols may be classified as *lyophilic* or *lyophobic*; the first ones are those that constitute bonds with the solvent and the second ones are those that do not constitute bonds with the solvent. If the solvent is water, lyophilic and lyophobic sols are, respectively, named *hydrophilic* and *hydrophobic sols*. The affinity of the *hydrophilic sols* for water is attributed to polar functional groups that are found on their surfaces. These groups include such polar groups as $-OH$, $-COOH$, and $-NH_2$. They are, respectively, named the *hydroxyl, carboxylic*, and *amine* groups. As established, the functional polar groups are found sticking out from the surface of the particle. Due to the affinity of these groups for water, the water is attached on the surface. This water is named *fixed water* and is attached on the surface and moves with the colloid. On the other hand, the *hydrophobic colloidal particles* do not have affinity for water; consequently, they do not contain any attached water. Habitually, inorganic colloidal particles are hydrophobic, and organic ones are hydrophilic [4]. As an illustration of an inorganic colloid is the case of clay colloids that produce turbidity in water, and an illustration of an organic colloidal particle is the colloids in domestic sewage [8].

Table 1. *Types of colloids [8].*

Dispersion Medium	Dispersed Phase	Common Name	Example
Solid	Solid	Solid sol	Coloured glass and gems, some alloys
Solid	Liquid	Solid emulsion	Jelly, gel, opal (SiO_2 and H_2O), pearl ($CaCO_3$ and H_2O)
Solid	Gas	None	Pumice, floating soap
Liquid	Solid	Liquid sol	Turbidity in water, starch suspension, ink, paint, milk of magnesia
Liquid	Gas	Foam	Whipped cream, beaten egg whites
Gas	Solid	Gaseous sol	Dust, smoke
Gas	Liquid	Gaseous emulsion	Mist, fog, cloud, spray
Gas	Gas	Not applicable	None

2.2. Zeta Potential

The repulsive property of colloids generates from the electrical forces that they possess [4]. At a near point from the colloid surface, the force is significant. It decreases near to zero at a high gap from the colloid. The electrical forces are generated because of the charges that the colloids have at their surfaces. These charges, named *primary charges,* are generated following one or both of two processes: (1) the rupture of the polar groups and (2) preferential adsorption of ions from water. The primary charges on hydrophobic colloids are attributed to preferential adsorption of ions from water [8].

The primary charges, on hydrophilic colloidal particles, are mainly attributed to the polar groups, e.g., the carboxylic and amine groups [4]. The phenomenon monitoring the charges apparition on these types of colloidal particles is shown in Fig. 1. The symbol R shows the colloidal particle entity. The colloidal particle is illustrated at the top of the drawing, without the influence of pH. By a convenient combination of the H^+ and OH^- being introduced to the suspension, the colloidal particle achieves ionisation of carboxylic and the amine groups. At this point, both ionised groups neutralise each other and the colloid is considered as neutral. This point is well known as the *isoelectric point*, and the related ion of the colloid is named the *zwitter ion*. When pH is augmented by introducing a base, the added OH^- will neutralise the acid extremity of the zwitter ion (i.e., the NH_3^+); the zwitter ion will disappear, and the entire colloidal entity will be negatively charged. The inverse situation is correct if the pH is decreased by the introduction of H^+ ions. The injected H^+ ions neutralise the base extremity of the zwitter ion (the COO^-); the zwitter ion disappears, and the entire colloidal entity will be positively charged. From this interpretation, a hydrophilic colloid can reach a primary charge of either negative or positive as a function of the pH [12]. The first electrical charges on a colloidal particle which, as we have established, could either be classified as positive or negative, attract ions of opposite charges from the suspension. These ions with opposite charges are named *counterions*. This is phenomenon shown in Fig. 2. When the primary charges are enough large, the attracted counterions can constitute a compact layer surrounding the first charges. This layer is named the *Stern layer*. In turn, the counterions have the possibility to attract their own counterions, the *coions* of the primary charges, constituting an additional layer. Because these coions constitute a continuous distribution of ions into the bulk of the suspension, they have tendency to be distributed; moreover, they may constitute a diffused layer. The second layer is well known as the *Gouy layer*. Consequently, the *Stern* and *Gouy* layers constitute an envelope of electric double layer surrounding the first charges. The totality of the electric charges in the Stern layer displaces with the colloidal particle; as a result, this layer is an attached layer. In the case of the Gouy layer, a fraction of the layer may displace with the colloidal particle by shearing at a *shear plane*. There is a possibility for this layer to shear off out of the boundary of the attached Stern layer evaluated from the surface of the colloidal particle. Consequently, certain variations of the charges in the layer displace with the colloid, at the same time others do not. This plane is shown in Fig. 2 [8].

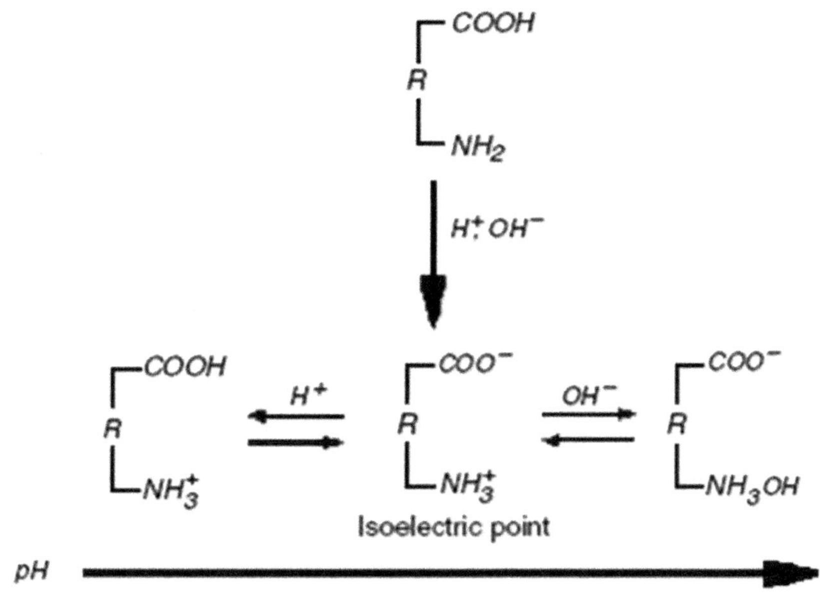

Figure 1. *Initial charges of a hydrophilic colloidal particle depending on pH [8].*

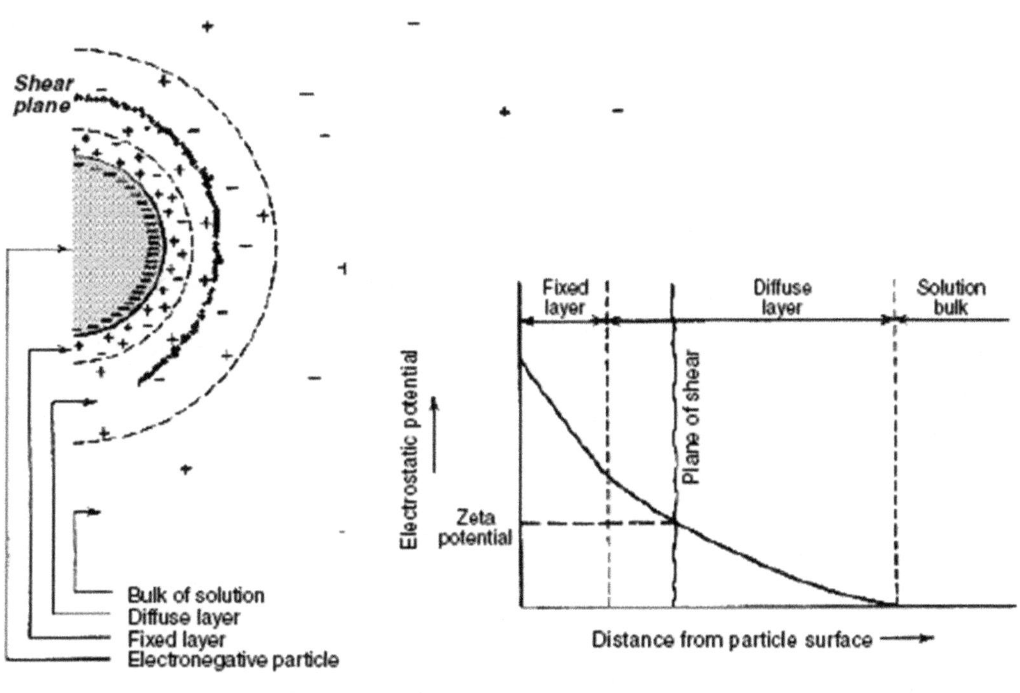

Figure 2. *Charged double layer surrounding a (negatively) charged colloidal particle (left) and change of electrostatic potential as a function of distance from colloid surface (right) [8].*

The electric charges possess an electrostatic potential. As shown on the right-hand side of Fig. 2, this potential is most important at the colloid surface and decreases to zero at the bulk of the suspension. The potential at a certain point from the surface at the birth of the shear plane is named the *Zeta potential* [4]. The greater this potential, the more important is the repulsive force and the more stable the colloidal particle [8].

3. Collision Mechanisms

Colloids are brought into contact following certain manners:

- Brownian diffusion (i.e., *perikinetic* agglomeration).
- Fluid motion (i.e., *orthokinetic* agglomeration).
- Differential sedimentation [8,13,14].

These mechanisms are illustrated in Fig. 3. As seen above, all particles in water are subject to random motion as a consequence of their thermal energy. This phenomenon is well known as *Brownian motion.* Due to this fact, collisions between colloids produce *perikinetic aggregation* [15]. It is easy to evaluate the frequency of the created collisions [16]. It is evident that Brownian (perikinetic) agglomeration has not the capacity to conduct to the formation of large aggregates. This is due to the decrease in particle content and the second-order nature of the process. Practically, flocculation processes are frequently performed under conditions where the suspension is under the action of certain form of *shear*, i.e., by stirring or flow [17]. When colloids are transported thanks to suspension, they may produce a great influence on the rate of colloid collision. This phenomenon is named *orthokinetic agglomeration* [15]. In fact, the primary theoretical concept to this phenomenon was the fruit of Smoluchowski's research, at the same time with his famous study on perikinetic agglomeration [18]. In the case of orthokinetic collisions, Smoluchowski studied the example of spherical particles in *uniform and laminar shear* [16]. Practically, such conditions are not found; however, the simple case constitutes an appropriate basic point. Fig. 4 illustrates the primary model for the Smoluchowski analysis of orthokinetic collision intensities. Two spherical colloids, of various sizes, are fixed in a uniform shear field. This significates that the velocity of the fluid varies linearly as a function of distance in only one direction, perpendicular to the direction of flow. The rate of variation of fluid velocity in the z-direction is du/dz. This is the *shear rate* and is attributed the symbol G. The centre of one colloid, radius a_j, is assumed to be fixed in a plane where the fluid velocity is zero, and colloids above and below this plane displace along fluid streamlines with various velocities, as a function of their situation. A colloid of radius a_i will *just* be in contact with the central sphere if its centre lies on a streamline at a distance $a_i + a_j$ from the plane where $u = 0$ ($a_i + a_j$ is the *collision radius*, as in the analysis of perikinetic agglomeration) [16]. The totality of colloids at points less than the collision radius will enter in collision with the central sphere, at rates that are function of their concentration and position (and consequently velocity) [14].

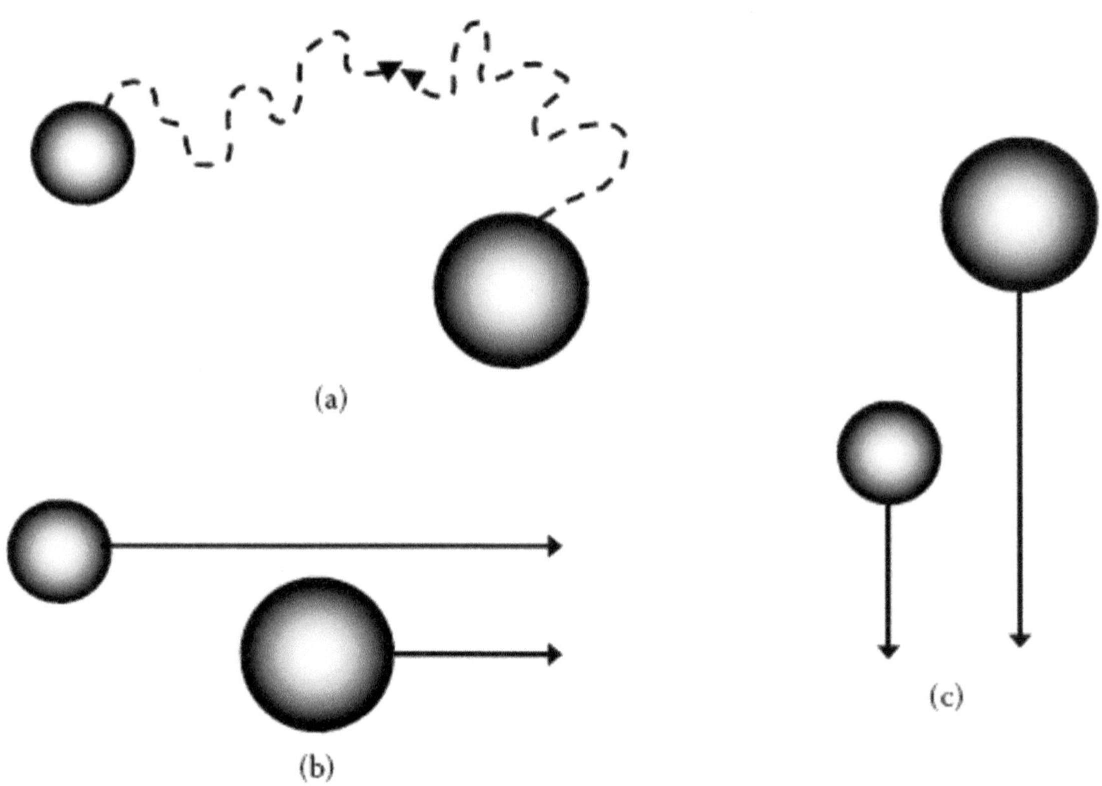

Figure 3*. Particle transport conducting to collisions by (a) Brownian diffusion, (b) fluid motion, and (c) differential sedimentation [14].*

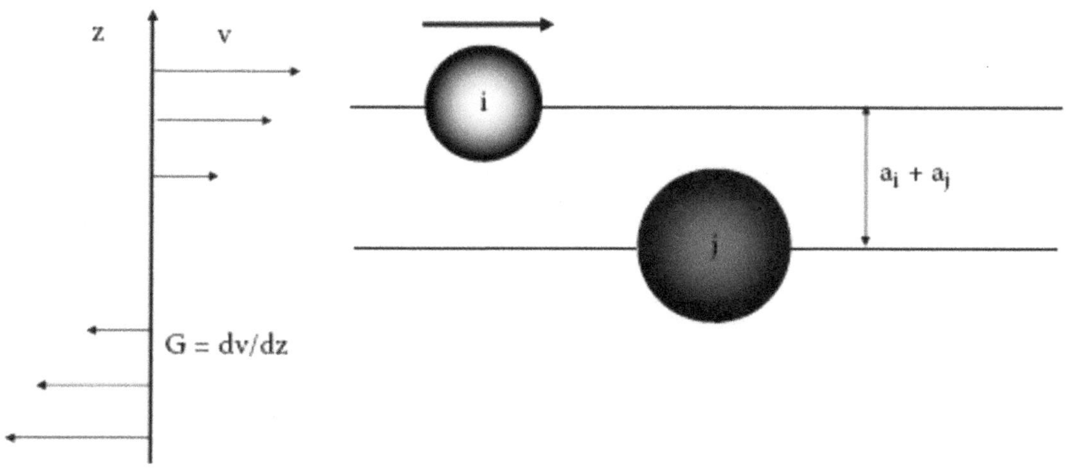

Figure 4. Model for orthokinetic agglomeration in uniform laminar shear [14].

On the other hand, there is an additional significant collision mechanism. This mechanism appears when colloids of different size (or density) are settling in water [15]. Evidently, larger and denser colloids will sediment faster and have the possibility to enter in collision with more slowly settling colloids during their settling [13,14].

4. Coagulation and Flocculation

4.1. Terminology

4.1.1. "Coagulation" and "Flocculation"

This section concerns processes in which suspended colloids constitute bigger agglomerates. These agglomerates can be eliminated with more efficiency by physical separation processes such as decantation and filtration [19-21]. Supposing that the colloids are originally stable, we may have two key steps in the agglomeration process. These fundamental stages are illustrated in Fig. 5:

- Destabilisation of colloids
- Collisions of colloids to build-up agglomerates [14].

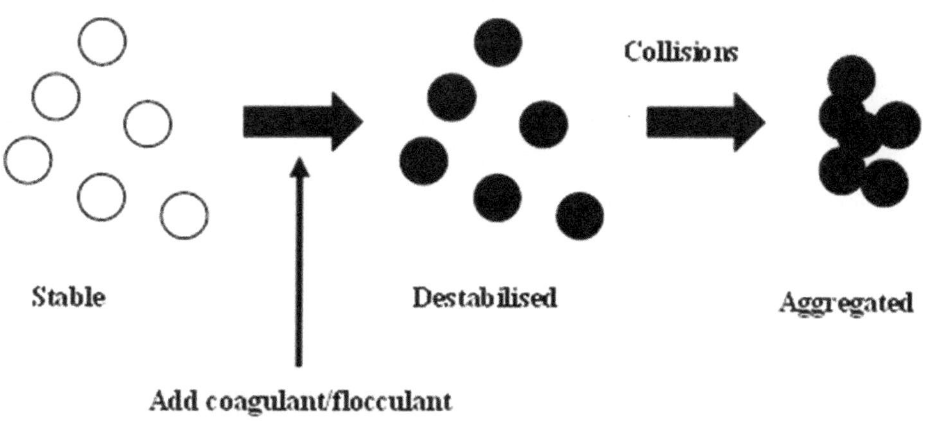

Figure 5. Destabilisation and aggregation of particles [14].

We will discuss the case of stable colloids; their stability is due to their surface charge and consequently electrical double-layer repulsion [22]. In such situation, destabilisation implicates an augmentation in ionic strength or a neutralisation of the colloid charge. Introducing salts to elevate ionic strength is not considered as a convenient choice and additional additives will be utilised, as discussed in the following paragraphs. The destabilisation stage aims to render the *collision efficiency,* α, as great as possible, in ideal conditions with $\alpha = 1$, until the point that every collision conducts to agglomeration [15]. In the case where colloids are completely destabilised, so that $\alpha = 1$, collisions are indispensable if agglomerates have to be constituted [16]. The colloid collision frequency is largely function of the particle content and of the mechanism by which collisions occur. For less concentrated water, where the frequency of the generated collisions may be insignificant, it may be easy for colloids to be completely destabilised but to demonstrate very little aggregation upon acceptable time values [16]. Due to the short-interval characteristic of interactions between particles, it is frequently easy to treat separately the destabilisation and collision phenomena [4]. That is mean that it is usually acceptable to suppose that the frequency of the generated collisions is not disturbed by interactions between colloids [14].

In the following paragraphs, the word "aggregation" will be utilised in a large sense, to significate any phenomenon whereby colloids aggregate together [23]. We may ask the question of other terms, particularly the largely employed words, i.e., *coagulation* and *flocculation* [24]. However, there is no wide acceptation concerning the manner, by which such words have to be utilised. Moreover, we may find at least two largely used conventions [14].

It is frequent to limit the word *coagulation* to situations where colloids are destabilised by salts or by charge neutralisation (CN) and the agglomerates (*coagula*) have tendency to be small and dense [25]. *Flocculation* is consequently limited to the examples in which polymer bridging is the key stage, agglomerates have tendency to become bigger, and more open structurally [26]. Due to the fractal nature of aggregates, it is frequent that bigger structures have tendency to be more open and less dense [27]. Therefore, the differentiation between small, compact coagula and larger, less dense flocs is an unavoidable result of the stronger interparticle binding in the case of polymers, conducting to larger agglomerates [17]. An additional complication is that *flocculation* is, in certain conditions, applied to situations where agglomeration produces in a *secondary minimum* [4]. An additional quite different adoption is frequently utilised in the field of water/wastewater technology [6]. Following this convention, *coagulation* is related to destabilisation, by the injection of convenient chemicals, and *flocculation* is related to the generation of agglomerates, frequently by certain form of fluid movement (i.e., orthokinetic aggregation) [28]. These are related to the two steps illustrated in Fig. 5 and could be considered as chemical and physical characteristics of the agglomeration phenomenon [14].

4.1.2. Destabilising Agents

Due to the first distinction established from *coagulation* and *flocculation*, the chemical products utilised to produce destabilisation of colloids may be named *coagulants* or *flocculants*, as a function of their manner of introduction and mechanism [29]. Consequently, coagulants would be inorganic salts, including those containing specifically adsorbing counterions, and flocculants would be long-chain polymers, which implicate bridging interactions [30]. Even if there are greatly several different types of destabilising chemicals, the wide proportion of those utilised practically are classified in just two categories:

- Hydrolysing metal coagulants
- Polymeric flocculants [31,32].

The type of these chemicals and their mechanism will be reviewed in the following sections [14].

4.2. Hydrolysing Metal Coagulants

The most largely utilised coagulants are based on aluminium and ferric salts, such as aluminium sulphate ("alum") and ferric chloride [33]. Initially, it was assumed that their mechanism was a consequence of the trivalent nature of the metals, producing Al^{3+} and Fe^{3+} ions in suspension, which are supposed to be very efficient in destabilising negatively charged particles [5]. In fact, this is a largely oversimplified consideration due to the fact that trivalent metal ions are easily *hydrolysed* in water, which has an important influence on their action as coagulants [14].

4.2.1. Hydrolysis of Metal Cations

In certain conditions, metal ions in solution are found as simple hydrated cations. This is the situation for alkali metal ions such as Na^+ and K^+. Due to the polar nature of water, such cations are *hydrated* at a certain level, which significates they are surrounded by some water molecules fixed by electrostatic attraction between the positive metal ion and the negative (oxygen) extremities of the water molecules. It is acceptable to resonate in matter of a *primary hydration shell*, where water molecules are directly related to the central metal ion and more strongly fixed water in a secondary hydration shell. In the case of the trivalent metal ions Al^{3+} and Fe^{3+}, it is established that the primary hydration shell is constituted of six water molecules in octahedral coordination (see Fig. 6 (a)). Due to the elevated positive charge on the central metal ion, electrons tend to be drawn in the direction of the metal from the water molecules, and this may conduct to the liberation of H^+ leaving a hydroxyl group fixed and a decreased positive charge for the metal, as illustrated in Fig. 6 (b). Since the phenomenon essentially implicates the splitting of water molecules, it is named *hydrolysis.* Since hydrolysis produces the liberation of H^+ into water, it is strongly dependent of pH. Indeed, elevated pH values favour dissociation and vice versa. Moreover, as each proton is liberated, the reducing positive charge renders additional dissociation more complicated. Consequently, with augmenting pH there is a sequence of hydrolysis equilibria, which can be presented as follows [14]:

$$M^{3+} \rightarrow M(OH)^{2+} \rightarrow M(OH)_2^+ \rightarrow M(OH)_3 \rightarrow M(OH)_4^- \quad (1)$$

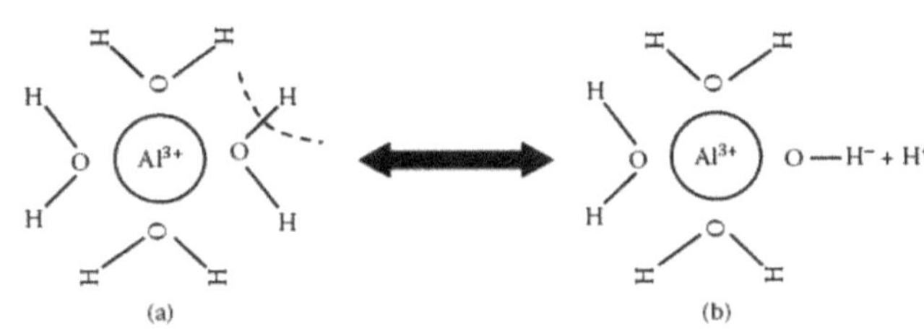

Figure 6. *Hydrolysis of Al^{3+} [(a) hydrated aluminium cation, only 4 of 6 H_2O molecules shown, (b) after release of H^+ to form $Al(OH)^{2+}$] [14].*

For convenience, H_2O molecules, in the hydration envelop, are not presented. Each of the steps in the hydrolysis phenomenon has a convenient equilibrium constant [14]:

$$M^{3+} + H_2O \leftrightarrow M(OH)^{2+} + H^+ \qquad K_1 \qquad (2)$$

$$M(OH)^{2+} + H_2O \leftrightarrow M(OH)_2^+ + H^+ \qquad K_2 \qquad (3)$$

$$M(OH)_2^+ + H_2O \leftrightarrow M(OH)_3 + H^+ \qquad K_3 \qquad (4)$$

$$M(OH)_3 + H_2O \leftrightarrow M(OH)_4^- + H^+ \qquad K_4 \qquad (5)$$

These are defined in the classical method, so that in the situation of K_2, for example:

$$K_2 = \frac{\left[M(OH)_2^+\right]\left[H^+\right]}{\left[M(OH)^{2+}\right]} \qquad (6)$$

where square brackets represent molar concentrations of the different species [14].

For Al and Fe(III), the uncharged hydroxide, $M(OH)_3$, has too poor solubility in water and is possibly to constitute a precipitate upon a certain interval of pH. This precipitation is crucial in the mechanism of hydrolysing metal coagulants. As well as the equilibrium constants listed earlier, a solubility constant for the metal hydroxide is also required, considering the following dissociation of the solid phase, $M(OH)_{3(s)}$:

$$M(OH)_{3(s)} \leftrightarrow M^{3+} + 3OH^- \qquad K_s \tag{7}$$

$$K_s = [M^{3+}][OH^-]^3 \tag{8}$$

If correct equilibrium were achieved, consequently the convenient solubility constants would be those for the stable crystalline forms such as gibbsite and goethite, in the case of Al and Fe, respectively. On the other hand, these are habitually constituted slowly (usually weeks or months). Considering the point of view of coagulation process, it is much more important to deal with the solubility constants (K_{sam}) of the amorphous precipitates that generate originally. However, these values are affected with some uncertainty and only evaluated values can be presented. They are frequently at least 100-fold bigger than values for the corresponding crystalline solids; consequently, the amorphous material is soluble. Table 2 presents certain values for the hydrolysis and solubility constants for Al and Fe(III) species in water at 25°C and at zero ionic strength; consequently, they are convenient for low salt concentrations, typical of many natural waters. The constants are shown in the conventional p*K* form (where $pK = -\log_{10}K$). Using these p*K* values, it is easy to evaluate, as a function of pH, the concentrations of the various dissolved hydrolysis products in equilibrium with the amorphous hydroxide precipitate. Due to uncertainties over the solubility constants for the amorphous precipitates, the results may not be very accurate, but they present a helpful indication of the relative significance of the different species over an interval of pH values. Fig. 7 is a *speciation* diagram illustrating the results of such evaluations for Al and Fe(III), considering data in Table 2. The total content of dissolved species, in equilibrium with the solid form, is really the solubility of the metal at a certain value of pH. From Fig. 7, it is obvious that there is a certain minimum solubility, which appears near neutral pH, for both metallic forms. It must be noted here that the minimum solubility of Fe(III) is greatly lower than that of Al and that the minimum is widely larger. Moreover, it is clear that, for of Al, the anionic form $Al(OH)_4^-$ (aluminate) is the predominant dissolved species near neutral pH. An additional manner of presenting the speciation values is to plot the mole fraction of each species as a function of the total dissolved quantity in equilibrium with the amorphous hydroxide. This has been performed in Fig. 8 for Al and Fe(III). These results illustrate great gaps between these metals. In the case of Al, the dominant species are Al^{3+}, at low pH (~ 4.5) and the aluminate ion, $Al(OH)_4^-$, at pH values more important than ~ 7. The intermediate species make only minor roles at pH values in the interval of ~ 4–6.5. For Fe(III), the different species are found upon a larger pH interval (about 8 units) and each hydrolysis product is predominant at certain pH values. This is the predicted behaviour for hydrolysis of metal ions. The cause that Al species are "squeezed" into a much narrower pH interval is assumed to be the consequence of a transition from octahedral coordination in $Al^{3+}\cdot 6H_2O$ to the tetrahedral $Al(OH)_4^-$. In the situation of Fe(III), octahedral coordination is fixed throughout. Moreover, it is important to indicate that the *soluble,* uncharged $Fe(OH)_3$ is the dominant dissolved Fe species in the pH interval 7–9 (even if the real content is only ~ 2×10^{-8} M). The corresponding Al species, $Al(OH)_3$, is often a minor dissolved component in *relative* terms, although it is at least 10 times more soluble than $Fe(OH)_3$ [14].

Table 2. *Equilibrium constants (pK values) for Al and Fe(III) hydrolysis and solubility of amorphous hydroxides (values for 25°C and zero ionic strength) [14].*

	pK_1	pK_2	pK_3	pK_4	pK_{sam}
Al^{3+}	4.95	5.6	6.7	5.6	31.5
Fe^{3+}	2.2	3.5	6	10	38

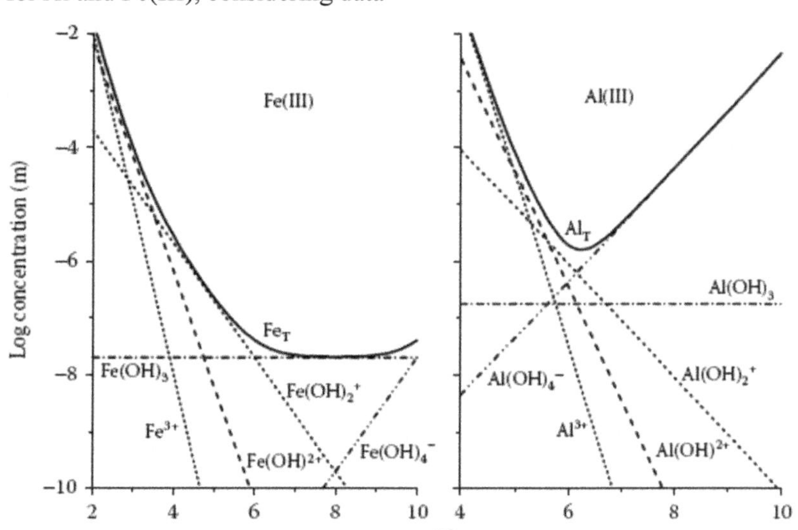

Figure 7. *Speciation diagrams for Fe(III) and Al(III) (only monomeric hydrolysis products indicated) [14].*

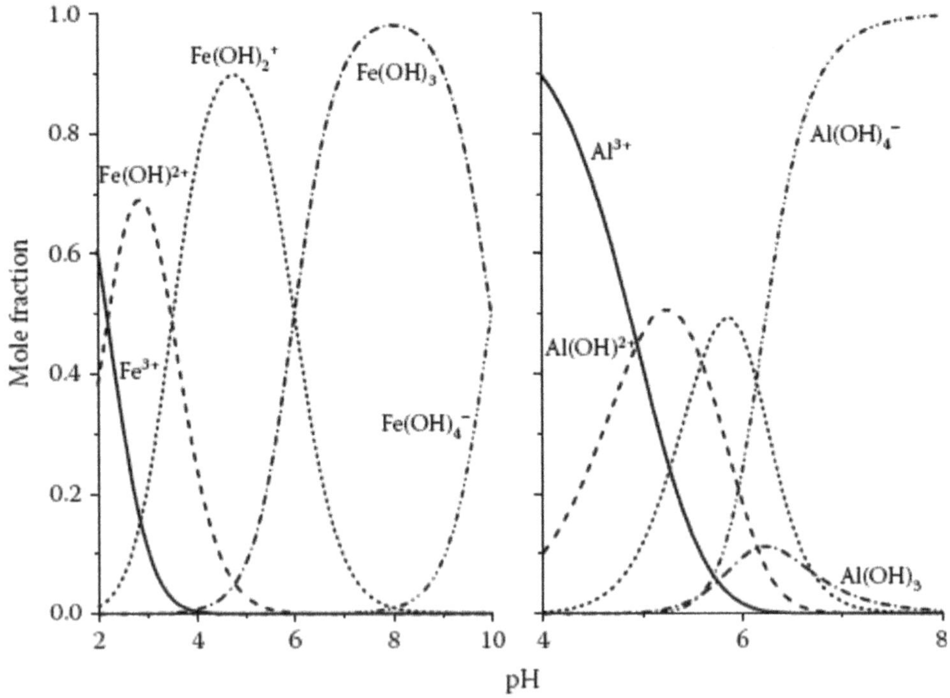

Figure 8. *Proportion (mole fraction) of hydrolysed Fe(III) and Al(III) species relative to total soluble metal concentration [14].*

Our review so far has not taken into consideration some difficulties. One of these difficulties is the action of different anionic species that may affect hydrolysis equilibria. As an example, it is established that fluoride constitutes strong complexes with Al and this generates most important aluminium solubility than would be expected based on the results in Fig. 8. Another remark is that only *monomeric* hydrolysis products have been taken into account, at the same time, under some conditions, *polynuclear* species can be significant. These constitute the subject of the next section [14].

4.2.2. Polynuclear Hydrolysis Products

In addition to the monomeric hydrolysis products discussed previously, there are several possible *polynuclear* forms that may be significant. For Al these include $Al_2(OH)_2^{4+}$ and $Al_3(OH)_4^{5+}$ [25], and there are similar forms for Fe(III) [34]. On the other hand, these may be unimportant at the low contents of the metals habitually utilised in coagulation [35]. Practically, only the monomeric products and the hydroxide precipitate may be significant. Polynuclear hydrolysis forms can be produced in some circumstances. The best encountered of these is $Al_{13}O_4(OH)_{24}^{7+}$ or "Al_{13}," which can be produced by certain neutralisation of aluminium salt solutions or by many different manners [25]. This tridecamer has the so-named keggin structure, consisting of a central tetrahedral AlO_4^{5-} unit surrounded by 12 Al octahedra with shared extremities. The tetrahedral and octahedral Al sites can be without difficulty distinguished in the ^{27}Al NMR spectrum. Under convenient conditions, Al_{13} constitutes rapidly and essentially irreversibly, remaining stable in aqueous solutions for long time. The tridecamer is assumed present largely in the natural aquatic medium (e.g., in acid forest soil water). Using coagulation data, additional polynuclear forms, such as the octamer, $Al_8(OH)_{20}^{4+}$, have been suggested [36]. In fact, there is no clear prove for the octamer and it is not likely to be important practically. There are several commercial chemicals based on prehydrolysed metal salts. For aluminium, a frequent example is the class of materials known as *polyaluminium chloride* (PACl), which may be formed by controlled neutralisation of aluminium chloride solutions [37]. It is possible that several of such chemicals contain significant contents of the tridecamer Al_{13}. For aluminium sulphate, it is hard to produce prehydrolysed species with elevated degrees of neutralisation since sulphate promotes hydroxide precipitation. The presence of small contents of dissolved silica can importantly ameliorate the stability, and the resulting chemical is named *polyaluminosilicate-sulphate* [14]. There are corresponding products containing polymerised iron species [38], even if these are not as largely utilised as PACl [39].

4.2.3. Action of Hydrolysing Coagulants

In fact, there are mainly two significant manners, in which hydrolysing coagulants may destabilise and coagulate particles with negative charge [14]. At low contents and upon appropriate pH values, cationic hydrolysis reactants may adsorb and neutralise the colloid charge, consequently producing destabilisation and coagulation [36]. At more important concentrations of metallic salt, hydroxide precipitation appears and this has a great action — forming the so-called *sweep* flocculation (*SF*) [40].

4.2.4. Charge Neutralisation (CN) by Adsorbed Species

At low contents of metal, only soluble species are detected (see Fig. 7). It is usually assumed that hydrolysed cationic species such as $Al(OH)^{2+}$ are more easily fixed on negative surfaces than the free metal ion and so may efficiently neutralise surface charge. In a general manner, CN with aluminium salts appears at low metal contents, habitually of the order of a few micromoles/L at ~ pH 7. It is known that, for many inorganic suspensions at pH 6, the quantity of Al required to neutralise the surface charge is near 5 μmoles per m^2 of particle surface (~ 130 μg Al per m^2). In fact, even at very low metal contents, the solubility of the amorphous hydroxide may be surmounted. In addition, in the area of neutral pH, cationic hydrolysis reactants represent only a small proportion of the total soluble metal, particularly for Al (Fig. 8). The fact that CN is frequently detected in such examples assumes that the efficient species might be colloidal hydroxide particles [12]. In the case of aluminium hydroxide, the point of zero charge is near pH 8; consequently, the precipitate particles should be positively charged at lower pH values. For ferric hydroxide, the point of zero charge is lower, near pH 7. Moreover, when the bulk solubility is not surmounted, it is likely that certain form of *surface precipitation* will appear because of nucleation at the surface. Really, it is not easy to differentiate between surface precipitation and the fixation of colloidal hydroxide particles that have been precipitated in solution [12]. A combination of these effects may be most possible practically and constitutes the basis of the *precipitation charge neutralisation (PCN)* model, which is shown schematically in Fig. 9. Whatever the precise nature of the charge-neutralising species, they are possible to be apt of charge reversal at higher dosages; in other words, this fact significates that there will be a characteristic *optimum dosage* at which coagulation is most performant [36]. At higher dosages, particles become positively charged and *restabilised.* The optimum dosage must be function of the particle concentration; however, practically the value is frequently low. In some cases, the optimum dosage interval can be limited; consequently, precise dosing control is required. Another disadvantage of relying on CN is that, for low particle concentrations, the collision rate and consequently the agglomeration rate will be low, and long periods may be required to give significantly large flocs [17]. Neutralising surface charge, by small-adsorbed species, does nothing to increase the collision rate, even if, of course, the collision *efficiency* can be importantly improved [14].

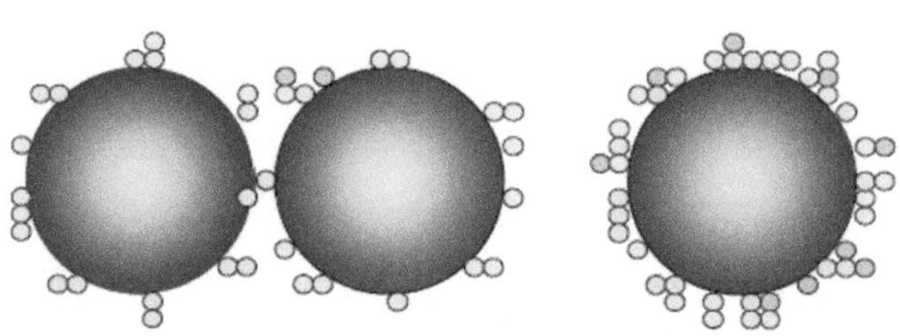

Figure 9. *Precipitation charge neutralisation (PCN) model, showing (a) CN and (b) charge reversal (restabilisation) of particles by precipitated hydroxide colloids [14].*

Some of the advantages announced for prehydrolysed coagulants are assumed to be a consequence of the presence of highly charged cationic species, such as $Al_{13}O_4(OH)_{24}{}^{7+}$. The fact that this ion carries 7 positive charges assumes that it would be very strongly adsorbed on negative surfaces and would be efficient in neutralising particle charge. If we assume that species such as Al_{13} can be more performant in charge neutralisation, it will be difficult to see how, at the optimum dosage, the coagulation rate could be more important than with other adsorbing cationic species [14].

4.2.5. "Sweep" Flocculation

Practically in water/wastewater treatment operations, metal coagulants are injected at dosages much higher than the solubility of the amorphous hydroxide and extensive precipitation appears. For some causes, that are not completely accepted, this can produce much more efficient separation than simple CN. The most possible interpretation is that the initial colloids are fixed, in a certain manner, into the growing hydroxide precipitate and are consequently eliminated from water. This enmeshment of particles is usually assumed as a *"sweeping"* action — consequently the term *"sweep flocculation"* (*SF*) [17]. The hydroxide precipitate could be considered as "bridging" particles together and consequently *"SF"* might be the more convenient term from one point of consideration [17,41]. In addition, in water/wastewater treatment, the formation of large hydroxide agglomerates needs some form of agitation; consequently, orthokinetic collisions are important, and this once again encourages the utilisation of the term "flocculation" [42]. The agglomerates constituted because of hydroxide precipitation are largely known as "flocs" [17]. However, it is confusing that the additives utilised are widely known as "coagulants" [14].

Moreover, *SF* usually conducts to faster aggregation than CN and produces stronger and larger flocs [17]. It is easy to understand the reason for the higher aggregation rate [14]. The generation of a hydroxide precipitate produces a big augmentation in the efficient particle concentration and consequently a most important collision rate, following Smoluchowski theory [15]. Hydroxide precipitates are constituted from large numbers of colloids, which constitute very soon after injection [12]. The agglomeration of these colloids produces low-density flocs, with a relatively big volume [23]. Following the theory of orthokinetic aggregation, the rate is directly proportional to the volume fraction of suspended particles, and this can be largely augmented by hydroxide precipitation [43]. This is the principal explanation why *SF* is so much more efficient than CN [44]. The flocs formed under *"sweep"* conditions are also stronger and consequently grow larger for the same shear conditions [23]. A great advantage of *SF* is the fact that it does not much depend on the nature of the colloids to be eliminated, whether bacteria, clays, oxides, or others [44]. For relatively less concentrated suspensions, the optimum coagulant injection is that which produces the most rapid hydroxide precipitation and is not practically function of the nature and concentration of suspended particles. The large volume associated with hydroxide flocs conducts to an important practical problem — the production of large quantities of sludge that requires to be removed of in certain mannert [23]. In a water treatment plant, most of the sludge generated is in bonds with metal hydroxide rather than the impurities eliminated from water. Even if there is habitually no important restabilisation in the case of *SF* and consequently no sharp optimum dosage region, overdosing is best avoided to limit the volume of sludge generated [37].

The mechanism of prehydrolysed coagulants, such as PACl, at typical injections also very possibly implicates hydroxide precipitation and *SF*, even if this point has not been widely studied [17,37]. There is prove that the nature of the precipitate differs from that produced with "alum" [14].

4.2.6. Overview

When hydrolysing coagulant is increased to a suspension of negatively charged colloids, four distinct zones are distinguished:

Zone 1: Very low dosage; colloids remain negative and consequently stable,

Zone 2: Dosage sufficient to produce CN and consequently coagulation,

Zone 3: Higher dosage generating charge reversal and restabilisation,

Zone 4: Still higher dosage producing hydroxide precipitation and *SF* [14].

Fig. 10 illustrates the results of a standard *jar test* manner, habitually utilised in water treatment plants. In this manner, a suspension is injected with various quantities of coagulant under standard mixing and decantation conditions [13]. Habitually, there is a brief *rapid mix* time instantaneously after injection. This is followed by a longer time of *slow stirring* during which flocs may be generated because of orthokinetic aggregation [17]. Then, these flocs are authorised to settle for a standard time, after which a sample of the supernatant water is aspired and its turbidity is evaluated [23]. This *residual turbidity* produces a good information of the intensity of elimination during decantation and consequently of the efficiency of the C/F process [13]. Fig. 10 illustrates that, at very low coagulant injection, the remaining turbidity is high, showing little or no sedimentation (*Zone 1*). As the dose is elevated, there is a narrow interval (*Zone 2*) where there is an important reduction in remaining turbidity. This is the area of CN by adsorbed species, and it is frequently found that the colloid charge (as measured, for instance, by electrophoretic mobility [45,46] or streaming current [47]) is near zero. At higher injections, remaining turbidity is once again high, showing restabilisation of the colloids because of excess adsorption and charge reversal (*Zone 3*). Finally, at still higher dosages, there is a significant removal in remaining turbidity due to hydroxide precipitation and *SF* (*Zone 4*). It should be indicated that the remaining turbidity in *Zone 4* is lower than that in *Zone 2*, illustrating that *SF* produces larger, faster-settling flocs than those generated by CN [23]. In addition, as mentioned above, there is no restabilisation after *Zone 4*. The behaviour illustrated in Fig. 10 is typical of aluminium salts near pH 7 [17]. With these conditions, the hydroxide precipitate is positively charged. At pH values near to the isoelectric point (around pH 8), *Zone 2* may not be clear and only *SF* is functioning [14].

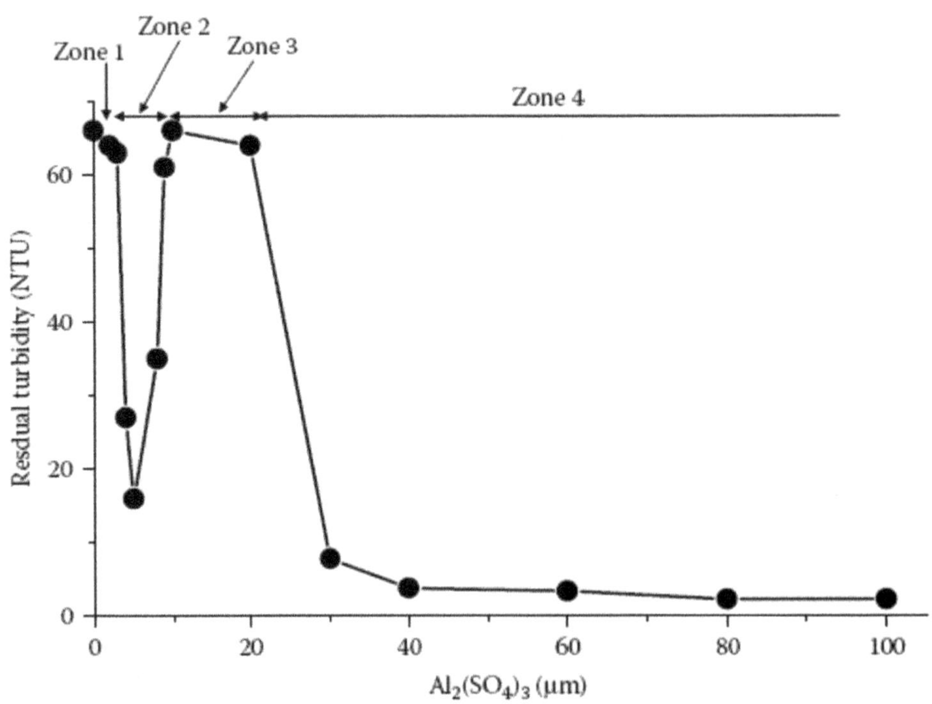

Figure 10. *Residual turbidity of kaolin suspensions after coagulation with aluminium sulphate over an interval of doses at pH 7 [14].*

4.2.7. Practical Aspects

Many important factors can importantly influence the efficiency of hydrolysing coagulants [17]. These comprise the effects of different anions and the effect of temperature [45]. Several frequent anions can constitute complexes with aluminium and iron (III) and can importantly influence hydroxide precipitation. An important illustration is sulphate, which is naturally present in water and may be introduced as aluminium or ferric sulphate during water treatment. Sulphate coordinates moderately strongly with Al, but the important influence is on the precipitation process. On the positive effect of the isoelectric point of aluminium hydroxide (i.e., below about pH 8), sulphate can adsorb on the precipitate and decrease its positive charge. This significates that the colloidal precipitate can agglomerate more rapidly to generate large hydroxide flocs [14,48].

Temperature has influences that are practically significant [49]. Particularly, at rather low temperatures, conventional aluminium coagulants have tendency to achieve less well for different causes [45]. Some prehydrolysed coagulants seem to be less influenced by low temperatures and are usually proposed for applications in cold areas [45]. Another advantage of prehydrolysed coagulants, such as PACl, is that, at efficient injections, they form less sludge than simple metal salts [37]. This may be partly attributed to the fact that they can be efficient at lower dosages [50]. Hydroxide flocs, as generated under *SF*, have tendency to be weak and are easily disrupted under high shear conditions [17]. In addition, the breakage can be irreversible at a certain level; consequently, flocs do not easily reform when the shear rate is decreased [14].

4.3. Flocculation Processes

The principal needs for efficient flocculation are as follows:

- Rapid mixing of coagulants (Fig. 11).
- Opportunity for collisions of destabilised particles and so flocculation [14].

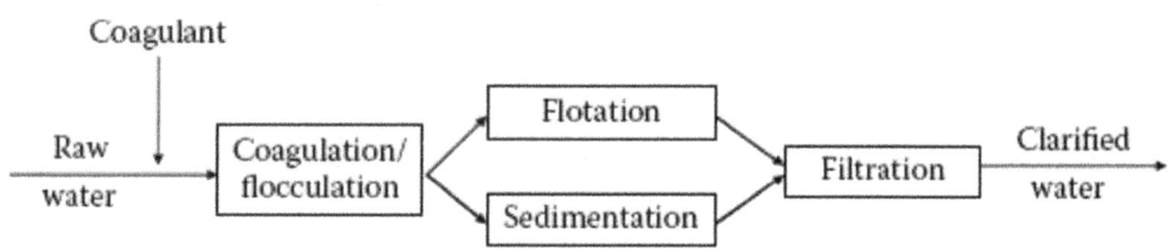

Figure 11. *Typical order of processes for particle separation in a water treatment plant [14].*

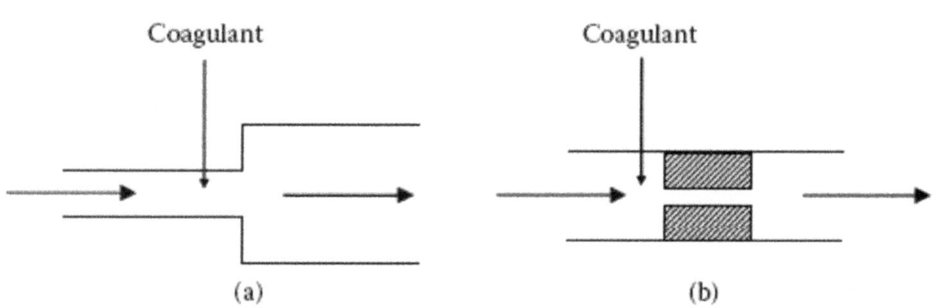

Figure 12. *Rapid mixing of coagulant by "in-pipe" methods [(a) widening pipe; (b) narrowing pipe] [14]Even if rapid mixing has long been established to have significant actions on flocculation phenomena and has been examined in some detail, it is possible that many examples of poor efficiency of practical flocculation units can be related to poor mixing [14,28].*

For the second step, some form of fluid motion has to be formed, which may be by mechanical stirring or flow (or both) [14,51].

4.3.1. Rapid Mixing

Mixing is necessary to help coagulant addition to disturb stability in the colloidal system [5]. For particles to aggregate they must collide, and mixing develops collision [15]. Brownian movement, the random motion affected to small particles by bombardment by individual water molecules, is always detected as a natural mixing force [52]. Nevertheless, additional mixing energy is required [53]. High intensity mixing, which distributes the coagulant and generates rapid collisions, is most performant [16]. The frequency and number of particle collisions are also significant in coagulation [15]. In a low turbidity water, the introduction of solids such as clay or the recycle of previously settled solids may be indispensable to augment the number of particle collisions [16]. Rapid mixing (sometimes called "flash mixing") is required to distribute the coagulant species among the particles in as short a period as possible [53]. In the case of coagulants that adsorb on particles and neutralise their charge, this can be significant [54]. Poor mixing can conduct to local overdosing of coagulant and consequently restabilisation of some particles [53]. For this cause, a short time of intense, turbulent mixing is preferable [55]. The high shear rates related to rapid mixing can also have an important role in the transport of coagulant species and can augment the rate of adsorption [53]. In the case of hydrolysing metal coagulants, upon conditions where hydroxide precipitation and *SF* are significant, the action of rapid mixing is not so obvious [44]. Nevertheless, it is established that hydrolysis rates are rapid and it is possible that rapid mixing conditions have some action in determining the relative rates of key processes such as adsorption and the formation of precipitates [52]. In principle, rapid mixing requires to be strong but of short period (i.e., few seconds) [53]. Otherwise, the nature of flocs generated subsequently can be influenced [23]. Prolonged times of intense mixing can conduct to the growth of small, compact flocs that grow slowly when the shear rate is decreased [55]. Rapid mixing may be realised in a flow-through stirred tank (a "backmix" reactor), even if this is an inefficient mixing device due to short-circuiting of flow [53]. It is not easy to achieve total and homogeneous distribution of introduced coagulant in a short period (i.e., less than one second) [54]. It is more frequent to inject coagulant at a point where there are turbulent conditions due to flow. This point may be in a channel — for example, where water flows over a weir — or in some kind of "in-pipe" mixer [56]. The latter way can implicate injecting coagulant at a point where the pipe either widens or narrows, as illustrated schematically in Fig. 12 [14].

4.3.2. Floc Formation

Frequently, growth of large flocs needs the application of velocity gradients or shear [57]. The major effects on flocculation rate are the particle (floc) size and content and the effective shear rate, G [58]. Higher shear rates generate improved particle collision rate but may decrease collision efficiency and provoke certain floc breakage [17]. A helpful compromise is a process known as *taper flocculation,* in which the effective shear rate is initially high, producing a rapid flocculation rate, and then progressively decreased so that large flocs can generate [23]. Practically, application of shear implicates the introduction of energy [59]. This can be realised in two manners: *mechanical* or *hydraulic.* Mechanical devices are typified by flow-through stirred tanks of different types, often known as *paddle flocculators* [60]. The paddles may rotate about vertical or horizontal axes, but in all conditions the power input to the water is a function of the drag force on the paddle and the rotation speed. The power introduction to the water could be theoretically evaluated, but it is not too hard to evaluate [14].

4.4. Enhanced Coagulation

Halogenated organics are generated if natural organic matter (NOM) enters in reaction with free chlorine or free bromine [61]. Free chlorine is frequently added to water directly as a primary or secondary disinfectant [62]. Free bromine is produced from the oxidation by chlorine of the bromide ion in the source water. Factors influencing the generation of these halogenated disinfection by-products (DBPs) comprise type and concentration of NOM, chlorine form and dose, time, bromide ion concentration, pH, organic nitrogen concentration, and temperature [63-65]. Because water treatment plants have been asked to control for total trihalomethans in the past, water treatment operators are likely familiar with some of the requirements that the Disinfectant/Disinfection By-Product (D/DBP) Rule implicate [66]. The main points of the DBP Rule and some of the main changes water supply systems are asked to satisfy with are published in literature [28]. Nevertheless, The D/DBP Rule imposes the utilisation of enhanced coagulation (EC) treatment for the removal of DBP precursors for surface water systems that have decantation capabilities [67]. The enhanced process implicates changes to the existing coagulation process such as augmenting the coagulant injection, decreasing the pH, or both sometimes [68-77].

Some authors [28] conducted bench, pilot, and demonstration scale studies to investigate arsenate removals during EC. The EC conditions in these researches comprised an augmentation of alum and ferric chloride coagulant injection from 10 to 30 mg L^{-1}, a reduction of a pH from 7 to 5.5, or both [28]. Results from these investigations established the following:

1. More than 90% arsenate removal can be realised upon EC conditions. Arsenate reductions above 90% were easily realised upon all conditions when ferric chloride was utilised.
2. EC using ferric salts is more performant for arsenic removal than EC using alum. With an influent arsenic content of 5 µg L^{-1}, ferric chloride realised 96% arsenate reduction with an injection of 10 mg L^{-1} and no acid introduction. When alum was utilised, 90% arsenate reduction could not be realised without decreasing the pH.
3. Reducing pH during EC increased arsenic removal by alum coagulation. With ferric coagulation, pH does not play an important role between 5.5 and 7.0 [28].

Nevertheless, post-treatment pH adjustment may be needed for corrosion control when the process is realised at a low pH [28].

5. Conclusions

The main conclusions can be summarised as follow:

1. Brownian movement, the random motion related to small particles by bombardment by individual water molecules, is always present as a natural mixing force. Even if rapid mixing has long been established to have great roles on flocculation phenomena and has been investigated at a certain level, it is possible that many examples of poor efficiency of practical flocculation units can be related to poor mixing.
2. *SF* conducts to faster agglomeration than CN and produces stronger and larger flocs: the generation of a hydroxide precipitate produces a big augmentation in the efficient particle concentration and so a greater collision rate. Hydroxide precipitates are generated from large numbers of colloids, which constitute very soon after injection. The flocs generated upon *"sweep"* conditions are also stronger and hence grow larger for the same shear conditions.

Acknowledgements

The present research work was undertaken by the Binladin Research Chair on Quality and Productivity Improvement in the Construction Industry funded by the Saudi Binladin Constructions Group; this is gratefully acknowledged. The opinions and conclusions presented in this paper are those of the authors and do not necessarily reflect the views of the sponsoring organisation.

References

[1] L. Rizzo, A. Di Gennaro, M. Gallo, V. Belgiorno, Coagulation/chlorination of surface water: A comparison between chitosan and metal salts, Sep. Purif. Technol. 62 (2008) 79–85.

[2] O.S. Amuda, I.A. Amoo, Coagulation/flocculation process and sludge conditioning in beverage industrial wastewater treatment, J. Hazard. Mater. 141 (2007) 778–783.

[3] J.H. Masliyah, S. Bhattacharjee, Electrokinetic and colloid transport phenomena, John Wiley & Sons, Inc., New Jersey, 2006.

[4] R.M. Pashley, M.E. Karaman, Applied colloid and surface chemistry, John Wiley & Sons Ltd, Chichester, England, 2004.

[5] J.E. Smith, Jr., R.C. Renner, B.A. Hegg, J.H. Bender, Upgrading existing or designing new drinking water treatment facilities, Noyes Data Corporation, New Jersey, 1991.

[6] A. Szyguła, E. Guibal, M.A. Palacín, M. Ruiz, A.M. Sastre, Removal of an anionic dye (Acid Blue 92) by coagulation–flocculation using chitosan, J. Environ. Manag. 90 (2009) 2979–2986.

[7] A. Rushton, A.S. Ward, R.G. Holdich, Solid-liquid filtration and separation technology, VCH, Weinheim, 1996.

[8] A.P.Sr. Sincero, G.A. Sincero, Physical–chemical treatment of water and wastewater, IWA Publishing & CRC Press, Boca Raton, 2003.

[9] F. Woodard, Industrial waste treatment handbook, Butterworth–Heinemann, Boston, 2001.

[10] W. Pietsch, Size enlargement by agglomeration, John Wiley & Sons Ltd, Chichester, UK, 1991.

[11] J.T. O'Connor, T. O'Connor, R. Twait, Water treatment plant performance evaluations and operations, John Wiley & Sons, Inc., New Jersey, 2009.

[12] S.A. Parsons, B. Jefferson, Introduction to potable water treatment processes, Blackwell Publishing Ltd, Oxford, UK, 2006.

[13] W. Lick, Sediment and contaminant transport in surface waters, IWA Publishing, CRC Press, Taylor & Francis Group, LLC, Boca Raton, 2009.

[14] J. Gregory, Particles in water, properties and processes, CRC Press (Taylor & Francis), Boca Raton, 2006.

[15] P. Dolejs, Considerations of the collision efficiency factor in coagulation of dissolved natural organic matter, In: Natural Organic Matter in Drinking Water, Origin, Characterization, and Removal, Workshop Proceedings, September 19-22, 1993, Chamonix, France, AWWA Research Foundation, USA, 165-172, 1994.

[16] D.N. Thomas, S.J. Judd, N. Fawcett, Flocculation modelling: A review, Water Res. 33 (1999) 1579-1592.

[17] E. Sharp, Natural organic matter coagulation, PhD thesis, School of Water Sciences, Cranfield University, Cranfield, UK, 2005.

[18] M.C. Sterling Jr., J.S. Bonner, A.N.S. Ernest, C.A. Page, R.L. Autenrieth, Application of fractal flocculation and vertical transport model to aquatic sol–sediment systems, Water Res. 39 (2005) 1818–1830.

[19] R.E. Weiner, R.A. Matthews, Environmental engineering, 4th Ed., Butterworth–Heinemann, New York, 2003.

[20] R.W. Crites, J. Middlebrooks, S.C. Reed, Natural wastewater treatment systems, CRC Press (Taylor & Francis), Boca Raton, 2006.

[21] K. Sutherland, Filters and filtration handbook, 5th Ed., Butterworth–Heinemann, Elsevier, Oxford, UK, 2008.

[22] T. Li, Z. Zhu, D. Wang, C. Yao, H. Tang, Characterization of floc size, strength and structure under various coagulation mechanisms, Powder Technol. 168 (2006) 104–110.

[23] M.I. Aguilar, J. Sáez, M. Lloréns, A. Soler, J.F. Ortuño, V. Meseguer, A. Fuentes, Improvement of coagulation–flocculation process using anionic polyacrylamide as coagulant aid, Chemosphere 58 (2005) 47–56.

[24] D.L. Russell, Practical wastewater treatment, John Wiley & Sons, New Jersey, 2006.

[25] X. Wu, X. Ge, D. Wang, H. Tang, Distinct coagulation mechanism and model between alum and high Al13-PACl, Colloids Surf. A: Physicochem. Eng. Aspects 305 (2007) 89–96.

[26] P. Cañizares, F. Martínez, C. Jiménez, C. Sáez, M.A. Rodrigo, Coagulation and electrocoagulation of oil-in-water emulsions, J. Hazard. Mater. 151 (2008) 44–51.

[27] J.-M. Siéliéchi, B.S. Lartiges, G.J. Kayem, S. Hupont, C. Frochot, J. Thieme, J. Ghanbaja, J.B. d'Espinose de la Caillerie, O. Barrès, R. Kamga, P. Levitz, L.J. Michot, Changes in humic acid conformation during coagulation with ferric chloride: Implications for drinking water treatment, Water Res. 42 (2008) 2111-2123.

[28] F.R. Spellman, Handbook of water and wastewater treatment plant operations, Lewis Publishers (CRC Press), Boca Raton, 2003.

[29] E. R. Alley, Water quality control handbook, 2nd Ed., McGraw-Hill, New York, 2007.

[30] S.P. Chesters, E.G. Darton, S. Gallego, F.D. Vigo, The safe use of cationic flocculants with reverse osmosis membranes, Desalin. Water Treat. 6 (2009) 144–151.

[31] The Nalco water handbook, 2nd Ed., F.N. Kemmer (Ed.), McGraw-Hill, New York, 1988.

[32] Y. Zeng, J. Park, Characterization and coagulation performance of a novel inorganic polymer coagulant—Poly-zinc-silicate-sulfate, Colloids Surf. A: Physicochem. Eng. Aspects 334 (2009) 147–154.

[33] I. Ivancev-Tumbas, B. Dalmacija, Effects of coagulation processes on aldehydes formation in groundwater treated with common oxidative agents, Water Res. 35 (2001) 3950-3958.

[34] A.I. Zouboulis, P.A. Moussas, F. Vasilakou, Polyferric sulphate: Preparation, characterisation and application in coagulation experiments, J. Hazard. Mater. 155 (2008) 459–468.

[35] S. Suarez, J.M. Lema, F. Omil, Pre-treatment of hospital wastewater by coagulation–flocculation and flotation, Biores. Technol. 100 (2009) 2138–2146.

[36] N.K. Shammas, Coagulation and flocculation (ch. 4), In: Physicochemical treatment processes (vol. 3), Handbook of environmental engineering, L.K. Wang, Y.-T. Huang, N.K. Shammas (Eds.), Humana Press Inc, New Jersey, 2005.

[37] C. Ye, D. Wang, B. Shi, J. Yu, J. Qu, M. Edwards, H. Tang, Alkalinity effect of coagulation with polyaluminum chlorides: Role of electrostatic patch, Colloids Surf. A: Physicochem. Eng. Aspects 294 (2007) 163–173.

[38] X. Xu, S. Yu,W. Shi, Z. Jiang, C. Wu, Effect of acid medium on the coagulation efficiency of polysilicate-ferric (PSF)—A new kind of inorganic polymer coagulant, Sep. Purif. Technol. 66 (2009) 486–491.

[39] B. Shi, G. Li, D. Wang, C. Feng, H. Tang, Removal of direct dyes by coagulation: The performance of preformed polymeric aluminum species, J. Hazard. Mater. 143 (2007) 567–574.

[40] X. Ntampou, A.I. Zouboulis, P. Samaras, Appropriate combination of physico-chemical methods (coagulation/flocculation and ozonation) for the efficient treatment of landfill leachates, Chemosphere 62 (2006) 722–730.

[41] A.T. Owen, P.D. Fawell, J.D. Swift, D.M. Labbett, F.A. Benn, J.B. Farrow, Using turbulent pipe flow to study the factors affecting polymer-bridging flocculation of mineral systems, Intern. J. Mineral Process. 87 (2008) 90–99.

[42] C.E. Capes, Particle size enlargement (vol. 1), In: Handbook of powder technology, J.C. Williams, T. Allen (Eds.), Elsevier, Amsterdam, 1980.

[43] D. Bouyer, C. Coufort, A. Liné, Z. Do-Quang, Experimental analysis of floc size distributions in a 1-L jar under different hydrodynamics and physicochemical conditions, J. Colloid Interface Sci. 292 (2005) 413–428.

[44] J. De Zuane, Handbook of drinking water quality, 2nd Ed., John Wiley & Sons, Inc., New York, 1997.

[45] F. Xiao, J. Ma, P. Yi, J.-C.H. Huang, Effects of low temperature on coagulation of kaolinite suspensions, Water Res. (2008) 2983-2992.

[46] F. Xiao, X. Zhang, J. Ma, Indecisiveness of electrophoretic mobility determination in evaluating Fe(III) coagulation performance, Sep. Purif. Technol. 68 (2009) 273–278.

[47] C.A. Walker, J.T. Kirby, S.K. Dentel, The streaming current detector: A quantitative model, J. Colloid Interface Sci. 182 (1996) 71-81.

[48] M. Kobayashi, Breakup and strength of polystyrene latex flocs subjected to a converging flow, Colloids Surf. A: Physicochem. Eng. Aspects 235 (2004) 73–78.

[49] A.M. Goula, M. Kostoglou, T.D. Karapantsios, A.I. Zouboulis, The effect of influent temperature variations in a sedimentation tank for potable water treatment — A computational fluid dynamics study, Water Res. 42 (2008) 3405-3414.

[50] A.L. Ahmad, S.S. Wong, T.T. Teng, A. Zuhairi, Improvement of alum and PACl coagulation by polyacrylamides (PAMs) for the treatment of pulp and paper mill wastewater, Chem. Eng. J. 137 (2008) 510–517.

[51] O.P. Prat, J.J. Ducoste, Simulation of flocculation in stirred vessels, Lagrangian versus Eulerian, Trans IChemE 85 (2007) 207–219.

[52] C. Coufort, D. Bouyer, A. Liné, B. Haut, Modelling of flocculation using a population balance equation, Chem. Eng. Process. 46 (2007) 1264–1273.

[53] P. Mavros, Flow visualization in stirred vessels: A review of experimental techniques, Trans IChemE 79 (2001) 113-127.

[54] H.-C. Kim, S. Lee, Pump diffusion flash mixing (PDFM) for improving coagulation process in drinking water treatment, Sep. Purif. Technol. 52 (2006) 117–125.

[55] A. Maes, S. Vreysen, N.N. Rulyov, Effect of various parameters on the ultraflocculation of fine sorbent particles, used in the wastewater purification from organic contaminants, Water Res. 37 (2003) 2090–2096.

[56] R.K. Thakur, Ch. Vial, K.D.P. Nigam, E.B. Nauman, G. Djelveh, Static mixers in the process industries - A review, Trans IChemE 81 (2003) 787-826.

[57] G.L. McConnachie, G.K. Folkard, M.A. Mtawali, J.P. Sutherland, Field trials of appropriate hydraulic flocculation processes, Water Res. 33 (1999) 1425-1434.

[58] S.A.A. Tabatabai, M.D. Kennedy, G.L. Amy, J.C. Schippers, Optimizing inline coagulation to reduce chemical consumption in MF/UF systems, Desalin. Water Treat. 6 (2009) 94–101.

[59] M. Zeidan, B.H. Xu, X. Jia, R.A. Williams, Simulation of aggregate deformation and breakup in simple shear flows using a combined continuum and discrete model, Trans IChemE 85 (2007) 1645–1654.

[60] C. Coufort, D. Bouyer, A. Liné, Flocculation related to local hydrodynamics in a Taylor–Couette reactor and in a jar, Chem. Eng. Sci. 60 (2005) 2179 – 2192.

[61] D.A. Reckhow, G. Makdissy, P.S. Rees, Disinfection by-product precursor content of natural organic matter extracts (ch. 6), In: Disinfection by-products in drinking water, occurrence, formation, health effects, and control, T. Karanfil, S.W. Krasner, P. Westerhoff, Y. Xie, (Eds.), ACS Symposium Series 995, ACS, Washinton, 2008.

[62] D. An, J. Song, W. Gao, G. Chen, N. Gao, Molecular weight distribution for NOM in different drinking water treatment processes, Desalin. Water Treat. 5 (2009) 267–274.

[63] Y.F. Xie, Disinfection by-products in drinking water, formation, analysis, and control, CRC Press LLC, Boca Raton, 2004.

[64] C.W.K. Chow, J.A. van Leeuwen, R. Fabris, M. Drikas, Optimised coagulation using aluminium sulfate for the removal of dissolved organic carbon, Desalination 245 (2009) 120–134.

[65] J. Kim, Fate of THMs and HAAs in low TOC surface water, Environ. Res. 109 (2009) 158–165.

[66] H. Wong, K.M. Mok, X.J. Fan, Natural organic matter and formation of trihalomethanes in two water treatment processes, Desalination 210 (2007) 44–51.

[67] T. Karanfil, S.W. Krasner, P. Westerhoff, Y. Xie, Recent advances in disinfection by-product formation, occurrence, control, health effects, and regulations (ch. 1), In: Disinfection by-products in drinking water, occurrence, formation, health effects, and control, T. Karanfil, S.W. Krasner, P. Westerhoff, Y. Xie, (Eds.), ACS Symposium Series 995, ACS, Washinton, 2008.

[68] USEPA (US Environmental Protection Agency), Enhanced coagulation and enhanced precipitative softening guidance manual, Office of water (4607), EPA 815-R-99-012, May 1999.

[69] D. Ghernaout, B. Ghernaout, A. Kellil, Natural organic matter removal and enhanced coagulation as a link between coagulation and electrocoagulation, Desalin. Water Treat. 2 (2009) 203–222.

[70] D. Ghernaout, D. Ghernaout, A. Saiba, Algae and cyanotoxins removal by coagulation/flocculation: A review, Desalin. Water Treat. 20 (2010) 133-143.

[71] D. Ghernaout, S. Moulay, N. Ait Messaoudene, M. Aichouni, M.W. Naceur, A. Boucherit, Coagulation and chlorination of NOM and algae in water treatment: A review, Intern. J. Environ. Monitor. Analys. 2 (2014) 23-34.

[72] D. Ghernaout, B. Ghernaout, M.W. Naceur, Embodying the chemical water treatment in the green chemistry—A review, Desalination 271 (2011) 1-10.

[73] D. Ghernaout, The hydrophilic/hydrophobic ratio vs. dissolved organics removal by coagulation – A review, J. King Saud Univer. – Sci. 26 (2014) 169-180.

[74] D. Ghernaout, The best available technology of water/wastewater treatment and seawater desalination: Simulation of the open sky seawater distillation, Green Sustain. Chem. 3 (2013) 68-88.

[75] D. Ghernaout, B. Ghernaout, On the concept of the future drinking water treatment plant: algae harvesting from the algal biomass for biodiesel production—a review, Desalin. Water Treat. 49 (2012) 1-18.

[76] D. Ghernaout, B. Ghernaout, Sweep flocculation as a second form of charge neutralisation—a review, Desalin. Water Treat. 44 (2012) 15-28.

[77] O. Bello, Y. Hamam, K. Djouani, Control of a coagulation chemical dosing unit for water treatment plants using MMPC based on fuzzy weighting, J. Water Process Engin. 4 (2014) 34-46.

19

Forms and Extent of Ferrugination in Soils of Georgia

Lia Matchavariani[1], Dali Nikolaishvili[2]

[1]Department of Soil Geography, Faculty of Exact & Natural Sciences, Iv Javakhishvili Tbilisi State University, Tbilisi, Georgia
[2]Department of Geomorphology & Cartography, Faculty of Exact & Natural Sciences, Iv Javakhishvili Tbilisi State University, Tbilisi, Georgia

Email address:
lia.matchavariani@tsu.ge (L. Matchavariani), dali.nikolaishvili@tsu.ge (D. Nikolaishvili)

Abstract: Ferruginization as one of the most characteristic processes in soils of Georgia, has been investigated both qualitatively and quantitatively mainly micropedologically, by polarizing microscopy on thin sections. A study was undertaken to determine some of pedological factors affecting the genesis of Fe-formations in soils series of West and East Georgia. This process is diagnosed by the presence of different forms of ferrous formations, concentration level or iron crystallization, ferruginized micro-zones of plasma impregnation with ferric hydroxides, etc. A map showing the distribution of the ferrugination process in the soils of Georgia was created.

Keywords: Ferum hydroxides, Concretions, Distribution, Georgia

1. Introduction

The movement of water in soil profiles, carry with it the soluble constituents, mostly iron. Subsequently, these solutes are deposited and segregated in neoplasms. Concretions of some kind are found in most soils, where they have been formed by secondary processes [1].

Origin and development of ferruginous concretions are widely distributed in many regions of the Earth, including humid subtropical of Georgia. Their distribution, morphology, physical speciation, chemical composition and other parameters were investigated by many authors in the 50-90-years of the last century as well as as in recent years [2, 3, 4, 5]. In details morpho-genetically typification of soil Fe and Fe-Mn concretions, with the use of complementary research methods at the macro-, meso-, micro- and submicro-levels have been studied in Georgia [6].

The humid subtropical soils of Georgia have the properties associated with the movement and/or transformation of iron in the profile (in respect of both, concretion-formation and impregnation of the basic soil mass) what is caused by the manifestation of a number of the elementary soil processes attributed to the group of hydrogen-accumulative (lateralization, plinthification), and metamorphic (ferrugination, ferrallitization, rubification) processes. The listed processes take place in the soils of Georgia at a more or less degree [7].

Lateralization – a hydrogen-accumulative process of allochthonic intra-soil ferrugination by forming strong concretionary or iron-clad interlayers of different structures. Lateralization is a process of both, old and modern ferrugination causing the exclusion of the significant amounts of iron and aluminum from the cycle leading to the formation of interlayers of different structures, different densities and correlations of Fe, Al, Si by accumulating the iron from the ground or soil waters during the lateral movement through the accumulative landscapes. A lateral layer may also be formed by the metamorphization of plinthite. Usually, a laterite with its unbroken structure in the soil profile has a binomial composition: the upper, concretionary (which can be formed independently from plinthite-formation) and lower – tabular (the result of a plinthite formation and solidification).

Plinthification - a hydrogen-accumulative process of transformation of the ferraliticized material by means of precipitation of the iron oxides from the rising ground waters on the kaolinite matrix.

Ferrugination - a metamorphic process of the iron liberalization from the minerals during their weathering and sedimentation in situ in the pores and cracks as autochthonic cutans of grains and micro-aggregates and clumps of hydroxides accompanied by brown or red coloration of the soil-forming rock.

Ferrallitization – a metamorphic process of intra-soil weathering of primary minerals by forming and relative accumulation in situ of the secondary clay of the ferralitic content. The leading component of a ferraliticized material is quartz, kaolin and minerals of the group of aluminum and ferric hydroxides.

Rubification (ferritization) – a metamorphic process of an irreversible coagulation and further crystallization of colloid ferric hydroxides in the soil profile as a result of intense periodic drying of soil in the dry and hot period of the year after they are introduced and precipitated in the wet period of the year.

Morphologically, the specifics of ferric neo-formations lie in that they are easily differentiated in the mixed mass with their content and structure. The most important criterion of the micromorphological diagnostics of Fe-formations is the degree of concentration or crystallization of substances [4]. Ferrugination on its turn, is also an important diagnostic sign of gleying.

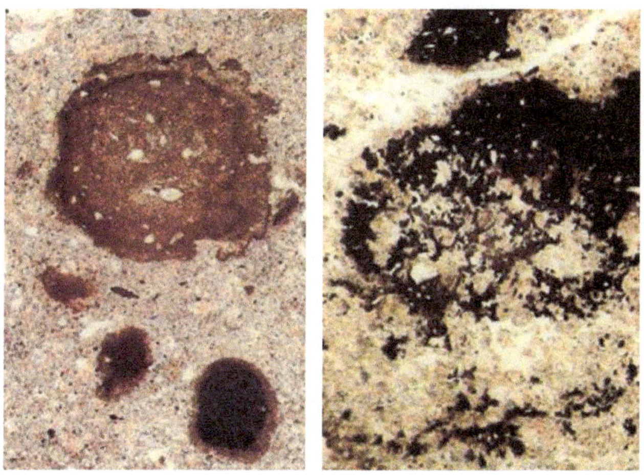

(a) (b)

Figure 1. Forms of iron (a) and manganese (b) compounds in soils, nic.II.

Neo-formations of iron compounds in fact, are found in all soils of humid territories [6, 7]. They are easy to recognize with their rusty-brown, reddish or yellowish-brown coloration in transmitted or reflected light; they are often isotropic except the crystals of goethite, hematite, etc.

Spots and flakes, dissipated in the basic mass, rarer than threads on the grains of the skeleton are the proof of the presence of less intense local processes of iron liberalization in situ. Separate or dissipated flakes can be the remains of organic-ferric complexes after their organic components are decomposed, or the result of Fe concentration with microorganisms, or the result of the local re-distribution of hydroxides.

Diffusion rings and strips are thin, continuous or discrete, single or complex (through concentric), ranging from the light rusty or orange through the blackish-brown color; they are always clearly distinguished in the basic mass. They are made of fine-dispersed Fe hydroxides, and organic-ferric complexes may also present. Sometimes, the diffusion rings are associated with the root conducting pores, i.e. the root excretions seem to cause the mobilization and re-distribution of hydroxides. The strips show the certain levels of ground waters, i.e. they are components of pseudofibers. The strips are profile traces of the near-root hydroxides concentrations, or they are formed in the periodically overwetted areas in terms of heterogeneous structure and grain size in the contact zone. Biogenic forms of hydroxides with low concentrations include ferruginized vegetation remains – ferruginous pseudomorphs in organic tissues.

Ferrous tubes or so called Rohrensteins (rhyzo concretions) are considered as a result of the further concentration of a substance in the near-root zones. As compared to diffusion rings, they are larger in size and have more impregnated basic mass along the root conducting pores of Fe or Mn compounds and humus substances. They are mostly typical to humus-accumulative horizons of soils of ground bogging.

Ferrous-manganous cutans, being the immediate product of illuviation, may be purely ferrous or manganese-ferrous. Often, they alternate with clay sinters. Purely manganese films, the manganese, easily recognizable by their black color with a metallic gleam, are rarer. Their typical forms are dendrites (fillers in thin, branched-out pores) and ferrous-manganese pseudomorphs on the vegetation remains.

Figure 2. Spectrum of genetic profiles Georgia's soils.

Concretions, micro-ortshteins, nodules – the forms of formations with more diversified content and appearance at high concentrations of ferrous substance. In addition to iron and manganese, these neo-formations include humus compounds and components of the basic mass. Many concretions include skeleton grains perhaps because they are formed through the segregation of hydroxides under favorable terms – skeletal micro-zones. Inclusion of the clay material in the concretion, papules in particular, is a clear evidence of the heterochrony of illuviation and concretion-formation. The morphology of the concretionary forms, particularly, of those inherited from the soil-forming specimens, is quite diversified.

Morphologically, the simplest of the concretionary forms is micro-ortshteins. Large micro-ortshteins can be fixed in a morphological profile as "leather coats". Micro-ortshteins are typical to elluvial horizons of soils with a texture-differentiated profile and can be considered a reliable criterion of surface gleying. Concretions can be of different forms; they contain Fe and Mn hydroxides, humus compounds and inclusions of skeletal grains. Concentric concretions are most often found in the upper horizons owing to the dynamics of the conditions of humidification and oxidation-reduction processes. The concretions with even distribution and high concentrations of hydroxides are sometimes called nodules.

The structure of ferrous-clay plasma is mainly described based on Kubiena's works [8], which contribution to micropedology has been evaluated in recent publications of G. Stoops [9, 10]. Initially, the ferrous-clay plasma is usually sub-divided into two types: plasma with the evenly distributed amorphous ferric hydroxides and plasma, in which amorphous ferric hydroxides present as clots. It is also classified by color: brownish-ocherous, yellow, red or greenish-grey plasma. The presence or absence of the crystallized forms of iron oxides is also considered.

Plasma of the first type. The brownish-ocherous plasma is a dense, fine-pored mass. Iron oxides in it are distributed diffusively. In Kubiena's opinion, such kind of plasma is mostly formed in terms of permanently humid subtropical climate. Yellow and red plasmas are typical of the soils in the subtropical zones with variable humidity. In the reflected light, the color of the yellow plasma has a brighter eggyolk yellow tint than in the transmitted light, while red plasma looks orange-red. Greenish-grey plasma is widely spread in gley soils. Usually, in case of irregular manifestation of the oxidation-reduction processes, a marble-like coloration occurs as spots of greenish, grayish, rusty-brown or ocherous color showing the different states of iron oxides.

Plasma of the second type includes: brown plasma with earthly appearance typical to the mountainous areas; loose, well-aggregated, highly porous, coagulated with ferric hydroxide, hardly transformable to a crystalline state; yellow with a red tint, typical to the soils of humid subtropics; includes many fine bright red coagulated lumps of ferric hydroxides; is characterized by a large number of Fe-concretions and spots, which are bright red in the reflected light.

2. Study Area and Methods

The object of the studies in fact, covers the entire soil spectrum of Georgia, which is characterized by a variety of soil types, spread on a small territory. Complicated geological structure, heterogeneity of relief, contrasting climatic conditions, the specificity of vegetation and biodiversity, comprehensively define a variety of soil cover of the country (Mountain Meadow; Brown Forest; Yellow-Brown; Mountain Chernozems; Raw Humus Calcareous; Chernozems; Cinnamonic; Meadow Cinnamonic; Grey Cinnamonic; Meadow Grey Cinnamonic; Salt Soils; Red Soils; Yellow Soils; Subtropical Podzolic; GleyPodzolic; Alluvial; Bog Soils).

The macro- and micromorphological descriptions were done in compliance with well-known guides and information manuals in micropedology [11, 12, 13]. In order to determine the common geographical regularities of formation and distribution of various types ferruginous segregations in soils of Georgia, the detailed micropedologic studies by using the systems of mutually complementary coupled macro-, mezzo-, micro- and sub-micromorphological methods, which clearly evidence the mechanism of formation of the genetic profiles of soils were used [6].

The taxonomic units of the national soil classification of Georgia (fig. 2), according to WRB [14] are systemized in following relation: Cambisols (Mountain Meadow, Meadow Grey Cinnamonic, Cinnamonic, Cinnamonic Calcareous, Meadow Cinnamonic); Umbrisols (Mountain Forest Meadow of West Georgia); Leptosols (Mountain Forest Meadow of East Georgia); Luvisols (Yellow Soils, Yellow Brown Forest); Faiozems (Mountain Meadow Chernozem like); Nitisols (Red Soils); Acrisols (Subtropical Podzols); Chernoziom (Vertisols); Kashtanozems (Grey Cinnamonic); Fluvisols (Alluvial Soils), etc.

3. Results and Discussions

Table 1. *The Ratio of Ferrugination among Profile-formation Process in maim Soils of Georgia.*

Soil Types	Soil-forming Processes				
	Fe	Ls	Gl	Cb	Hm
Brown Forest	++	++	+	-	++
Yellow-Brown	++	++	++	-	++
Mountain Chernozems	+	++	+	-	+++
Raw Humus Calcareous	+	++	-	+++	+
Chernozems	+	+	+	+++	+++
Cinnamonic	-	+	-	+++	+++
Meadow Cinnamonic	+	+	+	+++	+++
Grey Cinnamonic	-	+	-	+++	++
Meadow Grey Cinnamonic	+	+	+	+++	+
Salt Soils	+	++	++	+++	+
Red Soils	+++	+++	++	-	++
Yellow Soils	+++	++	++	-	++
Subtropical Podzolic	+++	+++	+++	-	+
Gley Podzolic	+++	+++	+++	-	+
Bog Soils	+++	+++	+	-	+
Alluvial	++	++	+	+ -	+

Fe – Ferrugination; Ls – Lessivage; Gl – Geying; Hm – Humification; Cb – Calcareousness.

In the soils of Georgia, the processes associated with the ferum hidroxides sedimentation, movement and transformation, were identified in many soils, particularly in the soils of the humid subtropics (Red Soils; Yellow Soils; Subtropical Podzolic; GleyPodzolic; Alluvial; Bog Soils); quite many signs of ferrugination are also seen in Brown Forest Soils as small micro-zones of segregation and impregnation of plasma. Partially ferrous extractions are seen in arid regions, as well. In the Salt Soils, sometimes there are iron extractions observed as small micro-concretions being an immediate result of the past phases of soil-formation. Depending on the kind of manifestation, it is easy to identify two forms of the ferrugination in the soils of Georgia: concretionary and plasma ferrugination, though in many soils of a humid subtropical type, it is manifested virtually in both forms - as concretionary formations (Fe and Fe-Mn) and zones of ferrugination impregnated in the basic mass (fig. 1). This is why the gradation of ferrugination for the soils of Georgia was basically done by considering the intensity of the process, manifestation: strongly ferruginated, moderately ferruginated and weakly ferruginated.

The Ratio of manifestation the ferrugination among other soil process, which are forming the main soil profiles of Georgia, is given in table 1.

Iron compounds as concretionary partings and/or deposited cemented layers in the profile, or zones of ferrugination are most common in the subtropical Podzols of Georgia (the most discussible and disputable soils in a genetic respect). The most informative source for a thorough study of the given question is the complex of micromorphological methods. Consequently, special detailed studies of Fe-formations were accomplished by using the complex of coupled micromorphological methods (including micro- and submicroscopy), and the last chapter of the present work is dedicated to this question.

Thus, the process of ferrugination in the soils of Georgia in a micromorphological respect, is diagnosed by the presence of different forms of ferrous formations, degree of concentration or crystallization of iron, character of ferruginized micro-zones or zones of plasma impregnation with ferric hydroxides, etc.

The map, showing the distribution of the ferrugination process in main soils of Georgia, has been drawn up (fig. 3). We have identified three stages of intensity the manifestations of this process: strong ferrugination, which is typical for Red soils, Yellow soils, Subtropical Podzolic and Gley Podzolic and Bog soils; moderate ferrugination, characterizes Brown Forest and Yellow-Brown soils and partly Alluvial Soils); manifestation of weak ferrugination observed sometimes in Mountain and Plain Chernozems, Meadow Cinnamonics and Raw Humus Calcareous soils, Meadow Grey Cinnamonic and Salt Soils. In Cinnamonic and Grey Cinnamonic soils epy ferrugination process practically no fixed.

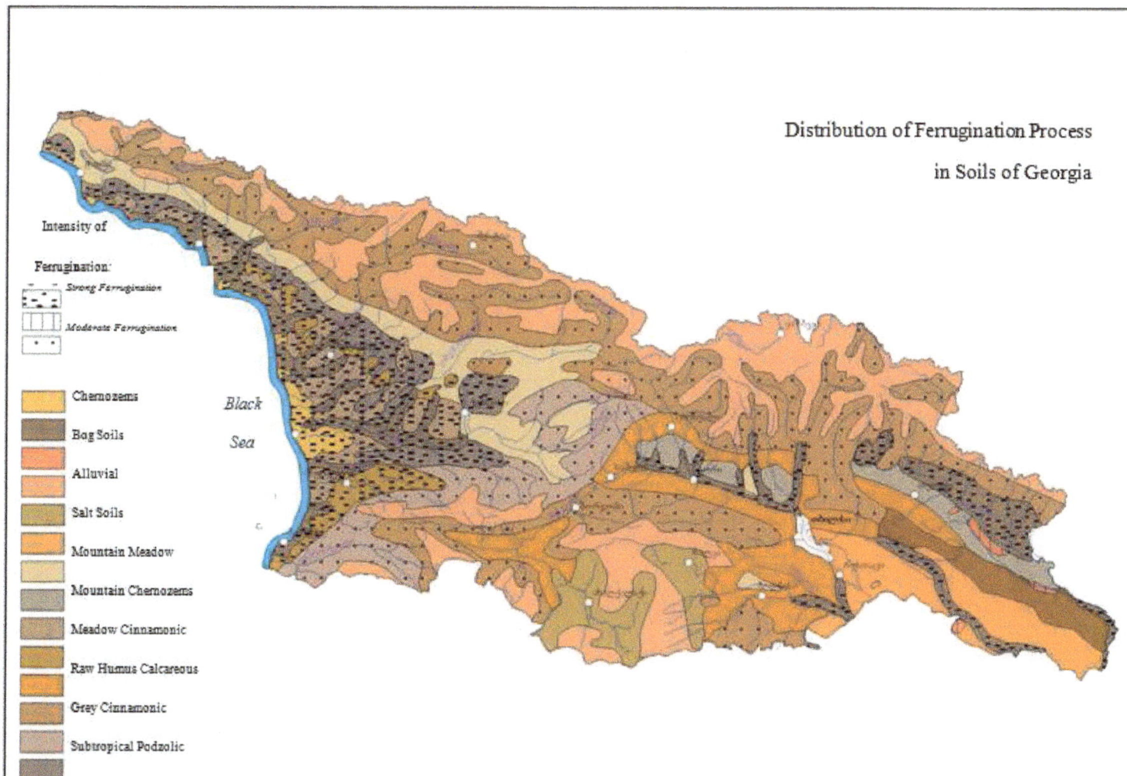

Figure 3. *Distribution of Ferrugination Process in Soils of Georgia.*

3. Conclusions

Ferruginization is one of the most characteristic processes in soils of Georgia, especially in the humid subtropical regions. The process of ferrugination in the soil profiles in a micromorphological respect, is diagnosed by the presence of different forms of ferrous formations, concentration level or iron crystallization, character of ferruginized micro-zones or zones of plasma impregnation with ferric hydroxides, etc. A map showing the distribution of the ferrugination process in the soils of Georgia was created by using software GIS.

References

[1] Buol S.W., Southard R.J., Graham R.C. and McDaniel P.A. Soil Genesis and Classification. ISBN 0-8138-2873-2. 5th Edition. Ames, Iowa: Iowa State Press, A Blackwell Pub. Co., 2003, 494 p.

[2] Latrille C., Elsassb F., Oort F.Van, Denaix L. Physical speciation of trace metals in Fe–Mn concretions from a rendzic lithosol developed on Sinemurian limestones (France). Geoderma, Elsevier, vol. 100, issues 1-2, 2001, pp. 127–146

[3] Gasparatos D.. Fe–Mn Concretions and Nodules to Sequester Heavy Metals in Soils. Environmental Chemistry for a Sustainable World, Springer, 2012, pp 443-474

[4] Gasparatos D. , Haidouti C. , Tarenidis D. Characterization of iron oxides in Fe-rich concretions from an imperfectly-drained Greek soil: a study by selective-dissolution techniques and X-ray diffraction. Archives of Agronomy and Soil Science, vol. 50, issue 4-5, 2004, pp. 485-493

[5] Jokova M., Kerchev G., Dilkova R. Relation between Waterlogging Degree and Behaviour of Mn and Fe Extractable Compounds along Depth of Textural Differrentiated soils. Почвознан., агрохим. и екол. (болг.), 2000, 35, №5, p.24-30

[6] Matchavariani L.G. Morphogenetic Typification of Concretions in Subtropical Podzolic Soils of Georgia. Eurasian Soil Science, Springer, ISSN: 1064-2293, 2005, vol. 38, #11, 1161-1172

[7] Matchavariani L.G., Geographic Paradigms of the Microfabrics of Major Soils of Georgia, Universal, Tbilisi, 2008, 300 p.

[8] Kubiena W. Micromorphological Features of Soil Geography. New Brunswick, New Jersey, 1970

[9] Stoops G. The "Fabric" of soil micromorphological research in the 20th century – A bibliometric analysis. Geoderma, 2014 - Elsevier. Volume 213, January 2014, 193-202

[10] Stoops G. Evaluation of Kubiëna's contribution to micropedology. At the occasion of the seventieth anniversary of his book "Micropedology". Eurasian Soil Science, vol.42, #6, 2009, Springer, 693-698

[11] Jahn R., Blume H.-P., Asio V.B. Students Guide for Soil Description, Soil Classfiication, and Site Evaluation. Halle, 2003.

[12] Bullock P., Fedoroff N., Jongerius A., Stoops G., Tursina T. Handbook for Soil Thin Section Description. Waine Research Publishing, Wolverhampton, Albrigton, UK, 1985

[13] FitzPatrick E.A. Micromorphology of Soils. Chapman and Hall, London, 1984

[14] World reference base for soil resources. FAO UN, World Soil Resources Reports – 103, Rome, 2006, 128 p.

20

Phenotype Variations of TAS2R38 Gene and Its Bioecological Significance

Rusudan Khukhunaishvili, Marina Koridze, Marina Nagervadze, Ciuri Khizrevanidze, Shorena Gabaidze

Department, of Biology, Faculty of Natural Sciences and Health Care, Batumi Shota Rustaveli State University, Batumi, Georgia

Email address:

rusudan_kh@yahoo.com (R. Khukhunaishvili), koridzemarina@gmail.com (M. Koridze), marinanagervadze@gmail.com *(M. Nagervadze)*
sopo.xozrevanidze@mail.ru (C. Khozrevanidze); gabaidzeshorena@mail.ru (S. Gabaidze)

Abstract: People have different feeling skills of the bitterness of compound phenylthiocarbamide (PTC). This trait has been shown to correlate with a number of dietary preferences and thus may have important implications for human health. There are two types of phenotypes in different ethno groups and populations according to this feature: PTC tester and PTC non-tester. Both phenotypes, besides some rare exceptions, are fixed in almost every population. We have studied the phenotypic variants of feeling the PTC bitterness spread in Georgian ethno group for the first time. Basing on the conducted experiments, about ¾ of the investigated population turned out to be sensitive towards PTC, and ¼ of the population – insensitive. We have analyzed the phenotypic structure of PTC feeling in the people who smoke in the population. The majority of the smokers (70%) turned out to be PTC non-tester and only 30% - PTC tester. It is important to study the phenotypic variants of feelings coded with TAS2R38 gene in the point of view of both – genetic structure of the population and bio-ecology in order to plan different kinds of preventive measures.

Keywords: PTC, Phenotype, Tester, Non-Tester

1. Introduction

In perceiving an environment by a human being, together with the other basic senses (sight, hear, smell, touch), the sense of taste also takes part. Different substances when penetrating into the mouth cavity cause the activation of receptor cells, which results in a specific feeling – sense of taste. The taste is based on the five basic senses (bitter, sweet, sour, salt and umami). The receptors of each category are activated with correspondent adequate irritants. It is important to feel bitter from them since they protect a person from receiving naturally toxic substances which, as a rule, have bitter taste [1]. In addition, differences in genetic variation in taste receptors between ethnic groups may contribute to differences in dietary intake patterns [2].

Feeling bitter of some substances by a person is characterized with polymorphism. Some people cannot feel the bitterness of this substance at all. One of these kinds of substances is phenylthiocarbamide (PTC). This substance cannot be found in the pure nature, though its analogies are included in the consistencies of some plants (cabbage, rape, broccoli). The taste of such plants is perceived differently by PTC tester and PTC non-tester.

The genetic basis of different feeling of PTC bitterness is TAS2R38 gene, which is localized in the 7th chromosome (7q) of the human being. The ratio of the phenotypes of this gene is an ethno-specific sign and reveals an association with different genetically determined and nonhereditary (ecological) factors. At present, the researchers recognize the bidominal structure of TAS2R38 gene haplotypes. Many researches in different ethnicities and races reveal large variations of phenotypic structure (from 55% to 85%) [3,4]. According to the existed data, PTC insensitive phenotypes are inclined to thyroid-deficient diseases of thyroid [5]. PTC insensitive phenotypes are dominant among the [6], and by the experiment conducted in order to connect with the body mass and food ration of the phenotypes, it was revealed a correlation (72% of the investigated people) between PTC insensitive phenotypes and overweightness of children [7]. Thus, it is important to study the phenotypic variability of the senses coded with TAS2R38 gene in the point of view of both – genetic structure of population and bio-ecology.

2. Study Area and Methods

We have pursued a research on the PTC bitterness sensitivity in the representatives of Georgian ethnicities of different ages and sexes. The inquired groups were randomly selected.

The research included 183 people. We have studied 115 female and 68 male people.

The research participants were inquired in our survey on the questionnaires drawn up by us. The questionnaire contained the questions on age, physical data (height, weight), blood group and Rhesus factor, undergone infectious diseases, endocrinal system pathologies (goiter, diabetes), smoking, etc.

The research was conducted by means of PTC standard testers, which contained 3.4 mg/kg of compound phenylthiocarbamide.

Basing on the data obtained via PTC testing, the people studied according to the sensitivity of bitterness was divided into PTC-tester and PTC non-tester phenotypes. We made notes of dominant, heterozygous and recessive genotypes with the following symbols: T – symbolically denotes the bitterness sensitivity; t – denotes non-sensitivity. Consequently, TT – is dominant homozygote according to the bitterness sensitivity; Tt – Heterozygote; tt – recessive Heterozygote.

The obtained results have been mathematically processed by a statistical method. We counted the allele concentration of PTC gene and the frequency of its propagation in the population with Hardy-Weinberg's formula:

$$q^2+2q(1-q)+(1-q),$$

which reflects the distribution of genotypes in a panmictic (freely hybridizable) population. In the formula, q denotes the T allele frequency of PTC gene in the population, and the frequency of the second t allele will be 1-q.

The authenticity of the obtained research results is proved by the formula:

$$M=\_+\sqrt{}\ p(100-p)/n$$

Where, P denotes the percentage data, n – the number people studied.

3. Results and Discussions

Despite the fact that the researchers recognize the bidominal structure of TAS2R38 gene haplotypes, many researches reveal large variations (from 55% to 85%) of phenotypic structure in different ethnicities and races [8]. The distribution of phenotypic variants of this gene occurs in a specific way in every population.

183 volunteers of Georgian nationality took part in the experiment. The inquired group included the people of both sexes aged from 18 to 70. According to the obtained results of the experiment, the part of the research objects felt the universal limited amount (3.4 mg/kg) of PTC that was perceivable for a person in a tester as very bitter, the part of them – as average bitter or slightly bitter, some of them could not feel the PTC bitterness at all. The PTC insensitive people fixed the flavor of paper or could not feel any flavor.

As we can see, according to the bitterness sensitivity rather wide range of research objects were revealed (the people who are being studied). We divided them conditionally into three categories:

- Dominant homozygote (TT) – clearly feels the bitterness of PTC;
- Heterozygote (Tt) – feels the PTC bitterness or slightly feels the bitter.
- Recessive homozygote (tt) – fail to feel PTC bitterness at all. They perceive the flavor of paper or have generally neutral feelings.

On the basis of the conducted experiments, 125 examined people of the studied population (183 people) turned out to belong to PTC tester phenotype, 58 – to PTC non-tester; consequently, the people bearing the PTC sensitive phenotype made up 68% of the research population, and PTC non-sensitive people – 32% of the research population (fig. 1).

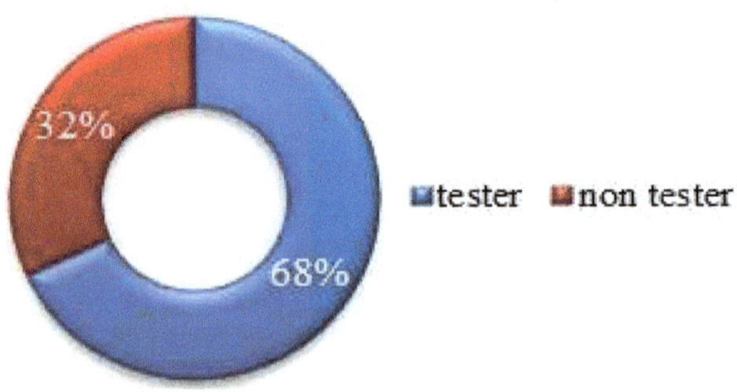

Figure 1. *Distribution of PTC tester and PTCnon-tester phenotypes in the population.*

From this, the recessive t - allele concentration of the coding gene of the bitterness sensitivity in the population equals to:

$$1-q=\sqrt{0,32}=0,57$$

Dominant T – allele concentration in the population equals to:

$$q=1-(1-q=1-0,56=0,43$$

Knowing the allele concentration enables us to calculate the distribution of the genotype frequency in the research population (fig. 2).

$$TT(q^2)=0,19;\ Tt(2q(1-q))=0,49;\ tt(1-q)^2=0,32.$$

The number of female representatives in the research amounted to 115 people. They composed the majority of the research population – 63%.

According to the experimental outcomes, 84 people turned out to be PTC tester, 31 – PTC non-tester. Consequently, the bearers of PTC tester phenotype were 73% of female inquired group and 27% were the people with PTC non-sensitivity (fig. 3).

Basing on the obtained results we can assume that in the female representatives of Georgian research ethnicities the sensitive phenotype towards the PTC bitterness equals to 73% of the population, insensitive – 27%, T and t allele concentration in the population – 0,48 and 0,52 and the frequency of genotypes of the sign of bitterness sensitivity is represented as following: dominant homozygote (TT) – 0,27; heterozygote (Tt) – 0,5; recessive homozygote (tt) – 0,23. As we can see, in the discussed group the inheritance of the bitterness sensitivity is the classical example of Mendel's first law. Essentially, the total obtained data coincide with the data obtained in the inquired population.

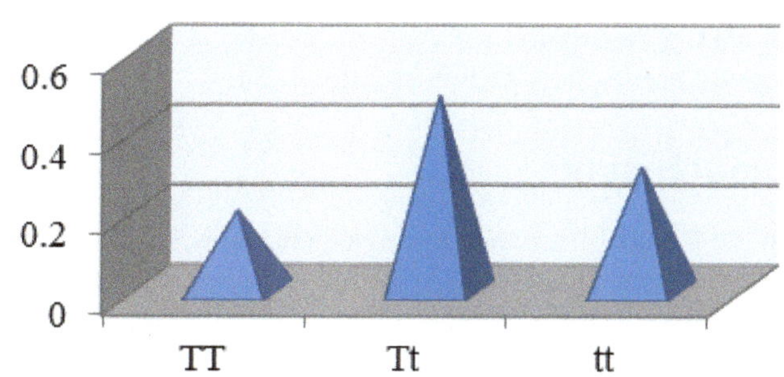

Figure 2. *Genotype frequency in the research population.*

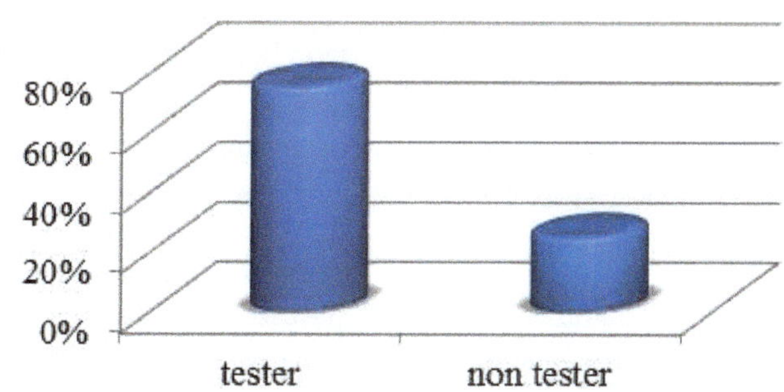

Figure 3. *Phenotypic variations according to female the sex.*

The male people in the inquired population were represented with 68 males. From this, 41 studied people turned out to be PTC tester, 27 – PTC non-tester. Consequently, the bearers of the PTC sensitive phenotypes made up 60% of the male inquired group; and the PTC non-sensitive people made up 40% (fig. 4).

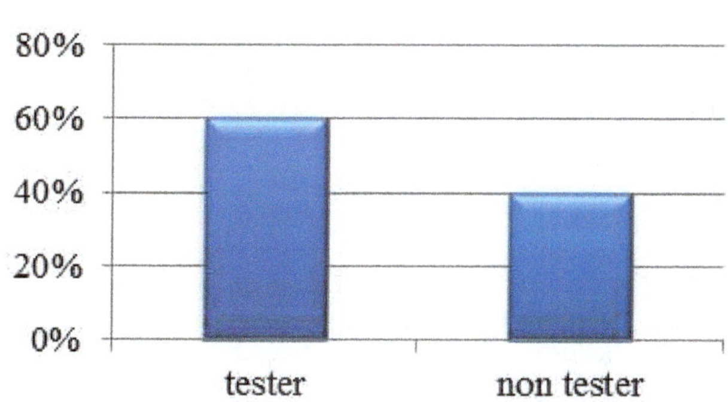

Figure 4. *Phenotypic variations according to the male sex.*

T and t allele concentration in the inquired group makes up 0,37 and 0,63 consequently, and the frequency of genotypes of bitterness sensitivity is represented as following: dominant homozygote (TT) – 0,13; heterozygote (Tt) – 0,47; recessive homozygote (tt) – 0,4.

Basing on the analyses of the obtained results, we can assume – according to the sexes in Georgian ethnicity, the structure of distributing phenotypic variations of PTC sensitivity is approximately equal. This fact has been corroborated by the equality of the frequency of the homozygotes in the population in almost every discussed cases and it does not change in connection to the sex (fig. 5).

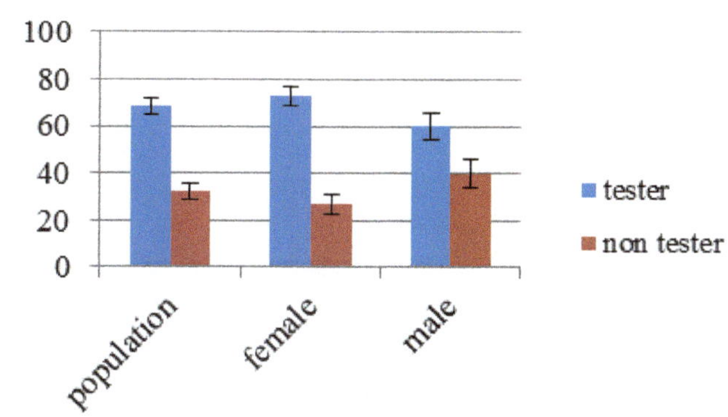

Figure 5. *PTC bitterness variations in the inquired population of Georgian ethnicity.*

There are rather controversial data in scientific references on the structure of the population of feeling PTC bitterness gene with respect to age. The majority of the authors considers the opinion that the ratio of the phenotypic variations of ethnicities and populations with respect to age is unchanged. The part of them thinks differently; they consider that according to the phenotypic expressions of this gene, varied ration is observed in different populations of different age groups. For instance, according to the current data of the molecular – genetic research on TAS2R16 gene, which was conducted with 941 individuals of 20-106 age groups in Italy, the high bitterness sensitivity was revealed among the aged people (dominant homozygote). These data increase from 35% to 55% in the aged people. The authors suggest that the PTC sensitivity could be one of the presumable markers for longevity [9].

We have studied phenotypic variations according to the ages towards PTC bitterness sensitivity in the randomly selected research populations. We divided the people studied during the analyses into three age categories: from 17 to 30, from 31 to 60 and above 60 age groups.

According to the conducted researches, the tester phenotype towards the PTC bitterness sensitivity made up 69±4,12 of the first study group; and non-tester phenotype – 31±4,12. The recessive t allele concentration in the first age group – 0,55, dominant T-allele – 0,45. The genotype frequency is represented as following: dominant homozygote (TT) – 0,2; heterozygote (Tt) – 0,49; recessive homozygote (tt) – 0,31 (fig. 6a).

In the second age group, the sensitive phenotype towards the PTC bitterness equals to 73% and non-sensitive phenotype –

27%. The concentrations of T and t – alleles in the study group equals to 0,52 and 0,48. The genotype frequency is represented as following: dominant homozygote (TT) – 0,23; heterozygote (Tt) – 0,5; recessive homozygote (tt) – 0,27 (fig. 6b)

In the third age group, 63% of the inquired people turned out to be the bearers of PTC tester phenotype, and 37% - non-tester. Consequently, in the third age group t-allele concentration is 0,61, T – 0,39. The genotype distribution frequency is: TT - 0,15; Tt- 0,48 and tt - 0,37.

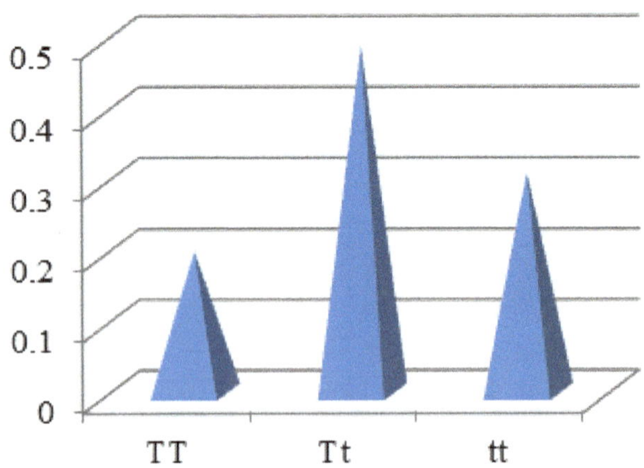

(a)

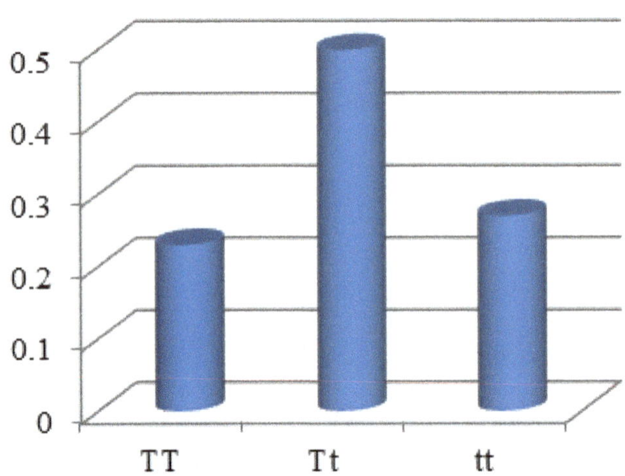

(b)

Figure 6. *Genotype frequency in the age groups of population: a - first age group; b - second age group.*

Towards the PTC sensitivity in the population studied by us, we have analyzed the frequency of phenotypic variants of the PTC bitterness sensitivity in the group addicted to cigarette. According to the experiments carried out in this regard, some phenotypes are associated with different diseases, especially with the high risk of thyroid cancer [10].

Based on the obtained data, in the group of smokers the tester phenotype having the ability of bitterness sensitivity was determined with 30% and 70% turned out to be PTC non-tester phenotype.

As a result of the conducted experiment and on the basis of the obtained results of the analysis of a group of smokers, we can assume that the majority of the smokers belong to PTC non-tester phenotype, and PTC tester are made up by about the quarter (30%) of the smoker group. As we can see, the phenotypic variations of PTC sensitivity in smokers are clearly different from the phenotypic structure of PTC sensitivity of the studied populations and generally European populations. At the same time, the obtained results coincide with the studies conducted by other authors in the cohorts of cigarette smokers, according to which the PTC non-sensitive phenotypes are fixed not only in the smokers but even in the majorities of their parents. The high percentage of the propagation of PTC non-tester phenotypes has been explained as if these people are less likely to feel the bitter taste of cigarettes. Due to this, the adults who have PTC non-tester phenotypes are more likely to become smokers in the future. Thus, PTC non-sensitive phenotype can be considered as a risk-group predisposed to this harmful habit.

4. Conclusions

On the basis of the research carried out in order to study the phenotypic variations of TAS2R38 gene in the Georgian ethnic group, we have found out that in the studied population the propagation frequency of the individuals having a phenotype with sensitivity towards the PTC bitterness equals to 68±3,14, non-sensitive phenotypes – 32±3,14. T allele concentration of TAS2R38 gene is 0,43, and t allele concentration – 0,57, consequently. The frequency of genotypes is TT - 0,19; Tt – 0,49; tt – 0,32;

According to a sex, the structure of the distribution of phenotypic variations of PTC sensitivity in the population is approximately equal.

The majority of the smokers in the studied population belong to PTC non-tester phenotype (70%). Consequently, this sign can be considered as a risk-group predisposed to cigarette consumption and during planning, preventive measures, it is recommended to consider the bearing of PTC non-tester phenotype.

References

[1] Kim UK, Drayna D. Genetics of individual differences in bitter taste perception: lessons from the PTC gene. Clin Genet. 2005 Apr;67(4), 275-280.

[2] Bibiana Garcia-Bailo, Clare Toguri, Karen M. Eny, and Ahmed El-Sohemy. Genetic Variation in Taste and Its Influence on Food Selection. OMICS A Journal of Integrative Biology 2009, v. 13(1), pp.69-80

[3] Drewnowski A., Rock CL.The influence of genetic taste markers on food acceptance. Am J Clin Nutr 1995, v.62, pp. 506–511

[4] O.V. Filiptsova, I.A. Timoshyna, Yu.N. Kobets, M.N. Kobets, I.S. Burlaka, I.A. Hurko. The population structure of Ukraine in relation to the phenylthiocarbamide sensitivity. Egyp MedHumGenet.. http://dx.doi.org/10.1016 /j.ejmhg. 2014.12.003. Published online January, 2015

[5] Facchini F., Pettener D., Rimondi A., Sichimbaeva K., Riva P., Salvi P., Pretolani E., Fiori G. Taste sensitivity to PTC and thyroid function (FT4 and TSH) in high- and low-altitude Kirghiz populations in the Pamir Hum Biol. 1994, v. 69(1), 97-106

[6] Cannon DS, Baker TB, Piper ME, Scholand MB, Lawrence DL, Drayna DT, McMahon WM, Villegas GM, Caton TC, Coon H, Leppert MF. Associations between phenylthiocarbamide gene polymorphisms and cigarette smoking. Nicotine Tob Res. 2005, v.7(6), 853-858

[7] Saraswathi YS, Mohsen Najafi, Vineeth VS, Kavitha P, Suttur S. Malini Association of phenylthiocarbamide taste blindness trait with early onset of childhood obesity in Mysore. Journal of Paramedical Sciences. 2011, vol.2 (4), 6-11

[8] Kim U, Wooding S, Ricci D, Jorde LB, Drayna D. Worldwide haplotype diversity and coding sequence variation at human bitter taste receptor loci. Hum Mutat. 2005, v. 26(3), 199-204

[9] Campa D, Rango F, Carrai M, Crocco, Montesanto A, CanzianF, Rose G, Rizzato C, Giuseppe P, Roberto Barale R. itter Taste Receptor Polymorphisms and Human Aging. PLoS One. 2012; 7(11): e45232.Published online November, 2012

[10] Shivaprasad H.S., Chaithra P.T., Kavitha P., Suttur S. Malini Role of phenylthiocarbamide as a genetic marker in predicting the predisposition of disease traits in Humans. J Nat Sci Biol Med. 2012, v. 3(1), 43–47

New Methods for Helicopter for Free Flight Inside Clouds and Precipitation Enhancement

Richard O. Claus[1], Tamara V. Tulaikova[2]

[1]NanoSonic Inc. Pembroke, Virginia, USA
[2]Wave Research Center at Prokhorov General Physics Institute, Moscow, Russia

Email address:
tulaikova@gmail.com (T. V. Tulaikova)

Abstract: The method of considerable decrease the CO_2 concentration in atmosphere based on precipitation enchainment is considered. Nowadays approaches for cloud reorganization are very important, because the sum of CO_2 and water vapor that collected in clouds makes 95% mass of greenhouse gases. Usage of helicopter for free flight inside natural clouds is considered in comparison with traditional airplanes. The method for anti-icing coating for helicopter and laboratory experiments with formulas for the cloud droplets adhesion/separation are presented. New technology for fast drying of air in medium near helicopter rotor is outlined.

Keywords: Acoustics, Helicopter, CO_2, Clouds, Precipitation Enhancement

1. Introduction and Background

Development of new technical methods for free atmosphere purification from greenhouse gases is very important in modern world and mitigation approach is necessary in order to combat the imminent dangers related to climate changes. Our climate is changing; this fact was proved by IPCC reports based on detailed world's science [1]. The global warming was caused by exceed of greenhouse gases, and annual CO_2 emission producing by world industry provide evident concentration increase in atmosphere at each year. The sum of $CO_2$and water vapor that collected in clouds makes 95% mass of greenhouse gases nowadays. Currently the role of clouds becomes important, so the methods of cloud reorganization are needed. Traditionally, humans tried use the natural clouds to produce a lot of water for agriculture at the necessary moments. The development of these methods are based on the fact that a typical cloud can contain thousands tons of water. Modern methods for precipitation enhancements have been developed in different countries around the world with wide practice [2]. The most popular current technologies for precipitation enhancement are based on airplanes those are sprinkling a chemical reagents into cloud. Note that a helicopter can serve to this purpose much better, because it is less speedy and more maneuverable in comparison with an airplane.

Our approach incorporates the possibility of stepwise CO_2 purification by cloud modification [3]. The first stage involves spraying of alkaline compounds, such as *KOH* or etc. inside cloud via an airplane or helicopter. The alkaline reagents significantly increase the solubility of CO_2 in water droplets due to increase the *pH* in cloud droplets, and rain droplets become saturated by atmospheric CO_2 during their gravitational fall from cloud to the ground. The CO_2 solubility significantly grows due to increase of *pH*, therefore ions concentrations of $[HCO_3^-]$ and $[CO_3^{2-}]$ increase in 10 and 100 times accordingly by each unit of *pH*. This method can be conducted in areas of the free atmosphere step by step, so its application on the global scale was analyzed for the whole Earth. The future technology can compensate for annual carbon emission by method application at 0.4 % – 0.08 % of our planet surface then pH = 10 – 11, [4]. The advantage of proposed approach is a good response in plant growth to the precipitation from modified clouds; one can see the strong vegetation growth by enriched watering at Fig.1 here and in [5]. To emulate the process of CO_2 absorption in water droplets during an indoor experiment, the similar changes were made in alkaline solution during long-time diffusion according to formula $L_D \approx (D \cdot t)^{1/2}$, there a diffusion coefficient in water is *D*, and *L* is a water layer depth. The

KOH was dissolved in water and resulting mixture was kept indoors during several hours $t = L_D^2/D$, as a result the solution has carbon ions of HCO_3^-, CO_3^{2-} and *K+*. The initial *pH*=12 was kept in these experiments. After saturation time *t*, the plants in brown pot were given the resulting solution and pure water was added to control blue pot (right side) with the same plants. These two watering processes were repeated regularly every day during one month. The experimental plants are beetroots (10 grains), carrots (20 grains), and parsley (2 grams) in each of plant pots. One can see a strong vegetation growth by enriched watering according to proposed method for all studied plants, Fig.1. Nowadays precipitation in industrial regions have led to soil deterioration up to *pH* = 4 or less.

Figure 1. *The indoor plants after 30 days correspondingly; plants in a right (brown) flower pot were watered by solution with carbon ions* CO_3^{2-}, HCO_3^- *and* K^+.

Other advantages of the method are based on a substantial joint use of surface of all small droplets in the current precipitation volume. The joint set of water droplets increases the available surface for gas/liquid interactions for fast and complete CO_2 absorption in the air volume under cloud. The process can be simulated in advance to take out from the atmospheric volume the desirable mass of carbon.

The idea of regional weather modification by precipitation enhancement was described earlier [6]. Nowadays the most popular methods are the sprinkling of hygroscopic particles or special solutions for warm clouds [7] and the introduction of glaciogenic substances into cold clouds [8] from an airplane. In addition to it, special acoustic devices could be used to accelerate the coalescence of water droplets in clouds. The background is the fact that acoustics provides high mobility for droplets, because they become involved in air vibrations inside sound waves with the best coalescence as a result. Previous acoustical experiments [9] and further analysis were carried out to show efficiency for acoustics inside clouds directly in the region with oversaturated water steam by using a modern airplanes or helicopter. The most effective is a joint utilization of two methods at the same time, which means that hygroscopic particles and acoustic influence would be directly applied inside one cloud area. This result depends on different areas of both considered methods. Dynamics of different hygroscopic particles show that only first 15 - 20 seconds demonstrate fast changes in the main cloud characteristics such as super saturation and spectrum dispersion. This is accompanied by an increase of the radii inside small drops from 0.01 - 0.1 μm up to 1 - 3 μm. So after the first major variations these parameters increase very slowly, hence significant changes in spectrum of cloud and precipitation increase up to 50 minutes [7]. Later the droplets should be triggered by sound waves in the form of vibrations to provide wide condensation and coalescence up to a drop size of 100 μm, and after this point gravity can predominate. Acoustic method was analyzed to obtain fast droplets coalescence inside nature clouds up to precipitation with big droplet sizes and gravity sedimentation [10]. Optimal acoustic generator is proposed; also optimal regimes of acoustical powers for typical cloud droplets are calculated in [10] and presented here at Fig.2. The Figure 2 demonstrates examples of the three acoustical ranges, as follows: (1) f = 20 Hz with lowest power Q = 175 W/m^2; (2) f = 50 Hz with Q = 800 W/m^2 and (3) f = 100 Hz with Q = 2500 W/m^2. We investigated the low frequency range from f = 50 - 100 Hz in order to provide droplet amplitude in a range L~ 0.5 – 2.5 mm for coalescence possibility for typical clouds. Additional analysis was performed and introduced the appropriate altitude changes relative to the main physical parameters in the model, as follows: air density (ρ_a), viscosity (η), sound velocity (C_a), and air temperature depending on km-altitudes, and an improvement decrease in necessary acoustic power was observed. Usage of sound source at the altitude of 2 – 6 km within the cloud by airplane or helicopter provides the high droplet coalescence to get precipitation.

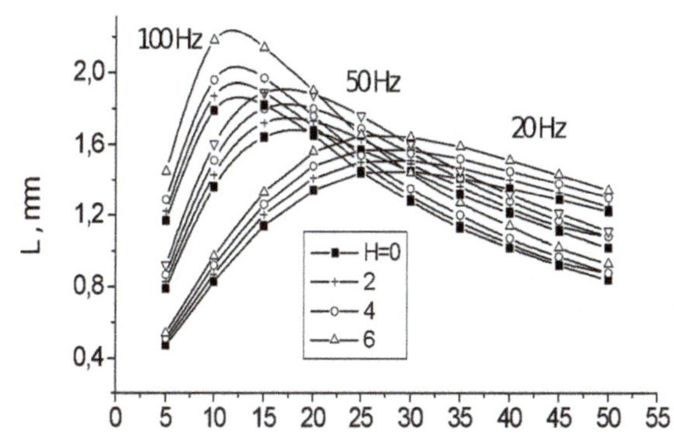

Figure 2. *Amplitudes L for vibrated droplets in acoustic field; droplet's radii are r = 5 – 50 μm and clouds are located at altitudes H = 0, 2, 4, or 6 km. Acoustic regimes are: f = 100 Hz and Q =2.5 kW/m²; f = 50 Hz and Q =800 W/m²; f = 20 Hz and Q = 175 W/m².*

The additional acoustic action can be effectively performed done by helicopter. The main advantage is the following. Rotor blade of the helicopter rotates with maximal angular velocity, so the linear velocity at the end of each blade reaches the speed of sound. Typical helicopters blade has a length of 10 meters and its partial speed varies linearly from 0 to 300 m/sec for further estimation here. Note that helicopter is moving forward with some speed, for example 100 m/sec; then supersonic motion of one blade is produced in small time periods. This effect takes place because there are the same directions of helicopter motion and one of its blades moving in the same direction with its own speed. It means that each of the blades overcomes the supersonic regimes and produces the shock wave. The shock wave has a steep front and bigger difference of pressure and local temperature comparable to surroundings. A shock way is a very best acoustical source for droplets in a cloud, such wave pushes the droplets to effective coalesce. Fast condensation inside the shock wave can be seen many times directly and also in photographs when airplane breaks the sound barrier. The helicopter has multiple advantages in comparison with airplane, and its icing immunity can be used successfully for meteorological applications, in particular for precipitation advantages. Techniques for anti-icing usage and improvement will be considered here in parts 3 and 4.

The modified precipitation droplets provide effective transport of CO_2 from the under-cloud atmospheric volumes to the ground and further more to soil, ground water and plants. Therefore, new systems for cloud management and regularly precipitation enhancement are necessary for recovery the atmosphere, soil, ocean and climate as a result.

2. The Precipitation Method for Carbon Transport from Atmosphere

The effect strongly relates to artificial changes of *pH* level in cloud water content. In nature also the ocean waters keeps significant masses of dissolved CO_2 due to its alkaline pH-level that is about $pH = 8.3$. There are known description for the insoluble and dissociated portions of the weak acid that remains after attaining the equilibrium of saturation for water by CO_2 as follows:

$$CO_2 + H_2O \Leftrightarrow H_2CO_3, HCO_3^- \Leftrightarrow CO_3^{2-} + H^+$$

$$H_2CO_3 \Leftrightarrow HCO_3^- + H^+ \quad (1)$$

The natural rain can't provide the carbon transport from atmosphere to the ground. In typical natural precipitation at the level $pH \approx 5.6$ the concentrations of the dissolved and dissociated portions are titled as follows: $[H_2CO_3]$, $[HCO_3^-]$, $[CO_3^{2-}]$, that are very small in this case. The gas solubility strongly increases with higher levels of *pH*. The effect in a cloud with 3 mkg / m^3 of NH_3 was shown by *Yunge* [11] and *Rasool* [12], so the medium has neutral $pH = 7$ but with appropriate increase for the concentration of dissolved ions. The partial concentrations are $[H_2CO_3] = 0.71$ mg/l; $[HCO_3^-] = 3.3$ mg/l, $[CO_3^{2-}] = 10^{-3}$ mg/*l*. The ratios of carbon in the first, second and third of the listed compounds are 0.1935; 0.1967; 0.20, correspondingly, that should be multiplied to concentration to find partial carbon masses and total one. The masses of carbon in 1 liter of water then $pH = 7$ are as follows: *0.1373* mg/*l*, *0.6491* mg/*L*, $2 \cdot 10^{-4}$ mg/*l*, and their sum is $C_{pH7} = 0.7866$ mg/l. The ionic concentrations of $[HCO_3^-]$ and $[CO_3^{2-}]$ increase in 10 and 100 times accordingly by each unit of *pH*. We performed similar calculations assuming that we increase *pH* in atmospheric water droplets up to 10, 11, etc [3-5]. The total mass of carbon in 1 liter of water with $pH = 10$ is $C_{pH10} = 849.1$ mg/*l*, see row 2 in Table 1.The carbon compound culd be measured in water using one of standard methods, see [14] or ets.

Let's assume the initial cloud modification to $pH = 10$, and precipitation layer $h_w = 10$ mm at surface $S = 1$ km^2, then it gives 10^7 liters. The total removed carbon mass will be $M = C_{pH!0} h_w S = 8.5$ tons in such modified precipitation water during one rain. Then $pH = 11$ the same rain will transport the 265 tons of carbon ions that is equal to 972 tons of CO_2 to the ground during this one rain.

Now, let's analyze the example volume of purified air in the atmosphere $U = 1$ km^3 under rain, where $h = 1$ km is the altitude of the cloud over land and the unit square of the surface is 1 km^2. The reduced CO_2 concentration was estimated by dissolving the carbon mass in alkaline rain using an assumption that the initial concentration of CO_2 was even under the cloud. The molar mass of CO_2 is 44, so the share of carbon in it is 0.2727. Using previous calculations, the atmospheric concentrations could be reduced to the following values: $C_{a-} = (C_{pH10}/0.2727) \cdot (h_w/h)$. Similar calculations with different *pH*-level, water layer h_w, and cloud altitude, *h*, show different possible warrants for utilization. For example, the concentration of removed CO_2 in this air volume should be approximately 420 ppm if the $pH = 10.3$ in rain water then $h_w / h = 10^{-4}$. If the height of the clouds over the land was greater than 1 km, for example $h = 6$ km, then all of the data of concentrations should be divided by 6 due to the increase in purified air volume. For this case and for a $pH = 10.8$, atmospheric purification will be complete again with removing concentration $C_{a-} = 403$ ppm, and the same precipitation layer is $h_w = 10^{-1}$ m. It is suggested the uniform CO2 vertical distributions, of cause the local effect is persists up to the moment of new injections or winds.

Key advantages for proposed method should be eliminated for alkaline rains in comparison with ocean water. First, due to small dimensions of precipitation droplets the falling time exceed the saturation time for the gas molecules inside precipitation droplets. The largest radius has been estimated as 3 mm up to drop disintegration. However, at the beginning of the precipitation the droplets have grown up to a radius of more then $r > 100$ μm the stated values for gravitational sedimentation. The fall velocity $V_g(r)$ can be approximated roughly using the stationary speed of droplet gravitation sedimentation as follows: $V_g(r) = \sqrt{2rgr_w / r_a}$, where ρ_a and $\rho_w$are the density for air and water respectively, and g is the acceleration due to gravity. The falling time for droplet can be estimated as $t_h \approx h / V_g(r)$ at the initial minimal altitude of $h = 1$ km. During CO_2 solubility, the next step demands estimations for the saturation time t_{aw} in falling droplets. The process of a concentration establishment described by the law: $t_{aw} \approx (4\pi r^3/3)/(K_{aw} 4\pi r^2) = r/(3K_{aw})$, the time depends on the ration of the aerated volume U_1 for a drop to the area of its surface S_1 in a gas flow [13].The process also depends on the gas exchange constant K_{aw}at the gas-liquid interface, the constant $K_{aw}(V)$ depends on the flow speed $V = V_g(r)$. In the past, the intensity of CO_2 gas exchange has been measured depending on the velocity of air flow V over the water surface [15-17] that gives $K_{aw}(5) \approx 6.4 \cdot 10^{-3}$ cm/ssthen V = 5m/s, and etc. Unifying the previous equation with measured constraint yields the saturation time t_{aw}= 6.3 sec, but falling time is t_{hl}= 257 sec for droplet radius r = 1 mm, and the similar for other sizes of droplets. Since $t_{aw} << t_h$ there is enough time for the saturation of typical rain droplets by CO_2 molecules during their flight to the ground from a 1 km altitude.

Analyses and calculations demonstrate the big total number of droplets in unit volume and their big summarized surface to contact with air during droplets flight. The analysis is based on the Marshall-Palmer approximation for the droplets spectrum with radii interval r_m = 0.05 - 3 mm during precipitation [3]. The volume of received water $U_l$was calculated as a sum of all of the falling droplets of the water layer I = 3 mm on a ground surface A = 1 m^2 during $T = 1$ hour. The calculations provide the total number of falling droplets in mentioned water unit, for a rain this number is N_r= 4.02·10^{11}, but for a drizzle $N_d \approx$ 6.76·10^{11} inside 3 liters of precipitation water. These results prove that a set of rain droplets have a large surface for gas/liquid interactions as compared to the ocean's plane surface. Then, each falling droplet runs at an air cylinder with a minimal volume length h and a ground area πr_m^2. The cylinder surface $S_m \approx h2\pi r_m$ means that air/water interface increases the contact for rainwater and purified air. The total sum of this surface for all droplets in considered water unit is S_r = 2·10^8 m^2 for rain, but the same units for a box of air volume has 4·10^3 m^2 with a vertical altitude of h = 1 km and a background A = 1 m^2. Important note is that due to the small concentration of CO_2 molecules (400 ppm), the probability of their collision with small water drops is low within clouds due to their small sizes. Therefore, a strong decrease in the concentration of CO_2 within clouds will not be achieved at the first step, and the alkaline reserve of a reagent introduced into the clouds will not be spent. The erosion of CO_2 as well as a significant reduction in air concentration will certainly occur in the sub-cloud volume via precipitation.

The *KOH* substance has hygroscopic properties such as that it absorb water around aerosol droplets due to condensation in the supersaturated media of clouds. The initial *KOH* aerosol droplets contain an alkaline saturated solution with a concentration n_0. Further analyses and calculations were performed in order to provide the average *pH*–levels in cloud water. The *KOH* reacts in water as follows:

$$KOH \rightarrow K^+ + OH^- \tag{2}$$

$$H^+ + OH^- \Leftrightarrow H_2O \ , \ K_w = [H^+] \cdot [OH^-]$$

Formulas for the quantity mass of *KOH* are the following:

$$n_{mol} = [OH^-] \tag{3}$$

The choice of a pH = 10, for example, demands that $[OH^-]$ = 10^{-4}, and the molar concentration now becomes10^{-4} mol/L according to last equation in (2). The mass concentration yields the value $n_{KOH}^{(pH10)}$ = 56·10^{-4} *g/L*, by taking into account the *KOH* molar mass of 56. The data can be used in the following manner: if the liquid water content in the medium of a unit cloud is W = 1 g/m^3, whereas 1 m^3 of an air cloud contains 1 cm^3 of pure water, and 5.6 mkg of masses should be introduced into a 1 m^3 of air within the cloud in order to ensure the appropriate pH= *10*. The cloud volume with 1 km^3 of volume and W = 1 g/m^3 need 5.6 kg to get pH = 10 in a volume about 1 km^3 of cloud. The 56 kg provides pH = 11 in the same cloud. For more accuracy the aerosol dynamics in cloud media was calculated in [3].

The proposed method can be applied for the whole Earth on the global scale. The surface area of our planet is A_E = 5,1·10^8 km^2 and the average annual layer of precipitation is h_w= 1,000 mm. Using algorithm and equations in [3-5] one can estimate the mass of removed atmospheric carbon (m_C^1) in 1 meter of precipitation water measured per surface 1 m^2, these are presented at 2nd row of Table 1, and pH = 10 - 11 is indicated at 1strow. The estimated masses of CO_2 in 1m^3 of water are presented at 3th row, so the carbon ratio in CO_2 is 0.2727. The calculated mass of formed carbon at the surface A_E of the whole planet is $M^A_C = m_C^1 \cdot A_E$ = 4.3·10^{14} kg then pH = 10, as listed at the 4th row of the Table 1. Hence the amount of carbon oxide is $M^A_{CO2} = M^A_C$ / 0.2727 and its value listed at the 5th row of the Table. In 2010 the global CO_2emission reached an amount of 3.06·10^{13} kg. For further calculations let's assume the global annual emission of CO_2 as the AE = 3.2·10^{13} kg. To compensate an annual CO_2emission from the ratio AE / M^A_{CO2} we estimate the minimal area at the Earth's surface, $P(A)$ %, to be used for proposed technology and the corresponding details are given at the 6th row of the Table. For our approach the method has to be applied on 2% - 0.08% of planet surface, the proposed method can completely compensate the annual carbon emission. To sum up let's estimate the mass of alkali to add for desired modifications in clouds. The mass *KOH* (Q_{KOH}) can be estimated using previous data for n_{KOH} = 5.6·10^{-3} mg/l then pH=10.Considering the precipitation layer of 1 m/year for the planet surface A its necessary percent, $P(A)$, can be estimated from the Table. The added mass of alkali at pH = 10 level is calculated here: $Q_{KOH}^{10} = n_{KOH} IAP(A)$ = 5.7·10^{10} kg. Note that similar estimations are presented at the bottom row of the Table 1. In comparison with the fertilizers for soil the added mass of alkali for clouds is relatively small.

Table 1. *The carbon mass (M^A_C) and CO_2 mass (M^A_{CO2}) at the Earth's surface A = 5.1·10^8 km^2; the minimal required surface P(A) and mass Q_{KOH} of KOH to compensate an annual carbon emission.*

1) pH	5.6	10	11
2) m_C^1, kg/m^3	1.6·10^{-4}	0.849	26.5
3) m_{CO2}^1, kg/m^3	5.9·10^{-4}	3.11	97.2
4) M^A_C, kg	8.2·10^{10}	4.3·10^{14}	1.35·10^{16}
5) M^A_{CO2}, kg	3·10^{11}	1.6·10^{15}	5·10^{16}
6) P(A), %	No possible	2%	0.06%
7) Q_{KOH} mil.tons		57	22.8

3. The Study for Helicopter Coating and Protection

The main principles for the laboratory measurements of surface properties for helicopter coating for its improvement are listed below. The purpose is to calculate forces acting on the cloud droplets of different sizes at different surfaces. The water drops were placed at different samples to study and make photographs of the contact area, and then photos were analyzed by computer. The angle θ and diameter d_k of contact of droplets with the surfaces were measured, these parameters are included in the following formula:

$$\sigma \cdot \cos\theta = \frac{F_1}{l} \tag{4}$$

here θ is a contact angle of surface and drop of water in contact point, F_1 is wetting force, l is a wetted perimeter. The surface tension of the water is σ = 75.50 mN/m for temperature T = 0°C. The force required to tear the drop from the surface is F_2, it can be calculated as follows:

$$\cos\theta \approx \frac{F_1}{F_2} \quad , \quad F_2 \approx l\sigma \tag{5}$$

The surface area of contact S_k and perimeter of this area l are calculated as follows:

$$S_k = \pi(d_k/2)^2 \text{ , } l = \pi d_k \tag{6}$$

Volumes for different drops in experiments were created by medical syringe with high accuracy. The largest drop has volume U_5 = 0.05 ml with the weight 0.05g. The smallest drop has the volume of U_1 = 0.01 ml and 0.01 g in weight. Also drops were produced and measured for the following volumes: U_2= 0.02 ml; U_3= 0.03 ml; U_4= 0.04 ml. One can estimate the radius of the drop from known formula for volume as follows:

$$r = \left(\frac{3U}{4\pi}\right)^{1/3} \tag{7}$$

From the formula (7) the free drop radii are r_1 =1.337; r_2= 1.684, r_3 = 1.928, r_4 = 2.122, r_5 = 2.229 mm respectively, and complete surface area $S = 4\pi r^2$ of droplet were calculated taking into account the droplet radius. Subtracting the surface area of $S$of the droplet and excluding its contact area we obtain the surface area S_a that air flow can blow up by the motion:

$$S_a \approx 4\pi r^2 - \pi(d_k/2)^2 \tag{8}$$

The series of measurements were done as well as calculations using the above procedure. A few typical photographs of water droplets on the surface which were analyzed are shown below in Fig. 3a, b.

The data for duralumin or Teflon surface are shown in Table 2a,b. One can see that large droplets can easily be eliminated from the surface because it takes small value of unified force $F_2/S_a$to do it.

Figure 3a. *Water droplets on duralumin surface.*

Figure 3b. *Droplets on Teflon, their masses are 0.01g; 0.02 g.*

Table 2a. *Droplet measurements on duralumin surface.*

M, G	θ, °	d_k,mm	F_2,mN	F_2/S_a, N/m²
0.01	60.2	3.42	0.811	61.1
0.05	68	6.5	1.542	47.13

Table 2b. *Droplet measurements on Teflon surface.*

M, G	θ, °	d_k,mm	F_2,mN	F_2/S_a, N/m²
0.01	144.77	1.35	*0.32*	15.2
0.02	146.59	2.03	*0.48*	14.9
0.03	152.21	2.48	*0.588*	14.0
0.04	161.15	2.80	*0.664*	13.2

Usually, clouds droplets have the radiuses r_i= 1 - 100 microns, such a small droplets can't be easily obtained in experiments, and further analysis aim to simulate the cloud medium. Let's use the algorithm for laboratory photos as at Fig.3a,b withlarge droplets on the surface to simulate small cloud droplets by approximation, Figure 4. Conversation to very small radii is shown in Fig. 4 by arrow. One can see that the ration is $r/r_k \approx 1$ for water/duralumin, so the area of contact of the micro drops is taken here equal to the cross section of the smallest droplets with of $S_{i(r)} = \pi(r_i)^2$ due to their good adhesion. The experimental measurement of the quantities F_2/S_a and F_2/S_kin Table 2a was used to calculate the droplet/duralumin dynamics. The force $F_2(r_i)$ for smallest droplets in clouds with radius r_i, measured in meters, can be calculated as follow:

$$(F_2/S_k) \cdot S_i(r) = 88.3 S_i(r) = F_2(r_i) \tag{9}$$

For hydrophobic Teflon surfaces, the contact area sure decreases with droplet mass. According to approximation at Fig.4 the small droplets with radius r exhibit much smaller radius in contact r_k on the Teflon-type surface, and ratio $r_k/r = 0.3$ of the contact radius to the radius of the drop decreased here. Simulation with the arrow on the experimental plot in Fig.4 indicates the greatest reliability of 0.3. The breakout force for small cloud droplets can be calculated using the coefficient of proportionality K:

$$F_2\,(r_i)= (F_2/\,S_k)\cdot K\cdot S(_ir)= 22.4\cdot S(_ir) \qquad (10)$$

$$K = \frac{S_k(r_i)}{S(r_i)} = \frac{4\pi r_i^2}{4\pi r_{k,i}^2} = \left(\frac{r_i}{r_{i,k}}\right)^2 \approx 0.3^2 = 0.1$$

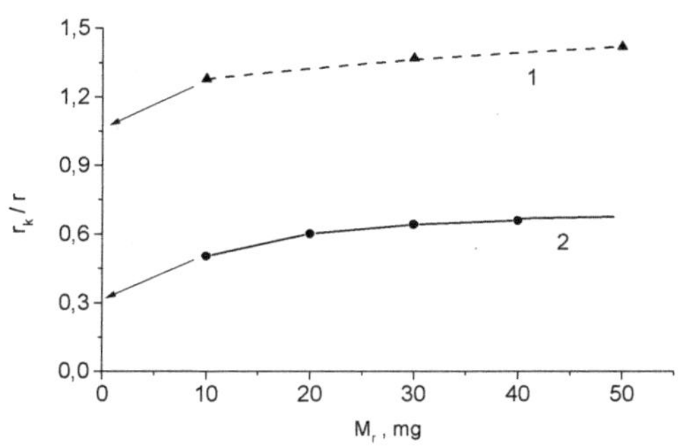

Figure 4. *The experimental ratio r_k/r of the contact radius as a function of decreasing droplet masses, M. The graph 1 displays duralumin surfaces, and the graph 2 is for Teflon.*

The results of calculation of forces required for separation of small droplets in typical cloud are shown in Table3.

Table 3. *The separation forces F_2 for droplets with radii r.*

, r,µm	F_2, N, in (9) for duralumin	F_2, N,from (10) for Teflon
1	$2.8\cdot10^{-10}$	$7\cdot10^{-11}$
2	$1.1\cdot10^{-9}$	$2.8\cdot10^{-10}$
10	$2.8\cdot10^{-8}$	$7\cdot10^{-9}$
100	$2.8\cdot10^{-6}$	$7\cdot10^{-7}$

Let's analyze different acted forces for tearing droplets from the surface on rotating helicopter blades. Firstly, the centrifugal force F_3acts on droplets of different sizes, r, on the blades at various distances from the axis of rotation L_b and creates the range of droplet speeds, respectively. Typically the velocity V at the end of the blade must not exceed 300 m/sec, so $V= 2\pi L_b\, f = 300$ m/s, with the rotor frequency is $f_3 = V/(2\pi L_b) = 4.775\ s^{-1}$, and centrifugal power for the droplet with mass, M, used for calculations:

$$F_3 = M(r)\omega^2 L_b = (4\pi r^3/3)\rho_w(2\pi)^2 f_3{}^2 L_b \qquad (11)$$

This relation show that the smaller droplets $r = 1$ micron cannot be removed by centrifugal force over the entire surface, even at a speed at the end of the helicopter blades $V_b = 300$ m/sec for both considered materials, because this value is too small, $F_3(L_b{=}10){=}3.76\cdot10^{-11}N$, see Table 3 for comparison with F_2. Droplets of radius $r = 2$ µm can be removed by centrifugal force from rotating blade with Teflon coating only from the 0.7-meters at the end of its length, because this force is $F_3(L_b{=}10){=}\ 3\cdot10^{-10}N$. Droplets with radius $r = 2$ µm can not be removed from rotating blade with duralumin surface. Droplets of radius $r = 10$ µm and more ones can be removed by centrifugal force for the length L_b= 10 m from rotating blade with Teflon surface in reality due to 5 times for exceeds of F_2, so the force $F_3(L_b{=}10){=}\ 3.76\cdot10^{-8}N$. There is partial water removing from duralumin and other no-Teflon coatings according to this algorithm.

In addition, the gravitational force $F_4 = M(r)g$ acts on a droplets. Calculations show that gravitational force is very small for all considered cloud droplets; this force can exceed the separation force only for big rainy droplets with radius r >1 mm.

The force created by flow of air is best for water/surface separation. For large drops on a moving surface one can use the following formula to estimate the drag force:

$$F_5 = \frac{\pi r^2 C\rho_w}{2}V^2 \qquad (12)$$

Taking into account speed the estimate for the Reynolds number here is $Re \sim 50$ (for $V = 100$m/s and $r = 10$µm), so coefficient was set $C = 0.4$ in (12). Air drag force in a form (12) can act effectively on the typical cloud droplets with r> 10 µm to provide the $F_5 > F_2$. For example then r = 10 µm, using equations (12) and (10) one can confirm the last statement for Teflon surface with speed V>0.3 m/s. These droplets ensure their separation from duralumin surface at speeds over 0.7 m/sec due to equations (12) with (9). In aerosol theory, in particular in the cloud theory of the motion of smallest particles of aerosol, it is assumed that the small droplets are moved by Stokes friction force into the air stream, this force can be determined by the formula:

$$F_6 = 6\pi r V \eta_a \qquad (13)$$

hereη_a is a dynamic viscosity of air. This force can tear the small cloud droplets from a moving surface effectively. The smallest droplets can be deleted from the body of the helicopter blades only then $F_2 > F_6$. It means that in practice the smallest droplets are present always at rotating blades of helicopter. For example, from equations (9) and (13) it follows that then r = 5 µm the surface speed should be larger then V= 147 m/s for duralumin surface. Also this means that half of rotating blades keep such droplets on its surface. On another hand, the Teflon coating was analyzed by comparison of forces in equations (10) and (13), so the speed should be V = 37 m/s and this means that more then 10% percent in a blade length will be free from there droplets.

Developed algorithm and calculations demonstrate that the helicopter body with hydrophobic Teflon-type coating does no hold the most of cloud droplets on its surface. Therefore hydrophobic coating provides a number of benefits and relieves the helicopter no-icing when driving in the clouds with greater conductivity. And additional to it, the smooth surface provides reflection of many droplets in comparison with rough one. The additional strong effect is the water vapor condensation and freezing at any moved surface in oversaturated medium at super cooled temperatures. The best way to stop this condensation is to dry of initial wet air as it is considered below.

4. New Approach for Helicopter to Form Dry Surroundings Around Itself

A method for reducing aircraft icing involves injecting the calculated amount of absorbent particles (e.g., salts) strictly in the space between the rotating blades [19]. Particles enlarge and get watered by cloud to safe big sizes of droplets. Then particles fall out of the field or screws under gravity or fall to the next blade due to the rotation of the ice-saline droplets. Afterwards particles are removed from the surface by centrifugal force or air stream, since these forces increase with increasing radius drops as $\sim r^3$, $\sim r^2$ respectively. These processes deliver the technological result of the complete removal of icing from the blades or/and tail rotor of the helicopter or prevention of its increase after the proposed system in flight. This method provides a significant reduction of energy consumption for anti-icing and easy technical implementation.

This technology protects the most important part of aircraft such as the main helicopter rotor providing the moving force. A helicopter is usually located in supersaturated and super-cooled areas of the cloud, so water vapor and droplets deposited on the metal can crystallize very fast. In comparison with airplanes, the greatest difficulty to affect helicopter flight is icing because helicopter flight altitude frequently coincides with the height of the clouds localization. Despite recent progress in designing methods for aircraft, this problem still is relevant, but the main advantages of helicopters compared to the plane is their good maneuverability in flight, in many respects due to the ability to hang in certain points of space. At low speeds the movement of the helicopter reduces the impact force of the flow of water drops in the air near the body and screw machine, which at high speeds can lead to failure. Therefore, the aim of the proposed method is to eliminate and prevent the icing of rotating blades of helicopters. To prove the efficiency of the proposed method the general principles of a mechanical device are used in analysis. Other systems related to the known methods and devices of this purpose are described in patents [20, 21, 22]. The most famous and widely used technologies include thermal method employing electric heaters of the front parts of rotating blades located inside them. It is widely used in the current system, but the main disadvantage of all electric systems is high power consumption of on-board power supply of the helicopter [20]. There are known the liquid systems, pneumatic, and other systems. The principle for the application of liquids for mentioned purpose, such as in the patent [23], is icing fluid. In practice for helicopters alcohol is used to act on the windshield of the cockpit. The principal disadvantage of mentioned liquid method is taking large mass of liquid required on board of the aircraft, so in practice it is impossible for the helicopters. The closest to the proposed approach is the method outlined in patent [21]. This method is too complicated; it comprises determining prior using radio measuring and active-passive radar meteorological sensing for vertical and horizontal dimensions recognition of the zones of supercooled clouds. The next step of preparations includes determining the average water content in the radial direction and other characteristics in order to detect the potentially dangerous areas of the cloud. The next step of this method involves an additional airplane fly firstly to do an active influence on these areas of potential helicopter icing by injecting into them the standard reagents that are AgI or CO_2 ice-forming solid reagents to get wet channel in cloud.

It is known that the liquid water content of clouds, W, is a basic physical parameter that determines the intensity of icing, I, measured in mm/min [25]. The mentioned value is proportional to the flight speed U, the coefficients of the surface material and its form β_1 and β_2; the icing intensity is $I \sim WU\beta_1\beta_2$.

According to the proposed method the small calculated amounts of salt having hygroscopic properties should be injected in the area between the rotor blades. The simple core experiment was performed with small helicopter's detail manufactured from duralumin material. The detail was placed in saturated solution of $NaCl$ to the time period more the one year. No considerable changes at the detailed surface were observed.

The physical mechanism is as follow. Each hygroscopic particle quickly increased in its size absorbing the water vapor around itself directly into the space between the blades of a helicopter rotor. It changes the properties in the falling droplets and air so that the icing of blades no available. Each particle of salt should be watered for the time before next rotating blades that is equal to $t_h = 1/f \approx 0.2$ sec, where f is a frequency of rotation of the rotor. The small salt particle is transformed into a large droplet size, absorbing all the excess moisture from the surrounding volume of air with radius L_{hp}. Large drops are safe, they are either fall out from the region under gravitational forces [21], or removed from the blades by centrifugal force or air stream, since these forces increase with increasing droplet radius as $r^3$and r^2 respectively. These saline droplets fall on the blade that releases salt due to the rotation of the screw of the helicopter and do not cause surface icing of blades. Note that the freezing point of an aqueous salt solution, which the hygroscopic salt materially reduced [26], for $NaCl$-26°C; for KCl-55°C. In the patent [22] the optimal composition of hygroscopic salts for anti icing agents is given for civiler application that could be used partially here, because magnesium metal and oxides were introduced in this mixture to prevent the corrosion for Fe.

Compared to [20] in the proposed method the main effect of anti-icing is achieved while reducing power consumption in a thousand times, which increases the reliability of piloting of helicopter. Compared to [22], the proposed method is significantly easier to use and require only a small amount of reagent 10 - 30 kg/hr. Compared to [22], this method provides accurate calculation of the required size and weight of hygroscopic particles related to cloud environments. Usage the proposed method in the industry delivers positive technical and economic effect and significantly increases the reliability of piloting a helicopter in bad weather.

Calculation of basic parameters of the proposed system is shown in example below. It was shown and calculated in [27] that it is sufficient to introduce a small amount, n, of hygroscopic salt for a substantial reduction of water content of the treated area. Estimations of watering hygroscopic solid particles such as salts $NaCl$, KCl show that particles from the initial range of $r = 10$ microns increased to 22 microns during the time between the release of salt into space and the arrival of the next blade due to rotation at an average water content of the cloud space of W = 1 g/m^3. As a result droplets of $NaCl$ will have concentration $\approx$10% and determine the concentration of solution droplets when the follow-blade propeller rotating at freezing temperature -13°C. In the techniques of active influence on the cloud developed now are a large number of hygroscopic particles, which can also be used for purposes of this method. Calculation of the minimal concentration, n, and summand mass, M, of hygroscopic particles to enter into rotating blade region allows for the diffusion of water supply to the salt particles from its individual volume of radius, L_{hp}. This calculation can be carried out by the next formulas:

$$L_{hp} = \sqrt{D / f} \tag{14}$$

$$n = \left\{ \left(\pi L_b^2 h\right) / L_{hp}^3 \right\} \tag{15}$$

$$M = \frac{4}{3}\pi r_s^3 \cdot \rho_s \cdot n \cdot K \cdot t_1 \tag{16}$$

Here, the coefficient of diffusion for small hygroscopic particles and air molecules in a turbulent air was D = 2 cm^2/sec, while the diffusion time in an individual particle volume between the arrival of the rotating blades t_h = 1/f = 0.21 sec. The acted are is ($\pi L_b^2 h$), then L_b = 10 m is a length of the blades of a helicopter, and h = 1 m is a thickness entrained air vane. The K (= 4) is number of blades. The ρ_s = 2 g/cm^3 is a density of salts, so L_{hp} = 6.5 mm is a characteristic diameter of the individual volume of each hygroscopic particle. The t_1 = 1 h is a time fly unit of one hour. In the example we get the estimated mass of salt M = *33 kg /h* that can be easy placed into helicopter in advance. The required mass may be reduced as a result of experiments taking into account long time of freezing brine.

Structurally, the system of delivering the air/salt mixture can be organized on the principle of dynamic sirens [3]. The stator is a stationary part with openings of the holes through which compressed air is periodically goes from central stator into the duct blade. The stator holes located outside of the rotating rotor holes that provide periodic coincidence holes, so compressed air can fall to the resonator with strong sound. The scope of the stator can have here a low power pump about 100 W and it is injected an air/salt mixture under pressure exceeding with 3 and to 5 atmospheres.This procedure provides the directional movement of the air. The resulting mass of hygroscopic particles emitted can easily be adjusted directly in the air in a controlled manner by varying the pressure in the volume of air/salt mixture during fly. The one implementation of the method to put particles is presented at Fig.5, where salty particles go away from small holes to opposite side in front of each blade. Another implementation is a common hole from additional tube located at the upper rotating blades to put particles to common area in particular in the back half of this spherical area.

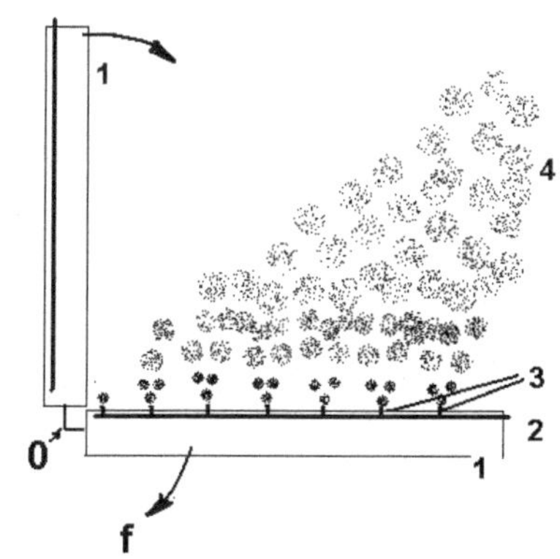

Figure 5. *Schematic representation of the method for removing ice from a helicopter rotor;*

1 - rotating the rotor blade at a frequency f, 0 is the place of rotation of the blades;
2 - duct for supplying the hygroscopic particles;
3 - holes in the duct;
4 - hygroscopic salt particles, flooded in the cloud space between the blades.

5. Conclusion

Helicopter is a very convenient instrument for precipitation enhancement, and additional proposed methods for reducing icing provide the freely fly inside natural clouds.

Acknowledgements

The authors are very grateful to Professors Thomas W. Choularton, Peter M. Cox, Yury A. Israel, Feodor V. Bunkin, Alexander Nadezhdinskii and Jean Charles Munch for useful consideration and to Dr. Svetlana Amirova for the help in a work.

References

[1] IPCC: Summary for Policymakers, In: "Climate Change 2014, Mitigation of Climate Change. Contribution of Working Group III to the Fifth Assessment Report of the Intergovernmental Panel on Climate Change", Edenhofer, O., R. Pichs-Madruga, Y. Sokona, E. Farahani, S. Kadner, K. Seyboth, A. Adler, I. Baum, S. Brunner, P. Eickemeier, B. Kriemann, J. Savolainen, S. Schlömer, C. von Stechow, T. Zwickel and J.C. Minx (eds). Cambridge University Press, Cambridge, United Kingdom and New York, NY, USA, 2014.

[2] R.T. Bruintjes. "A review of cloud seeding experiments to enhance precipitation and some new prospects". BAMS, v.80, 1999, pp.805-820.

[3] T. Tulaikova, S. Amirova. "The effective possibility for atmosphere CO_2 purification", 2012, Lap-Lambert, Germany

[4] Svetlana Amirova, Tamara Tulaikova. "One possibility for atmosphere CO_2 purification to get climate recovery". Science Discovery in Science PG. Special Issue: New Technical Ideas for Climate Recovery. Vol. 3, No. 1-2, 2015, pp.1-6.

[5] T. Tulaikova, S. Amirova. "The method for effective CO_2 purification in the atmosphere". Global Journal of Science Frontier Research. Vol. 15-H, No.1, Version 1.0, 2015, pp. 1-9.

[6] I. Langmuir, I. and V. Schaefer. "Improved methods of conditioning surfaces for adsorption" J. Am. Chem. Soc., 59, 1937, pp.1762-1763.

[7] A.S. Drofa. et al. "Formation of cloud microstructure: the role of hygroscopic particles". Izvestiya. Atmospheric and oceanic physics, v.42, 2006, pp. 355-366.

[8] S.M Shmeter, G.P. Beryulev. "Efficiency of cloud and precipitation modification with hygroscopic aerosols". Rus. Meteorology and Hydrology, v.2, 2005, pp. 43-60.

[9] Mednikov, A.P. "Acoustic coagulation and precipitation of aerosols". New York, Consultants Bureau, 1965.

[10] T.V Tulaikova, A.V. Michtchenko, S.R. Amirova. "Acoustic rains". Moscow, Physmathkniga-press, 2010

[11] H. Yunge. "Chemical compounds and radio-activity in atmosphere". Clarendon, Oxford, 1965.

[12] S.I Rasool, ed. "Chemistry of the lower atmosphere". Plenum, New York, 1973.

[13] E.K. Bjutner. "The influence of pollution of the ocean surface to transport for heat, gas and humidity". Hydromet-press, Leningrad, 1985.

[14] Edwin A. Roehl JR., Ruby C. Daamen, John B. Cook. "Estimating seawater intrusion impacts on coastal intakes as a result of climate change". Journal - American Water Works Association,. http://dx.doi.org/10.5942/jawwa2013.105.0131

[15] H.C. Broecher, J. Peterman. "The influence of wind on CO_2 exchange in a wind-water tunnel including the effect of minelayers". J. Mar. Res., V. 36, 1978, pp. 595-610.

[16] R. Wanninkhof, M. Knox. "Chemical enhancement of CO_2 exchange in natural water". Limnol. Oceanogr., v. 41, 1996, pp. 689-68.

[17] S. Kobayashi, T. Oguchi, S. Tanelli, E. Im. "Backscattering enhancement on spheroid-shaped hydrometeors: Considerations in water and ice particles uniform size and Marshall-Palmer distributed rains". Radio Science, V. 42, doi: 10.1029/2006RS003503.

[18] "The Pike. Method for determination of wet solder materials". GOST 23904-79, from 11.29.1979, Moscow, Russia.

[19] T.V. Tulaikova. "A method for the elimination and prevention of icing of the rotor of the helicopter in the clouds". Patent Ru2541548 from 24.07.2013.

[20] G.G. Blutshtein, E.V. Busigin, S.G. Golberg, V.V. Nam. "Electro heating anti-icing system, for helicopters for example". Patern RU 2226481 from 19.07.2002.

[21] V.D. Stepanenko, G.G. Shukin at ell. "The method for decrease the icing for aircraft". PatentRU 2228883, from 11.01.2002.

[22] Yu. S. Saphrigin, Yu. V. Buksha at ell. "The method of anti icing substrat". Patent RU 2212428, from 25.04.2002.

[23] R Heikky, Yu.-P. Nieminen. "The anti icing liquid for aircraft and related roads". Patent RU 2191199, from 26.08.1998.

[24] A.A. Chernikov, G.P. Beriulev, V.P. Korneev. "The method of active action for clouds". Patent RU 2295231 from 07.07.2005.

[25] I.P. Mazin. "Physical fundamentals of aircraft icing". Moscow: Gidrometeoizdat, 1957.

[26] E.V. Zielinska, E.Y. Voronin. "Theoretical aspects of the use of hydro resources". Moscow: The Academy of Natural Sciences, 2009.

[27] Y. S. Sedunov. "Physics of liquid-phase in the atmosphere". Leningrad: Gidrometeoizdat, 1972.

22

Natural and Anthropogenic Dynamics of Vegetation in the Aral Sea Coast

Dimeyeva Liliya

Institute of Botany & Phytointroduction /Ministry of Education & Science, Almaty, Republic of Kazakhstan

Email address:
l.dimeyeva@mail.ru

Abstract: Natural dynamics (primary successions) is studied in the dry seabed of the Aral Sea. Long-term studies of vegetation have identified three types of primary successions: psammosere, halosere and potamosere (sere of shrubby riparian vegetation). They differ by soil texture and salinity, patterns of temporal dynamics, and stages, selected on a basis of ecological-physiognomic features of dominant plants. Late seral stages were identified for succession types: psammophytic shrub (*Calligonum* spp, *Astragalus brachypus*, etc.) for psammosere; haloxerophytic and xerophytic dwarf semishrubs (*Anabasis salsa, Artemisia pauciflora, A. terrae-albae*) for halosere. There is a change of a dominant plant and succession dynamics in late seral stages in potamosere (*Tamarix* spp. → *Calligonum* spp, *Haloxylon aphyllum*, *Artemisia terrae-albae*). Anthropogenic dynamics of vegetation (secondary successions) depends on factors of disturbance. There is a set of anthropogenic factors causing degradation of vegetation cover: (1) agricultural: overgrazing, haymaking, plowing, clearing trees and shrubs; (2) linear structures (paved and dirt roads); (3) water management: construction and operation of hydraulic structures, fluctuation in river runoff and the sea level, disturbance in the natural flooding regime; (4) fires; (5) recreations. The leading factors in the region are connected with water management and irrigation. To identify the dynamics of plant communities and potential degradation trends, there are series of shifts (successional series): hydro-, xero-, halo-, psammo- seres characterizing moisture content and edaphic environments. Changes in the hydrologic regime of the Syrdarya river, building of the hydraulic structures lead to reduction of hydromorphic vegetation and a change it into halophytic desert. The construction of the Kokaral dam and restoration of the Small Aral Sea has led to the rehabilitation of wetlands and plant diversity. Hydrogenous succession facilitated a gradual recovery of populations of rare species listed in the Red Data Book of Kazakhstan, IUCN, Europe (*Scirpus kasachstanicus, Nymphoides peltatum, Salvinia natans, Typha minima*).

Keywords: Aral Sea, Primary, Secondary Successions

1. Introduction

Vegetation dynamics is the gradual directional irreversible changes caused by both internal and external factors [1; 2]. V. Sukachev [3; 4] distinguished the following types of successions in natural vegetation dynamics: syngenetic, endoecogenic, exogenic.

Autogenic successions caused by internal factors, allogeneic ones are related to external causes. Division successions on the primary and secondary was first introduced by F. Clements [5]. Primary succession is the establishment of plants on land that has not been previously vegetated. Secondary succession is the invasion of a habitat by plants on land that was previously vegetated, but destroyed by natural or human disturbances [6]. Primary and secondary successions refer to autogenic [2; 7]. They are controlled by four basic models [8; 9]: facilitation, tolerance, inhibition, and neutrality.

Anthropogenic successions are a special case of allogeneic (exogenic). After termination of the external factor allogenic succession is replaced by autogenic restoring. Classification of anthropogenic successions is based on the agents that cause them: technogenic, fire, irrigation, recreation, etc.

The desiccation of the Aral Sea has started in the early 60s of the last century. Primary succession covers a territory of 60,000 square kilometers. About 70% of the desiccated area are represented by salt desert. Long-term study has demonstrated that the primary succession rate is rather slow, especially within the dry seafloor of 2000s. Vegetation change sequence, trend of plant colonization and probable final stages depend on the sea sediments (substratum). The dry seabed (Aralkum) is a natural laboratory for studying the process of climate aridization, desertification, formation of primary plant communities and ecosystems. The remoteness from populated areas, inaccessibility to transport, and unsuitability for grazing to a minimum reduced anthropogenic impact. Therefore, the primary succession of the Aral coast can be seen as a classic example of autogenic ecological succession of the coastal plains in the time of marine regressions started from pioneer plants of the initial stage to the late seral plant communities.

The flora of the Aral Sea coast (in the limits of Kazakhstan) comprises 414 vascular plant species of 192 genera and 43 families. There are 24 (5.8%) endemics of Kazakhstan in the species list [10]. The dry seabed has been colonized by 344 species (83%) [11], among them are 19 endemics. The flora of the wetlands of the Syrdarya river delta front and the surrounding area consists of 112 vascular plant species of 82 genera and 35 families [12]. Four plant species are listed in Red Data Book of Kazakhstan [13]: *Nymphoides peltatum* (S.G.Gmel) O. Kuntze, *Atriplex pratovii* Suchor., *Scirpus kasachstanicus* Dobroch., *Tulipa borszczowii* Regel. In the flora of the wetlands are found European threatened species [14]: *Salvinia natans* (L.) All., *Typha minima* Funck.

2. Study Area & Methods

Study on natural dynamics was carried out in the dry seabed and the ancient marine terraces of the Aral Sea. Anthropogenic dynamics was studied in the Syrdarya river delta.

Traditional methods of field geobotany [15; 16] were used for identifying patterns of natural and anthropogenic vegetation dynamics, including: landscape and environmental profiling; geobotanical description of main plant communities. Collection of herbarium specimens was conducted simultaneously with the study of vegetation cover in all habitats. The Vegetation Description Form was used to record vegetative and environmental characteristics of the 100 (200 sq. m) sample plots. The form included the following information: GPS coordinates, elevation, topography, soil type, percent cover, species list, degree and type of disturbance, etc.

The study of plant successions is based on direct and indirect methods [17]. To study anthropogenic vegetation dynamics method of anthropo-dynamic series (transects) [18] was used. The spatial position of plant communities and their successional status is determined by a distance from the source of disturbance and the impact intensity [19]. The assessment of the degree of anthropogenic transformation is determined according to the criteria of vegetation disturbance: species composition, dominant of plant community, projective coverage, abundance, productivity, vitality, habitus, status of litter, presence of indicators of transformation, coverage of biological crust [20-23]. To identify the degree of disturbance is necessary to know the characteristics of undisturbed plant communities. Condition of a soil cover is also taking into account.

3. Results & Discussions

First investigations on the formation of vegetation cover in the dry seabed of the Aral Sea in Kazakhstan were started in the late 1970s the last century, when the sea level desiccation has become a disaster [24; 25]. In that time were already identified the main types of vegetation (Halophyta, Psammophyta, Potamophyta) and schemes of anticipated changes depending on different factors [26]. The initial stages of primary succession (syngenesis, *sensu* Sukachev[27]) have been described in detail by Kurochkina and Wucherer [28; 29]. Physiological aspects of primary successions were investigated by Pankratova [30; 31]; the dynamics of aboveground productivity of plant communities along spatial and temporal gradient – by Dimeyeva [32]. Psammophytic succession of the south-eastern coast of the Aral Sea was described in detail by Wucherer and Breckle [33]. Causes influencing primary successions such as time, exogenic factors, and plant sources were revealed [34].

Three types of primary successions: psammosere, halosere and potamosere (sere of shrubby riparian vegetation) have identified and described in 2007 [35; 36]. They differ by soil texture and salinity, patterns of temporal dynamics, indices of plant diversity and stages, selected on a basis of ecological-physiognomic features of dominant plants. Assessment of distribution and structure of vegetation in the Aral Sea ancient marine terraces helped to identify the late seral and sub-climax stages [37]. The age of the ancient marine terraces is over 3000 years [38].

The sandy belt encircles the whole Aral Seashore. At first stages of the sea regression, plant colonization all over the coastline was by psammosere.

Currently on coastal sands have been formed psammophyitic shrub vegetation (*Calligonum* spp, *Astragalus brachypus, Eremosparton aphyllum, Atraphaxis spinosa*), often with the participation of desert tree *Haloxylon aphyllum*.

Five stages of succession have been found for the psammosere:

(1) Stage of annual halomesophytes and halomesoxerophytes (*Salicornia europaea, Suaeda crassifolia, S. acuminata*) on marshy solonchak soils;
(2) Stage of annual halomesoxerophytes (*Atriplex pratovii*) on coastal solonchaks and coastal solonchakous soils with blown sand cover;
(3) Stage of psammoxerophylous grass *(Stipagrostis pennata*) on slightly saline hummocky sands;
(4) Stage of psammomesoxerophilic shrubs (*Calligonum* spp., *Eremosparton aphyllum, Astragalus brachypus*, with *Haloxylon aphyllum*) on non-saline hummocky sands;

(5) Stage of sub-climax xerophytic dwarf semishrub (sagebrush, wormwood), and haloxylon (*Artemisia terrae-albae, A. arenaria, Haloxylon aphyllum*) on plain and undulating sands, and late seral mixed psammophytic shrub (*Calligonum* spp., *Astragalus brachypus, Salsola arbuscula*) communities on hummocky sands.

Psammophytic succession is subject to the facilitation model. Within 2-3 years salt-marsh grounds keep moisture to support the growth of succulent halophytes. Gradually, the groundwater level decreases, surface horizon of sandy soil loses moisture after that wind started to change ideally smooth.

relief to undulated. Such conditions become favorable for the *Atriplex pratovii* colonization. Sustainable desalinization takes place after 10 years of continental development [39]. By this time, communities of *Stipagrostis pennata* and *Eremosparton aphyllum* are formed. The Soil becomes more stable, litter is accumulated, in plant communities appear psammophytes of continental sands (*Chondrilla brevirostris, Linaria dolichoceras, Astragalus lehmannianus, Artemisia arenaria*). Gradually exogenous factors lose their importance, thus endogenic processes start playing a leading role.

The severe salinity and heavy texture of soils are the main features of vegetation development in halosere. This is the most widely spread succession type. The halophytic succession passes several stages from annual hyperhalophytic saltwort communities to the haloxerophytic and xerophytic dwarf semishrub communities.

Five stages of the halosere have been identified:

(1) Stage of annual halomesophytes and haloxeromesophytes (*Salicornia europaea, Suaeda acuminata, S. crassifolia*) on marshy solonchaks;
(2) Stage of annual haloxeromesophytes and halomesoxerophytes on coastal solonchaks;
(3) Stage of halomesoxerophilic shrubs (*Halostachys belangeriana*) on coastal solonchaks and coastal solonchakous soils;
(4) Stage of halomesoxerophilic dwarf semishrubs (*Halocnemum strobilaceum*) on coastal solonchaks and coastal solonchakous soils;
(5) Stage of haloxerophilic and xerophilic dwarf semishrubs: (a) anabasis (*Anabasis salsa*), black sagebrush (*Artemisia pauciflora*) on desert solonetz; (b) kochia (*Kochia prostrata*), sagebrush (*Artemisia terrae-albae, A. scopiformis*), suaeda (*Suaeda physophora*) on brown desert solonetzic soils.

Halosere represents a succession, the development of which is a model of tolerance in conditions of changes in groundwater level, salinity of soil horizons and redistribution of salts in the soil profile. The environment determines the selection of salt-tolerant species.

Potamosere is typical for the formation of the tugaic (floodplain) shrub vegetation. In the dry seabed of the Aral Sea communities of *Tamarix* genera (*T. laxa, T. elongata, T. hispida, T. ramosissima*) belong to this type. Tamarisk thickets in river valleys pertain to desertified tugai [40].

Tamarisk community development takes place mostly in the surf belt, when the sea stays for a long time at the same elevation mark. As the sea desiccates, there can be formed several belts parallel to the coast line. In the composition of plant communities are observed meadow-riparian species (*Aeluropus littoralis, Calamagrostis dubia, Alhagi pseudalhagi Karelinia caspia, Argusia sibirica,* etc.), psammophytes (*Salsola paulsenii, Coryspermum aralo-caspicum, Stipagrostis pennata*), halophytes (*Limonium otolepis, Salsola nitraria*) and ephemerals (*Hyoscyamus pusillus, Senecio noeanus, Strigosella circinata*). Vegetation development follows the facilitation model, when the predominating species create proper conditions for smaller plants normally associated with the dominant species.

The formation of plant communities related to tugaic vegetation in the dry sea floor is largely depends on ecological conditions and biological features of the dominant species. The root system of tamarisk is closely associated with groundwater table. Tamarisk can grow when the ground water level is 1.5-5 m.

Roots can go down to 7-8 m deep caused by groundwater decrease. In the transition to automorphic ground conditions, riparian shrubs lose the dominate role and are replaced by desert species: haloxylon, sagebrush, wormwood [41]. In the area of ancient marine terraces, tamarisk becomes ecological relict. By reducing groundwater level tamarisk community is changed by late seral desert species.

For the potamosere, three stages have been detected:

(1) Stage of dense shoots of tamarisk (*Tamarix* spp.) on marshy solonchaks in the surf belt;
(2) Stage of formation of tamarisk communities; stabilization of species composition and ecological conditions (during the life time of the dominant);
(3) Stage of change in vegetation type - tugaic shrub by desert psammophytic shrub (*Calligonum* spp., *Astragalus brachypus, Ammodendron bifolium*), haloxylon (*Haloxylon aphyllum, H. persicum*), and sagebrush, wormwood (*Artemisia terrae-albae, A. arenaria*).

Anthropogenic dynamics of vegetation (secondary successions) depends on factors of disturbance. There are several anthropogenic factors causing destabilization of vegetation cover and the environment in the coastal area and the Syrdarya river delta.

Agricultural factors: livestock breeding (well-worn paths, overgrazing in watering places); plant breeding (haymaking, plowing, discharge drainage water to the river); clearing trees and shrubs. Grazing is a major factor impacting negatively on floodplain vegetation. Due to overgrazing the diversity of grass species reduces; the process of fodder plants (*Phragmites australis, Calamagrostis epigeios, Alhagi pseudalhagi*) disappearance provides vacant niches for occupation of toxic and unpalatable species (*Xanthium strumarium, Peganum harmala, Sphaerophysa salsula,* etc). Desert rangelands, in general, are characterized by a medium degree of disturbance with a strong degree in a distance of 5 km around wells and villages. Some areas of shifting sand are strongly desertified, it is connected mainly with overgrazing, wind erosion and shrub- and woodlands clearing. Vegetation of sandy dunes with psammophytic shrubs, tamarisk and haloxylon communities are moderately disturbed by grazing. Communities of annual and perennial saltworts in salty areas are weakly changed due to their poor properties as forage. Each rangeland type has its own series of degradation depending on initial species composition and soil. However, strong degree of desertification causes convergence of different vegetation types to formation weed aggregations with dominance *Ceratocarpus arenarius, Heliotropium argusioides, Euphorbia seguierana, Peganum harmala, Cousinia affinis, Iris tenuifolia, Salsola nitraria* in the sandy deserts and *Ceratocarpus urticulosus, Anabasis aphylla, Climacoptera brachiata* – in the clay deserts.

Reeds in the area are traditionally mowed for hay. Denudation of the coastal zone may disrupt the natural functioning of ecosystems, reduce their role as natural filters and shelter for waterfowl.

Fires can occur under the influence of both natural and anthropogenic factors. Reed stand is frequently subjected to intentional burning by humans with expectation of young shoots emergence in replace of burned out old grass. The fires involve floodplain forest and shrubs as well. The functioning of ecosystems becomes disturbed. Reed communities recover quickly, though the restoration of tugaic woodlands takes a long time. Mature tugaic trees cannot recover: they are replaced by shrubs.

Recreation is connected with the spontaneous development of tourism on the coastal area of the Small Aral Sea and campings of fishermen on the river banks. Irregular activity leads to accumulation of garbage and the total destruction of vegetation on the coastal strip. Mesophytic grass and forb meadows (*Calamagrostis epigeios, Glycyrrhiza glabra, Lythrum salicaria, Althaea officinalis*) are trampled down, damaged woodlands, mainly young shoots of *Salix songarica, S. wilhelmsiana, Elaeagnus oxycarpa*.

Linear structures (paved and dirt roads), especially new dirt roads, laid out along the shoreline of the Small Aral Sea and the Syrdarya river, causing disturbance of the vegetation or completely destroy it in the area of 10-20 m away the rut of roads.

The leading anthropogenic factors in the region are connected with water management and irrigation. To identify the dynamics of plant communities and potential degradation trends, there are series of shifts (successional series of plant communities): hydro-, xero-, halo-, psammo-seres characterizing moisture content and edaphic environments [42]. Alteration in hydrologic regime to less watering leads to reduction of hydromorphic vegetation and a change it into halophytic desert; ecological ranges become shortened with less floristic and phytocoenotic diversity. Successions of vegetation have mixed natural-anthropogenic origin: climax stage is not reached [43]. Change the ecological situation due to human impact in the Syrdarya river valley and the delta has began in 40th of the XX century. By the 1960s the rate of water flow use in the river basin approached to the unit. Series of dams in upper and middle stream completely stopped spring-summer floods in 1971, decreased run-off volume, increased mineralization. Anthropogenic factors are leading exogenous factors which control vegetation dynamics in the Syrdarya river valley and delta. Natural plant communities have practically disappeared, especially in irrigation regions. Natural vegetation cover disturbed in all elements of landscape structure. Earlier described plant communities have preserved as separate fragments differing by more simple spatial structure and incomplete floristic composition.

Disturbance of agricultural technologies and crop rotation led to secondary salinization of soils, 60% of irrigated lands affected by high salinity in 2000 [44]. Due to deterioration of the ecological conditions part of agricultural fields have abandoned. Only halophytes (*Halocnemum strobilaceum, Halostachys belangeriana, Suaeda microphylla*, etc.) can grow in fallow lands.

The Hydrological regime of delta lakes has an influence on wetlands [45]. Decreasing of water level changes hydrophyte vegetation (*Typha angustifolia, Scirpus lacustris, Phragmites australis*) into halophytic meadows (*Aeluropus littoralis, Juncus gerardii, Crypsis schoenoides*) and shrublands (*Tamarix hispida, T. laxa, Lycium ruthenicum*). New wetlands have formed along irrigation canals. Plant communities are distributed by short ecological ranges according to changes of soil moisture and degree of salinity: *Typha angustifolia, T. laxmannii, Scirpus littoralis* → *Calamagrostis epigeios, C. pseudophragmites* → *Karelinia caspia, Suaeda acuminata, S. microphylla, Saussurea amara*, etc.

Tugai forest in a middle stream of the river valley has decreased by 70-75% [21]. Natural communities of turanga (*Populus pruinosa, P. diversifolia)* has preserved by several fragments, shrublands (*Salix songarica, S. wilhelmsiana*) – in the lower flow. River-channel straightening is one of limiting factors for distribution of the floodplain forest.

In 2005 the Aral Sea was divided into the Small and Large Aral due to completion of the Kokaral Dam. After construction of the dam the Small Aral Sea level raised by 12 m. Mineralization of the sea water decreased from 23 to 5.2 (7.5) g/l. Restoration of the Small Aral Sea has led to the rehabilitation of wetlands and biodiversity [46; 47; 48]. Hydrogenous succession facilitated to a gradual recovery of populations of rare species listed in Red Data Book of Kazakhstan, IUCN, Europe (*Scirpus kasachstanicus, Nymphoides peltatum, Salvinia natans, Typha minima*).

The newly developing vegetation in the Syrdarya delta front is subject to various degrees of disturbance. The conditions of the aquatic vegetation, including beds of *Nymphoides peltatum, Potamogeton perfoliatus, P. crispus, P. pectinatum, Myriophyllum spicatum*, and submerged thickets of *Zostera noltii* correspond to their natural state.

The grassy marshes of reeds, *Typha angustifolia*, *Scirpus littoralis, S. lacustris*, *Bolboschoenus planiculmis, B. maritimus* covering the area between the ducts of the Syrdarya mouth, are not disturbed due to inaccessibility. The emerging communities of mixed grass meadows with seedlings of willow on the flats are not transformed. Young floodplain forests are impaired weakly and moderately. Middle-aged willow (*Salix songarica, S. wilhelmsiana)* forest reveals a severe level of disruption due to grazing and recreation. The cattle are grazed in the coastal thickets of reeds and cattail and the grass is mowed for hay.

Heavily disturbed soil and vegetation cover is observed near the dam. The soils of these areas are completely destroyed mechanically till the parent rock. In such areas the original soil is replaced by backfilling and pavement. The vegetation is represented by sparse aggregations of annual saltworts interspersed with single bushes of tamarisk. Plant communities grown in adjacent to the road areas and downstream of the dam are heavily disturbed.

4. Conclusions

The study of natural and anthropogenic vegetation dynamics is an important task for environmentally unstable the Aral region. Assessment of the modern state of vegetation cover helps us to identify the processes of desertification and determine the main goals to reduce the negative consequences of the ecological crisis. Principles of management of natural and anthropogenic vegetation dynamics are based on a determination of successional status of plant communities. The study of primary successions, the sequence of stages is a scientific basis for the phytomelioration. Acceleration of primary plant succession is achieved by introduction of late succession species of competitive (C) strategy (*Haloxylon aphyllum*) at the initial stages [49]. A study of secondary successions allows us to identify the opportunities for natural revegetation after termination of human impact, or to choose the methods and species-phytomelorants to speed up the recovery of degraded pastures or create green shelter belts around settlements [50].

Assessment of restoration potential of hydromorphic ecosystems after construction of the Kokaral Dam showed that over a period of 6-8 years, wetlands were rehabilitated [47; 48]. The wetlands maintain high biodiversity of ecosystem components; therefore in 2012 they were included in the Ramsar List of globally important wetlands. National Important Plant Area site "The Syrdarya River Delta" was described in the river delta front and the deltaic lake Raimkol [51].

However, anthropogenic factors have a significant impact on vegetation and biota, therefore, the conservation of the most valuable sites and giving them the status of nature reserves will preserve them for future generations. In the Aral Sea region already exists "Barsa-Kelmes" Nature reserve, where ecosystems of northern deserts are protected, including the desiccated floor of the Aral Sea [52]. Taking into account the significance for biodiversity conservation it is proposed to establish the nature protected territory "Cluster area of the Syrdarya delta front" within the area of the "Barsa-Kelmes State Nature Reserve" [47]. In the buffer zone adjacent to a protected natural area is important arrange monitoring over the rate of anthropogenic load and implement ecologically-friendly nature management.

References

[1] T.I. Rabotnov, Phytocoenology. Moscow: Moscow University, 1978, 384.

[2] B.M.Mirkin, L.G. Naumova, and A. Solomesh, Modern science about vegetation. Moscow: Logos, 2001, 264.

[3] V.N. Sukachev, Plant communities (Introduction to phytocoenology). Leningrad-Moscow: Kniga, 1926, 240.

[4] V.N. Sukachev, "An idea of development in phytocoenology", Soviet botany, vol. 1-3, 1942.

[5] F.E. Clements, Plant succession: an analysis of the development of vegetation. Washington D.C., 1916, 512.

[6] M. Pidwirny, Plant Succession. Fundamentals of Physical Geography. 2nd Edition. Date Viewed. http://www.physicalgeography.net/fundamentals/9i.html, 2006.

[7] M. Begon, J.L. Harper and C.R. Townsend, Ecology: individuals, populations and communities. Sunderland, Mass.: Sinauer Associates, 1986, 876.

[8] J. Connel and R. Slatyer, "Mechanisms of succession in natural communities and their role in community stability and organization", Am. Nat., vol. 111: 982, pp. 1119-1144, 1977.

[9] D.B. Botkin, "Causality and succession", in Forest succession: concepts and application, N.-Y., 1981, pp. 36-55.

[10] L.A. Dimeyeva, I.V. Pankratova, "Peculiarities of the Aral Sea and the Caspian Sea marine plains flora", Izvestia of Penza Pedag. University. Natural sciences, vol. 3, pp. 23-27, 2011.

[11] L.A. Dimeyeva, S.W. Breckle and W. Wucherer, "Flora of the Aralkum desert (in the limits of Kazakhstan", Bulletin of NAS RK. Biol. and Medical series, vol. 6, pp. 25-31, 2008.

[12] L.A. Dimeyeva and Z. Alimbetova,"Flora and vegetation of the cluster area "The Syrdarya river delta front" of Barsa-Kelmes nature reserve", in Proceedings of the conference dedicated to the 40th anniversary of Mangyshlak experimental botanical garden, Aktau, Kazakhstan, pp. 153-157, 2012.

[13] The Red Data Book of Kazakhstan. Plants, Almaty: ArtPrint, vol. 2:1, 431, 2014.

[14] Convention on the conservation of European wildlife and natural habitats. Annex I - Strictly protected flora species. Bern, 19.IX, 1979.

[15] B.A. Bykov, Geobotany. Alma-Ata: Nauka, 1978, 288 .

[16] Field Geobotany. Moscow-Leningrad: Nauka, vol. 1-5, 1959-1976.

[17] V.D. Alexandrova, "Study of vegetation cover changes", in Field Geobotany, Moscow-Leningrad: Nauka, vol. 3, 1964, pp. 300-447.

[18] L.Ya. Kurochkina, G.B. Makulbekova and G.K. Bizhanova, "Methods of study and mapping of anthropogenic changes of vegetation", Problems of deserts development, vol. 3, pp. 3-9, 1983.

[19] T.I. Rabotnov, Phytocenology. Moscow: Moscow University, 1983, 296.

[20] E.I.Rachkovskaya, N.P. Ogar and O.V. Marynich, "Factors of anthropogenic transformation and their influence on steppe vegetation in Kazakhstan", Steppe bulletin, vol. 5, pp. 22-25, 1999.

[21] N.P. Ogar, Vegetation of the river valleys in semiarid and arid regions of continental Asia. Theses of DSci in biology. Almaty, 1999, 47.

[22] D. Tongway, Rangeland soil condition assessment manual. Canberra: CSIRO, 1994, 69.

[23] L.A. Dimeyeva, "About additional criteria of assessment and recovery of anthropogenic ecosystems", Arid ecosystems, vol. 10:22, pp. 112-120, 2004.

[24] L.Ya. Kurochkina, "Botanical investigations in the Aral Sea basin", Problems of deserts development, vol., 2, pp. 9-17, 1979.

[25] W. Wucherer, "Primary colonization of the dry sea floor of the Aral Sea", Problems of deserts development, vol. 3, pp. 9-18, 1979.

[26] G.B. Makulbekova, "Vegetation of Bosay transect and its changes", Problems of deserts development, vol. 5, pp. 31-36, 1979.

[27] V.N. Sukachev, "About some key issues of phytocoenology", in Problems of botany, Moscow-Leningrad: AS of SSSR, vol. 1, p. 449-464, 1950.

[28] L.Ya. Kurochkina and W.Wucherer, "The development of V.N. Sukachev ideas about syngenesis", in Problems of dynamics of biogeocoenoses, Moscow: Nauka, 1987, pp. 5-27.

[29] W. Wucherer, Vegetation development on the new habitats in the desert. Alma-Ata: Gylym, 1990, 216.

[30] I.V. Pankratova, "Characteristics of sand dune ecosystems in the north-eastern coast of Barsakelmes island", in Proceedings of Nature Reserve "Barsa-Kelmes", Almaty: Tethis, vol. 2, pp. 45-80, 2007.

[31] I.V. Pankratova, "Water regime of phytocoenoses in the sandy coast of the Aral Sea", Arid ecosystems, vol. 13:33-34, pp. 101–112, 2007.

[32] L.A. Dimeyeva, "Productivity dynamics of phytocoenoses in the dry seabed of the Aral Sea", News of Nation. Acad. of Sc. Biol. Series, vol. 5, pp. 17-24, 1994.

[33] W. Wucherer and S.W. Breckle, "Psammophytic succession at the south-east coast of the Aral Sea", in: Botanical geography of Kazakhstan and Middle Asia, E.I. Rachkosvkaya, E.A. Volkova and V.N. Khramtsov (Editors), Saint Petersburg: Boston-Specter, 2003, pp. 340-347.

[34] W.Wucherer, S.-W. Breckle and A. Buras, "Primary succession in the Aralkum", in Aralkum – a Man-Made Desert, S.W.Breckle, W.Wucherer, L.A. Dimeyeva and N.P. Ogar (Editors), Springer Heidelberg Dordrecht London New York, 2012, pp.161-198.

[35] L.A. Dimeyeva, "Primary successions in the new Aral Sea shore", Basic and Applied Dryland research, vol. 1, pp. 1-16, 2007. http://www.badr-online.de

[36] L.A. Dimeyeva, "Mechanisms of primary successions in the Aral Sea coast", Arid ecosystems, vol.13: 33-34, pp. 91-102, 2007.

[37] L.A. Dimeyeva, "Ecological-historical stages of Priaralie vegetation forming", Bulletin of Moscow Society of Naturalists. Biol. Series, vol. 100:2, pp. 72-84, 1995.

[38] M.E. Gorodetskaya, "About the Aral Sea terraces", Geomorphology, vol. 1, pp. 46-55, 1978.

[39] V.M.Borovsky, N.F. Mozhaitseva, R.H. Kievskaya, V.A. Kornienko and V.P. Bogachev, "Formation of soil cover in the dry seabed of the Aral Sea", in The condition of the waters and the dry seabed of the Aral Sea, Alma-Ata: Nauka, 1983, pp. 43-91.

[40] S.A. Nikitin, Arboreal and shrub vegetation of deserts in USSR. Moscow: Nauka, 1966, 255.

[41] Z.A. Mailun, "Tugaic vegetation – Potamophyta", in Vegetation cover of Uzbekistan, Tashkent: Fan, vol. 2, pp. 303 –375, 1973.

[42] N.M. Novikova, Principals of conservation of botanical diversity in delta plains of Turan. Theses of DSci in geography. Moscow, 1997, 104.

[43] V.S. Zaletaev, "Irrigation desertification and destabilization of environment as ecological influence of irrigation in arid zone", Arid ecosystems, vol. 1:1, pp. 22-26, 1995.

[44] G.V. Geldyeva, T.I. Budnikova, I.B. Skorintseva, T.A. Basova, R.Yu. Tokmagambetova and R.V. Plohih, Landscape assurance schemes to combat desertification in the Syrdarya river valley. Almaty, 2004, 236.

[45] L.A. Dimeyeva,"Geobotanical aspects of irrigation desertification in the Syrdarya river valley", in Geographic problems of sustainable development: theory and practice, Almaty, 2008, pp. 444-454.

[46] M.Horikawa, L.A. Dimeyeva, T. Oyabu, I. Tsuyama, Yu. Morimoto and N. Ishida, "Evaluation of wetland changes in lower Syrdarya region, the Aral sea, and possibilities of ecological rehabilitation", in Proceedings of Nature Reserve "Barsa-Kelmes", Almaty: Tethis, vol. 2, pp. 95-103, 2007.

[47] Biodiversity of wetlands in the delta front of the Syrdarya river. M.O. Ospanov and K.Zh. Stamkulova (Editors), Almaty, 2012, 62.

[48] Monitoring of Ramsar wetlands in the Syrdarya river delta. M.O. Ospanov and K.Zh. Stamkulova (Editors), Almaty, 2014, 104.

[49] L.A. Dimeyeva, N. Ishida, Z. Alimbetova and G. Satekeev, "Establishment of forest plantations in saline bare lands of the Aral Sea coast", in Proceedings of Int. conf. "IFAS – 20 yeas on a way of collaboration", Almaty, Kazakhstan, pp. 145-151, 2013.

[50] L.A. Dimeyeva, "Restoration of the Aral Sea coastal rangelands", in Proceedings of VII Int. Rangeland Congress, Durban, South Africa, pp. 1148-1151, 2003.

[51] L.A. Dimeyeva, B.M. Sultanova, Z. Alimbetova, "Important plant area – "The Syrdarya river delta", in Proceedings of Int. conf. dedicated to 75 anniversary of Barsa-Kelmes nature reserve, Aralsk, Kazakhstan, pp. 9-17, 2014.

[52] L.A. Dimeyeva, N.P. Ogar, Z. Alimbetova and S.W. Breckle, "Nature conservation in the Aral Sea region: Barsa-Kelmes as an example", in Aralkum – a Man-Made Desert, S.W. Breckle, W.Wucherer, L.A. Dimeyeva and N.P. Ogar (Editors), Springer Heidelberg Dordrecht London New York, pp. 315-341.

Controlling Coagulation Process: From Zeta Potential to Streaming Potential

Djamel Ghernaout[1, 2, *], Abdulaziz Ibraheem Al-Ghonamy[2], Mohamed Wahib Naceur[1], Ahmed Boucherit[1], Noureddine Ait Messaoudene[2], Mohamed Aichouni[2], Ammar Abdallah Mahjoubi[2], Noureddine Ali Elboughdiri[2]

[1]Department of Chemical Engineering, University of Blida, Blida, Algeria

[2]Binladin Research Chair on Quality and Productivity Improvement in the Construction Industry, College of Engineering, University of Ha'il, Ha'il, Saudi Arabia

Email address:

dj.ghernaout@uoh.edu.sa (D. Ghernaout), djamel_andalus@hotmail.com (D. Ghernaout)

Abstract: This review concerns the using of Zeta and streaming potentials in the coagulation process control. Coagulation process is usually explained by charge neutralization mechanism. The negative charge may be quantified by Zeta potential or streaming potential measures. Prior to the advent of streaming current monitors, Zeta meters were the primary instruments for measuring electrokinetic properties as related to coagulant dose. Both instruments measure the potential and indirectly the particle surface charge, but use very different methods. Even if the on-line streaming current monitor can provide coagulation process optimization when properly installed, maintained, and interpreted, jar tests experiments and Zeta meters remain indispensable.

Keywords: Electrophoretic Mobility, Zeta Potential, Streaming Potential, Streaming Current, Streaming Current Detector, Coagulation Control

1. Introduction

Conventional water technology implies the treatment of extremely complex colloidal dispersions through coagulation/flocculation process [1]. Coagulating such colloids is achieved through controlling their surface electrical charges. This action is realized by adsorption of ions of opposite charge to the colloid surface [2,3]. This adsorption decreases the mutual electrostatic repulsion between dispersed colloids that prevents their aggregation into flocs [4-6].

This mechanism is called charge neutralization control [7-10]. Controlling the resultant electric charge of the colloid is the main key to the successful steps implicated in aggregation of colloids in raw waters [5,11]. In 1966, the streaming current detector was invented [12]. Prior to this device, there was no on-line, efficient tool to control a characteristic proportional to the value of charge neutralization other than to control its result (in other words, coagulation as studied in simple jar tests) [5,13,14].

In this review paper, we discuss the evolution from Zeta potential to streaming potential as a mean for coagulant control. As these parameters are related to the colloid charge, we also briefly discuss the mechanism of charge neutralization.

2. From Zeta Potential to Streaming Potential

For over 30 years, with a large success streaming current monitors have been used for on-line coagulant and polymer control in water and wastewater plants [15]. Before 25 years, users have not at their possession an appropriate mean to check-up the on-line unit operation and/or response to process variations. As with any control device, when the reading or output varies, users have to know if it is due to a process variation, or if something is not correct with the control device. Moreover, if the technician makes a chemical pump adjustment and the on-line streaming current device doesn't respond correctly, there may be issues with the pump, an empty feed tank, etc., and nothing bad with the device. Users have to know something is not correct and have quick troubleshooting methods. The laboratory streaming current monitor gives them that mean. The laboratory unit is also a great tool for users to see how changes such as pH, alkalinity, turbidity, mixing time, and chemical dosages affect the streaming current reading [15].

On the other hand, jar tests have been largely used to determine the coagulant dose. Jar tests have disadvantages in that they are time consuming, intermittent, and are subject to variations in user's visual observations [15]. To ameliorate coagulant demand determinations, Zeta meters have been used to increase jar tests results quality [16]. Zeta potential is an indirect evaluation of particle surface charge [16]. Before the use of streaming current monitors, Zeta meters were the first instruments for evaluating electrokinetic properties as related to coagulant dose [17]. Both tools measure charge, but use different mechanisms [15].

The Zeta meter (Fig. 1(a)) uses a stereoscopic microscope to control particle motion through a glass cell named an *electrophoresis cell*. Electrodes in each end of the cell create an electric field. Colloids having a positive charge (+) will move toward the anode (-), and negative colloids (-) toward the cathode (+). Colloid velocity and direction both of them are related to its Zeta potential. The Zeta potential relating to right coagulant dose changes from water to water. It is crucial to keep in mind that a Zeta potential of zero is difficult even if the optimum dose is fixed. The aim is to decrease repulsive forces until particle collisions cause agglomeration, and reach the correct particle size/density conducting to the best flocculation, settling, and filtration. In fact, this "correct charge" value may be negative, zero, or positive, but frequently negative. Zeta potential is usually expressed in mV. An usual correct potential (and consequently the electric charge) might be in the range -10 to -5 mV [15].

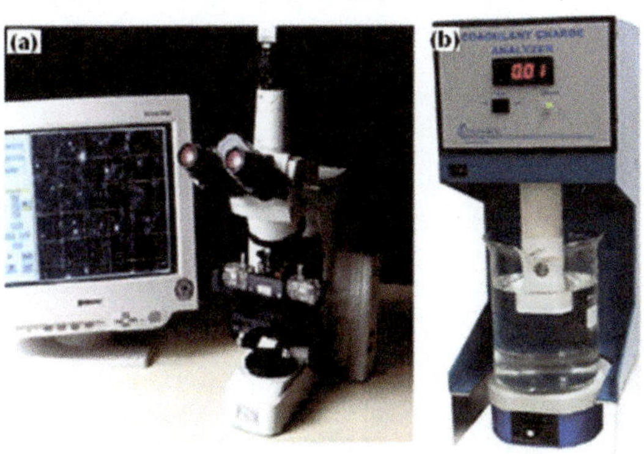

Fig. 1*. Zeta meter (a) and (b) streaming current monitor [15].*

In fact, the streaming current monitor (Fig. 1(b)) is a charge evaluating instrument. The charge evaluated is the net ionic and colloid surface charge (positive and negative) in the sample being examined. However, streaming current is in relation to Zeta potential by:

$$SC \; \alpha \; \frac{ZP \times D}{N} \tag{1}$$

Where *SC* is the streaming current, *ZP* is the Zeta potential, *D* = dielectric constant, and *N* = viscosity of the fluid. The treated sample flows into the chamber where it is drawn into the bore during the piston upstroke and is expelled on the downstroke. Colloids in the sample are temporarily fixed on the piston and cylinder surfaces [5]. As the water is moved back and forth, mobile counter-ions surrounding these particles are moved downstream to the electrodes. This reciprocating motion of like charges causes an alternating current to be created which is known as streaming current. This current is amplified, through the convenient electronics, and read as the streaming current value [15].

In fact, streaming current is another mean to evaluate Zeta potential. However, the measured numerical values are not identical. For example, a Zeta potential of -10 mV is not -10 streaming current units. This is not a problem in practical use due to the fact that the optimum treatment program is always empirically determined for each plant. When it is question to remove turbidity, the aim is to reach optimum particle size/density. When it is question to remove organics, as this the case of *enhanced coagulation*, optimum dosage may not even be in relation with particle collisions and floc birth [18]. This constitutes a real advantage over Zeta meters, which rely on visual observation of colloids [15].

Moreover, *optimum* coagulation chemistry is generally known as "the lowest dosage of all chemicals that results in the desired removal of contaminants with the lowest total operating costs." The on-line streaming current monitor can assure this optimization when conveniently installed, maintained, and interpreted [15].

In their study, Wäsche *et al.* [19] compared Zeta potential, as evaluated with the electrokinetic sonic amplitude method, to streaming potential as evaluated with the particle charge detection method. The measurements have been realized on identical suspensions of α-Al_2O_3, SiC and Si_3N_4 with a solids volume fraction of 1% and 10%. Zeta potential and streaming potential are related by a linear relation [20]. For this relationship, the correlation coefficients are better than 0.99 for the suspensions with 1% solids loading. For the suspensions with 10% solids loading, correlation coefficients are slightly lower due to the influence of increasing viscosity. Regardless of lower correlation at the higher solids loading, the isoelectric pH is constant within the limits of accuracy of the pH measurements [19].

A good review on the history of electrokinetic phenomena is presented by Wall [21]. An updated review on the measurement and interpretation of electrokinetic phenomena may be found elsewhere [16].

3. Colloidal Particles and the Double Layer Model

3.1. Colloidal Particles

Turbidity is constituted by suspended particles in water in

the size range of approximately 0.01 to 100 µm in size [22]. The larger portion can generally be removed by settling. The smaller particles, with sizes of < 5 µm, are called *colloidal particles*. Colloids are characterized by slow sedimentation velocities and so cannot be easily removed by settling. The behavior of colloids in suspension is greatly affected by their electrostatic charge [23]. This colloidal charge appears because of the uneven surface characteristics of the colloids and is usually negative. The charge on each colloid will be in repulsion with others and avoid important flocculation from producing. Charge neutralization is the main goal of coagulation [22,24,25].

3.2. The Double Layer Model

The double layer model is often called to interpret the distribution of electric charges around each colloid [22,26].

For practical interests we will suppose a negatively charge colloid. Nearest to the negative surface of the colloid there is a layer of strongly fixed positive ions – this is called the *Stern layer*. Further positive ions are still fixed to the colloid but are repelled by the Stern layer [22]; in the same manner, negative ions are attracted to the positive ions but repelled by the colloid [22]. A dynamic equilibrium of negative and positive charges constitutes, outside the Stern layer, a layer known as the *diffuse layer*. The presence of positive ions in the diffuse layer proportionally decreases as the distance from the colloid increases until beyond a certain point the ion concentrations are the same as the equilibrium in the water [22].

The strongly fixed positive ions near the surface and the charged layer surrounding this is where the name *double layer* originates from [26]. The surface and Stern layer together are actually termed a *double layer*. Moreover, the surface with a diffuse layer constitute a double layer. However, the surface plus a Stern layer plus a diffuse layer constitute a triple layer model.The point just outside the Stern layer is usually known as the *shear plane* since movement of the colloid through the suspension shears the diffuse layer away [22,23,27-29]. In fact, in some cases, the shear plane can be quite displaced beyond the Stern plane.

The thickness of these layers is a function of the concentration of ions in suspension. At any point from the surface, the charge density is equal to the difference between the concentration of positive and negative ions at that point (Fig. 2) [22,30].

The potential at the separation between the Stern layer and diffuse layer is named the *Zeta potential*. Evaluating Zeta potential is beneficial as it is a direct determination of the amount of energy required to bring separate particles together [22].

Zeta potential can be evaluated quite easily. This is done by using a microscope to observe turbidity particles inside a thin chamber called an *electrophoresis cell*. An electric field is applied along the cell, if the particles are negatively charged they will move towards the positive end of the cell, carrying their Stern layer ions with them. The average velocity with which these particles move depends on the Zeta potential value, as it is this that determines the net electrical force operating on the particle and its Stern layer ions [22].

In fact, the relationship between Zeta potential and this average velocity is not always straightforward to calculate, as it is a function of the fluid's viscosity, dielectric constant, conductivity and temperature. Consequently, Zeta potential, which should have units of mV, is often expressed in terms of electrophoretic mobility. This has rather confusing units in velocity per electric field strength, typically µm/s per V/cm [22]. Other disadvantages of measuring Zeta potential in this way include:

- It can only be measured for particles that are large enough to be detected and tracked through a microscope.
- It requires a laboratory technician to operate the apparatus and to observe the particles through a microscope. Consequently, Zeta potential measurement is slow, labor intensive and cannot be performed online.

Zeta potential measurement has useful applications in laboratory studies of coagulation/flocculation processes, but is of little use for online automation [22].

4. Streaming Potential

A streaming current and streaming potential are two interrelated electrokinetic phenomena studied in the areas of surface chemistry and electrochemistry (Table 1) [22,31]. They are an electric current or potential which originates when an electrolyte is driven by a pressure gradient through a channel (or capillary) or porous plug with charged walls [31,32]. Streaming currents in well-defined geometries are a sensitive method to determine the Zeta potential of surfaces [31].

Table 1. *Streaming current and streaming potential [22].*

	Streaming current and streaming potential
1.	A streaming current meter is an instrument for measuring the electric charge that exists on small, suspended particles in liquid. A streaming current meter is the only on-line instrument that can be utilized to evaluate coagulated particle stability for the feedback control of coagulant dosage.
2.	The streaming current meter is also referred to as the streaming current detector or streaming potential detector as well as streaming current monitor and streaming current analyzer in various works. The word 'detector' is utilized in earlier literature instead of the words 'meter', 'monitor' or 'analyzer' because earlier streaming current meters were considered to only provide a qualitative indication of the presence and sign of charge, rather than a useful quantitative measurement. The words 1: 'Meter', 2: 'Monitor' and 3: 'Analyzer' all mean the same thing in this context, except that the implication of sophistication increases in this order.

4.1. Origin of the Streaming Current

Adjacent to the channel walls, the charge-neutrality of the liquid is violated due to the presence of the electric double layer (Fig. 2) [30] or in other terms the overall free energy is equilibrated due to opposing electrical and osmotic pressure forces: a thin layer of counter-ions attracted by the charged surface [31]. The transport of counter-ions along with the pressure-driven fluid flow produces a net charge transport: the streaming current. The reverse effect, creating a fluid flow by applying a potential difference, is named *electroosmotic flow* [31].

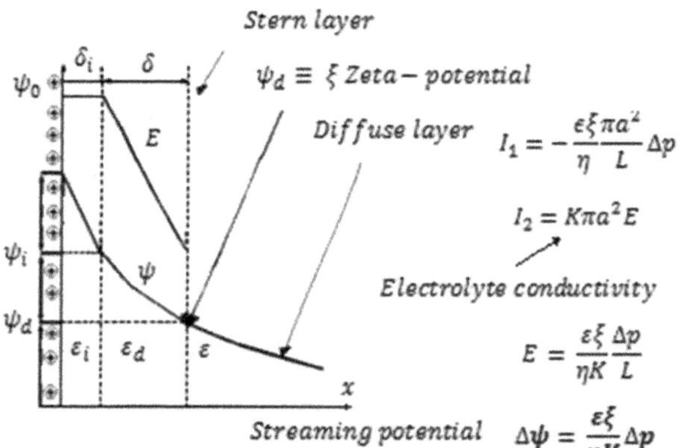

Fig. 2. Electric double layer [30].

4.2. Measurement Method

A typical device to determine streaming currents consists of two reversible electrodes placed on either side of a fluidic geometry across which a known pressure difference is applied. When both electrodes are held at the same potential, the streaming current is evaluated directly as the electric current flowing through the electrodes. Alternatively, the electrodes can be left floating, allowing a streaming potential to build up between the two ends of the channel. A streaming potential is defined as positive when the electric potential is higher on the high pressure end of the flow system that on the low pressure end [31].

The value of streaming current observed in a capillary is usually in relation with the Zeta potential through the relation:

$$I_{str} = -\frac{\epsilon_{rs}\epsilon_0 a^2}{\eta}\frac{\Delta P}{L}\zeta \quad (2)$$

The conduction current, which is equal in magnitude to the streaming current at steady state, is:

$$I = K_L a^2 \frac{U_{str}}{L} \quad (3)$$

At steady state, the streaming potential built up across the flow system is calculated by:

$$U_{str} = \frac{\epsilon_{rs}\epsilon_0\zeta}{\eta K_L}\Delta P \quad (4)$$

Where:

- I_{str}: streaming current under short-circuit conditions, A
- U_{str}: streaming potential at zero net current conditions, V
- I_c: conduction current, A
- ε_{rs}: relative permittivity of the liquid, dimensionless
- ε_0: electrical permittivity of vacuum, F·m^{-1}
- η: dynamic viscosity of the liquid, kg·m^{-1}s^{-1}
- ζ: Zeta potential, V
- ΔP: pressure difference, Pa
- L: capillary length, m
- a: capillary radius, m
- K_L: specific conductivity of the bulk liquid, S m^{-1}.

The above equations assume that:

- the double layer is not too large compared to the pores or capillaries (i.e.,κa>> 1), where κ is the reciprocal of the Debye length,
- there is no surface conduction (which typically may become important when the Zeta potential is large, e.g., $|\zeta| > 50$ mV),
- there is no electric double layer polarisation [31,33].

4.3. Streaming Current Isochoric and Non-isochoric. Seismo-Electric Effect

Streaming potential and streaming current occur in porous bodies when liquid is being pumped through them. These electrokinetic phenomena have been known for almost 200 years [34]. They are largely used for evaluating Zeta potential in porous bodies. The important field of application so far is in the paper industry. Streaming potential is frequently the preferred manner of realizing this measurement. However, the streaming current method does not require conductivity for calculating Zeta potential, which is an advantage. One of the important areas of the streaming current and streaming potential application is *seismo-electric effect*. This phenomenon is detected in underground rocks when sound waves propagate through. This seismo-electric effect is largely used for characterizing underground formations with regard to water or oil presence [34].

4.4. The Streaming Current Effect

The streaming current (or streaming potential) effect is a reverse manifestation of the electrophoresis effect. This occurs when a particle is mechanically moved through the fluid, or the fluid moved past the particle, conducting to a separation of charges pushing a potential to exist [22,35].

A commonly occurring situation is that water with negatively charged particles is forced through a filter. The negatively charged particles become lodged in the filter, while the mobile positively charged ions are swept downstream. This separation of charges produces an electrical potential to exist across the filter, called the *streaming potential*. If electrodes are inserted upstream and downstream of the filter, then the electrical potential can be determined. The upstream electrode is negative and the downstream electrode is positive [22].

The electrical potential results in a current being conducted upstream through the water to remove the charge separation, this is the streaming current. The electrical potential therefore is a function of the conductivity of the liquid [22].

Streaming current is a more important quantity than streaming potential, as it does not depend on the water's conductivity. Streaming current can be determined whenever the water with suspended charged particles is forced through thin capillaries or other barriers to the particle's motion. However, useful determinations are extremely hard to directly make from this effect, since the current generated is extremely small and easily obscured by potentials that exist for other reasons and electrical paths through the environment [22].

5. Using a Zeta Potential Meter or a Streaming Current Meter

Zeta potential is a parameter determined in the laboratory; its measurement requires a technician to use a microscope to evaluate it [36]. Provided it these measurements are done correctly, they are very repeatable for different water parameters. Streaming current is an online measurement. A streaming current monitor runs continuously and only occasionally requires manual input. However, streaming current does not 100% directly relate to Zeta potential and the desired streaming current set point must be manually evaluated and if plant conditions significantly vary [36].

Moreover, even if the concept of Zeta potential was established more than 100 years ago [37], a limitation on the Zeta potential mean is that it is not a continuous on-line measuring device [36,37]. The streaming current detector was introduced in response to this disadvantage. Its main benefit is rapid detection of plant upsets. In fact, streaming current is another mean to evaluate Zeta potential, but it is really in relation with the Zeta potential of a solid surface, such as the walls of a cylindrical tube, and not the Zeta potential of the colloid. Streaming current is well used as an on-line controller of Zeta potential, but it is only an indication since the value is not scaled. In other words, as shown above, a change in 10 streaming current units does not correspond directly to a Zeta potential change of 10 mV. Moreover, the zero position is frequently shifted importantly from true zero. It was proposed that a streaming current monitor must be usually calibrated using a Zeta meter [37]. If cost is not a problem then this will conduct to an extremely accurate handling of the Zeta potential in the coagulated water. Practically, it is much easier and almost as precise to utilize jar tests to calibrate the streaming current monitor. Drinking water utilities frequently utilize streaming current monitors in their plants and utilize Zeta potential meters in their central laboratories [36,38].

Furthermore, there are updated versions of Zeta potential measuring devices (e.g., Malvern Zetasizer® [39]) based on the Doppler velocitometry. In fact, Zeta potential is evaluated by applying an electric field through the dispersion. Colloids within the dispersion with a Zeta potential will migrate toward the electrode of opposite charge with a velocity proportional to the magnitude of the Zeta potential. This velocity is evaluated using the phenomenon of laser Doppler anemometry. The frequency shift, or phase shift, of an incident laser beam produced by these displacing colloids is evaluated in terms of the particle mobility, and this mobility is related to the Zeta potential by inputting the dispersant viscosity, and the application of the Smoluchowski or Huckel theories [39].

5.1. Finding the Optimal Flocculation Point with a Streaming Potential Instrument

Researchers such as Wang [40] affirmed that once the optimal operating conditions are established using laboratory tests, the streaming current devise can be called to fix them by controlling the variation of the charge density of the colloids. However, these devices have not been helpful for evaluating the optimal operating conditions since they are not able to find the neutral charge point, which is theoretically the optimal chemical dosage point. As an illustration, if you were to take a bucket of water from the river, you could determine its pH value. You would then have a starting point for adjusting the pH to the desired point. The same would be correct for temperature, conductivity, etc. Theoretically, you have to be able to determine the net charge density of the water as well, providing a starting point for adjusting it for optimal coagulation. On the other hand, traditional streaming current devices can only determine the relative change of the charge density in the suspension; they cannot measure the neutral charge point, which is the baseline determination for chemical optimization [40]. However, this is not correct if a titration is realized using the coagulant chemical while determining streaming current. Table 2 shows how the laboratory streaming current monitor is utilized [15].

Table 2. *How the laboratory streaming current monitor is used [15].*

	How the laboratory streaming current monitor is used
1.	Using a simple titration procedure, the laboratory streaming current monitor is utilized to correlate "best jar" with a quantified charge value and optimum plant performance. This provides a quick, easy and visible way for operators to understand the relationship between water quality and streaming current.
2.	Once this correlation is determined, jar tests usually becomes less frequent or unnecessary. If the plant has an on-line streaming current monitor, the laboratory unit serves as a periodic verification of its operation just like a laboratory pH-meter, or turbidity meter does for on-line instruments.
3.	The laboratory streaming current monitor can also be a useful tool for comparing strength/activity of different coagulants/polymers prior to full scale trials.

5.1.1. Streaming Current Technique

As shown above, in a streaming current device, water is pushed through by a piston in the cylinder (Fig. 3) [40-42]. The relative motion between the piston and the cylinder walls produces a charged stream, which generates a potential difference (voltage) through the electrodes in the cylinder. The signal is transformed to a streaming potential signal. The streaming current method has to be able to determine the real net charge density in the sample suspension, not only the relative variation of the charge density [40]. This is the same case as in the determination of a voltage with a voltmeter, which determines voltage magnitude as well as its polarity.

For the same concept, the streaming current device has to be able to evaluate the net charge magnitude and polarity. If the suspension being examined is negatively charged, a streaming current device should show the negative polarity as well as the intensity on a linear scale. The same device has to indicate a positive signal for positively charged suspension. When suspension is electrically neutral (positive and negative charges are exactly equilibrated), the device has to indicate the value of zero [40]. Fig. 4 shows a streaming current device measurement for suspension samples with different charge distributions [40,43].

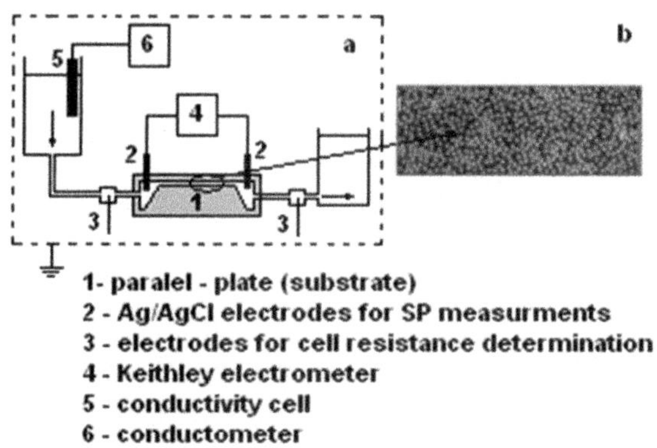

Fig. 3. A schematic view of the parallel-plate channel cell used for streaming potential (SP) measurements of particle covered surfaces, part "a". In part "b" the micrograph of deposited particles is shown (polystyrene latex particles coverage 0.25, diameter 800 nm) [41].

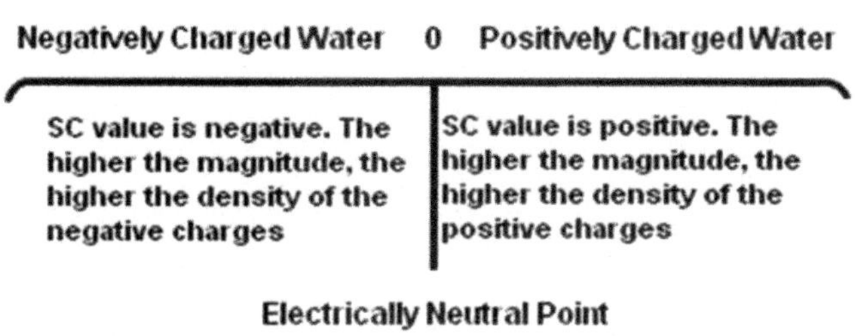

Fig. 4. Charge distributions versus streaming current (SC) value [40].

5.1.2. Instrument Should Determine Neutral Charge Point and Change in Charge Density

Theoretically, a streaming current instrument should be able to determine the neutral charge point as well as the variation in charge density on a linear scale [40]. This gives a basic point for chemical optimization and also gives a baseline for comparing results from different coagulants, water sources, and seasonal variations. Empirical data can be registered and consulted as conditions vary, allowing scientific interpretation of process conditions [40].

Each application is different, since are the requirements for coagulant dosage optimization. Although the electrically neutral point is in theory the optimal point, it is not obligatory the desired operating point for all applications. Traditional streaming current instruments are habitually "zeroed" once a desired operating point has been measured, but they cannot be used to detect the neutral charge point [40].

5.1.3. New Instrument Provides Two Measurement Signals

It was developed an online streaming potential instrument, the streaming potential system, which gives two measurement signals: an absolute signal and a relative operating signal [40]. The absolute signal determines the net charge density with charge polarity and magnitude, in which the neutral charge point is always shown as zero. The relative signal is adjustable so that the zero point can be set to the desired operating point [40]. Users can utilize the absolute signal to establish the theoretical operating point. The relative signal is utilized when the desired point is different from the theoretical point [40].

A streaming potential system was monted at Crooked Creek Water Reclamation Facility in Norcross, Ga., in July 1998. This online instrument was utilized to control centrate charge. Fig. 5 shows the strong relationship between the streaming potential system signal and polymer dosage.

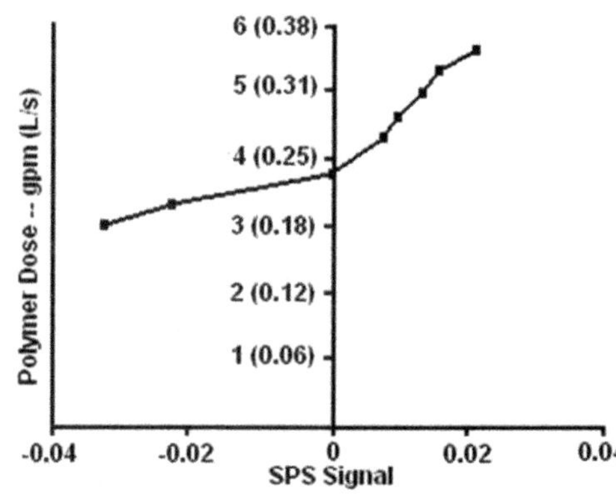

Fig. 5. Relationship between the streaming potential system (SPS) signal and polymer dosage [40].

5.2. Mechanism of Operation of the Streaming Current Detector

Barron *et al.* [5] studied the mechanism of operation of the streaming current detector by the use of well characterized surfaces and dispersed colloidal particles. The response of the instrument in the presence of various electrolytes and with the addition of colloids of known electrokinetic properties shows that the apparatus is measuring a quantity that demonstrates contributions from the charge on both the walls and piston of the apparatus and from colloidal material associated with the surfaces. Analysis of the output of the instrument in terms of the magnitude of the current is poorly taken into consideration by theory. The efficiency of adsorption of colloids, polymers and other simple reagents to the walls and/or piston and the ease of removal of these substrates was shown to be critical to the response of the device [5].

El-Gholabzouri *et al.* [44] compared the Zeta potential obtained from streaming current and electrophoretic mobility determinations. The electrokinetic experiments were realized using two anionic polystyrene latexes whose surface charge density and average particle diameter were -9.8 and -3.4 μC cm^{-2} and 0.65 and 1.2 μm, respectively. For the conversion, two theories were utilized: the Levine *et al.* theory for the streaming current and O'Brien–White for the electrophoretic mobility measurements. This way most of the phenomena taking place at the interface are taken into consideration. The Zeta potentials obtained this manner compare well for the largest latex with the lowest surface charge. However, there are substantial discrepancies for the smaller latex sample with the higher surface charge density. These discrepancies are interpreted in terms of the effect of the surface conductance on the electrokinetic signal [44].

Dentel *et al.* [45] utilized a streaming current detector in continuous flow experiments to study its response to different conditions. Buffered water and kaolinite systems were coagulated with aluminium sulphate (alum) at various doses. The dose was incremented in a stepwise manner while monitoring streaming current detector output, or alternated between samples with and without alum added. It was established that the streaming current detector responded rapidly and reversibly to variations in sample characteristics over a large interval of coagulant doses. Readings were similar to those realized from batch streaming current detector determinations. However, when extremely high doses of alum or ferric chloride (2×10^{-3} M as Al or Fe) were injected, the streaming current detector signal became erratic. Extent and characteristics of this effect differed following the coagulant utilized, delay time between the coagulant addition and streaming current detector sampling, coagulant dose and whether or not an ultrasonic cleaning accessory was used. Several different cleaning methods were shown to alleviate the difficulties, but some cleaning agents themselves interfered with streaming current detector output [45].

Childress and Elimelech [46] utilized a streaming potential analyzer to study the effect of solution chemistry on the surface charge of four commercial reverse osmosis and nanofiltration membranes. Zeta potentials of these membranes were examined for aqueous solutions of various chemical compositions over a pH interval of 2 to 9. In the presence of an indifferent electrolyte (NaCl), the isoelectric points of these membranes range from 3.0 to 5.2. The curves of Zeta potential versus solution pH for all membranes show a shape characteristic of amphoteric surfaces with acidic and basic functional groups. Results with salts containing divalent ions ($CaCl_2$, Na_2SO_4, and $MgSO_4$) show that divalent cations more readily adsorb to the membrane surface than divalent anions, especially in the higher pH interval. Three sources of humic acid, Suwannee River humic acid, peat humic acid, and Aldrich humic acid, were utilized to study the effect of dissolved natural organic matter on membrane surface charge. Other solution chemistries implicated in their investigation comprise an anionic surfactant (sodium dodecyl sulfate) and a cationic surfactant (dodecyltrimethylammonium bromide). Results illustrate that humic substances and surfactants readily adsorb to the membrane surface and markedly effect the membrane surface charge [46].

In 1974, Cserfalvi *et al.* [47] established that because of optical reasons, electrophoretic measuring methods utilized for the measurement of the electrokinetlc potential of dispersed colloids, cannot be applied to macromolecular colloids of low concentration.

Recently, Adamczyk *et al.* [41] discussed recent developments in the field of streaming current/streaming potential determination for surfaces covered by particles, including polyelectrolytes and proteins. They examined the new theoretical results enabling us measure quantitatively the relationship between the streaming potential and surface coverage of colloids. They discussed and explained experimental results achieved for model systems of monodisperse polymeric particles in terms of the theoretical approach. They also pointed out a major utility of the streaming potential measurements for determination of submonolayer coverages of particles and proteins [41].

6. Charge Neutralization as the Key Mechanism in Coagulation/Flocculation

In coagulation process, the importance of charge neutralization mechanism is established in several studies using clay suspensions, organic matters (natural organic matter and dyes) solutions, micro-algae and microorganisms suspensions since the beginning of the coagulation/flocculation investigations. In fact, the net negative charge on colloids in most raw waters and the coagulant dose are typically imposed by the concentration of natural organic matter. For example, dissolved or sorbed humic materials at neutral pH have ~ 10 meq of negative charge per g of natural organic matter. The most highly charged clays have a negative charge as high as 1 meq per g of clay. Thus, for equal concentrations of natural organic matter and total suspended solids, the minimum coagulant demand for the natural organic matter would be at least 10 times the coagulant demand for the total suspended solids [16].

As an illustration, in 1966, Cardwell [48] suggested a technique for studying adsorption in which the detection is realized by a streaming current method. This method makes use of an instrument performing continuous measurements for studying an adsorption process. A study has been accomplished of the adsorption of cetyl dimethyl benzyl ammonium chloride by a monoionic sodium bentonite. The adsorption was followed in two parts: the amount adsorbed to the cationic exchange capacity of the clay and the amount adsorbed from this point to the critical micelle concentration of the quaternary salt [48].

Kam and Gregory [49] studied the anionic charge carried by aquatic humic substances which have a major part in their interaction with metal ions and other cationic species. In fact, removal of such substances by coagulation/flocculation process can be, at least partly, achieved by charge neutralization. They examined the charge densities of a commercial humic acid and an aquatic humic extract by following their interactions with a series of synthetic cationic polyelectrolytes. The methods utilized were colloid titration by spectrophotometry and streaming current detection, and flocculation determined by color removal and by an optical monitoring method. For a given cationic polyelectrolyte, all four methods gave charge densities for humic substances which were in good correlation. However, great differences in the apparent humic charge density were detected, depending on the charge density of the cationic polyelectrolyte utilized. With low charge density polyelectrolytes, the apparent anionic charge of the humic substances was detected to be low. With higher polyelectrolyte charge densities, the apparent humic substance charge density augments and arrives to a limiting value when the polyelectrolyte charge is bigger than ~ 3 meq/g. This shows a non-stoichiometric interaction between the anionic sites of the humic substances and the cationic charges of the low-charge polyelectrolytes. Optimum flocculation of humics is produced with less cationic charge in the case of low-charge polyelectrolytes than those with higher charge density. However, the degree of removal was greatly better in the latter case. In all cases, the molecular weight of the cationic polyelectrolytes (over a range from about 50,000 to 15 million) is shown to have no effect on the results [49].

Wu *et al.* [50] studied the interactions of silica microspheres by light scattering in presence of alum and polyaluminium chloride (PACl) with various OH/Al ratios. They examined coagulation behaviors using different coagulant dosages at constant pH (6.5) and salt concentration (0.01 mol/L). Depending on the measurement of size distribution and Zeta potential, charge neutralization was suggested to be the basic coagulation step for all the coagulants while other distinct coagulation steps were also implicated based on *in situ* formed or preformed hydrolyzing products. Precipitate coverage and *sweep* flocculation were called for alum, contrasted to polycation patch and bridge aggregation for PACl [50].

Recently, Xu *et al.* [7] focused on the effect of coagulant dose and pH on the coagulation behaviors of nano-Al_{13} polymer ($Al_{13}O_4(OH)_{24}{}^{7+}$) with respect to the treatment of low specific UV absorbance surface water. The elimination of fine particles and natural organic matter was studied under different coagulation conditions. PACl and traditional aluminum coagulant, alum, were utilized for purpose of comparison. Floc size, strength and fractal structure were examined under different coagulation conditions. The results demonstrated that fine particles, and hydrophobic natural organic matter with large molecular weight, could be easily eliminated by the three coagulants in a wide dose range (5–17 mg/L). While the proportion of hydrophilic natural organic matter in the removed dissolved organic matter was bigger in Al_m and Al_{13} species coagulation at low and moderate dose (lower than 11 mg/L). Neutral and alkaline favored the particle elimination, while acidic pH participated in better hydrophilic natural organic matter elimination. The floc strength study demonstrated that the floc strength constituted by Al_{13} under different coagulation conditions was in the following order: pH 6.5 (15 mg/L) > pH 4.0 (9 mg/L) > pH 8.0 (15 mg/L). For $Al_2(SO4)_3$ coagulation, the order was as following: pH 4.0 (9 mg/L) > pH 6.5 (15 mg/L) > pH 8.0 (15 mg/L). Moreover, the fractal dimension of flocs indicated that Al_{13} species conducted to compact aggregates at low concentration and pH [7].

An interesting study was performed by Ye *et al.* [51], they studied the coagulation behavior of PACl with various basicities (*B* = OH/Al values) under different alkalinities. They get better insights into the coagulation mechanisms implicating interactions between hydrolyzed Al(III) products and colloids. Jar tests were utilized to measure the coagulation efficiencies, including Zeta potentials, residual turbidities and pH values. An optical control technique of photometric dispersion analyzer was used to follow the coagulation dynamics. The experimental results illustrate that the traditional coagulant such as alum implicates a rapid hydrolysis after dosing and the *in situ* formed hydrolysis products can destabilize the kaolin particles by precipitation charge neutralization. The preformed polymeric species in PACls show a relatively high stability after dosing and can constitute "electrostatic patches" on clay particle surfaces. These patches play an important role in "electrostatic patch coagulation". Increasing alkalinity increases both precipitation charge neutralization and electrostatic patch coagulation areas. When alkalinity is low alkalinity, electrostatic patch coagulation with high Al_b contained in PACls works better than precipitation charge neutralization coagulation; increased alkalinity ameliorates the efficiency of traditional coagulant due to *sweep* flocculation. When alkalinity is higher, more coagulant is required to realize complete charge neutralization. The stoichiometric relationships between the dosage and alkalinity are different following the *B* values of PACls [51].

7. Indecisiveness of Electrophoretic Mobility Determination and Effects of Low Temperature on Coagulation

It is well established that the magnitude of electrophoretic mobility is frequently utilized to represent the repulsion force between colloids in an aqueous system. A generally accepted idea is that colloids will coagulate efficiently when the suspension electrophoretic mobility is near to zero, which is frequently achieved by dosing trivalent cations such as Fe(III) [52,53]. Concerning Fe(III) coagulation, however, Xiao *et al.* [53] found that colloids were firstly enmeshed by voluminous ferric hydroxide precipitate or flocs, and the enmeshment of colloids by flocs was not dependent of the suspension electrophoretic mobility during Fe(III) coagulation. A suspension with either highly negative or positive electrophoretic mobility could be well coagulated and clarified if the solution pH and Fe(III) concentration favored the formation and fast growth of ferric hydroxide precipitate. For an efficient Fe(III) coagulation, the electrophoretic mobility was not obligatory zero [52].

In addition, Xiao *et al.* [54] compared coagulation of kaolinite suspensions at low temperatures with that at an ambient temperature of 22°C, and the process was studied with regard to the coagulation rate and chemical characteristics of coagulation. Experiments utilizing a photometric dispersion analyzer proved that coagulation of kaolinite suspensions could be considered as a two-phase process: low temperature importantly decreases the coagulation rate of the first phase but not that of the second one. On the other hand, results illustrated that low temperature did not serve to impede the hydrolysis of aluminum [Al(III)] within 1 min of alum addition. The measurements of electrophoretic mobility show that destabilization of kaolinite particles by hydrolyzed Al species was not decreased by low temperature within 1 min of alum addition. Slow coagulation at low temperature is attributed to the lowered coagulation rate but not the altered chemistry aspect of Al(III). Moreover, the change in settled turbidity after 20-min flocculation as a function of coagulant dosage was more significant in the cold due to the low coagulation rate.

Consequently, Xiao *et al.* [52,54] confirmed that Zeta potential (or electrophoretic mobility) is not an important determinant when a *sweep* coagulation mechanism dominates.

8. Practical Considerations

As concluded by Dempsey [16], plant operators should take into consideration the coagulation mechanism at their facility and the diagnostic tools that can be utilized to ameliorate treatment efficiency or to respond to variations in water quality. As an example, the city of Milwaukee WI changed from alum to PAC1 during the winter of '1992-93. This conducted to a decrease in coagulant dose from ~30 ppm alum to ~5 ppm PAC1. It is possible the destabilization mechanism was varied from *sweep* flocculation to charge neutralization but jar tests (rather than electrokinetic tests) were retained for determination of convenient coagulation conditions. Spring '1993 brought unusual weather conditions, culminating with a significant freeze between two thaws. These conditions are frequently related to rapidly deteriorating water quality due to run-off, sieving of contaminants behind a slush dam followed by release during a thaw, and the effects of cold water on the position of the $Al(OH)_{3(am)}$ solubility diagram. Moreover, Milwaukee suffered some critical equipment and chemical failures. Hundreds of thousands were sickened by *cryptosporidiosis* [16]. It is not possible to tell today whether the system failed in part because of overdosing or underdosing. It is likely that the problems could have been easily analyzed and treated if Zeta meters or streaming current monitors had been utilized [16].

Furthermore, Dempsey [16] presented a strategy for application of Zeta meters for control of charge neutralization. Finally, coagulant control in water treatment [55-62] remains difficult to be well modeled and more progress is needed in coagulation process comprehension and quantification.

9. Conclusions

This review discusses the common approaches of measuring electrokinetic properties of colloidal systems—mainly Zeta potential, streaming potential, and streaming current—with respect to optimization of coagulation/flocculation processes in water treatment. From this review, three main conclusions are drawn:

1. The electric double layer plays the major determinant role of colloidal interactions and colloid stability. The charge and potential of a range of surfaces and colloidal particles have been measured successfully in order to quantify the electric double layer for various systems. Surface charge may be determined by charge titration, whereas Zeta potential is generally determined by electrokinetic techniques (e.g., electrophoresis, streaming potential, electroosmosis). The distinction between the surface potential and the Zeta potential is that the latter applies at the plane of shear, which may be up to 2-3 water diameters away from the surface.
2. From a user's point of view, streaming current assures a good on-line tool for optimizing and controlling coagulant dose. With the development of laboratory streaming current monitors, users can verify on-line operation and evaluate coagulant dose quicker than performing jar tests, which remain important. The combination of on-line and laboratory units help to characterize the best coagulation for a varying water source.
3. Zeta potential or electrophoretic mobility is not an important factor when *sweep* coagulation dominates, which is the common case in the real-world coagulation processes.

Acknowledgements

The present research work was undertaken by the Binladin Research Chair on Quality and Productivity Improvement in the Construction Industry funded by the Saudi Binladin Constructions Group; this is gratefully acknowledged. The opinions and conclusions presented in this paper are those of the authors and do not necessarily reflect the views of the sponsoring organization.

References

[1] D. Ghernaout and M.W. Naceur, Ferrate(VI): In situ generation and water treatment – A review, Desalin. Water Treat., 30 (2011) 319–332.

[2] B. Ghernaout, D. Ghernaout and A. Saiba, Algae and cyanotoxins removal by coagulation/flocculation: A review, Desalin. Water Treat., 20 (2010) 133–143.

[3] W. Yu, J. Gregory and L.C. Campos, Breakage and re-growth of flocs formed by charge neutralization using alum and polyDADMAC, Water Res., 44 (2010) 3959-3965.

[4] R.J. François, Strength of aluminium hydroxide flocs, Water Res., 21 (1987) 1023-1030.

[5] W. Barron, B.S. Murray, P.J. Scales, T.W. Healy, D.R. Dixon and M. Pascoe, The streaming current detector: a comparison with conventional electrokinetic techniques, Colloids Surf. A: Physicochem. Eng. Aspects, 88 (1994) 129-139.

[6] S.R. Gray and C.B. Ritchie, Effect of organic polyelectrolyte characteristics on floc strength, Colloids Surf. A: Physicochem. Eng. Aspects, 273 (2006) 184–188.

[7] W. Xu, B. Gao, Y. Wang, Z. Yang and X. Bo, Role of Al_{13} species in removal of natural organic matter from low specific UV absorbance surface water and the aggregates characterization, Chem. Eng. J., 171 (2011) 926– 934.

[8] C. Huang and H. Shiu, Interactions between alum and organics in coagulation, Colloids Surf. A: Physicochem. Eng. Aspects, 113 (1996) 155-163.

[9] J. Oh and S. Lee, Influence of streaming potential on flux decline of microfiltration with in-line rapid pre-coagulation process for drinking water production, J. Membr. Sci., 254 (2005) 39–47.

[10] H. Liu, C. Hu, H. Zhao and J. Qu, Coagulation of humic acid by PACl with high content of Al_{13}: The role of aluminum speciation, Sep. Purif. Technol., 70 (2009) 225–230.

[11] S. Byun, J. Oh, B.-Y. Lee and S. Lee, Improvement of coagulation efficiency using instantaneous flash mixer (IFM) for water treatment, Colloids Surf. A: Physicochem. Eng. Aspects, 268 (2005) 104–110.

[12] W.F. Gerdes, A new instrument -- The streaming current detector, Paper presented at the 12th National ISA Analysis Instrument Symposium, pp. 181-198, Houston, Texas, May 11-13, 1966.

[13] (a) S. Xia, X. Li, Q. Zhang, B. Xu and G. Li, Ultrafiltration of surface water with coagulation pretreatment by streaming current control, Desalination, 204 (2007) 351–358.

(b) H.N. Jang, D.S. Lee and S.O. Ko, The effects of coagulation with MF/UF membrane filtration in drinking water treatment, Desalin. Water Treat., 19 (2010) 138–145.

[14] S.K. Dentel, A.V. Thomas and K.M. Kingery, Evaluation of the streaming current detector--I. Use in jar tests, Water Res., 23 (1989) 413-421.

[15] Laboratory streaming current monitor defines best coagulation chemistry, http://www.chemtrac.com/products/cca3100/documents/Laboratory_SCM.pdf (accessed on 29/06/11).

[16] B.A. Dempsey, Coagulant characteristics and reactions (Ch. 2), In: G. Newcombe and D. Dixon (Eds.), Interface science in drinking water treatment, Elsevier, 2006.

(a) W.E. Neuman, Optimizing coagulation with pilot filters and zeta potential, J. Am. Water Works Ass. 73 (1981) 472-475.

(b) A.V. Delgado, F. González-Caballero, R.J. Hunter, L.K. Koopal and J. Lyklema, Measurement and interpretation of electrokinetic phenomena, J. Colloid Interface Sci., 309 (2007) 194-224.

[17] (a) E.L. Bean, S.J. Campbell, F.R. Anspach, R.W. Ockershausen, C.J. Peterman, Zeta potential measurements in the control of coagulation chemical doses, J. Am. Water Works Ass. 56 (1964) 214-227.

(b) M. Elimelech, W.H. Chen and J.J. Waypa, Measuring the zeta (electrokinetic) potential of reverse osmosis membranes by a streaming potential analyzer, Desalination, 95 (1994) 269-266.

[18] D. Ghernaout, B. Ghernaout and M.W. Naceur, Embodying the chemical water treatment in the green chemistry—A review, Desalination, 271 (2011) 1–10.

[19] R. Wäsche, M. Naito and V.A. Hackley, Experimental study on zeta potential and streaming potential of advanced ceramic powders, Powder Technol., 123 (2002) 275–281.

[20] P.R. Johnson, A comparison of streaming and microelectrophoresis methods for obtaining the ζ potential of granular porous media surfaces, J. Colloid Interface Sci., 209 (1999)264–267.

[21] S. Wall, The history of electrokinetic phenomena, Current Opinion Colloid Interface Sci., 15 (2010) 119–124.

[22] (a) D. Edney, Introduction to the theory of the streaming current meter, http://www.accufloc.com/downloads/SCMTheory.pdf (accessed on 29/06/11).

(b) W. Olthuis, B. Schippers, J. Eijkel and A. van den Berg, Energy from streaming current and potential, Sensors Actuators B, 111–112 (2005) 385–389.

[23] (a) G. Lagaly, Colloid clay science (Ch. 5), In: F. Bergaya, B.K.G. Theng and G. Lagaly (Eds.), Handbook of clay science, Developments in clay science (Vol. 1), Elsevier, 2006.

(b) T. Ishibashi, Coagulation mechanisms: An electron microscopic study using aluminum sulfate, J. Am. Water Works Ass. 72 (1980) 514-518.

[24] D. Ghernaout, B. Ghernaout and A. Kellil, Natural organic matter removal and enhanced coagulation as a link between coagulation and electrocoagulation, Desalin. Water Treat., 2 (2009) 203–222.

[25] J. Zhang and J. Buffle, Kinetics of hematite aggregation by polyacrylic acid: Importance of charge neutralisation, J. Colloid Interface Sci., 174 (1995) 500-509.

[26] M. Zembala, Electrokinetics of heterogeneous interfaces, Adv. Colloid Interface Sci., 112 (2004) 59– 92.

[27] F. Mietta, C. Chassagne and J.C. Winterwerp, Shear-induced flocculation of a suspension of kaolinite as function of pH and salt concentration, J. Colloid Interface Sci., 336 (2009) 134–141.

[28] M.A. Yukselen and J. Gregory, The reversibility of floc breakage, Int. J. Miner.Process., 73 (2004) 251– 259.

[29] I.G. Droppo, K. Exall and K. Stafford, Effects of chemical amendments on aquatic floc structure, settling and strength, Water Res., 42 (2008) 169-179.

[30] K. Sadlej, Streaming potential and streaming current of a particle covered surface, Part 2: Virial expansion and simulations, http://fluid.ippt.gov.pl/seminar/text/sadlej280109.pdf, 28/01/09 (accessed on 29/06/11).

[31] Streaming current, http://en.wikipedia.org/wiki/Streaming_current (accessed on 28/06/11).

[32] Encyclopedia2, Streaming current,http://encyclopedia2.thefreedictionary.com/Streaming +current%2Fpotential (accessed on 28/06/11).

[33] M.-S. Chun, S.-Y.Lee, and S.-M. Yang, Estimation of zeta potential by electrokinetic analysis of ionic fluid flows through a divergent microchannel, J. Colloid Interface Sci., 266 (2003) 120–126.

[34] (a) Dispersion technology, Inc.,Streaming Current isochoric and non-isochoric. Seismoelectric effect, http://www.dispersion.com/streaming-current.html (accessed on 29/06/11).

(b) A.S. Dukhin, V.N. Shilov, The seismoelectric effect: A nonisochoric streaming current 2. Theory and its experimental verification, J. Colloid Interface Sci., 346 (2010) 248–253.

[35] C.A. Walker, J.T. Kirby and S.K. Dentel, The streaming current detector: A quantitative model, J. Colloid Interface Sci.,182 (1996) 71–81.

[36] Accufloc, Should I use a zeta potential meter or a streaming current meter? http://www.accufloc.com/zeta_sc.html (accessed on 28/06/11).

[37] Zeta-meter,Everything you want to know about Zeta Potential, http://www.zeta-meter.com/everything.pdf (accessed on 28/06/11).

[38] C. Igarashi and K. Nishizawa, Effect of turbulent flow condition on streaming potential measurement using cylindrical cells, Colloids Surf. A: Physicochem. Eng. Aspects, 303 (2007) 235–240.

[39] (a) M.J. Jaycock, Assumptions made in the measurement of zeta-potential by streaming current/potential detectors, Paper Technol., 36 (1995) 35-38.

(b) Malvern Zetasizer®,Zeta potential measurement using laser Doppler electrophoresis (LDE), http://www.malvern.com/LabEng/technology/zeta_potential/zeta_potential_LDE.htm (accessed on 20/12/11).

(c) L. Li, H. Zhang, G. Pan, Influence of zeta potential on the flocculation of cyanobacteria cells using chitosan modified soil, J. Environ. Sci. 28 (2015) 47-53.

[40] J. Wang, Finding the optimal flocculation point with a streaming potential instrument, J. Am. Water Works Assoc., November (2001) 54-56, http://www.sentrolsystems.com/AWWA_paper.pdf (accessed on 28/06/11).

[41] Z. Adamczyk, M. Nattich and M. Zaucha, Electrokinetics of particle covered surfaces, Current Opinion Colloid Interface Sci., 15 (2010) 175–183.

[42] K. Sadlej, Streaming potential and streaming current of a particle covered surface, Part 1, http://fluid.ippt.gov.pl/seminar/text/sadlej210109.pdf, 21/01/09 (accessed on 29/06/11).

[43] T.J. Day, On-line control of coagulants, polymer dosing with streaming current technology, http://www.waterworld.com/index/display/article-display/200746/articles/waterworld/volume-20/issue-3/editorial-focus/on-line-control-of-coagulants-polymer-dosing-with-streaming-current-technology.html, 21/06/11 (accessed on 28/06/11).

[44] O. El-Gholabzouri, M.A. Cabrerizo and R. Hidalgo-Àlvarez, Comparative electrophoretic mobility and streaming current study for ζ-potential determination, Colloids Surf. A: Physicochem. Eng. Aspects, 159 (1999) 449–457.

[45] S.K. Dentel, A.V. Thomas and K.M. Kingery, Evaluation of the streaming current detector--II. Continuous flow tests, Water Res., 23 (1989) 423-430.

[46] A.E. Childress and M. Elimelech, Effect of solution chemistry on the surface charge of polymeric reverse osmosis and nanofiltration membranes,J. Membr. Sci., 119 (1996) 253-268.

[47] T. Cserfalvi, T. Meisel and E. Pungor, Determination of the electrokinetic potential of dispersion and macromolecular colloids by measurement of the streaming potential. I., Electroanalytical Chem. InterfactalElectrochem., 53 (1974) 365-370.

[48] P.H. Cardwell, Adsorption studies using a streaming current detector, J. Colloid Interface Sci., 22 (1966) 430-437.

[49] (a) S.-K. Kam and J. Gregory, The interaction of humic substances with cationic polyelectrolytes, Water Res., 35 (2001) 3557–3566.

(b) L.H. Mikkelsen, Applications and limitations of the colloid titration method for measuring activated sludge surface charges, Water Res., 37 (2003) 2458–2466.

(c) S. Bratskaya, A. Golikov, T. Lutsenko, O. Nesterova and V. Dudarchik, Charge characteristics of humic and fulvic acids: Comparative analysis by colloid titration and potentiometric titration with continuous p*K*-distribution function model, Chemosphere, 73 (2008) 557-563.

[50] X. Wu, X. Ge, D. Wang and H. Tang, Distinct mechanisms of particle aggregation induced by alum and PACl: Floc structure and DLVO evaluation, Colloids Surf. A: Physicochem. Eng. Aspects, 347 (2009) 56-63.

[51] (a) C. Ye, D. Wang, B. Shi, J. Yu, J. Qu, M. Edwards and H. Tang, Alkalinity effect of coagulation with polyaluminum chlorides: Role of electrostatic patch, Colloids Surf. A: Physicochem. Eng. Aspects, 294 (2007) 163-173.

(b) Y.-L. Cheng, R.-J. Wong, J.C.-T. Lin, C. Huang, D.-J. Lee, A.S. Mujumdar, Water coagulation using electrostatic patch coagulation (EPC) mechanism, Drying Technol. 28 (2010) 850-857.

[52] F. Xiao, X. Zhang and C. Lee, Is electrophoretic mobility determination meaningful for aluminum(III)coagulation of kaolinite suspension?, J. Colloid Interface Sci., 327 (2008) 348-353.

[53] F. Xiao, X. Zhang and J. Ma, Indecisiveness of electrophoretic mobility determination in evaluating Fe(III) coagulation performance, Sep. Purif. Technol., 68 (2009) 273-278.

[54] F. Xiao, J. Ma, P. Yi and J.-C.H. Huang, Effects of low temperature on coagulation of kaolinite suspensions, Water Res., 42 (2008) 2983-2992.

[55] S.K. Dentel, Coagulant control in water treatment, Crit. Rev. Environ. Control, 21(1991) 41-135.

[56] O. Oriekhova, S. Stoll, Investigation of FeCl3 induced coagulation processes using electrophoretic measurement, nanoparticle tracking analysis and dynamic light scattering: Importance of pH and colloid surface charge, Colloid Surface A 461 (2014) 212-219.

[57] E.A. López-Maldonado, M.T. Oropeza-Guzman, J.L. Jurado-Baizaval,A. Ochoa-Terán, Coagulation–flocculation mechanisms in wastewater treatment plants through zeta potential measurements, J. Hazard. Mater., 279 (2014) 1-10.

[58] D. Ghernaout, S. Moulay, N. Ait Messaoudene, M. Aichouni, M.W. Naceur, A. Boucherit, Coagulation and chlorination of NOM and algae in water treatment: A review, Intern. J. Environ. Monitor. Analysis, 2 (2014) 23-34.

[59] D. Ghernaout, The hydrophilic/hydrophobic ratio vs. dissolved organics removal by coagulation – A review, J. King Saud University – Sci., 26 (2014) 169-180.

[60] D. Ghernaout, The best available technology of water/wastewater treatment and seawater desalination: Simulation of the open sky seawater distillation, Green Sustain. Chem., 3 (2013) 68-88.

[61] D. Ghernaout, B. Ghernaout, Sweep flocculation as a second form of charge neutralisation—a review, Desalin. Water Treat. 44 (2012) 15-28.

[62] J. Duan, J. Wang, T. Guo, J. Gregory, Zeta potentials and sizes of aluminum salt precipitates – Effect of anions and organics and implications for coagulation mechanisms, J. Water Process Eng. 4 (2014) 224-232.

Resources and Threats of the Eco-Tourism Development in Georgia

Giorgi Khomeriki[1], Giorgi Meladze[2]

[1]Vakhushti Bagrationi Institute of Geography at Iv.Javakhishvili Tbilisi State University, Tbilisi, Georgia
[2]Dep. of Geography, Iv. Javakhishvili Tbilisi State University, Tbilisi, Georgia

Email address:
gogihom@hotmail.com (G. Khomeriki), meladzeg@gmail.com (G. Meladze)

Abstract: Ecological tourism can play a certain role in solving the problems of environment protection and sustainable and safe development of the country. The article considers some specific problems of the eco-tourism development in Georgia. The results of the study of the tourism potential in the regions in 2009-2014 revealed that: Georgia has substantial means to develop eco-tourism. This is true not only for the protected areas, but also for individual natural and geographical areas of the different regions (historical areas) of the country. The distinctive feature of the country's eco-tourism potential is particularly diversified relief forms and natural eco-systems, in particular, intense variation of the landscapes and great contrasts between them (seaside and mountain, humid subtropical and arid zone, river gorges and steppe landscapes alternate over small areas), widely distributed endemic biological species, and traditional ethnographic, farming and domestic cultural forms survived in the historical regions (mostly in the mountains) of the country. The principal eco-tourism value is the natural eco-systems being only insignificantly modified due to industrial changes and maintaining their "natural originality". In addition, numerous remnants of historical and cultural heritage give them a certain value. They naturally merge with the environment and not dominate over it, but add to its beauty. The threats to the environment are mostly associated with an increasing anthropogenic "aggression", destruction of the traditional mode of nature management meaning certain degree of responsibility and deficient environmental laws and mechanisms of their execution. The problem is further aggravated by the fact that due to the limited territories, the eco-systems of the country are particularly sensitive to technogenic pressure. Concrete eco-tourism projects must inevitably envisage preliminary geo-ecological works.

Keywords: Eco-Tourism, Resources, Threats, Visitors, Georgia

1. Introduction

In recent years, ecological tourism has been actively developing in Georgia, which is important for the perspectives of the rational use of resources and positioning of the country on the world tourism market. The article considers some specific problems of the ecotourism development in Georgia.

Tourism in Georgia started to develop in the XIX century and was associated with the curative and health-improving properties of the natural environment and aesthetic values of the country. However, it should be said that in the past, tourism with its incomes greatly was behind other branches of economy.

The economy of modern Georgia is still in its transformation phase, and the objective of Georgian economics is to identify and concentrate on the sectors of economy capable of making the country competitive. In solving this problem, the trends of the international markets and the country's own resources should be considered. In this respect, the ecotourism potential of Georgia's natural environment becomes increasingly important. Fortunately, unlike many developed industries, there are undisturbed ecological systems, natural landscapes and natural sights survived in Georgia being an object of increasing interest on the international tourist market. In Georgia there are fourteen types of landscapes and seventy one genera [1]. According to the landscape diversity, Georgia is in 12th place in the world. But with regards to the amount of landscape types per square km, Georgia is in the first place. In other areas that are the equal in size to Georgia we may find only one type of landscape, while in Georgia we have twenty two, so ten times more than in other regions [2]. In spite of the fact that

Georgia possesses a small territory (69.7 thousand sq. km), it has approximately all the landscape types represented in the world, except savannas, deserts and rainforests. In Georgia we have approximately all types of landscape zones, which range from humid, dry subtropical valleys to high mountain alpine and glacier zone [3]; There is the sharply expressed vertical zonation of landscapes. The country's highest peak is mount Shkhara (5203 m), while the lowest point is located between Poti and village Kulevi (1.2-2.3 m below sea level) [4].

On the background of the aggravated natural threats worldwide, we can assume that these assets of our natural environment will be more valuable, while their protection is going to be a more urgent issue in the future. We agree that the most important resources serving as a support for the competitive strategy of Georgia are natural conditions and natural resources of the country, including eco-tourism potential. In this respect, the promotion of ecotourism development seems one of the most important directions of the sustainable development of the country.

It is known that in the historical past, Georgia had natural areas and objects protected by the state or society. For instance, as far back as in the XII century, Queen Tamar of Georgia issued a royal decree to protect certain territories of the country, and in the seventeenth century, the Code of Laws issued by King Vakhtang VI of Georgia established the territory of Korughi as a protected object, where cutting trees, farming and even walking was prohibited, and this territory was protected by guards [5]. It is also known that there were so called *Khati* (place of worship) forests in the mountainous regions of Georgia in the past, which were protected by the local communities. Virtually, these forests were reserves. In 1912, Lagodekhi Reserve was established in Kakheti. It was the first official reserve in Georgia. In the following years of the XX century, there were other reserves, managed reserves and protected areas established in Georgia.

Adoption of the Law Concerning the System of Protected Territories by the Parliament of Georgia in 1996, lying the foundation to the establishment of the environmental protection standards and categories based on the recommendations of International Union for Conservation of Nature (IUCN) [6] in Georgia was an important fact. From this time onward, a system of territories protected by the state with a single management started to develop, with the promotion of ecological tourism as one of its priorities.

2. Study Area & Methods

The study covers the protected areas of Georgia, with the following National Parks being the principal eco-tourist destinations: Algeti, Borjomi-Kharagauli, Vashlovani, Tbilisi, Tusheti, Kolkheti, Machakhela, Mtirala, Kazbegi and Javakheti. The eco-tours are partially admitted to other areas, such as Tusheti and Kintrishi protected landscapes as well as managed reserves, where in addition to eco-tours, other tourist activities are also admitted.

The total area of the National Parks of Georgia is 2767 sq. km, making almost 4% of the territory of the country. The National Parks are established in different corners of Georgia. The objects they protect (endemic biological varieties, unique landscapes, undisturbed ecological systems and historical-cultural monuments) are vary significantly in nature. On the one hand, the National Parks include the eco-systems typical to wet subtropics, e.g. the eco-systems in Mtirala or Kolkheti National Parks, and on the other hand, they cover the eco-systems typical to arid landscapes, e.g. the ones found in Vashlovani National Park. Borjomi-Kharagauli National Park is distinguished for a unique eco-system of coniferous forests, while Tusheti National Park is known for a rare variety of endemic biological species (the National Park includes 230 Caucasian endemic species, with 11 of them found only in Georgia), and accommodation, historical and ethnographic values of the local settlements. All other National Parks also have distinctive features.

The study is based on the expeditions organized in the protected areas and perspective eco-tourist areas (Kvemo Kartli, Erusheti Ridge, Guria mountainous area, Javakheti volcanic mountainous area, etc.) in different years. In addition to the direct observations, the study uses literary, statistical and cartographic methods. The article presents the primary results of the study of the ecological tourism development in Georgia. The next stage will be dedicated to the geo-ecological and tourist-geographical study of the individual protected areas.

3. Results & Discussions

Ecotourism is a relatively new branch in Georgia. Its origination and development are associated with the protected areas.

At present, there are 10 national parks, 14 state reserves, 18 managed reserves and 2 protected landscapes in Georgia, with different kinds and degrees of tourist activities permitted.

Tourism in the protected areas of Georgia is characterized by an increasing trend (fig. 1). According to data of the Agency of Protected Areas, the number of ecotours and visitors in the protected areas over the recent years has been featuring a dynamic increase. In particular, the total number of visitors increased from 12,2 thousand to 417,8 thousand – more than 34-fold increase - over the period of 2008-2014. The greatest boost in the number of visitors was featured in 2011 (with 240% growth). As for 2012, it was characterized by a slight fall of the same number. However, in the following years, the rate of increase in the number of visitors recovered.

The 2014 reduction (to +1.9 %) of the increase rate of total annual number of tourists in Georgia was not reflected in ecotourism (+19,2%). This could be explained by the fact that the main generating countries for ecotourism – Israel, Germany, Poland, Russia and Ukraine – were affected less by the stricter visa regulations imposed by the Georgian government in 2014 that made the visa procedures more complicated. The regulation changes were directed more at countries visitors from which represented insignificant part of the ecotourism field – namely Iran, Egypt, Iraq and China.

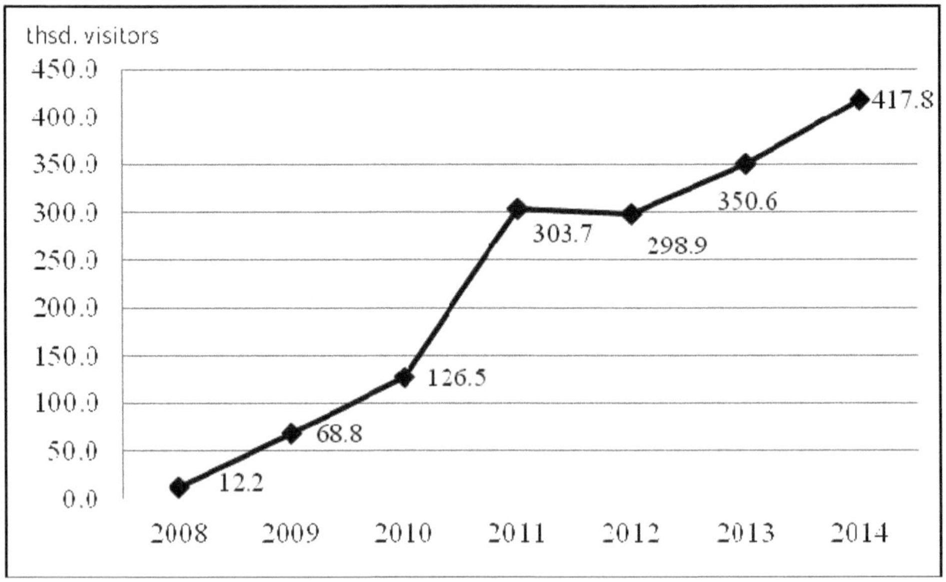

Figure 1. *Total number of visitors to the protected areas of Georgia*

Source: Georgian National Tourism Administration

Similar changes are reflected in income dynamics (fig. 2). Revenue from ecotours in protected areas over 2008-2014 increased 23-fold (55,7 thousand to 1318,1 thousand GEL). As the given figures show, the rate of income growth until 2011 (48% of the average annual increase) significantly fell back the rate of growth of the number of visitors (500% of average annual increase). However, from 2011, the rate of income growth tended to increase rapidly and overcame the rate of growth of the visitors' number. This can be explained by the establishment of new territories and improvement of the reception facilities and infrastructure for the eco-tourists in the protected areas in 2011.

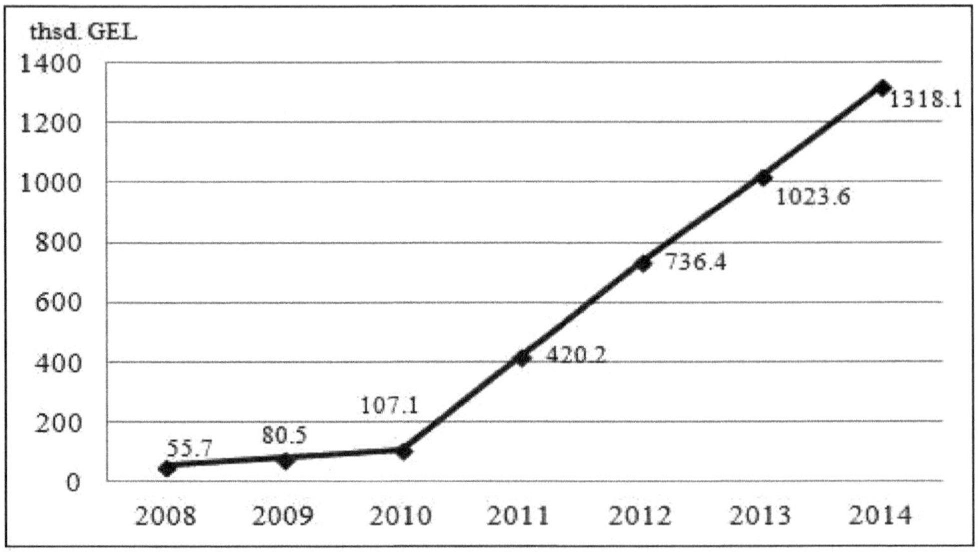

Figure 2. *Size of Incomes from the Protected Areas of Georgia*

Source: Agency of Protected Areas.

As per the structure of visitors to the protected areas, several countries worldwide are remarkable in this respect. In 2014, most tourists traveled from Israel (16,9 thousand people), followed by Russia (15,1 thousand people), Ukraine (13,6 thousand), Poland (13,1 thousand) and Germany (5,8 thousand).

The visitors are mostly young people up to 35 years of age, mostly traveling in groups of 4 to 6. The distribution of visitors in different protected areas is very interesting (fig. 3).

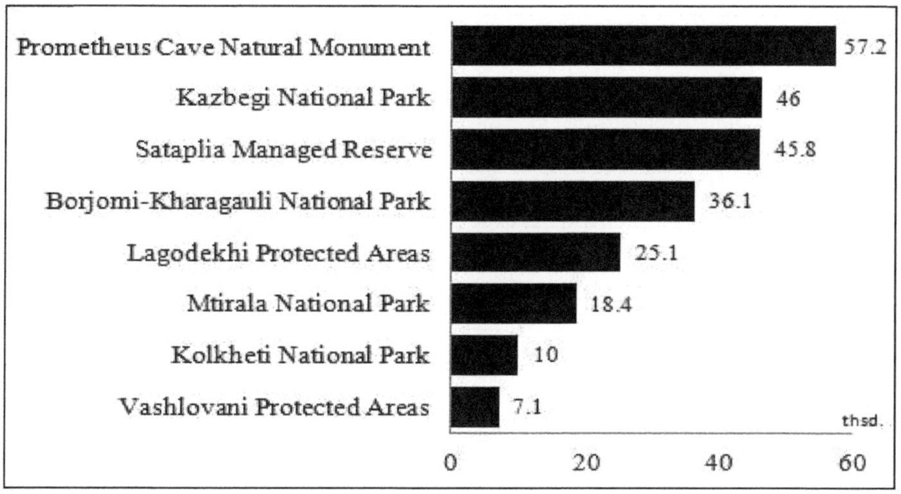

Figure 3. *Distribution of Visitors by Protected Areas by 2014*

Source: Agency of Protected Areas.

As the figure 3 shows, outstanding destinations are Prometheus and Sataplia karst caves. In our opinion, the major reason for this is the best equipment and facilities of these destinations and large and efficient private investments in them [7].

It should be noted that the system of tourist reception, accommodation and service still fails to meet the visitors' demands. The same is true with cognitive ecological programs [6]. Recovering these deficiencies is essentially important to attract eco-tourists and create the perspectives to increase incomes in the future.

The trend of increasing number of ecotourists in Georgia is first of all, caused by the increasing interest in Georgia on the world tourist market as in a country with fresh and important tourist and recreational resources. The number of tourists to the country is characterized by a general increase. As per the data of the Ministry of Internal Affairs of Georgia, the number of visitors to the country has steadfastly increased in recent years (+23,4% per year). In 2014 the rate of increase in total (tourist as well as non-tourist) arrivals to Georgia was reduced to +1,9% (The number of the total arrivals exceeded 5 mn 493 thousand) [8]. This reduction was mostly caused by external factors – mainly the unstable political and economic situation in Ukraine and Russia. Despite this the rate of increase in tourist visits was reduced to a far less degree - the number of tourists (2 mn 218 thousand) represented a reduction to +7,4% increase rate compared to the previous year [8].

The country has all three components of the ecological tourism potential: nature diversity, variety of historical and cultural heritage and ethnographical diversity. However, there are problems, which are to be solved to develop the perspectives of ecotourism.

The number of tourists is expected to continue to grow in the future, and the stress on the natural eco-systems and traditional economic and ethnographic properties of local communities will increase consequently. Besides, as it is known, these major resources of ecotourism are very sensitive to excessive anthropogenic impact.

Therefore, when planning and organizing the eco-tours, the limits of the admissible number of tourists to the protected areas are to be taken into account by considering the specifics of each eco-system. This problem can be solved by accomplishing ecological, geo-ecological and touristic-geographical studies.

Developing the ecotourism in a qualitative rather than quantitative respect is particularly important. In particular, the guest reception facilities, such as accommodation, catering, recreation, entertainment and cognitive facilities and infrastructure must be provided in greater numbers and modernized to be an organic part of the surrounding environment and not surpass it; an increase of per tourist income should be promoted, and it is also important to prolong the touristic season.

Development of the local market of ecotourism is an important reserve. The specific weight of domestic travelers is still very little. The problem here is the high prices of service incompliant with the purchasing ability of the population [6].

In the experts' opinion, the major deficiency of tourism is the underdeveloped touristic business in Georgia in the peripheries of the country, as well as inflexible management, poor service, non-proficiency in foreign languages, outdated material-technical base and infrastructure, lack of modern standards and high service prices not corresponding to the service quality [6, p.7].

The major goal of establishing and developing a system of protected territories is the protection and maintenance of bio-ecological values of the natural environment, such as bio-diversity and endemic and rare biological species. As it is known, the Caucasus was named by the IUCN among 34 world hotspots, particularly rich in species and with great bio-diversity and was named by the World Wildlife Fund among 200 key eco-regions of the world [6. p.21].

Villages and communities of Georgia are also an important eco-tourist resource. Remote and often abandoned mountain villages, as well as villages with the survived elements of the traditional Georgian life and farming and ethnic villages survived at some places of the country (e.g. villages of German Lutherans, dukhobors, Daghestani or Greek people and other nationalities) are a subject of the keenest interest. As the incoming tour-operators state, such destinations are in great demand. On the other hand, due to a number of respects.

The threats to the nature of Georgia, the major riches of the country, is the aggravated ecological problem, such as reduced forest areas, active erosion processes, devastated bio-diversity, fertile soils occupied with industrial and construction sites, accelerated deforestation and active natural calamities, are to a great extent associated with the irrational and irresponsible nature management. The situation can be rescued through active environmental measures, with the expansion of the system of the protected territories as one of the major measures.

The nature of Georgia has significant resources to expand the eco-tourist areas. There are a number of undisturbed natural eco-systems survived in Georgia and there are areas with ecologically valuable objects of different scales. There are forms of organic unity of society and nature still found in our natural environment. The historical monuments and traditional villages scattered in the nature add to the attractiveness of the country. At this point, there is one conceptual issue of how to treat ecotourism. Is it cognition only, or is it also visiting the environment being in an organic co-existence with humans?

Another principal opinion concerns the role of the non-living nature in the selection of the protected areas. Surely, the rare species of living nature of Georgia need protection, but it does not seem fair to us to underestimate the role of non-living forms of the eco-systems, such as relief forms, soil, watercourses, karst landscapes or climatic conditions in the eco-systems. In fact, the geographical specifics of these elements of the eco-systems result in the diversity of the biological species. It is the non-living nature subject to a much more intense negative industrial and anthropogenic stress with consequent influence on the whole eco-systems. In our opinion, when creating the new protected areas in the future, the necessity for protecting the non-living nature should be considered. This opinion is based on the observations made during the complex geographical expeditions organized in the regions of Georgia in recent years.

During our expeditions, the primary study of the ecological complexes and rare relief forms revealed the objects, which can be formed as protected areas and consequently, become an active ecotourism objects. In our view, such objects can be:

- Individual cliffy systems of various hypsometric heights and originations, in particular mountain ridges, e.g. Birtvisi, etc.
- Large karst formations, e.g. karst massif of Racha, which is very large, but in fact remains unstudied [9].
- Mountainous zone in Guria: mountain forests, mouths and upper reaches of the Rivers Gubazeuli, Bzhuzhi and Supsa, Bakhmaro and Gomismiti Resorts, including the Reserve of Pontic oak. This zone is subject to heavy anthropogenic impact and the eco-systems here degrade rapidly.
- Gorges in Kvemo Kartli: the canyons of the Rivers Khrami (near Samshvilde), Chivchavi and Karabulakhi [10].
- Javakheti volcanic mountainous area with impressive megalith monuments formed (Shaori and Abuli fortresses, monastery cut in Samsari cliff) [11].
- Mountainous zone in Samegrelo: Egrisi Ridge, the River Khobistskali Gorge, Lake Tobavarchkhili.
- Erusheti Mountainous zone in Samtskhe.
- Davitgaredja, etc.

4. Conclusions

Ecotourism in Georgia is characterized by increased rates. The number of visitors and incomes grow rapidly. The major factor contributing to the growing interest in this kind of tourism and ecological activity is the important natural resources of the country, in particular, diversified landscapes densely clustered in different areas, surviving undisturbed eco-systems and endemic biological species mostly found in the protected areas.

Some of the negative factors are the underdeveloped specific material-technical and scientific-information bases for tourist reception and the inflexible pricing system.

The prospects for the ecotourism development are associated with the establishment of the new visitor service objects in the protected areas and perfection of the existing ones, as well as the extension and perfection of the research and cognitive programs.

Extension of the system of the protected areas with the ecologically valuable but threatened eco-systems and non-living nature objects, which are not a part of the system at present, represents a significant potencial.

With regards to state policy, it is important to ensure a well-organized development of ecotourism in the country based on the reasonable strategy, by strictly observing the ecological norms and considering scientifically developed recommendations.

References

[1] N.Beruchashvili. Landscape Map of Caucasus. Tbilisi, Scale 1:1,000,000.

[2] N.Beruchashvili.Diversity of Georgia's Landscapes and Geographical Analysis of Landscapes Diversity of the World. Proc. of the First National Conference: Biological and Landscape Diversity of Georgia. Tbilisi, 2000, 221-250

[3] Nikolaishvili D., Matchavariani L., Demetrashvili O. Methodology of Determining the Agri-tourism Potential on Georgia's Example. Journal of Tourism Challenges and Trends - Rural Tourism, vol. IV, No 1, 2011, 59-76

[4] Nikolaishvili D., Donadze T., Meladze G., Trapaidze V., Dvalashvili G., Sharashenidze M., Tigishvili T. Evaluating the Touristic Potential of Georgia's Landscapes. Forum geographic. Vol. XI. Issue 2 (12-12-20112), 2013, 145-154.

[5] Protected Areas of Georgia. Tsignis saxelosno, Tbilisi, 2007, 5.

[6] Assessment of Ecotourism Potential in Georgia. IUCN, Tbilisi, 2008.http://www.greengeorgia.ge/sites/default/files/Ecotouris m %20Assessment%20in%20Georgia.pdf

[7] G.Khomeriki, D.Maisuradze, K.Tsikarishvili, T.Khucishvili. Carst Caves of Georgia as the Significant Tourist Resource. Geography of Georgia, TSU, N 6, Tbilisi, 2007, pp. 69 -78.

[8] http://www.police.ge/ge/useful-information/statistics/sazghvri s- kvetis-statistika

[9] M. Gongadze, G. Lominadze, G. Khomeriki. Karst of the Racha Range and its Touristic and Recreational Potential. Collected papers of Institute of Geography of Georgia. New series, N 6 (85). Tbilisi, 2014. 53-57.

[10] G. Khomeriki, V. Neidze, M. Tutberidze, M. Gongadze, G.Lominadze, N.Chikhradze. Tourist-Recreational Potential of Kvemo Kartli (Geographical Analyses. Collected papers of Institute of Geography of Georgia. New series, #N4 (83). Tbilisi, 2012, 237-246.

[11] V. Neidze, M. Tutberidze, G. Khomeriki. Tourism in Djavakheti: Development Factors. Collected papers of Institute of Geography of Georgia. New series, N5 (84). Tbilisi, 2013. 240-245.

New Methodologie Eco-Friendly Using Calcined Eggshell Meal Catalyst for Synthesis of Benzimidazoles, Benzoxazoles and Benzothiazoles

Abdeljabar Haboub[1], Meryem Hamlich[1], Souad Harkati[1], Yassine Riadi[2], Rachid Slimani[2], Mina Aadil[3], Anouzla Abdelkader[1], Saïd Lazar[2], Mohamed Safi[1, *]

[1]Laboratoire Materials, Catalysis & Natural Resource Development, URAC University of Hassan II Mohammedia-Casablanca, BP 146, Mohammedia, Morocco
[2]Laboratory Biochemistry, Environment & Agri URAC University of Hassan II Mohammedia-Casablanca, BP 146, Mohammedia, Morocco
[3]Laboratoire of Physical Chemistry and Bio-organic Chemistry, University of Hassan II Mohammedia-Casablanca, BP 146, Mohammedia, Morocco

Email address:
mohamedsafi@yahoo.fr (M. Safi)

Abstract: The Calcined Eggshell Meal (CEM) doped is a new solid support has been employed as a catalyst for efficient synthesis of benzimidazoles, benzoxazoles and benzothiazoles. Taking into account environmental and economical considerations, the handling of CEM used here presents many advantages such as simple and convenient procedure, easy purification and shorter reaction time and enhanced recycling possibilities, which are now well established in fine organic synthesis.

Keywords: Calcined Eggshell Meal, Heterogeneous Catalysis, Benzimidazoles, Benzoxazoles, Benzothiazoles, Recyclable Catalyst

1. Introduction

The benzimidazoles, benzoxazoles and benzothiazoles structural motifs play very important roles in numerous pharmaceutical molecules with a wide range of biological properties including antibacterial [1], antibiotic [2], anticonvulsant [3], anticancer [4], antihypertensive [5], antifungal [6], anti-inflammatory [7], immunosuppressant agents [8] and antiviral effects [9]. Consequently, the synthesis of the heterocyclic nucleus has gained great importance. Many methods for the synthesis of benzimidazoles [10-19], benzoxazoles [20-24] and benzothiazoles [25-29] have been discovered and reported.

However, all of these procedures suffer from one or more of the following disadvantages such as long reaction times, low yields of the products, harsh reaction conditions, the use of excess amounts of reagents, tedious workup procedures, and co-occurrence of several side reactions. In addition, some of the catalysts and reagents are expensive, toxic, and air sensitive. Therefore, there is still a need to search for better catalysts that could be superior to the existing ones with regards to toxicity, handling and operational simplicity.

Nowadays, many organic reactions have been devised in such a way that the reagents are deposited on various inorganic solid supports. These reagents are becoming a major area of chemical research due to the wide variety of organic transformation that can be carried out under mild experimental conditions and the continuing drive for cleaner and more efficient processes [30].

For our own part, we recently reported that Animal Bone Meal (ABM), a source of biogenic apatite and a cost-effective material, could be used as catalyst in several reactions [31].

Eggshell Meal represents a source of biogenic calcium carbonate. Considering good catalytic properties of synthetic calcium carbonate powder towards organic synthesis, the main aim of this study was to evaluate Calcined Eggshell

Meal (CEM) as original heterogeneous catalyst in organic synthesis. The natural support CEM has employed as catalyst in Knoevanegel reaction [33].

A brief and easy method to prepare the CEM as catalyst in synthesis of benzimidazoles, benzoxazoles and benzothiazoles has been developed. As far as we know, it is the first report of the use of this natural source to perform organic reactions with low cost and high efficiency.

In continuation of our effort to explore the catalytic activity of CEM for useful organic transformation we herein report that Lewis acids doped CEM catalyzed a simple and environmentally benign synthesis of benzimidazoles, benzoxazoles and benzothiazoles, from the reaction of *o*-phenylenediamine, *o*-aminophenol and *o*-aminothiophenol with different aldehydes (Scheme 1).

1 X = O, S, NH

2 R = H, Cl, NO_2, F CH_3O, CH_3

3 X = NH
4 X = O
5 X = S

Scheme 1. *Synthesis of benzimidazoles 3, benzoxazoles 4 and benzothiazoles 5 catalyzed by CEM, $ZnCl_2$/CEM and $CuBr_2$/CEM.*

2. Experimental

2.1. Preparation and Characterisation of CEM Catalyst

Eggshells were collected, then washed several times with tap water and left in open air for several days to get rid of odours. Later, they were transferred to the oven at 80°C for drying. The dried Eggshells were crushed and milled into different particle sizes in the range 45-200 µm, calcined at 800°C for 2h. The residue was washed with water and was used after drying 24h at 80°C. The residue was washed with water and was dried overnight at 100°C in a conventional draying oven and then calcined at a heating rate of 2°C/min to 400°C and kept at this temperature for 4h. The resulting material was denominated CEM.

The chemical composition of CEM shows a high yield of Ca (61.95%) compared to small amounts of Mg (0.79%), Si (0.65%), Na (0.64%), P (0.47%), Al (0.17%) and Cl (0.12%). The X-ray diffraction of our Catalyst is shown in Figure 1 show a main peak appeared at 2θ=34.13°. In addition, this spectrum shows several other small peaks at 2θ = 18.07°, 23.08°, 28.72°, 29.43°, 36.03°, 39.47°, 48.55°, 50.85°, 57.48°, 60.74°, 62.62°, 64.30°, 71.80° and 81.88°. This spectrum confirms the presence of calcite, syn and portlandite. Analysis of FT-IR spectrum of CEM gave some bands at 3600 cm^{-1} that are assigned to -OH stretching modes and the stretching and folding of carbonate group has been assigned to peaks at 1500 cm^{-1} and 865 cm^{-1}. The specific surface area of the CEM calculated by the BET (Brunauer-Emmett-Teller) method is 62.42 m²/g.

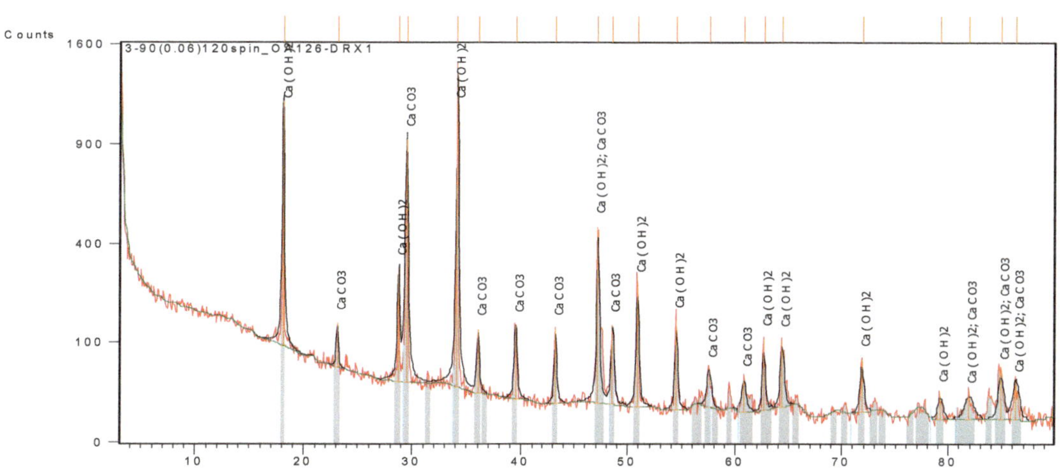

Figure 1. *X-ray diffraction of CEM.*

2.2. Reaction Procedure

A mixture of *o*-phenylenediamine 1a, *o*-aminophenol 1b or *o*-aminothiophenol 1c (1mmol), different aromatic aldehydes 2 (1,1mmol), and a pinch of catalyst (100mg) in 10mL toluene were stirred for 10min and refluxed further at 110°C for an appropriate time (Table 1) which was monitored by TLC (*n*-hexane/EtOAc : 2/1) and HPLC. After completion of the reaction, catalyst was recovered from the filtrate, concentrated on a rotatory evaporator and chromatographed on a silica gel column to offer pure product.

3. Results and Discussion

3.1. Optimization of the Heterocyclic Construction

In the initial experiment, we investigated various conditions in the model reaction of *o*-phenylenediamine 1a with benzaldehyde 2a in the presence of CEM/$ZnCl_2$ (Scheme 2) and the results were summarized in Table 1.

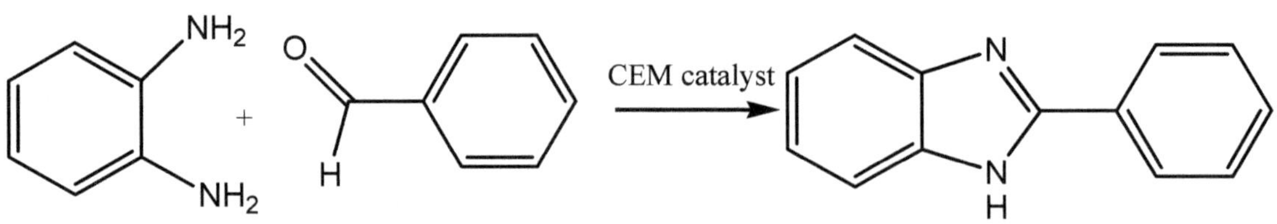

1a **2a** **3a**

Scheme 2. *Synthesis of benzimidazoles 3a catalyzed by $ZnCl_2$/CEM.*

The effect of various solvents (such as $CHCl_3$, CH_3OH, CH_3CN, THF, H_2O, 1,4-dioxane and toluene) on the model reaction was conducted in the presence of CEM/$ZnCl_2$. The result indicated that the solvents had a significant effect on the product yield. The use of, $CHCl_3$, CH_3CN and 1,4-dioxane as solvent gave moderate yields (Table 1, Entries 1, 3 and 6), The use H_2O as solvent gave no reaction(Entry 5), but the good yields was observed with THF and free solvent (Entries 4, 13). The best conversion was observed when the reaction was performed in toluene at 110°C (Entry 10). A mixture of starting materials and final heterocyclic was obtained. The best solvent remains toluene, but it should be pointed out that heterocyclic construction also requires thermal, aprotic and apolar conditions.

Without catalyst in refluxing toluene, condensation of benzaldehyde 2a with *o*-phenylenediamine 1agave low yield and required long reaction times (Entry 12). No condensation was observed in the absence of air [33] prevents the required oxidative step and benzimidazole were not detected (Entry11).

In air and in the presence of $ZnCl_2$/CEM, reactions between 2a and *o*-phenylenediamine furnished the desired cycloadducts 3a after 3h in good to excellent yields (Entries 8,9,10). In addition, temperature and time play a role, reactions performed in refluxing toluene showed a better efficiency.

Table 1. *Optimization for the synthesis of 3a catalyzed by $ZnCl_2$/CEM.*

Entry	Solvent	T (C°)	Time(h)	Yield (%)[a]
1	$CHCl_3$	50	3h	61
2	CH_3OH	25	3h	83
3	CH_3CN	25	3h	49
4	THF	66	3h	81
5	Water	100	3h	NR[c]
6	Dioxane	100	3h	61
7	Toluene	70	3h	26
8	Toluene	100	3h	76
9	Toluene	110	2h	78
10	Toluene	110	3h	96
11[b]	Toluene	110	24h	NR[c]
12	Toluene	110	24h	34[d]
13	free	100	3h	80

[a] Yields in pure isolated products; [b] No Air; [c] No Reaction; [d] No catalyst.

3.2. Extension of the Methodology

For these purpose, two other Lewis acid catalysts were prepared by impregnating CEM with $ZnCl_2$ or $CuBr_2$. Heterocyclic construction efficiency was also performed using CEM alone and doped CEM analogs.

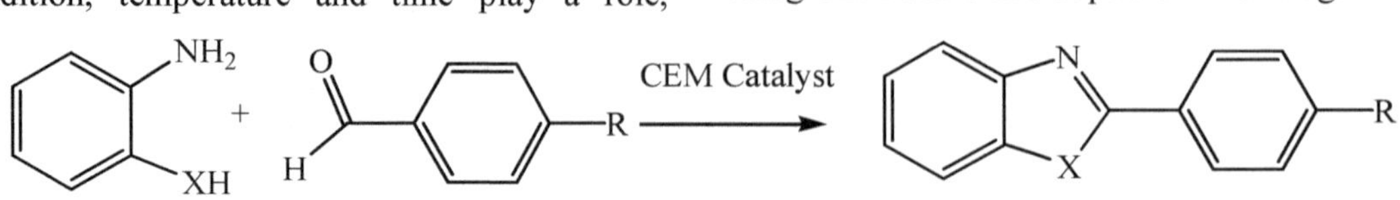

1a X = NH
1b X= O
1c X=S

2a-f

3a-f X =NH
4a-f X= O
5a-f X=S

Scheme 3. *Synthesis of benzimidazoles 3, benzoxazoles 4 and benzothiazoles 5 under CEM, CEM/$ZnCl_2$ and CEM/$CuBr_2$.*

In air, a slight excess of aldehyde 2a-f (1.1 equiv.) was used in the presence of amino derivatives 1a-c and CEM catalysts at 110°C to build a small library of fused heterocyclic of type 3-5 (Table 2). Reaction progress was monitored by HPLC (C18, CH_3CN /H_20 70/30, λ=254 nm, Flow rate : 0,3mL/mn). After completion of the reaction, the crude product was purified by column chromatography and recrystallization to afford pure benzimidazoles 3, benzoxazoles 4 and benzothiazoles 5. Physical and spectral characterization of the products was confirmed by comparison with available literature data [31, 33-37].

With CEM alone, the reaction proceeded smoothly throughout heterocyclic construction and the yield was lower (Table 2, Entries 9, 10). An electro-donating group on aldehyde 2 decreased the efficiency of the heterocyclic building. Reaction yields were between 9 and 94% (Table 2).

Under similar conditions, the use of CEM doped with MX_2 Lewis acids the products were isolated in moderate to excellent yields (52-98%). Degradation was not observed and after 3h, heterocyclic 3-5 were obtained after an easy purification step. Reaction yields and speed slightly decreased when methoxy or methyl benzaldehydes 2d and 2e (Entries 4-5, 10-11 and 16-17) were employed. The conversion obtained with CEM, $ZnCl_2$/CEM and $CuBr_2$/CEM clearly showed the positive effect of the impregnating process $ZnCl_2$ doped CEM.

Table 2. *Synthesis of benzimidazoles 3, benzoxazoles 4 and benzothiazoles 5 under CEM, $ZnCl_2$/CEM, and $CuBr_2$/CEM catalysis.*

Entry	Reagent	X	R	Yield %, CEM	Yield % $ZnCl_2$/CEM	Yield % $CuBr_2$/ CEM
1	3a	NH	H	86%	94%	55%
2	3b		Cl	79%	85%	44%
3	3c		NO_2	90%	65%	82%
4	3d		OCH_3	46%	84%	47%
5	3e		CH_3	81%	98%	96%
6	3f		F	46%	53%	82%
7	4a	O	H	50%	52%	52%
8	4b		Cl	94%	67%	66%
9	4c		NO_2	32%	76%	39%
10	4d		OCH_3	9%	59%	80%
11	4e		CH_3	72%	62%	72%
12	4f		F	84%	87%	48%
13	5a	S	H	90%	93%	93%
14	5b		Cl	90%	95%	97%
15	5c		NO_2	20%	54%	58%
16	5d		OCH_3	90%	66%	82%
17	5e		CH_3	92%	66%	98%
18	5f		F	90%	92%	94%

[a] Yields in pure isolated products; t = 3h; T= 110°C.

The recycling performance of the CEM/$ZnCl_2$ in the model reaction was also investigated. After completion of the reaction, the isolated CEM/$ZnCl_2$ was washed with CH_2Cl_2 and reactivated at 400°C for 2h, and can be reused for times, after for recycles the catalytic activity of CEM/$ZnCl_2$ was almost the same as that of fresh catalyst (Table 3). Catalytic processes and reduced environmental problems (lower energy, valorization of waste as natural and reusable catalyst, non toxicity of the catalyst, reduced amount of solvent).

Table 3. *Studies on the reuse of CEM/$ZnCl_2$.*

Entry	Run	Yield % 3a
1	fresh	99 %
2	Recycle I	96 %
3	Recycle II	95 %
4	Recycle III	92%
5	Recycle IV	91 %

[a] Each reaction was carried out as described in reference [36]

4. Conclusion

In summary, we have developed a simple and efficient one-pot synthetic approach for the synthesis of benzimidazoles, benzoxazoles and benzothiazoles by the condensation of various *o*-phenylenediamine, *o*-aminophenol or *o*-aminothiophenol and aromatic aldehydes in the presence of CEM/$ZnCl_2$. This methodology offers very attractive features such as reduced reaction times, high yields and atom efficiency, as well as easy product separation. The operational simplicity of the procedure is also attractive from the green chemistry point of view.

References

[1] I. Yildiz-Oren, I. Yalcin, E. Aki-Sener, N. Ucarturk, Eur. J. Med. Chem. 39 (2004) 291.

[2] X. Song, B.S. Vig, P. L. Lorenzi, J. C. Drach, L. B. Townsend, G. L. Amidon, J. Med. Chem. 48 (2005) 1274.

[3] (a) W. G. Bywater, W. R. Coleman, O. Kamm, H. H. Merritt, J. Am. Chem. Soc. 67 (1945) 905. (b) A. Benazzouz, T. Boraud, V. Dubédat, A. Boireau, J. M. Stutzmann, C. Gross, Eur. J. Pharmacol. 284(1995) 299.

[4] (a) D. Kumar, M. R. Jacob, M. B. Reynolds, S. M. Kerwin, Bioorg. Med. Chem. 10 (2002) 3997 (b) M. R. DeLuca, S. M. Kerwin. Tetrahedron Lett, 38 (1997) 199 (c) S. Sato, T. Kajiura, M. Noguchi, K. Takehana, T . Kobayasho, T. Tsuji, J. Antibiot. 54 (2001) 102 (d) M. Ueki, K. Shibata, M. Taniguchi, J. Antibiot, 51 (1998) 883.

[5] K. Kubo, Y. Inada, Y. Kohara, Y. Sugiura, M. Ojima, K. Itoh, Y. Nishikawa, T. Naka, J. Med. Chem, 36 (1993) 1772-1784

[6] M. Yamato, J. Pharm. Soc. Jpn. 112 (1992) 81

[7] P. A. Thakurdesai, S. G. Wadodkar, C. T. Chopade, Pharmacology online, 1 (2007) 314-329

[8] (a) M. R. Grimmett, A. R. In: Katritzky, C.W. Rees, Scriven E.F.V. (eds) Comprehensive Heterocyclic Chemestry II, Vol.3, (1996), Elsevier Science Ltd, Oxford. (b) M. R. Grimmett, Imidazole and Benzimidazole Synthesis, Academic Press, San Diego (1997).

[9] X. Song, B. S. Vig, P. L. Lorenzi, J. C. Drach, L. B. Townsend, G. L. Amidon, J. Med. Chem. 48(2005) 1274

[10] S.T. Huang, I. J. Hsei, C. Chen, Bioorg. Med. Chem. 14 (2006) 6106-6119.

[11] J. L. Girardet, L. B. Townsend, J. Org. Chem, 64 (1999) 4169-4172.

[12] C. M. Yeh, C. L. Tung, C. M. Sun, J. Comb. Chem, 2 (2000) 341-348.

[13] J. J. Chen, V. Wie, J. C. Drach,. L. B. Townsend, J. Med. Chem. 43 (2000) 2449-2456.

[14] D. Tumelty, K. Cao, C. P. Holmes, Org. Lett. 3 (2001) 83-86.

[15] J. Mann, A. Baron, Y. Opoku-Boahen, E. Johansson, G. Parkinson, L. R. Kelland, S. Neidle, J. Med. Chem. 44 (2001) 138-144.

[16] B. Raju, N. Nguyen, G. W. Holland, J. Comb. Chem, 4 (2002) 320-328.

[17] H. Akamatsu, K. Fukase, S. Kusumoto, J. Comb. Chem. 4 (2002) 475-483.

[18] C. E. Hoesl, A. Nefzi, R. A. Houghten, J. Comb. Chem. 5 (2003) 155-160.

[19] D. Vourloumis, M Takahashi, K. B. Simonsen, B. K. Ayida,. S. Barluenga, G. C. Winters, T. Hermann, Tetrahedron Lett. 44 (2003) 2807-2811.

[20] D. F. Shi, T. D. Bradshaw, S. Wrigley, C. J. McCall, P. Lelieveld, I. Fichtner, M .F. G. Stevens, J. Med. Chem. 39 (1996) 3375-3384.

[21] X. Beebe, D. Wodka, T. J. Sowin, J. Comb. Chem, 3 (2001) 360-366.

[22] A. Hari, C. Karan, W. C. Rodrigues, B. L. Miller, J. Org. Chem. 66 (2001) 991-996.

[23] R. S. Pottorf,. N. K. Chadha, M. Katkevics, V. Ozola, E. Suna,. H. Ghane, T. Regberg, M. R. Player, Tetrahedron Lett. 44 (2003) 175-178.

[24] F. Chen,.C. Shen, D. Yang, Tetrahedron Lett. 52 (2011) 2128-2131.

[25] M. S. Chua, D. F. Shi, S. Wrigley, T. D. Bradshaw, I. Hutchinson, P. N. Shaw, D. A. Barrett, L. A. Stanley, M. F. G. Stevens, J. Med. Chem. 42 (1999) 381-392.

[26] E. Kashiyama, I. Hutchinson, M. S. Chua, S. F. Stinson, L. R. Phillips, G. Kaur, E. A. Sausville, T. D. Bradshaw, A. D. Westwell, M. F. G. Stevens, J. Med. Chem. 42 (1999) 4172-4184.

[27] I. Hutchinson, M. S. Chua, H. L. Browne, V. Trapani, T. D. Bradshaw, A. D. Westwell, M. F. G. Stevens, J. Med. Chem. 44 (2001) 1446-1455.

[28] W. Leng, Y. Zhou, Q. Xu, J. Liu, Macromolecules, 34 (2001) 4774-4779.

[29] I. Hutchinson, S. A. Jennings, B. R. Vishnuvajjala, A. D. Westwell, M. F. G. Stevens, J. Med. Chem. 45 (2002) 744-747.

[30] G. H. Ponser, Angew. Chem. Int. Ed. Engl. 17 (1978) 487-496.

[31] (a) Y. Riadi, R. Mamouni, Y. Abrouki, M. ElHaddad, N. Saffaj, S. ElAntri, S. Routier, G. Guillaumet, S. Lazar, Lett. Org. Chem. 7 (2010) 269-271. (b) Y. Riadi, R. Mamouni, R. Azzalou, R. Boulahjar, Y. Abrouki, M. ElHaddad, S. Routier, G. Guillaumet, S. Lazar, Tetrahedron Lett. 51 (2010) 6715-6717. (c) Y. Riadi, R. Mamouni,. R. Azzalou, M. ElHaddad,. S. Routier, G. Guillaumet,S. Lazar,. Tetrahedron Lett. 52 (2011) 3492-3495. (d) Y. Riadi, Y. Abrouki, R. Mamouni,.M. ElHaddad, S. Routier, G. Guillaumet,.S. Lazar, Chem. Cent. J. 6 (2012) 60. (e) Y. Riadi, Y. Abrouki, S. ElAntri, R. Mamouni, M. ElHaddad, S. Routier, G. Guillaumet, S. Lazar, Int. J. Chem. 34 (2013) 1152-1156.

[32] Y. Riadi, R. Slimani, A. Haboub, S. Elantri, M. Safi, S. Lazar, Mor. J. Chem. 2 (2013) 24-28.

[33] L. Songnian, Y. Lihu, Tetrahedron Lett. 46 (2005) 4315-4319.

[34] V. R. Devalla, K. Ethirajulu, J. Chem. Soc. (1995) 1497-1501.

[35] K. Bougrin, A. Loupy, M. Soufiaoui, Tetrahedron 54 (1998) 8055-8064.

[36] A. B. Naidu, G. Sekar, Synthesis. (2010) 579-586.

[37] S. L. Balaji, R. P. Umesh, R. M. Jyotirling,.A. M. Ramrao, Bull. Korean Chem. Soc. 31 (2010) 2329-2332.

26

Impact of Global Warming on the Vegetation Durable and Distribution Area of Crops in the Humid Subtropical and Mountainous Regions of Georgia

Giorgi Meladze, Maia Meladze

Institute of Hydrometeorology at the Georgian Technical University, Tbilisi, Georgia

Email address:

meladze.agromet@gmail.com (G. Meladze); meladzem@gmail.com (M. Meladze)

Abstract: The surface air temperature in the Eastern and Western parts of Georgia in conditions of global warming, has increased by 0.2-0.5°C, respectively. By 2030-2050, the average increase in the annual temperature may reach 1-2°C. Therefore, the preliminary evaluation of the impact of such an increase on the agrarian sector is important. The increasing and decreasing trends of the duration of the vegetation period and sums of active temperatures and atmospheric precipitations were revealed for the vegetation period in humid subtropical zone of Adjara in West Georgia and mountainous regions of Samtskhe-Javakheti in East Georgia, according to the trend. The increasing trend of the sum of temperatures in the humid subtropical zone of Adjara has reached 52°C for half a century (1956-2005). In terms of such a temperature increase, the sum of active temperatures is going to exceed 100°C by 2050 what will be beneficial for the ripening of citrus fruits and high productivity of other crops in terms of moderate soil productivity. The increasing trend for the sum of active temperatures in Samtskhe-Javakheti region in the above-said period reached 125°C according to the trend, and may double by 2050 reaching 3192°C. The given sum of active temperatures will improve the productivity of cereals, fruit-trees, vegetables, vine (early, with moderate ripeness) and other crops, in terms of relevant irrigation. According to the current (basic) and future scenario (2020-2050) in Adjara region, by considering 1°C increase in air temperature, there are four micro-zones allocated (up to 500 m a.s.l.) to grow appropriate crops. For Samtskhe-Javakheti region Under the current (basic) and future scenario, the increase in temperature by 2°C, there are four micro-zones allocated (by 500 m gradation), to grow appropriate crops.

Keywords: Crops, Active Temperature Sum, Scenario

1. Introduction

The data analysis of monitoring over the global climate change evidences an increasing trend in surface air temperature. According to the IPCC report, the average annual air temperature has increased by 0.6°C. The researchers prove that global warming is mainly associated with the anthropogenic impact, and therefore, human activities must be reasonably and purposefully planned. In particular, the concentration of carbon dioxide (CO_2) in the atmosphere must be reduced, as it has like „greenhouse effect" resulting in the process of warming in the lower layers of the atmosphere [1].

At the end of the 20th century, the concentration of carbon dioxide increased by 10% [2], and this trend continues in the 21th century. By 2030, the increase may double what will lead to the temperature increase by 2-3°C [3, 4]. The said problem must be solved by the joint efforts of the world countries. Otherwise, if the climate warming continues for many years, the ecological balance established for many centuries may be negatively influenced. Global climate warming has spread in the Western and Eastern regions of Georgia. The surface air temperature has increased by 0.2-0.5°C, respectively [5, 6, 7]. The greater temperature increase is observed in east Georgia as compared to west Georgia. Such an increase must be taken into account in the future because the temperature increase may continue and lead to the 1-2°C increase in the average multi-year temperature by 2030-2050. Consequently, we must have some preliminary awareness of the outcomes of such an impact on the economic branches of the country, and agrarian sector in particular, which is a relatively vulnerable branch.

2. Study Area & Methods

Khelvachauri municipality (humid subtropical region of Adjara) is located 15 m a.s.l. and Akhaltsikhe municipality (mountainous regions of Samtskhe-Javakheti) - to 1000 m a.s.l. Sums of active temperatures here is 4400-3000°C (respectively) and atmospheric precipitations makes 1360-3080 mm (respectively).

In conditions of global warming, we must identify the sums of active temperatures and atmospheric precipitations, as well as the increasing and decreasing trends of the duration of the vegetation period (in days) in the vegetation period by using the trend method. This allows identifying their changes in time and space. The zones of spreading and normal growing and development of the crops must be identified according to the future scenarios. With this purpose, based on the observation data of Khelvachauri municipality weather station, the subtropical region of Adjara (West Georgia) and according to the mountainous region of Samtskhe-Javakheti of East Georgia (Akhaltsikhe municipality weather station), the materials of current (basic) meteorological observations were developed (1956-2005). The future scenarios (2020-2050) were developed in a similar way calculated with model ECHAM4 and A2 future scenario [6].

3. Results & Discussions

According to the current (basic) and future scenario (at the increase temperature by 1°C) in the humid subtropical region of Adjara, it was established that the transition of average daily air temperature above 10°C in spring is observed on 24.III (current) and the transition below 10°C in autumn is observed on 3.XII (current). Under the scenario the temperature increase by 1°C, the transition of average daily air temperature above 10°C in spring is observed on 16.III and the transition below 10°C in autumn is observed on 9.XII. So, in case of the increase scenario (at the increase temperature by 1°C), the transition above 10°C in spring starts approximately 8 days earlier and ends 6 days later in autumn than in case of the current (basic). Thus, the duration of the vegetation period (in days) increases from 254 days (current) to 268 days (scenario), i.e. by 14 days. Such an expanded period will allow the agricultural workers and farmers to accomplish the soil cultivation works, fertilization, sowing cereals and vegetables, planting seedlings, etc. 8 days earlier in the spring. The duration of vegetation in autumn 6 days longer will promote sowing the autumn crops within the terms envisaged by agricultural methods, gaining lossless harvesting, etc.

In the mountainous region of Samtskhe-Javakheti, the date of the temperature transition above 10°C in spring is 22.IV (current), and the transition below 10°C in autumn is 15.X. At the increase temperature by 2°C (future scenario), the transition above 10°C in spring will occur on 11.IV, and the transition below 10°C in autumn will occur on 24.X. The duration of the vegetation period (in days) in Samtskhe-Javakheti region also increases. In particular, in spring, according to the scenario at the increase temperature by 2°C, the transition of the temperature above 10°C starts 11 days earlier and ends 9 days later than with the current (basic). Therefore, the duration of the vegetation period increases from 176 days (current) to 196 days, i.e. by 20 days. The given number of days will allow the agricultural workers and farmers to accomplish spring and autumn works. For the humid subtropical zone of Ajara, the sums of active temperatures above 10°C between the above-said dates are calculated, with the sum of the basic temperature of 4400°C, and it is 4650° under scenario (at the increase temperature by1°C). The sum of temperatures under the current (basic) will increase by 250°C what will have a favorable impact on the development of citrus cultures, particularly, on full ripening and quality of the fruit at the locations with a less sum of temperatures. Besides, this will allow gaining high-quality harvest of cereals, tea, vine, fruit, technical cultures (tung, laurel, tobacco) and other cultures in terms of proper soil humidity.

The sums of active temperatures are similarly calculated in Samtskhe-Javakheti region between the above-said dates and they make 2880°C (current) on average, and - 3510°C under the scenario, at the increase temperature by 2°C (2040-2050). Such sum of temperatures is beneficial for the areas with crops lacking normal development due to the low sum of temperatures. The impact of global warming on the variation of the sums of active temperatures in Adjara humid subtropical zone is shown as a trend line (Fig. 1).

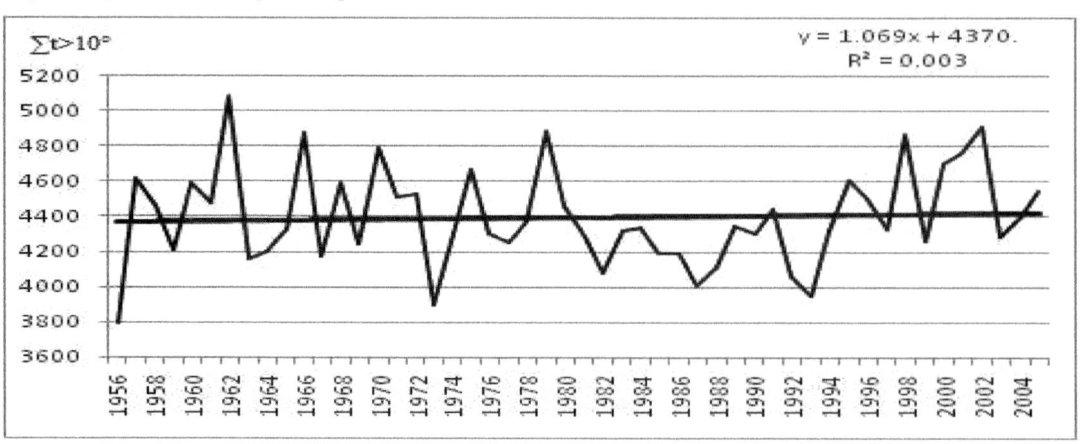

Figure 1. *Dynamics of the course and trend line of the sum active temperatures (Khelvachauri)*

The trend line (Fig. 1) clearly shows the increase of the sum of temperatures in the vegetation period. In the initial period (1956), it is 4371°C and reaches 4423°C at the end of 2005. In this period (50 years) it was revealed that the sum of active temperatures increased by 52°C. If the temperature increase continues at the same rate, by 2050, it will be a little more than 100°C, what will be favorable to gain a high-quality and biologically ripe fruits of citrus cultures. In addition, the productivity of other agricultural crops will improve in terms of optimal soil humidity.

In Samtskhe-Javakheti mountainous region (Fig. 2) the trend line clearly shows the tendency of the increasing sum of active temperatures.

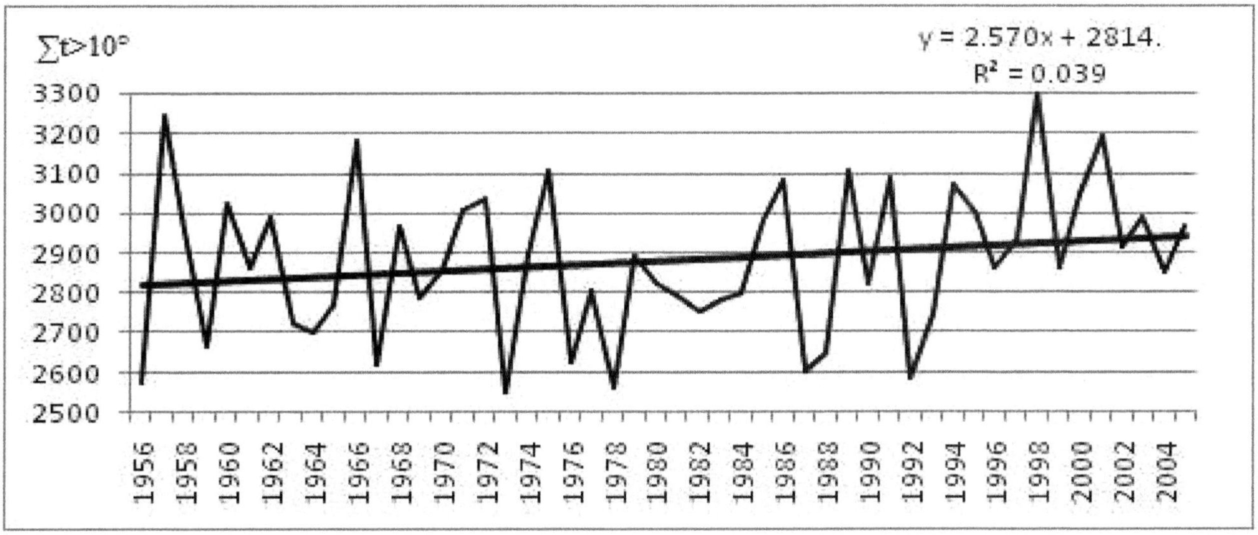

Figure 2. *Dynamics of the course and trend line of the sum active temperatures (Akhaltsikhe)*

In the initial period (1956), it is 2817°C and reaches 2942°C by 2005. The sum of active temperatures in this period increased by 125°C. If the trend of the temperature increase continues in the vegetation period, it may double by 2050 and reach 3192°C. Such sum of temperatures will improve the productivity of cereals, fruit, vine (early, of average ripeness), vegetables and other crops in terms of proper irrigation measures.

In conditions of global warming, the areas where the global warming will not have negative outcomes must be selected to grow crops. Consequently, the degree of change of the sums of active temperatures and distribution areas of crops depending on the height from the sea level under current (basic) and future (2020-2050) scenario by 1 and 2°C increase in temperature is assessed and taken into account. In this connection, the distribution areas of crops depending on the agro-climatic zones were allocated [8, 9]. A zone including the Black Sea coastline and spreading at about 500 m a.s.l., is identified in the humid subtropical region of Adjara. At every 100 m height of this zone, there are four micro-zones of distribution of citrus cultures identified and the active temperature sums by 1°C temperature increase are fixed [8]. The sum of temperatures in the micro-zone at 100 m a.s.l. is 4210°C. The given sum of temperatures ensures full ripening of early and late tangerine every year, and ripening of early orange and grapefruit for 6 or 7 times and late ones for 4 or 5 times in every 10 and more years. The sum of temperatures at 200 m a.s.l. is 4070°C, and therefore, full ripening of early tangerine is expected every year and that of late tangerine is expected for 8 or 9 times in every ten or more years. Full ripening of early orange is possible once or twice in every ten years. The sum of temperatures at 300 m a.s.l. is 3920°C. In this micro-zone, full ripening of early tangerine is likely for 4 or 5 times and 2 or 3 times in ten or more years for late tangerine varieties. The sum of active temperatures in the micro-zone at 400 m a.s.l. is 3780°C. This sum of temperatures does not ensure full ripening of the citrus fruits. As for the lemon, its fruit is desirable to harvest in terms of the sum of temperatures of 3800-4000°C. In such terms, the lemon fruit is of a light green, partially yellowish color with high content of vitamin „C". Therefore, lemon can be grown in every micro-zone, provided it is protected against the frost in winter [8].

According to the developed future scenario, at the increase in air temperature by 1°C, the sum of temperatures in the micro-zone at the height of 100 m is 4340°C and it is 4210°C at the height of 200 m. In terms of such temperatures, full ripeness of early and late citrus varieties will be possible every year as compared to the above-mentioned (basic) micro-zones (100-200 m a.s.l.). The sum of temperatures at 300 m a.s.l. is 4080°C. In this micro-zone, full ripening of tangerine is possible every year, while full ripening of early orange is possible for 4 or 5 times and that of late orange and grapefruit is possible for 3 or 4 times in every ten years. The sum of active temperatures at 400 m a.s.l. decreases and makes 3920°C. In terms of such a sum of temperatures, full ripening of early tangerine is possible for 7 or 8 times, that of late tangerine is possible for 5 or 6 times in ten years, while full ripening of early orange is possible once or twice in every ten years. In conditions of global warming under the future scenario in temperature increase by 1°C, the sum of active temperatures is increased, as mentioned already. Therefore, full ripeness of tangerine fruit in the humid subtropical zone of Ajara region is possible at the height of 300 to 400 m. Orange and grapefruit are an exception (with full ripeness once or twice in every ten years). According to the future scenario, as a result of the increased sum of temperatures (150-200°C), the productivity and quality of other agricultural crops (cereals, tea, vine, vegetables, nut, kiwi (actinidia), technical and essence varieties) will improve in terms of optimal soil humidity.

In the vegetation period, the atmospheric precipitations, along with the sums of active temperatures, play an important role in the growth and development of crops and gaining the harvest. In this connection, the sums of atmospheric precipitations in the given period (1956-2005) was determined in Adjara region and the relevant dynamics were drafted according to the trend (Fig. 3).

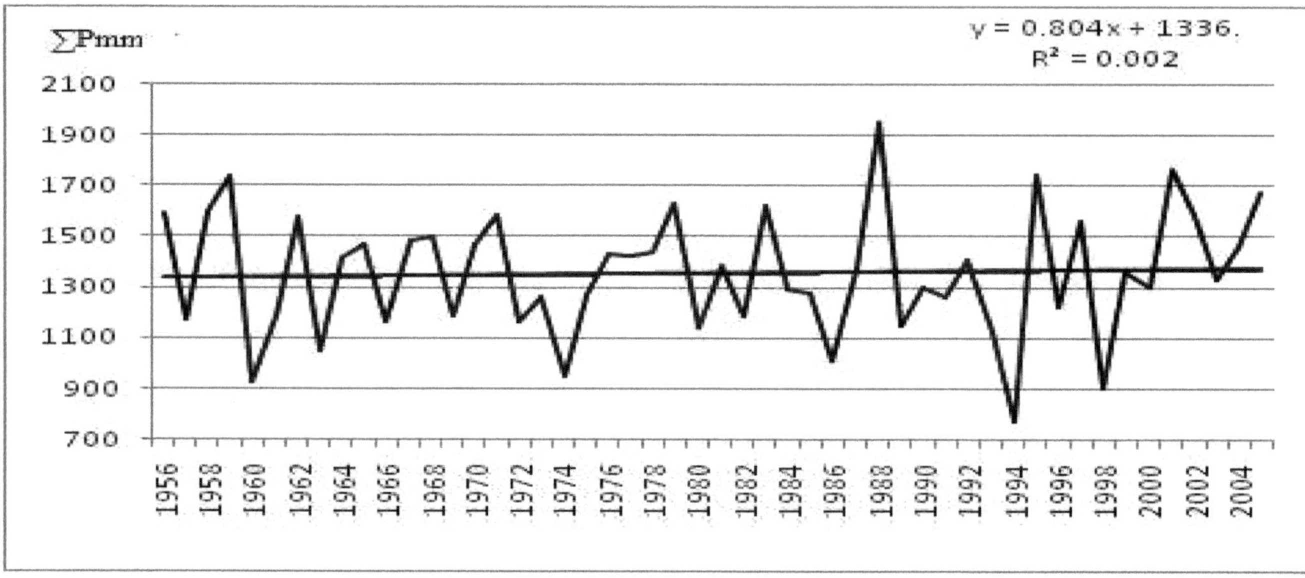

Figure 3. *Dynamics of the course and trend line of the atmospheric precipitations (Khelvachauri)*

In the initial period (1956), the sum of atmospheric precipitations was 1337 mm and reached 1377 mm after 50 years (in 2005). The tendency of increase of the atmospheric precipitations is 40 mm, showing an increase of 8 mm in every ten years. If this increasing tendency continues, by 2050, the amount of precipitations may double in the above-said zone what will slightly improve the soil humidity.

In Samtskhe-Javakheti mountainous region of East Georgia, there were 4 agro-climatic zones allocated in connection to global warming. The first zone spreads to 1000 m a.s.l., and the sum of active temperatures here is 2800°C (current). According to the scenario, in temperature increase by 2°C, it is - 3320°C. The sum of temperatures is surely increased what will improve the productivity and quality of the cereals (spring wheat, maize, etc.), vine (early varieties) and fruit. This will also boost the harvest of vegetables and potato, with the second harvest possible to gain and the vine (late varieties) will be possible to distribute, but with high-quality full quality ripeness only for 7 or 8 times in every ten years. The second zone spreads at about 1500 m with the sum of active temperatures of 2000°C (current). Under future scenario (at the increase temperature by 2°C), the sum of active temperatures is 2450°C (showing an increase of 450°C). Surely, in such terms, it will be possible to expand the distribution area of fruit trees, berries, cereals, vegetables, potato and other cultures, and production area of cattle succulent root crops. The third zone spreads at about 2000 m with the current (basic) sum of active temperatures not exceeding 1200°C, and is 1580°C under scenario (at the increase temperature by 2°C). The increased sum of temperatures will promote the growth and development and high productivity of autumn and spring crops, early pomaceous fruit: apple, pear, berries: black and red currants, willow-leaved sea-buckthorn, raspberries, vegetables, potato, as well as cattle succulent root crops and pastures and hay-making meadows. The fourth zone spreads at about 2500 m with quite decreased sum of active temperatures, with the current (basic) it is 1030°C at the height of 2100 m and it is 390°C at the height of 2500 m. The latter sum of temperatures is not suitable even for growing grass as a pasture. Under the future scenario (at the increase temperature by 2°C), the sum of active temperatures will increase to 1400°C at the height of 2100 m and it will be 700°C at the height of 2500 m. Such sum of temperatures as compared to the current, creates better conditions for such cultures to spread as early wheat (autumn and spring), barley, vegetables, potato, berries (black and red currants), willow-leaved sea-buckthorn, chokeberry and above-mentioned cattle succulent root crops at the height of 2200-2300 m, and pastures and hay-making meadows at the height of 2300-2500 m.

Normal growth and development of these cultures in the vegetation period besides the warmth, depends on the atmospheric precipitations, while Samtskhe-Javakheti mountainous region is less provided with the latter. Therefore, the precipitations in the vegetation period in different years (1956-2005) were assessed by identifying and drafting the dynamics of the course of atmoshperic precipitations on the background of global warming (Fig. 4).

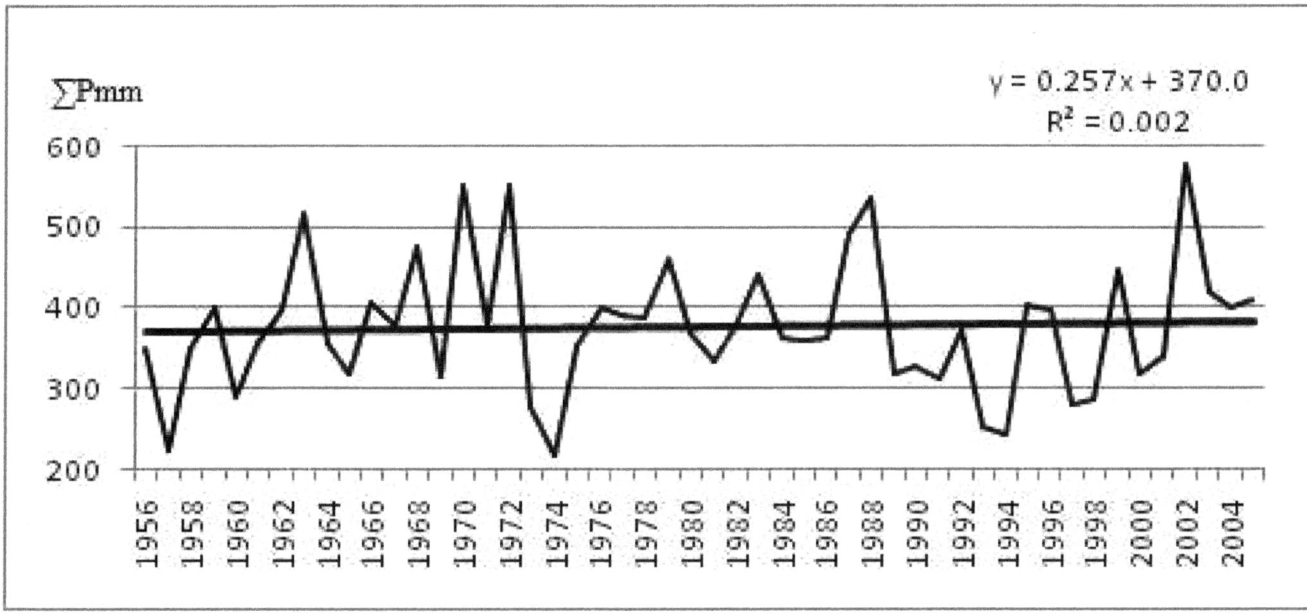

Figure 4. *Dynamics of the course and trend line of the atmospheric precipitations (Akhaltsikhe)*

In this period (in 1956), the sum of atmospheric precipitations was 370 mm according to the trend, and it is 383 mm after 50 years (in 2005). As one can see, an increasing trend of the precipitations was fixed (by 13 mm). By 2040-2050, this figure may double (making 26 mm), but cannot create optimal conditions for the growth and development of crops. Therefore, a stable productivity can be achieved by providing the proper soil humidity where this is possible.

The nomogram of heat provision in the vegetation period in case of the current (1956-2005) and future scenarios (2020-2050) by 1°C temperature increase was designed for Adjara humid subtropical zone [10] by using the sum of the above-given temperature data and relevant agro-meteorological method. The similar nomogram of heat provision following the current (1956-2005) and future scenario (2020-2050) by 2°C temperature increase was designed for Samtskhe-Javakheti mountainous region [11]. By using the nomograms, the provision (%) of active temperature sums with the current and future scenario by 1°C and 2°C increases in every ten years and more was calculated (Table 1).

Table 1. *Provision by active temperature sums (above 10°C)*

Regions	Year	Provision by active temperature sums, %						
		95	**90**	**70**	**50**	**30**	**10**	**5**
Adjara	1956-2005 Current (basic)	4080	4150	4290	4400	4550	4880	4920
	2020-2050 Scenario, 1°C- increase	4370	4430	4540	4650	4810	5020	5070
Samtckhe-Javakheti	1956-2005 Current (basic)	2470	2540	2720	2880	2980	3170	3240
	2020-2050 Scenario, 2°C - increase	2960	3090	3350	3510	3650	3900	4010

4. Conclusion

The results of the study evidence that the global warming has a clear impact on the increase of active temperatures sums. In particular, in case of temperature increase by 1°C under the scenario, it increases by 250°C on average, and in case of temperature increase by 2°C , it increases by 500°C or more. Consequently, the distribution area of the plants will expand and the zones will spread higher: the humid subtropical zone of Adjara will rise by 100-150 m and Samtskhe-Javakheti mountainous zone will rise up to 200-300 m as compared to the present areas of distribution. This suggests that the agricultural workers and farmers will be able to rationally spread the profitable crops. The vegetation period in the humid subtropical zone of Adjara under scenario by the 1°C temperature increase, will be prolonged by 14 days, and it will be prolonged by 20 days in Samtskhe-Javakheti mountainous zone according to the scenario by 2°C temperature increase. These indicators will help realize the agro-technical measures under optimal conditions.

References

[1] H.S. Mavi, G.I. Tupper. Agrometeorology: Principles and Applications of Climate Staties in Agriculture. Binghamton, HY USA. The Haworth Press Ins., Food Products Press®, 2004, p. 364.

[2] Human Development Report. Fighting Climate Change: Human Solidarity Divided Word. Published for the United Nations Development Programme (UNDP), 2007-2008, p. 31.

[3] Impacts of Desertification and Drought and other Extreme Meteorological Events. Prepared by Gathara S.T. Geneva, Switzerland, 2006, p. 88.

[4] H.P. Das, T.I. Adamenko, K.A. Anaman, R.G. Gommes and G. Johnson. Agro-meteorology Related to Extreme Events. World Meteorological Organization - Geneva - Switzerland, 2003, p. 152.

[5] K.Tavartkiladze, E.Elizbarashvili, D.Mumladze, J.Vachnadze. Empirical Model of the Change of Overground Air Temperature Field. Tbilisi, 1999, p. 106.

[6] Georgia's Second National Communication to the UNFCCC. Ministry of Environmental Protection and Natural Researches - UNPP, Tbilisi, 2009, p. 237.

[7] K.Tavartkiladze, N.Begalishvili, T.Tsintsadze, A.Kikava. Influence of Global Warming on the Near-Surface Air Temperature Field in Georgia. Bulletin of The Georgian National Academy of Sciences, vol.6, № 3, 2012, pp. 55-60.

[8] G.Meladze, M.Meladze. Influence of the Climatic Global Warming on Agrometeorological Zone of Humid Subtropics of Georgia. Geography of Georgia, TSU, №6-7, 2008, pp. 95-101.

[9] G.Meladze, M.Tutarashvili, M.Meladze. Influence of Global Warming on Changes of Agroclimatic Zones. Proceedings of Hydrometeorology Institute, vol.115, 2008, pp. 97-104.

[10] G.Meladze, M.Meladze. Change of Agroecological Zones under the Future Scenario (2020-2050 year) in the Conditions of Global Warming. Collection of Papers of International Conference. Geographical Society of Georgia, I.Javakhishvili Tbilisi State University, 2011, pp. 28-32.

[11] G.Meladze, M.Meladze. Scenarios of Distribution of Zones Agricultural Crops and Reception of Two Yields in the Conditions of Global Warming (2020-2050, on an example of Dedoplistskaro). International Scientific Conference - Pressing Problems in Hydrometeorology and Ecology. Proceedings of Hydrometeorology Institute, vol.117, 2011, pp. 70-81.

Study and Preservation of Wild Relatives of Cultivated Plants of the Northern Tien Shan (Within Kazakhstan)

Gulnara Sitpayeva

Institute of botany and phytointroduction, Almaty, Kazakhstan

Email address:
sitpaeva@mail.ru

Abstract: Modern biodiversity declining, including wild relatives of cultivated plants (WRCP) is caused by anthropogenic pressure and represents a serious threat to the world community. Since 2013 in the Institute of Botany and Phytointroduction CS MES RK the state scientific and technical program: "Botanical variety of wild relatives of cultivated plants of Kazakhstan as a source of enrichment and preservation of the agrobiodiversity gene pool for realization of the Food programme" has been realized. In 2014 the WRCP species of the natural flora of the mountain regions of Southeast Kazakhstan were the objects of researches within 4 floristic areas – the southern hillside of the Zhungar, Ile and Kungey Alatau; Ketmen, Terskey Alatau and Shu-Ile mountains. Results of researches: the Northern Tien Shan botanical diversity was revealed; mobilization of reproductive material was made; features of species structure of WRCP were studied, their phytocenosis characteristics were obtained; resource and breeding-genetic assessment of WRCP was made; their phytopathogens were revealed; passports of species of WRCP of the studied territory were made. The ridges of the Northern Tien Shan belong to the regions possessing a rather high degree of botanical diversity and due to it, they are considered to be the important objects of WRCP' concentration in Kazakhstan. The taxonomical analysis demonstrated that WRCP of the studied regions are presented by 46 families with 148 genera including 269 series. The greatest species variety of WRCPwas presented in the following families: *Poaceae*Barnhart – 66 species; *Rosaceae*Juss. – 38, *Fabaceae*Lindl. – 28 species. The hierarchical order of the first three leading is *Poaceae*, *Rosaceae*, *Fabaceae*. In 2014 the species share of 5 families increased: *Asteraceae*, there were 11 representatives and it became 17, *Lamiaceae* (9 – 13), *Alliaceae* (8 – 9), *Caprifoliaceae* (4 – 7), *Polygonaceae* (5 – 12).

Keywords: Wild Relatives of Cultivated Plants, Kazakhstan

1. Introduction

This article presents the results of research being carried out in the framework of the scientific and technical program "The botanical diversity of wild relatives of cultivated plants of Kazakhstan as a source of enrichment and genetic conservation of agro-biodiversity for implementation of the Food Program "(2013-2015) developed and implemented by the staff of RSE "Institute of Botany and Phytointroduction" CS MES RK (as the parent organization conducting research). The concept of crop wild relatives (WRCP) is interpreted by us like a number of other authors [1] quite extensively. Under WRCP we understand the natural flora species evolutionarily genetically close to cultivated plants and, above all, the species used by humans for introduction, introduction to the culture, as well as in crosses to produce improved cultivars. From this definition it follows that WRCP include all the most genetically similar species of the genus to which the cultivated species relate.

According to Kamelin R.V. the total number of cultivated plants is estimated at about 3,000 species of seed plants, that is about 1% of their composition in the world flora [2]. As you know, Vavilov N.I. identified 8 major centers of origin and resettlement of cultivated flora. A large part of the territory of Kazakhstan applied by him to the Central Asian Center of origin of cultivated plants. In turn Sinskaya E.N. proposedto allocate 5 major geographic regions (each with its own subdomains) of the historical development of cultural flora [3]. In her understanding Kazakhstan is situated within the Mid-South-West Asian subregion of the Old Mediterranean area (which is fully consistent with the Central Asian center of Vavilov N.I.). And according to the views of Wolfe E.V. the main part of the territory of Kazakhstan should be assigned to the Eastern Mediterranean subregion of the Mediterranean region [4] – one of the 16 floristic regions into which this author shared the globe.

The role of biological diversity, including WRCP in human life is difficult to overestimate. For example, in the documents of the Food and Agriculture Organization (FAO) it is said: "Among all of the Earth's riches the most important for all nations and generations are plant genetic resources" [5]. However, according to FAO experts about 75% of the global genetic diversity of agricultural crops has been lost in the course of the twentieth century. The current loss of biodiversity, including diversity of the WRCP is caused, as a rule, by human factors and represents a serious threat to the international community. [6] The strategic objectives of conservation and balanced use of biological diversity is *ex situ* conservation of wild relatives of cultivated plants in the regions of their origin; preservation of the gene pool of local landraces of crops; creation of the bank of germplasm of rare and endangered plant speciesю

2. Study Area & Methods

This work is directed to the realization of the above mentioned objectives; the results of the implementation of this program within the Kazakh part of the Northern Tien Shan (Chu-Ili mountains, ridges: Ile and Kungey Alatau; Ketmen and Terskey Alatau and the southern slopes of Dzhungar Alatau) are presented in this manuscript.The areas studied in 2014 are extremely mountainous regions of the Northern Tien Shan. The highest species diversity of flora is concentrated in the mountains. It is due to the peculiarities of geographical location, altitude, orientation of the ridge, as well as soil and climatic conditions of mountain areas. In addition, the common history of the geological formation determines the formation of species composition and originality.

While studying the flora of the WRCP of the Northern Tien Shan the classical botanical research methods were used. The main floristic method used in the field researches was route-reconnaissance method. The cameral treatment of raw material was carried out in strict accordance with all the requirements for collection, drying, sanitary processing and storage of herbarium specimens in the Herbarium (AA). The fundamental floristic reports covering the territory of Kazakhstan were used for identification of the collected material [7-10].

Vegetation was studied using traditional methods of field geobotanical studies [11-17], including geobotanical description of the main plant communities and landscape-ecological profiling. Field studies were carried out by a combination of detailed-routing method and the method of landscape ecological profiling using topographic base, space images. At points fixed on the ground by GPS device, on special geobotanical forms the detailed description of the main geobotanical prevailing over the area and plant communities was carried out. To assess the degradation degree of vegetation the variety criteria were used [18-21].

For an inventory of commercial arrays of the WRCP, development of the recommendations on the optimal mode of laying and rational use of their resource base the traditional resource keeping techniques of [22] and geobotanical studies were used with experimental plots in the key points. All obtained data were processed statistically and summarized in inventory. By a similar scheme data on stocks of raw materials in key areas possessing commercial overgrown were also processed.

Research on the diversity of micromycetes parasitizing on the WRCP included collecting herbarium material for research, description of the symptoms of defeat, mycological study, and inventory of diseases. While the territory inspection the following damages of plants were noticed: fungal coat; pustules; spot (necrosis, ulceration); mummification and so on. The identification of fungi was carried out with the help of domestic and foreign determinants [23-28].

Besides for the unification of registration methods of the results of field studies, as well as the presentation of the data in the report for all the executors of the program (employees of 5 State Botanical Gardens) general guidelines were developed. In particular, there were developed: "Description form of viewpoint and collecting material (herbarium and seeds)"; "The scheme of the geobotanical description of the community"; "The form of describing particular thickets of WRCP species" (due to the determination of the resource stocks). The reminder for collecting and primary processing of seeds and the form of the "Statement of acceptance and transfer of seed samples"

To summarize the data for each species classified as wild relatives of cultivated plants Passport of the WRCP species was developed including the following characteristics: taxonomic affiliation (family, genus, species); life form (annual, herbaceous perennial, etc.); morphology description (root, stem and so on.); terms of phenological phases (flowering and fruiting); especially environmental confinement; general distribution; distribution in Kazakhstan; the place of actual registration (co-ordinates of growing place); phytocenotic role of the species (dominant, component, etc.); fungal diseases affecting this species; seed characteristics (size, shape, and so on).

Thus, according to the described methods the novelty of the research program is a comprehensive approach to the study of the WRCP of Kazakhstan: from their detection and comprehensive (floral, phytocenotic, resource and so on.) analysis to their saving in nature, in living and herbarium collections, seed banks, as well as studying their diseases and creating the electronic information database (DB).

3. Results & Discussion

So, according to N.I. Vavilov [29] the greatest of diversity of cultivars and forms of cultivated plants is concentrated in the mountainous areas. He regarded them as "natural experimental stations of new race formation" for flora and fauna, as well as for humans. The taxonomic analysis of species composition showed that the WRCP of the studied regions were represented by 46 families with 148 genera uniting 269 species. The species composition of the leading 10 families of the WRCP combines 207 species taxa (according to "Fig. 1").

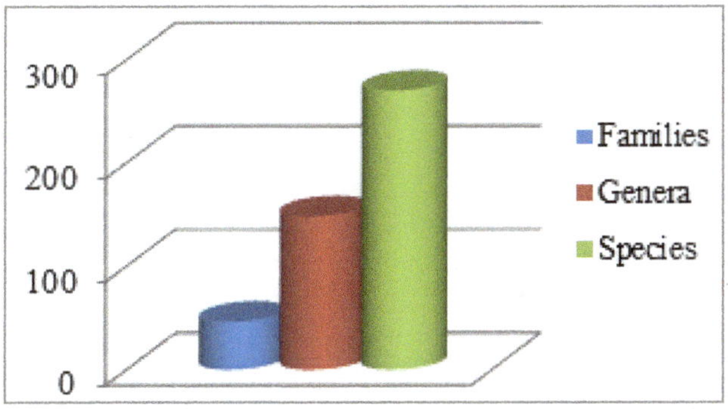

Figure 1. *Quantitative indicators of taxonomic compositionof the WRCP collected in 2014*

The quantitative distribution between families was unequal. The following families were characterized by the most specific content of wild relatives: *Poaceae* Barnhart – 66 specis, *Rosaceae* Juss. – 38, *Fabaceae* Lindl. – 28, *Asteraceae*Dumort. – 17, *Lamiaceae*Lindl. – 13, *Polygonaceae*Juss. – 12, *Chenopodiaceae* – 10; *Alliaceae* J. Agardh – 9, *Brassicaceae*Burnett. – 7and *Caprifoliaceae*Juss. – 7. The other families were represented only by 62 species of the relatives. Among them: *Apiaceae*Lindl. was represented by 5 species, *Asparagaceae*Juss. and*Ranunculaceae* – each by 4, *Grossulariaceae*DC. и*Solanaceae* – by 3. The families *Amaranthaceae, Berberidaceae, Betulaceae, Cannabaceae*Endl., *Hypericaceae*Juss., *Iridaceae, Moraceae* Link, *Malvaceae* Juss., *Nitariaceae,Plantaginaceae, Rhamnaceae*Juss. and*Urticaceae* possess by 2 species. The other 11 families (*Aceraceae*Juss., *Anacardiaceae* Lindl., *Asphodeliaceae, Crassulaceae, Cupressaceae, Dipsacaceae, Ephedraceae* Dumort., *Limoniaceae, Paeoniaceae, Valerianaceae* Batsch, *Viburnaceae)* – onlyby 1 species.

The distribution of the wild relatives of cultivated plants by genera was also unequal. The highest number of species the following genera contain: *Allium* L. (9 species), *Rosa* (8), *Lonicera*L.(7), *Artemisia* L. (6), *Agropyron, Leymus, Poa, Rumex*and *Cotoneaster* (5). In general, the composition of these genera was 55 species. The genera contain in their composition by 4 species: *Glycyrrhiza, Elymus, Hordeum, Crataegus*and *Asparagus.* The genera containing by 3 species of the wild relatives: *Bromus, Eremopyrum, Festuca,Rheum, Clematis, Sasola, Amoria*and *Medicago*. The genera: *Elytrigia, Helictotrichon, Koeleria, Melica, Phleum, Atraphaxis, Fragaria, Malus, Rubus, Sanguesorba, Sorbus, Urtica, Amaranthus, Inula, Lactuca, Berberis, Betula, Lepidium, Kraschennikovia, Elaeagnus, Alhagi, Caragana, Hedysarum, Vicia, Ribes, Hypericum, Iris, Mentha, Salvia, Linum, Nitraria, Morus, Plantago*and*Calamagrostis*possess by 2 species in their composition. The other genera were presented only by 1 species. In general this group unites 112 species. The other genera of the WRCP were presented only by 1 species.

Search and collection of wild relatives of cultivated plants in 2014 was carried out in a volume of 4 floristic areas. The distribution of species by region and unevenly (as shown in "Fig. 2").

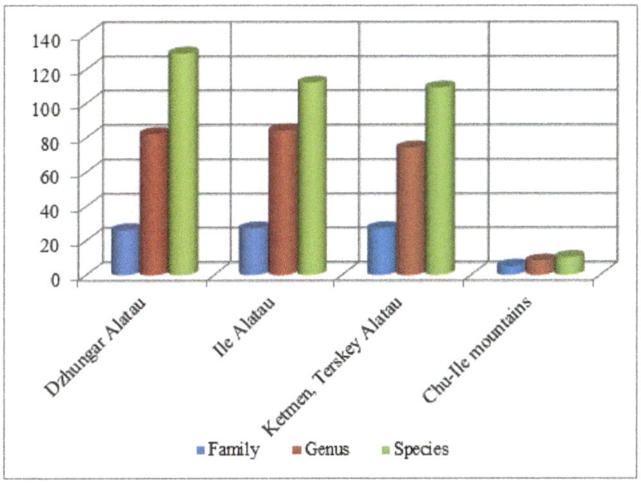

Figure 2. *Distribution of the WRCP according to floristic regions*

Quantitative indicators of the number of families of the WRCP were about the same: 27 families – in the Ile Alatau, and 26 – in Dzhungar Alatau and KetmenTerskey-Alatau. There were only five families in the Chu-Ili Mountains.

Generic spectrum showed similar data: the largest number of generic taxa were accounted for the Ile Alatau (84 species), Dzhungar Alatau (82) and KetmenTerskey-Alatau (74). In the Chu-Ili Mountains the number of genera was 8. The distribution of the number of members collected DSKR was unequal among floristic regions. To establish a set of species growing within one or more of the studied regions, the comparative analysis of their geographic range were carried out.

Thus, it for the floristic areas Dzhungar Alatau and Ketmen-Terskey Alatau the greatest number of common species of wild relatives was revealed – 50 species.Such species as *Achilleamillefolium, Achnatherumsplendens, Agropyroncristatum, Anisantha tectorum, Anthriscussylvestris, Armeniacavulgaris*, Bromus squarrosus, *Betulatianschanica, Brachypodiumpinnatum, Betulapendula*and others. Some lower rates of common species of the relatives were identified for the Ile Alatau and for the Ketmen-Terskey Range – 35 species represented by the following representatives: *Achilleamillefolium, Agrostis gigantea, Armeniaca vulgaris,Calamagrostisepigeios, Cotoneaster melanocarpus, Dactylisglomerata, Elytrigiarepens, Leonurus turkestanicus, Lonicerahispida, Melica transsilvanica, Melilotusofficinalis, Origanumvulgare, Phleum phleoides, Plantagolanceolata*and others.

The number of wild relatives marked for Ile Alatau and Dzhungar Alatau drops a little till 32 species:

Crataegus songorica, Dactylisglomerata, Daucuscarota, Deschampsia caespitosa, Elaeagnus oxycarpa, Ephedra equisetina, Glycyrrhiza uralensis, Hypericum perforatum, Elytrigiarepens, Glycyrrhizaglabra, Glycyrrhiza uralensis, Glycyrrhiza uralensis, Hypericum perforatum, Loniceratatarica, Menthalongifolia, Rosa alberti, Rosa *beggeriana, Rosa fedtschenkoana, Rosa platyacantha, Salvia deserta, Ziziphora clinopodioides*and others.

For the floristic regions that are geographically distant from each other the number of common species of the relatives naturally drops. Since the number of common species for the Chu-Ile Mountains and Ketmen-Terskey Alatau it was 6; for the DzungarianAltatau and Chu-Ile mountains there were 5 representatives.

Despite the fact that geographically Ile Alatau and the Chu-Ile Mountains are adjacent territories, but they have a different altitude and climatic parameters. These features are reflected in the minimum quantitative composition of the common WRCP one can find there only 7 species:*Elytrigiarepens, Eremodaucuslehmannii, Hypericumperforatum, Aegilops cylindrica,Poa pratensis* and others. And only 3 species of the WRCP can be found within the whole studied floristic regions: *Elytrigiarepens,Hypericum perforatum* and *Salvia deserta.*

In 2014 during the implementation of the research work the points of growth of the wild relatives of cultivated plants of the number of the species included into the Red Book of Kazakhstan [30] were marked. For instance such species as *Malussieversii, Armeniacavulgaris* Lam., *Rheum wittrockii*Lundstr., *Paeoniahybrida*Pall., *Aflatuniaulmiflora, Pistaciavera* and *Vitisvinifera.*

Within the study the state of plant communities was analyzed involving the WRCP. It was revealed that in various parts the vegetation was transformed in varying degrees. In general the degree of disturbance was medium. In some places there were revealed strongly transformed areas. Steep slopes were slightly violated.

The main factors of transformation were: plowing in the foothills, changes of the river flow, canals, grazing, haying, recreation and road digression. It was noted that in the Ile Alatau the natural vegetation was not left in the steppe zone. Focusing the flowering ornamental species of tulips, irises and other ephemeroids early in spring leads to increased recreational pressure on vegetation. Within the especially protected territories (Ile-Alatau National Park, Almaty reserve, museum "Tamgalytas") natural ecosystems with WRCP were preserved.

As a result of the resource researches in the territories of 24, 25 and 25a, floristic regions (the ranges Ketmen, Terskey, Kungey and the southern slopes of Dzhungar Alatau) 93 species of the WRCP of Kazakhstan were revealed and the reserves of air-dried and fresh raw material were noted, 16 species from the following families: *Rosaceae* Juss. (7 species), *Lamiaceae* Lindl. (3), *Berberidaceae*Juss. (2*), Asteraceae*Dumort. (1), *Elaeagnaceae*Juss. (1), *Polygonaceae* Juss. (1), *Ranuncullaceae*Juss. (1).

The identified in the territory of the Ketmen ridge and southern macroslope of the Dzhungar Alatau ridge the stocks of dog rose, hawthorn Korol'kova, cotoneaster multiflori, old-man's-pepper, common origanum, mint longifolia and Ziziphoraclinopodioides can be recommended for implementation of stocking up in limited quantities and for the needs of the local network of pharmacies. For industrial stocking up the thickets of medicinal and food plants was identified, such species as: *Aconitum leucostomum*Worosch. and*Rumextianschanicus* A. Los. (hr. Kungei Alatau), *Berberissphaerocarpa* (Ketmen) and Hippophae*rhamnoides* (southern slopes of the ridge Dzhungar Alatau).

The natural reserves of raw materials of the "endangered species" were taken into account: *Armeniaca vulgaris* Lam., *Málussiéversii* and *Berberisiliensis*. They should be used as reserve plots for the collection of seed and planting material.

Among 16 identified resource species of the WRCP of Southeastern Kazakhstan 25% of the species were provided by the raw material base and were suitable for commercial preparations. Other species of the WRCP (75%), for which even the limited supply of raw materials in a small area were taken into account, represent a unique genetic material of biological diversity, formed under the conditions of the studied floristic regions of South-eastern Kazakhstan.

During the cameral treatment of the herbarium material collected during expeditions and revision of the herbarium funds of IBP it was found that the WRCP in the study area there were 437 taxa of parasitic fungi. Rusts occur in a quantity of 47 species (11%), smut fungi – 44 species (10%). Ergot pathogen is one species of fungus (0.2%). Among the total number of detected phytopathogens 72 species of fungi (17%) cause leaf spots of the host plant, 30 species (7%) – *Septoria blotch.* 3 species cause scab pathogens (1%), leaf curl 2 species (1%), choke diseases of cereals – 2 species (1%), rot of fruit and roots – 36 (8%), stem rot – 32 species (5%). The remaining diseases account for 39% or 185 species of fungi pathogens.

The result of breeding and genetic studies of wild apricot forests of the western part of the Ile Alatau (the Zhetyzhol ridge) 20 forms of apricot with valuable genetic traits were chosen and described. The passports for selected forms of the *Armeniaca vulgaris* were made. It is noted that the release forms are characterized by high polymorphism on fruit size, color, ripening, consistency pulp, taste, etc. Also from natural ecosystems of the surveyed gorges 5 plant species were selected: *Morusnigra, Morusalba, Rosa platyacantha*Schrenk and *Loniceramicrophylla* Willd. exSchultes..

During the survey of the southern macro slope of the Dzhungar Alatau forms of the species from 5 genera were selected: hawthorn, buckthorn, barberry, dog rose and cotoneaster. It was revealed that the main anthropogenic impacts on natural ecosystems, determining the degradation of fruit forests are grazing and planting of alien tree species by the Forestry employees.

Total there were selected, marked and certified plus trees of *Armeniaca vulgaris* (20 copies), *Morus alba* (2) *Morusnigra* (2), *Cerasustianschanica*(3), *Rubus caesius* (2), *Loniceramicrophylla* (2), *Crataeguskorolkowii* (2), *Hippophaerhamnoides* L. (4), *Berberisiliensis* (11) *Rosa platyacantha* (8), *Rosa laxa* (12), *Rosa iliensis* (1), *Rosa beggeriana* (5), *Cotoneaster melanocarpus* (1) *Cotoneaster meultiflorus* (2), *Cotoneaster megalocarpus* (1).

The analysis of the intraspecific variability of generative organs (inflorescence, fruit and seed) in 15 natural populations of *Berberisiliensis* was carried out. To identify form diversity of the dimensional features the "method for detecting interference distributions" was proposed and justified.With this method there were revealed 53 forms of *Berberisiliensis* by gradations of sizes of metamers of the inflorescence axis, by the abundance of flowers in the brush, by the length of stalks, height and the maximum diameter of the fruits, the length and thickness of the seeds.

Form diversity of *Berberisiliensis* was individual and populationly was very specific. When analyzing the types of flower brushes which combining the characteristics of abundance shaped flowers, and peduncle length metameresit was determined that more than 50% of the identified 86types were populationally specific and 25% of the types were implemented only in two populations. However, two types of inflorescences brushes were revealed, each of them was required in each of the populations of this metapopulation. These types of indicators were implemented in 10-12 populations. Individual, populational, metapopulational variability of traits and their ontogenetic variability due to age, plants and fruit cluster position in the structure of the plant were demonstrated. Based on the results of the described researches the recommendations for effective conservation and breeding use of *Berberisiliensis* were done.

4. Conclusion

Thus, study of the WRCP of Kazakh part of the Northern Tien Shan revealed their botanical diversity in this area containing 269 species from 148 genera and 46 families; it was made the mobilization of reproductive material of the WRCP – 893 samples. The peculiarities of the species composition of the WRCP were studied; their phytocenotic characteristics were obtained; resource and selection-genetic evaluation of the WRCP was done; their phytopathogens were revealed; the passports of the WRCP growing in the target territory were compounded.

Acknowledgments

This work was made possible due to the state funding of Republic of Kazakhstan in the frame of the scientific-technical program No. 0199"Botanical variety of wild relatives of cultivated plants of Kazakhstan as a source of enrichment and preservation of the agrobiodiversity gene pool for realization of the Food programme", 2013-2015.

References

[1] O.N. Korovina, "Nature genepool of wild relatives of cultivated plants of the USSR Flora ant its protection,"Leningrad, 1986, P. 126.*Published in Russian.*

[2] R.V. Kamelin, "Great selection of humanity dawn," Barnaul, Azbuka, 2005, 126 pp.*Published in Russian.*

[3] E.N. Sinskaya, "N.I. Vavilov doctrine about the historical and geographical centers of the cultivated flora development," In "Questions of geography of cultivated plants and N.I. Vavilov," Moscow-Leningrad, Nauka, 1966, pp. 22-31.*Published in Russian.*

[4] E.V. Wolf, "The cultural flora of the Globe," 1987, Leningrad, 324 pp.*Published in Russian.*

[5] AReportontheCommissiononPlantGenetic Resources, Rome Second Session, March 16-21, (1987).http://apps3.fao.org/wiews/Morocco/Paper3.jsp

[6] Agenda 21, 1992. http://univdhaka.academia.edu/SahiduzzamanSohel/Papers/369556/Agenda_21_Rio_Earth_Summit-1992.

[7] "Flora of the USSR," 1934-1965, Moscow-Leningrad, Publish House of the Academy of Sciences of the USSR, Vol. 1-30. *Published in Russian.*

[8] "Flora of KazSSR," 1956-1966,Alma-Ata, Publish House of the Academy of Science of KazSSR, Vol. 1-30. *Published in Russian.*

[9] "The illustrated determinant of Kazakhstan plants," edited by V.P. Goloskokov, 1972, Alma-Ata: Publish House "Nauka of KazSSR," Vol. 1, 2. *Published in Russian.*

[10] "The determinant of Plants of Middle Asia and Kazakhstan," 1969-1987, Tashkent: Fan, Vol. 1-9. *Published in Russian.*

[11] "Fieldgeobotany," 1959-1976, Moscow-Leningrad: Nauka, Vol. 1-5. *Published in Russian.*

[12] B.A. Bykov, "Geobotany," 1978, Alma-Ata: Nauka, 288 pp.*Published in Russian.*

[13] A.M. Berlyant, "Geoinformationmap-making," 1997, Moscow. 60 pp.*Published in Russian.*

[14] Z.V. Karamysheva, E.I. Rachkovskaya, "Experience of large-scale geobotanical map-making (for flora of the southwestern part of Central Kazakhstan Upland). Principles and methods of geobotanical map-making," 1963, Moscow-Lenungrad, 6 pp.*Published in Russian.*

[15] B.V. Vinogradov, "Airmethods of studying the flora of arid zones," 1960, Leningrad, 360 pp.*Published in Russian.*

[16] V.B. Sochava, "Geobotanical map-making. Plant classification as an ierarchi of the dynamic systems," 1972, Leningrad, pp. 33-38. *Published in Russian.*

[17] "Methodoly of the state evaluation and map-making of the ecosystems in extreme conditions," 1993, Puschino, 202 pp.*Published in Russian.*

[18] "Methodical recommendations on estimation and map-making of the modern condition of the ecosystems of the Mongol People Republic," 1989, Ulan-Bator, 107 pp.*Published in Russian.*

[19] E.I. Rachkovskaya, N.P. Ogar, O.V. Marynich, "Factorsoftheanthropogenictransformationandtheirinfluenceupontheflora ofKazakhstansteppes,"J. Steppebulletin, Novosibirsk, Vol. 5, pp. 22-25.*Published in Russian.*

[20] O.V. Marynich, "Anthropogenictransformationofthesteppeflora (for Central Kazakhstan)," thesis of the candidate of biological sciences, 1999, Almaty, 29 pp.*Published in Russian.*

[21] D.Tongway, "Rangeland soil condition assessment manual," 1994, Canberra: CSIRO, 69 p.

[22] "Methods of the determination of medicine plants,"1986, Moscow, 50 pp. *Published in Russian.*

[23] U. Braun, "A monograph of the *Erysiphales* (powdery mildews)," Nova Hedwigia, 1987, Vol. 89, pp..1–700.

[24] U. Braun, R.T.A. Cook, "Taxonomic manual of the Erysiphales (powdery mildews)," 2012, RBC-KNAW Fungal biodiversity centre. Utrecht, the Netherland, 707 pp.

[25] K. Seifert, G. Morgan-Jones, W. Gams, B. Kendrick, "The Genera of Hyphomycetes" 2011,Utrecht: CBS-KNAW Fungal Biodiversity Centre. [CBS Biodiversity Series no. 2.],997 p.

[26] V.A. Bilai, E.Z. Koval, "Aspergillus. The determinant,", 1988, Kiev: Naukovadumka, 204 pp.*Published in Russian.*

[27] D.ASatton, A. FotergillandM. Rinaldi, "Determinantofpathogenicandconditionallypathogenicfungi," 2001, Moscow, 468 pp. *Published in Russian.*

[28] M.A. Klich, "Identification of common *Aspergillus*species," 2002, CentraalbureauvoorSchimmelcultures, 116 p.

[29] N.I. Vavilov, "Origin centres of cultivated plants," 1926, in Proceedings on applied botany and breeding, Vol. 16, No. 2, pp. 3-48. *Published in Russian.*

[30] The Red book of Kazakh SSR. Part 2," 1981, Alma-Ata: Nauka, 262 pp. *Published in Russian.*

An Eco-Friendly Paradigm for the Production of Potentially Anti-Parasitic Drugs

Younes Abrouki[1, *], Abdelkader Anouzla[1], Hayat Loukili[1], Rabiaa Lotfi[1], Ahmed Rayadh[1], My Abdellah Bahlaoui[1], Driss Zakarya[1], Mohamed Zahouily[1], Rachid Mamouni[2], Saïd Sebti[3], Adil Mandil[1]

[1]Department of Research, Faculty of Science and Technology, University Hassan II, Mohammedia, Morocco
[2]Laboratory Organic Chemistry, Team Bio-Organic Chemistry Applied, Ibn Zuhr University, Agadir, Morocco
[3]Department of Chemistry, Faculty of Science Ben M'Sik, University Hassan II, Casablanca, Morocco

Email address:
abroukiyounes@yahoo.fr (Y. Abrouki)

Abstract: A cleaner production of various potentially anti-parasitic drugs based bis-(arylmethylidene)cycloalkanones have been developed using a heterogeneous catalysis technology in aqueous media without organic solvent. The higher activity and selectivity of our solid-state synthetic phosphate catalyst coupled with their ease of use and reduced environmental problems makes them attractive alternatives to homogeneous basic reagents.

Keywords: Green Synthesis, Heterogeneous Catalysis, Anti-Parasitic Drugs, Bis-(Arylmethylidene) Cycloalkanones, Synthetic Phosphate, Aqueous Media

1. Introduction

Chagas' disease and Leishmaniasis are endemic protozoan diseases recognized as epidemiological priorities by the World Health Organization (WHO). Chagas' disease is found mainly in Latin America and it is estimated to affect 12 to 14 million people. Leishmaniasis affects 12 million people in 88 countries, 72 of which are developing countries [1].

Chagas' disease [2] and Leishmaniasis [3] are caused by the trypanosomatid parasites *Trypanosoma cruzi* and *Leishmania* sp, respectively.

There are limited drugs available for the treatment of both diseases, and the currently used chemotherapeutic agents are still inadequate due to their extreme toxicity [4]. In addition, drug resistant parasite strains have become more prevalent. The search for more effective and less toxic drugs to treat both diseases have became mandatory [5].

In search for new drugs with improved antiparasitic profiles, the brazilian authors are showed that the bis-(arylmethylidene)cycloalkanones presented a activity against amastigotes forms of *Trypanosoma cruzi* and promastigotes forms of Leishmania amazonensis [6].

Cross-aldol condensation [7] is an important protocol for the synthesis of various bis-(arylmethylidene)cycloalkanones. This reaction is classically carried out using strong acid or base [8]. Indeed, different organometallic complexes, Lewis acid are found to be able to catalyze this reaction [9]. However, most of the above-mentioned process generated a large amount of toxic waste and involved tedious separation procedures since the reactions were homogeneous and the catalysts were destroyed or discarded on the work-up procedure [10].

Recently, because of increased environmental concerns and strong economic driver in chemical research and industry, great efforts have been made on heterogeneous catalysis, which allows the catalyst for easy isolation and recycling. Thus, solid catalysts have drawn considerable attention in organic synthesis due to their advantages such as environmental compatibility, reusability, high selectivity, simple operation, non-corrosiveness and low cost [11].

In continuation of our ongoing program to develop new environmentally process in organic synthesis, we describe in this paper, a environmentally friendly method for the synthesis of various bis-(arylmethylidene)cycloalkanones-based potentially anti-parasitic drugs by Cross-aldol condensation catalyzed by synthetic phosphate in aqueous media.

2. Materials and Methods

2.1. Chemicals and Instrumentations

All commercial reagents and solvents were used without further purification. X-ray diffraction (XRD) patterns of the catalysts were obtained on a Philips 1710 diffractometer using Cu-Kα radiation. Surface areas were determined at 77 K using a Coulter SA 31000 instrument with an automated gas volumetric method employing nitrogen as the adsorbate. NMR spectra were recorded on a Bruker ARX 300 spectrometer. Mass spectra were recorded on a VG Autospec spectrometer. FTIR spectra were recorded on an ATI Mattson-Genesis Series spectrophotometer using the KBr disc method.

2.2. Preparation and Characterization of Catalysts

The fluorapatite (FAP) have been prepared by reaction between diammonium phosphate, calcium nitrate and ammonium fluoride in presence of ammonia [12]. The FAP obtained was calcined at 900°C before use. The structure of this catalyst was confirmed by X-ray diffraction, infrared spectra and chemical analysis. The surface area of calcined FAP was determined by the BET method and found to be $S=15.4\ m^2\ g^{-1}$. The total pore volume was calculated by the BJH method ($V_T=0.058\ cm^3\ g^{-1}$).

The modified fluorapatite (Na/FAP) has been prepared by impregnation of the FAP with a solution of sodium nitrate followed by calcination at 900°C [13]. The surface area of the new catalyst Na/FAP was determined by the BET method as $5.4\ m^2\ g^{-1}$ and the total pore volume obtained by the BJH method is $0.0032\ cm^3\ g^{-1}$.

2.3. General Procedure

The general procedure for synthesis of bis-(arylmethylidene)cycloalkanones is reported in Figure 1, as follows: To a flask containing a mixture by taking 1:2 mole ratios of cyclopentanone 1 and arylaldehydes 2 in aqueous media, the Na/FAP catalyst was added and the mixture was stirred at reflux during 3 hours.

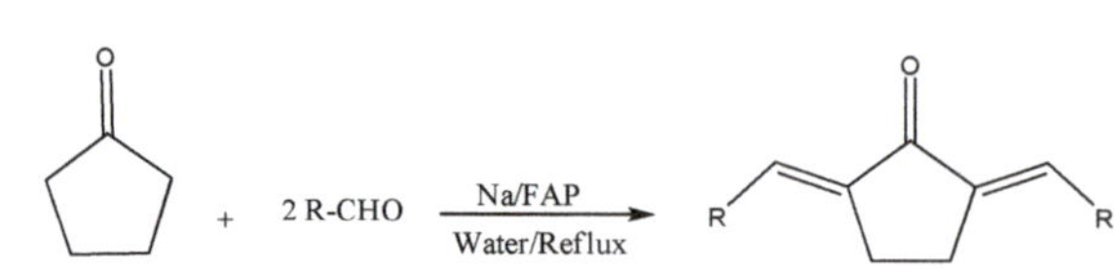

1 2 3

Figure 1. Synthesis of Potentially Anti-Parasitic Drugs.

Hot water (2×20 mL) was added, followed by simple filtration. After filtration and extraction with hot water, the solution was concentrated and purified by silica gel chromatography (*n* hexane/ethyl acetate: 7/3). Na/FAP was reactivated by drying at 150 °C or, alternatively washed with acetone and calcined at 900 °C for 1 h. The crude product 3 was identified by melting point, ^{1}HNMR, ^{13}C NMR, and IR spectroscopies.

3. Results and Discussion

After synthesis and characterization of fluorapatite and Na/FAP, we tested these two materials as catalysts in the synthesis of bis-(arylmethylidene)cycloalkanones in aqueous media, which tends to a green synthesis strategy, going from modernization of organic synthesis toward sustainability. The cross-aldol condensation between cyclopentanone 1 and benzaldehyde 2a, were chosen as model substrates to determine suitable reaction conditions for synthesis of bis-(arylmethylidene)cycloalkanones. The optimal reaction condition for synthesis of this product was obtained by central composite design [14] and it's reported in Figure 2.

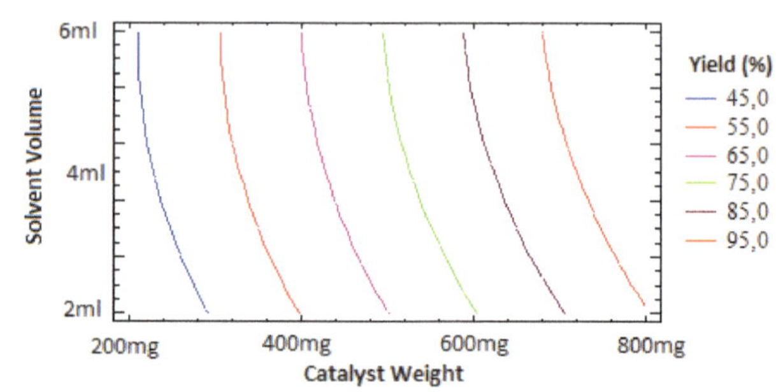

Figure 2. Optimization of reaction condition for cross-aldol condensation by central composite design.

We studied the influence of several parameters on a model cross-aldol condensation, specifically: (i) optimal catalyst preparation conditions (Na/FAP), (ii) quantity of catalyst, and (iii) kinetics of the reaction.

The investigation of central composite design showed that the optimum model reaction conditions are: catalyst weight = 700 mg, reaction time = 3 h and solvent volume = 5 mL. To determine the scope and limitation of this reaction, the optimum condensations were applied to other substrates as shown in table 1.

Table 1. Synthesis of various bis-(arylmethylidene)cycloalkanones.

Entry	R-CHO	Product (*)	Yield (%)
3a	C_6H_5CHO	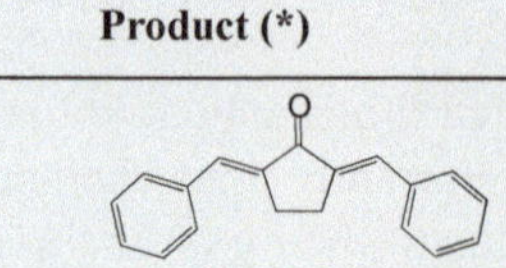	92
3b	4-Cl-C_6H_4CHO	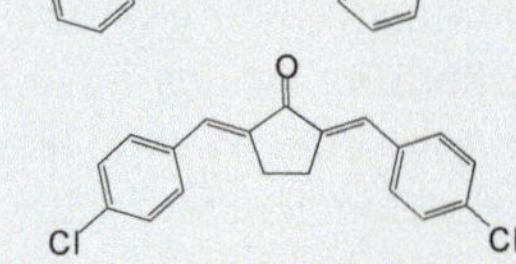	83
3c	4-MeO-C_6H_4CHO	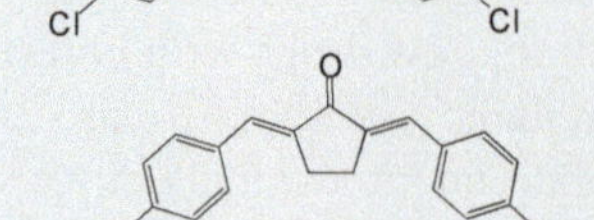	72

Entry	R-CHO	Product (*)	Yield (%)
3d	$3-NO_2-C_6H_4CHO$	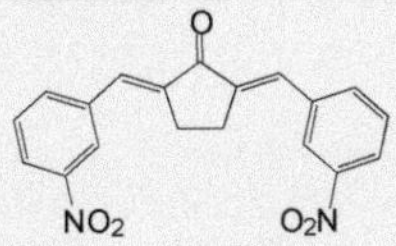	64
3e	PhC_2H_2CHO	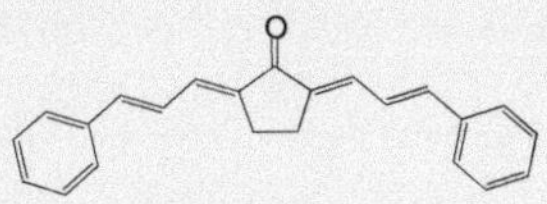	80
3f	C_4H_3CHO	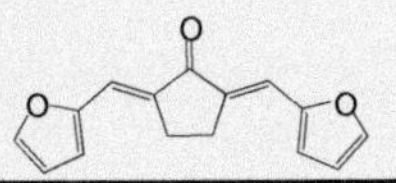	93

* The analytical data were identical to those reported in the literature.

In this regard, the preparation of potentially anti-parasitic drugs was carried out under reflux conditions by taking 1:2 mole ratios of cyclopentanones and arylaldehydes in water in the presence of modified fluorapatite Na/FAP catalyst to give the desired bis-(arylmethylidene)cycloalkanones in excellent yields. In general, the use of FAP as the heterogeneous catalyst in the crossed-aldol condensation has allowed the isolation of our products in low yields with slow reaction time. Under similar conditions in the absence of synthetic phosphate FAP or Na/FAP catalyst, only the starting material was regenerated, thus highlighting the role of this catalyst. When the reactions were examined in different solvents such as methanol, toluene, chloroform, water and THF, we found that water is the best solvent due to high yields and shorter reaction times. Moreover, using water as the reaction medium made the separation process much easier with high purity of final product than using organic solvents. However, when toluene, THF, chloroform, and methanol were used as solvents, a homogenous mixture of product and starting materials was obtained at the end of the reaction. After filtration of the final product, synthetic phosphate catalyst was recovered quantitatively by simple filtration and regenerated by calcination for each new reuse. The table show that the comparison of the results obtained by our method with some of those reported shows the efficiency of this method because of shorter reaction times, high yields, and application of an inexpensive and readily available catalyst. A simple procedure and the use of water as a solvent are expected to contribute to the development of more benign crossed-aldol reactions. The modified synthetic phosphate Na/FAP catalyst is a new, inexpensive, and attractive solid support which can contribute to the development of catalytic processes and reduced environmental problems and it is the first use of this material as the catalyst.

4. Conclusions

In conclusion, we have developed a clean and easy method for the synthesis of potentially anti-parasitic drugs based bis-(arylmethylidene)cycloalkanones using an inexpensive, reusable, easy to handle, noncorrosive, and environmentally benign catalysts, fluorapatite and especially sodium-modified fluorapatite. The use of these catalysts offers diverse advantages including simplicity of operation due to the heterogeneous nature of reaction, easy workup, high yields, and catalyst reusability.

References

[1] WHO, World Health Organization. Available from: http://apps.who.int/tdr/svc/publications/tdrnews/tdr-news-issue-84 (accessed 02.22.10).

[2] J.A. Urbina, Parasitological cure of Chagas disease, Oswaldo Cruz. 94 (1999) 349-355.

[3] S.A. Grevelink, E.A. Lerner, Leishmaniasis, J. Am. Acad. Dermatol. 34 (1996) 257-272.

[4] S.L. Croft, G.H. Coombs, Leishmaniasis - current chemotherapy and recent advances in the search for novel drugs, Trends Parasitol. 19 (2003) 502-508.

[5] R.B. Oliveira, A.B.M. Vaz, R.O. Alves, D.B. Liarte, C.L. Donnici, A.J. Romanha, C.L. Zani, Arylfurans as potential Trypanosoma cruzi trypanothione, Mem. Inst. Oswaldo Cruz 101 (2006) 169-173.

[6] F.P. Braga, V.P. Alves, R.S. Ferreira, R.B. Fradico, P.S. Lage, M.C. Duarte, T.G. Ribeiro, A.S. Júnior, A.J. Romanha, M.L. Tonini, M. Steindel, E.F. Coelho, R.B. Oliveira, Synthesis and evaluation of the antiparasitic activity of bis-(arylmethylidene) cycloalkanones, European Journal of Medicinal Chemistry 71 (2014) 282-289.

[7] Y. Riadi, R. Mamouni, R. Azzalou, R. Boulahjar, Y. Abrouki, M. El Haddad, S. Routier, G. Guillaumet, S. Lazar, Animal bone meal as an efficient catalyst for crossed-aldol condensation, Tetrahedron Letters 51 (2010) 6715-6717

[8] P. Salehi, M. Dabiri, M.A. Zolfigol, M.B. Fard, Silica sulfuric acid as an efficient and reusable reagent for crossed-aldol condensation of ketones with aromatic aldehydes under solvent-free conditions, J. Braz. Chem. Soc. 15 (2004) 773-776.

[9] M.S. Abaee, M.M. Mojtahedi, S. Forghani, N.M. Ghandchi, M. Forouzani, R. Sharifi, B. Chaharnazm, A Green, inexpensive and efficient organocatalyzed procedure for aqueous aldol condensations, J. Braz. Chem. Soc. 20 (2009) 1895-1900.

[10] M.S. Abaee, M.M. Mojtahedi, R. Sharifi, M.M. Zahedi, H. Abbasi, K. Tabar- Heidar, Facile synthesis of bis(arylmethylidene)cycloalkanones mediated by lithium perchlorate under solvent-free conditions, Iran. J. Chem. Soc. 3 (2006) 293-296.

[11] Y. Abrouki, M. Zahouily, A. Rayadh, B. Bahlaouan, S. Sebti, A natural phosphate and doped-catalyzed Michael addition of mercaptans to α, β -unsaturated carbonyl compounds, Tet. Lett., 43, (2002), pp. 8951-8953.

[12] M. Zahouily, Y. Abrouki, A. Rayadh, S. Sebti, H. Dhimane, M. David, Fluorapatite: efficient catalyst for the Michael addition, Tet. Lett., 44 (2003), 2463-2465.

[13] A Solhy, A Smahi, H El Badaoui, B Elaabar, A. Amoukal, A. Tikad, S. Sebti, D.J. Macquarrie, Efficient hydration of nitriles to amides catalysed by sodium nitrate modified fluorapatite, Tet. Lett., 44 (2003), 4031-4033.

[14] Y. Abrouki, A. Anouzla, H. Loukili, R. Lotfi, A. Rayadh, M.A. Bahlaoui, S. Sebti, D. Zakarya, M. Zahouily, Central composite experimental design applied to the catalytic carbon-sulfur bond formation by fluorapatite catalyst, American Journal of Applied Chemistry, 1 (2013), 22-27.

29

North American Multi-Purpose Dear Introducent Plants in Western Georgia Humid Subtropical Conditions

Mariam Metreveli[1, 2], Avtandil Meskhidze[1], Feride Tchaidze[2], Julieta Jakeli[2], Dali Beridze[1]

[1]Batumi Shota Rustaveli State University, Biology Department, Faculty of Natural Sciences and Health Care, Batumi, Georgia
[2]Batumi Botanical Garden, Batumi, Georgia

Email address:
metrevelim@list.ru (M. Metreveli), meskhidze.a@gmail.com (A. Meskhidze), feride_tchaidze@mail.ru (F. Tchaidze), juli.jakeli@mail.ru (J. Jakeli), dalidali59@gmail.com (D. Beridze)

Abstract: The article deals with bioecological peculiarities and application of exotic plants *Liquidambar styraciflua L., Liriodendron tulipifera L.* and *Magnolia grandiflora L.* introduced in Adjara or the South Colchis Black Sea littoral in the conditions of Transcaucasia, namely, Western Georgia humid subtropics. Namely, the following has been studied peculiarities of growth and development, adaptation, reproduction and natural rehabilitation, Antimicrobial Action of Extracts Obtained from Leaves. The multifold application recommendations are given on.

Keywords: South Colchis, Exotic Plants, Bioecological Peculiarities, Adaptation, Antimicrobial Action

---

1. Introduction

We have set as our aim to study some bioecological peculiarities of North American exotic plants Liquidambar styraciflua, Liriodendron tulipifera and Magnolia grandiflora introduced in Adjara Black Sea littoral in the conditions of Western Georgia humid subtropics.

The Black Sea littoral of Georgia is distinguished by the diversity of the local as well as exotic, introduced flora. Especially interesting in this respect is the Ajara, namely, Batumi coastline [1, 2, 3]. Adjara or South Colchis floristic region occupies one of the significant places in the Caucasus by vegetation diversity, high specific share of endemism and uniqueness of forest ecosystems. With historical past of vegetation, the specific composition of floristic complexes, systematic structure the territory of Adjara belongs to Colchis or East Euxeinos province of the sub-Mediterranean region of the Mediterranean world. Its floristic diversity is characterized by the flora of the tertiary period Mediterranean autochthonic origin [4, 5]. At present Ajara is mentioned as Southern Kolkheti (Colchis).

At the same time, it is remarkable that the introduction of plants began here as early as the end of the 19th century and today, many of them have been implemented into decorative gardening, subtropical agriculture and other fields of economy. The origins of distribution and implementation of exotic plants in practice take place from the Batumi Botanic Garden that is included in the unique botanic gardens worldwide. It is century old and is arranged according to the principle of phyto-geographical sections. The Garden represents over 2 000 woody plant species characteristic to all the continents of the world, including North American moody introduce the number of which reaches 217 species for today. A number of scholarly works have been dedicated to the study of the adaptation of exotic plants along the Georgian Black Sea littoral [6, 7, 8, 9, 10].

The nature of Georgia is quite diverse and each of its regions is characterized by specific soil-climatic conditions. Diversity of Georgian climate is preconditioned, on the one hand, by its location at the northern border of subtropical zone between the Caspian and Black Seas, and on the other hand, by the particular complexity of its relief – Caucasus ranges of various heights and directions. Atmospheric processes developed with moderate as well as subtropical zones participate in the formation of climate in Georgia. Adjara is situated in the southwestern part of Georgia and is distinguished by specific physical-geographical peculiarities that are determined by a number of reasons. Significant among them is the location of Adjara in southern latitudes that stipulates abundance of solar energy, and the climatic impact of the Black Sea that is more revealing along the coastline. It is along the coastline that Liquidambar styraciflua, Liriodendron tulipifera and Magnolia grandiflora grow, within the height of 0-250 meters. These areas are distinguished by subtropical climatic conditions, directly border the sea and are protected by the mountains from three sides. The thermal regime is preconditioned by the geographical locations, the direct closeness to the sea. The average annual air temperature in 14-150°C; the average temperature of the coldest month January is 6-70°C; the temperature of the warmest months – July and August – is 220°C. The lowest temperatures are -8 and -90 C. The highest temperature is 360°C. The average quantity of annual precipitation is 2500 mm, whereas general – 2500-3500 mm. The relative air humidity is 80-82%. There are mostly red soils. It rarely snows, but when it does, it sometimes reaches 1 m and because of wetness it in most cases destructively affects some of the exotic plants [9, 11].

The natural habitats of our exotics are the mixed-type forests of the central, eastern, southern and southeastern states of eastern USA where the climate is mostly of the Atlantic-monsoon type of moderate and subtropical zone [12, 13, 14].

We have studied the growth, development, reproduction and natural rehabilitation peculiarities of Liquidambar styraciflua, Liriodendron tulipifera and Magnolia grandiflora, as well as defined their frost-resistance and winter-resistance qualities. For the purpose of defining their role in environmental improvement, we have studied the antimicrobial action on the example of protistocidal activity. An experiment is started for the determination of bactericidal and fungicidal actions of extracts obtained from the leaves. As it is well known, plants can produce biologically active substances of complex chemical composition that represent one of the preconditioning factors of plant immunity; they have antimicrobial activity and destroy or prevent development of pathogenic microorganisms and simple unicellular organisms (sometimes, multi-cellular organisms too). These biologically active substances are known as phytoncides [15, 16]. Volatile phytogenic compounds conduct health-improvement effect of the environment – they are one of the important regulators of air physical-chemical composition. Application of antimicrobial action of plant phytoncides or "natural antibiotics" is available in ecology, medicine, agriculture and other fields. As far as the species under study represent important garden and park plants with a grand crown and leaves, we got interested in antimicrobial activities of extracts obtained from their leaves.

2. Materials and Methods

For the determination of the growth and development peculiarities of the North American exotic plants *Liquidambar styraciflua, Liriodendron tulipifera* and *Magnolia grandiflora* along the Adjara coastline of the Georgian Black Sea littoral applying corresponding methodology, we have studied the growth, leaf-bearing, flowering, fruit development, seed production and dissemination processes of the solid age specimens growing in the Batumi Botanical Garden. Each decade of every month during the whole year we were observing the phases of bud swelling and opening, complete leaf-bearing and leaf-falling, sprout growth beginning and ending, generative buds swelling and flowering starts and ending, fruit knotting, development and seed dissemination. For the determination of the adaptation degree of the exotics we have fixed the degree of their natural rehabilitation in new environments, namely, the degree of self-regeneration through seeds and vegetation [17, 18]; we have also studied the impact of one of the limiting factors for exotics' adaptation along the Ajara coastline of the Georgian Black Sea littoral – periodical low temperatures and the thickness of snow [19]; as far as all the three exotic plants are the rare species for the organization of parks and forests and their role is interesting in environment improvement, we have studied the antimicrobial action of their leaf extract. We have obtained the final results of prosticidal action on the basis of several repeated experiments. In the case of determination of prosticidal action, the simple unicellular organism – *Paramecium caudatum* - was used. The experiment was based on the Boris Tokin method [15, 16] according to which prosticidal action is determined by the time of complete destruction of infusorians under the action of leaf extract. The degree of bactericidal and fungicidal activity is in the process of study, for which we use the strains of fungus and bacterial pathogenic cultures, including quarantine pathogens, existing at the Institute of Phytopthology and Biodiversity of Batumi Shota Rustaveli State University. For the comparative analysis, we have included in the experiment the worldwide acknowledged plant species of strong antiseptic action: onion *(Allium cepa L.),* garlic (*Allium sativum.*), myrtle (*Myrtus communis l.),* geranium (*Pelargonium roseum Ait.*); we have studied the peculiarities of graft and seed reproduction with control and bio-stimulator variants at the Batumi Botanical Garden orangery and nursery [20, 21].

3. Results and Discussions

3.1. Growth, Reproduction and Adaptation Peculi-arities

3.1.1. Growth and Development Peculiarities

Below are given the results of several years' observations on the growth phases of the exotics under study. The process of each phase is given in calendar dates (table 1).

Liquidambar styraciflua L. - Family *Altingiaceae Lindl.* [22].

They are high deciduous trees, with straight, elegant stem and broad pyramid-shape crown. Centenary trees in Adjara are usually 30-35 m high with stem diameter of 1-1,5 m. Massive swelling of vegetative buds begins in late February – early March; buds open from second-third decade of March; apical growth starts in the third decade of March and lasts till July in most cases, but sometimes ends at the end of June or lasts till first decade of August inclusive. The average growth comprises 8,5 cm – 20,5 cm. On shoots some callus-like growths can develop. Leafing stage begins in late March – early April. Leaf color changes from the first decade in

November, leaf falling begins from the second decade in November and continuing till the third decade in December. The average size of 5-7 lobed leaves is 15-20 cm. The plant massively flowers from the third decade in March to the first decade of May during which period the plant is beautified by masculine inflorescence in the form of long apical brushy bunches and single hanging glomes. Fruit ripening starts from the third decade of October and ends in November. Dissemination from the spherical long-stem hanging collective fruit made up of numerous small boxes ends by the third decade of December. Diameter of ball-shape collective fruits is 2, 5 – 4 cm. There are very small 1-2 seeds in each small box. Empty fruits remain on the trees till spring. Fruit-bearing is abundant.

The comparison of flowering and fruit-bearing dates in the habitat is given in the table 2.

It should be noted that along our coastline the dates of vegetative and generative development of North American and East Asian liquid ambars coincide, i.e. they begin development from March and continue till the third decade of November total comprising 8-9 months. They flower in March-April and give fruit in October-November-December. They are characterized by vegetative and mixed-type bud development. When in spring the mixed-type bud open, first the flowering process takes place, followed afterwards by leafing and apical growth processes.

Liquidambar styraciflua L. is considered to be one of the most beautiful decorative plants inasmuch as it is effective due to beautiful, symmetrical, wide, pyramid-shed crown and changing leaf colors, especially in autumn when it is colored with orange, red and purple leaves and very beautiful, hanging, spherical, round cone-like fruit. Liquidambar resinous or amber tree is also interesting for the purpose of using its walnut-like texture wood and composition of aromatic “ambra liquida” balsamic compounds in medicine and perfumery[23].

Liquidambar styraciflua L. is a species protected by IUCN Red List and is the category of LC (Least Concern)[24,25,26].

Liriodendron tulipifera L.–Family *Magnoliaceae Juss.*

It is a huge, tall deciduous tree-plant, growing about 60 m high with 3-4 m diameter. It has grown in the territory of the Batumi Botanical Garden since 1913. Centenary specimens reach here 35 m height, stem diameter – 160 cm, crown projection – 24x20 m. Swelling of vegetative buds begins in March and sprout growing (shooting) - in early April. Sprout growing generally ends at the end of July whereas on the upper branches it continuous till mid August. The average length of the shoots annual growth is 25 cm. Leaves are 4 – 6 - lobed. By the end of July the length of the grown leaves reaches up to 15 cm. Leaves become yellow and begin to fall gradually. In the first half of November all the leaves have already obtained beautiful golden-yellow colouring. The massive falling of leaves begins in the second half of November and ends in December. Flowering begins from the second half of April till the end of May, sometimes including the first half of June. Flowers appear on the mixed-type shoots of the current year. These shoots begin growing earlier than the vegetative shoots and end very soon, remaining short, about 5 cm long. Vegetative buds develop at the top. They soon swell, grow and give first flowers already in the second or third decade of April, in the period of massive leafing of the plant, in conditions of 14-15^{0}C temperature. Flowers are single, androgynous, arranged at the top of the shoot, 6 cm diameter; 6 crown petals are yellow-greenish with an orange-reddish spot at the stem. They resemble tulips in shapes, hence their name. They yield cucumber-like aroma and are melleous. In the eastern part of the USA it is a very mellous plant. Flowering is annual. Respectively, fruit-bearing is annually too, and is quite abundant. Fruit is of 3, 5 – 5 cm long, cone-like collective fruit directed upwards to short shoots, and consists of a stem and numerous 4 cm long alates with one wing and four faceted seeds that fall after ripening. Seed ripening begins at the end of October and in November massive ripening proceeds. Ripen “cones” open in single wings that can be spread by wind. By the end of December all the seeds are disseminated already and only the stem remains on the tree where seeds were placed. These stems remain all through winter, sometimes up to the appearance of new leaves.

The comparison of flowering and fruit-bearing dates in the habitat is given in the table 2.

Like the south-eastern region of the USA, Liriodendron is considered as a marvelous decorative and important garden-park plant along the Ajara Black Sea littoral with the size of its crown, projection, grace, especially colors of autumn leaves and beautiful flowering. It is a valuable mellous plant for bee keeping. It is also well known for its precious wood. Interested farmers cultivate it for this purpose and get desirable results. The roots, bark and leaves can be used for medicinal purposes (especially due to *tulipiferine* consistency)[23,27].

Magnolia grandiflora L.–Family *Magnoliaceae Juss.*

Magnolia grandiflora, is an evergreen tall tree. In the habitat it reaches 30 m height, stem diameter is 1, 5 m. In the conditions of Ajara the height of the centenary ones reaches 22-25 m, stem diameter – 1-1, 5 m; the crown is wide-pyramid-shape, is characterized by dense branching, massively fuzzy vegetative buds are produced at the shoot tops and leaf bays. Flower buds are created at the top of the previous year shoots, they are also fuzzy. The leaves are big, leather, whole-edged, shiny, 12-25 cm long; leaves live 2 – 2, 5 years. Two-year old leaves become yellow and fall in spring (April-May), when new leaves appear. During the first intensive period of leaf-falling the 50-60% fall is observed in two-year leaves. There is massive leaf-falling in May. In summer leaf-fall decreases (15%), in autumn 40% of the rest of the 2, 5-year leaves fall. In the hot summer conditions the leaf-falling increases. Young leaves have soft, light green plates. By the early June the leaves become leather and obtain dark-green coloring. Thus, *M. grandiflora* has 2 well-expressed periods of leaf falling – spring and autumn. During the first years of life *M. grandiflora* grows rather slowly by in 10 years time, especially in conditions of fertile soil, it grows very fast and yields annual shoot growth up to 0, 5 – 0, 6 m.

The growth and development of this magnolia is very

interesting because of its being an evergreen plant. Vegetation begins significantly late, in April. Simultaneous to this phenophase leaves open and shoot growing starts. The entire leafage conducts 29/V – 5/VI; new leaves open until 10-15 July when shoot growth finishes. Sometimes shoot growth lasts till July-August and ends in the production of apical bud. Mostly, in August II growth appears that is produced from terminal buds opened in the upper branches. One-year shoots get branched in mid June and produce II stage shoots from accessory vegetative buds that long continue growing and exceed mother shoots 3-4 times. Next year, generative buds and vegetative accessory buds are produced at the apex of short generative shoots. They can open the same year, but the main shoot does not grow any more. Vegetative shoot is long and produces vegetative buds at the apex as well as leaf bosoms. Next year it will be the lateral shoots that continue growth and not the main-parent ones. Shoot growth is 26-35 cm in average. Shoot wooding begins from 25/VI-10/VII and ends by 20-30 August. The growth and flower buds are quite shaped by the end of July. At the beginning of October all the parts of the flower can be distinguished in the vegetative buds.

Generative buds swell a bit later than vegetative ones. At the beginning of May buds greatly increase in size and begin to develop for 20-35 days, ending in flowering (end of May - beginning of June). The diameter of the apical, single milky-white, androgynous flowers arranged in three circles and consisting of 6-9 crown petals, is 25-30 cm. During bad weather conditions small bugs inhabit the flowers by means of which the pollination process is conducted. *M. grandiflora* flowers do not open all together, but in succession, from May to the end of October. Massively flowering takes place in June-July; afterwards it slows down and then restarts again. Single flowers can open even in November along the Adjara coastline.

Like vegetation, flowering is a long process. Flowers and ripe seedy fruits can be found in trees simultaneously. Sometimes, in October, on older specimens, you can see together buds, developed buds, flowers, fruits and ripe seeds. 12 cm long and 6 cm wide collective fruits are straight-standing "multi-leaves" in the form of big "cones". Fruits are knotted all through the summer, sometimes in autumn too. From fruit knotting up to its ripening 150-160 days pass. Fruit ripens in October-November. There are 20-25 seeds in one cone. Seeds are black, bean-shape, oblong, rather thick, 3-6 mm in diameter, 6-15 mm in length, covered with bright red fleshy jacket. The massive fall of the seed takes place after their complete ripening. After seed falling "cones" gradually fade away and fall, but sometimes empty, dry cones will remain in trees until the next year vegetation period. Satisfactory fruiting is developed in well-lighted trees, whereas the badly-lighted ones. In the Adjara coastline conditions the plant bears fruit annually, in most cases massively.

The comparison of flowering and fruit-bearing dates in the habitat is given in the table 2.

Magnolia grandiflora, is a highly decorative plant for the Adjara littoral. It is a unique garden-park with its elegant crown, leaves similar to *Ficus elastic* leaves, beautiful flowers, cone-like fruit and red-jacket seeds hung on threads in ripening season, as well as adaptation to city conditions. It is also a very interesting plant in medicine and perfumery for containing essential oils in all the organs and alkaloids, glycosides and essential oils in the leaves [23].

Table 1. *Phenological Cycle (by months) of Liquidambar styraciflua, Liriodendron tulipifera, Magnolia grandiflora in the Humid Subtropical Conditions along Adjara Black Sea Littoral.*

№	Species	Growth		Leafing		Flowering		Fruiting		Leaf falling	
		start	end	start	end	start	end	start	end	start	end
1	Liquidambar styraciflua	III, IV	VII	IV	V	III	V	X	XI	XI	XII
2	Liriodendron tulipifera	III, IV	VII	IV	VI	IV	VI	X	XI	XI	XII
3	Magnolia grandiflora	IV	VIII	IV	VI	V, X	VIII, XI	X	XI	III, X	V, XII

Table 2. *Flowering and Fruiting Dates of Liquidambar styraciflua, Liriodendron tulipifera, Magnolia grandiflora in Habitat and along Adjara Black Sea Littoral.*

№	Species	Flowering in habitat	Fruiting in habitat	Flowering in Adjara	Fruiting in Adjara
1	Liquidambar Styraciflua	May	October	March–May	October– November
2	Liriodendron Tulipifera	April–May (June)	October	April–May (June)	October– November
3	Magnolia Grandiflora	May–August	October	May–July/ August; October–November	October– November

3.1.2. Peculiarities of Reproduction and Natural Rehabilitation

One of the signs of adaptation of the exotic plants to the new environmental conditions is the degree of reproduction and natural rehabilitation with seeds as well as vegetative way.

For the determination of this peculiarity we have conducted collecting and planting of all three types of seeds. Also, taking stem grafts and rooting with using test variants and growth stimulators. As growth stimulators, we have used 0, 01% of indoleacetic acid and 2% sucrose solutions. For graft rooting we used mixed substrate of peat, sand and humus. Grafting results are shown in the table 3 whereas the seed reproduction results are discussed in the text.

In case of *Liquidambar styraciflua,* the seed taken in autumn was planted in spring (March) and it gave better results rather than planting right after seed taking. In this occasion the developmental degree comprised 40-75%, while the degree of development for the seeds taken in autumn was about 30%. Seeds grow rather slowly; they begin branching from 3-4 years, i.e. this is the time when cork-like sprouts appear. The height of 3-year-old plants is 20-40 cm. According to the sources, liquidambars begin reproductive development rather late, at the age of 10-15 years. However, out of seedlings obtained by us for years, the 6-7-year old specimens gave flowers and even fruits. Graft reproduction of liquidambar stems did not give any results in our experiment.

It should be noted that on the territory of the Batumi Botanical Garden as well as all over the littoral, there are numerous self-seeding forms around the centenary liquidambars.

Out of the seeds obtained from *Liriodendron tulipifera* fruits, only 5% appears to be of normal development. Respectively, we received low percent of seed germination in our experiment, and that was only in spring, the first decade of March, and not in autumn. Seedlings obtained from spring sowing grow and develop very well. According to the sources, liriodendrons begin reproductive development rather late, at the age of 20-25 years. However, out of seedlings obtained by us for years, the 10 – 12-year old specimens gave flowers and even fruits. Summer graft reproduction gave good results. The degree of rooting, increased by 30% after using bio-stimulators.

On the territory of the Batumi Botanical Garden as well as all over the littoral, there are numerous natural self-seeding and vegetative forms around the centenary liriodendrons. Abundance of self-seeding can be explained by seed development in great numbers. Among them, relatively old-aged specimens produce flowers and fruits.

We have processed the seeds of *Magnolia grandiflora* well before sowing, that means that, first of all they were rid of red fleshy jackets and washed. Part of the seed is planted right after taking in autumn and it gives 40-45% of growth. The other part is kept in cold conditions with stratification till spring. The seed stored in damp peat moss and placed in polyethylene sacks are in the refrigerator till spring. Seeds planted after 4-4,5 months of stratification show 30% of growth. Seedlings are characterized by unusually slow growth: the height of 6-year-old seedlings is 2, 5 cm that gives flowers at this age. Rooting of spring grafts shows very good results and the degree of rooting using bio-stimulators increases to 40-45%.

Centenary exotics of *Magnolia grandiflora* do not develop self-seeding in the subtropical climatic conditions of Adjara Black Sea littoral. However, there are numerous specimens received after rooting from branches bent to the earth that creates an association to the mangrove forests. \

Table 3. *Results of Vegetative Graft Rooting.*

№	Species	Number of grafts	Degree of rooting in %		
			Test version	Indoleacetic acid 0,01%	Saccharin 2%
1	Liquidambar tyraciflua	100	0	0	0
2	Liriodendron tulipifera	100	30	60	65
3	Magnolia grandiflora	100	35	75	80

3.1.3. Peculiarities of the Adaptation to New Environment

One of the signs of the complete adaptation of exotic plants to the new environmental conditions is the resistance towards limiting factors. For the Adjara humid subtropical climate such factors are: periodical frosty winter (-8-9^{0}C), periodical snow up to 0,5-0,8 m high and strong wind. According to the works of the local scientists as well as our observations we can conclude that during the century all the three exotics have managed to overcome critical dangers from the above given factors positively. Only sometimes, young sprouts of liriodendrons are characterized by fragility and the plant gets mechanically damaged by heavy snow. IN general, all the three species are characterized by high ability of adaptation. There is one colorful huge specimen of liriodendron at the Batumi Botanical Garden which several years ago got rooted out during the tempest only because it was growing at the edge of the valley. However, it entirely stretched over the valley and created a live bridge, the roots and apical branches were covered with earth and it got successfully rooted again. The tree has still been alive for over 30 years already. Its branches took vertical direction; they grow, produce flowers and fruits and together with specimens received from the apical branches create the association of children trees grown from the tree stretched over the valley.

3.2. Determination of Antimicrobial Action of Extracts Obtained from Liquidambar Styraciflua, Liriodendron Tulipifera, Magnolia Grandiflora Leaves

For the purpose of defining antimicrobial activity of the extracts obtained from the leaves of the species under study, we have included in the analysis fungal and bacterial pathogens obtained in pure culture during phytopathological research at the Institute of Phytopathology and Biodiversity. However, we have not finished the experiments yet due to insufficient number of repetitions. They are in the process and we refrain from making any conclusions about the results. At this stage we give the results of protistocide action below.

We determined protistocide action according to the time of the complete destruction of *Paramecium caudatum* upon the impact of volatile phytoncide compounds of the extract obtained from leaves [15, 16]. The material for analysis was taken in spring, summer and autumn periods. The extract was obtained by squeezing well-crushed leaves.

Experiments were conducted during 3 years in three calendar dates of different phases of vegetative and generative development: the last decade of May, the first decade of August and the first decade of October. According to the obtained result, we conclude that the activation of phytoncide actions increases from spring to summer while afterwards decreases from summer towards autumn. Exotics are characterized by the highest phytoncide activities in August (table 4).

Table 4. *Protistocide activity of the species under study.*

№	Species	Last decade of May		Necessary time to kill infusories (in minutes)
		Exposition time		
		start	end	
1	Liquidambar styraciflua	14.15.00	14.22.40	7.40
2	Magnolia grandiflora	13.35.40	13.46.50	11.10
3	Liriodendron tulipifera	14.28.40	14.40.40	12.00

Table 4. *Continue.*

№	Species	First decade of August			First decade of October		
		Exposition time		Necessary time to kill infusories (in minutes)	Exposition time		Necessary time to kill infusories (in minutes)
		start	end		start	end	
1	Liquidambar styraciflua	13.45.00	13.51.30	6.30	13.50.10	13.58.55	8.45
2	Magnolia grandiflora	14.24.20	14.33.50	9.30	14.05.10	14.16.25	11.15
3	Liriodendron tulipifera	14.10.15	14.21.30	11.15	13.40.15	13.53.20	13.05

The studied species can be divided in accordance with the strength of phytoncide activity as follows: the species of the strong phytoncide activity - *Liquidambar styraciflua;* quite strong - *Magnolia grandiflora;* the species of the medium phytoncide activity - *Liriodendron tulipifera.*

In the case of our experiment the simple organisms not only hamper their development but are completely destroyed. For a certain period after the destruction all the organelles of their cells are distinctly shown and later their lysis takes place.

For the purpose of comparison, we have included in the experiment plant species known as universal antiseptics [15, 16]: onion (*Allium cepa*), garlic (*Allium sativum*), myrthe (*Myrtus communis*), and geranium (*Pelargonium roseum*) (table 5). We have determined their protistocide activities in the given situation. As a result, we have found out that protistocide activities of liquidambars approximate to the protistocide activities of onions and cherry-laurel in this experiment.

Table 5. *Activity of plants known with strong antiseptic properties in the given experiment.*

№	Species	Necessary time to kill infusories (in minutes)
4	Allium cepa L. - onion	5.35
5	Allium sativum L. - garlic	7.05
6	Myrtus communis l. - myrtle	11.05
7	Pelargonium roseum Ait. - geranium	9.00

4. Conclusion

On the basis of the obtained results we can conclude that North American exotics *Liquidambar styraciflua L., Liriodendron tulipifera L., Magnolia grandiflora L.* are characterized by complete adaptation to the local humid subtropical climatic conditions along the transcaucasian Black Sea littoral, namely, in Western Georgia, Adjara or South Kolkheti (Colchis). They grow well, develop, give flowers and fruits abundantly, propagate and undergo self-rehabilitation – develop self-seeding forms (except Magnolia) and vegetative shoot forms (except liquidambars). They are characterized by high antimicrobial action on the basis of protistocidal activity of volatile phytogenic compounds extracted from the leaves. They represent highly-decorative garden-park species in terms of using in decorative gardening and city landscape gardening as well as important forest and woody culture, especially in case of liriodendron; also, they are noteworthy plants for medicine and perfumery.

Finally, on the basis of scientists' recommendations these widely spread species along the Georgian Black Sea littoral can easily be included in the gold fund of decorative gardening and become symbols of exotic flora of the Georgian coastline. Magnolia is characterized by low branch rooting and new specimen's growth only in Batumi, like its habitat. Nomenclature

References

[1] Pilipenko F. (1978) „Invasive Trees and Shrubs along the Caucasian Black Sea Littoral". "Nauka" Publishing, Leningrad.

[2] Hinkul C. G. (1936) Plant Introduction and Naturalization in Humid Subtropics of the USSR, Batumi Botanical Garden Bulletin. №1, Batumi Subtropical Botanical Garden Publishing, 3-44.

[3] Manjavidze D., Tsitsvidze A., Matinyan A. (1965) Perspective Varieties of Woody Plants Introduced at Batumi Botanical Garden for Forestry and Forest Management, Tbilisi, "Metsniereba" publishing, 67 p.

[4] Kolakovskiy A. (1961) «Vegetative World in Colchis", MSU publishing, Moscow, 459 p.

[5] Shetekauri Sh. (1988) Biotops of Petrophytic Flora of the High Mountains Caucasus, Bull. Georgian Academy of Sciences, 159: 3.

[6] Trees and Shrubs at Batumi Botanical Garden (2007) Part II (The Angiosperms). Editor in Chief – Vano Papunidze. Batumi, 425 p.

[7] Trees and shrubs of Batumi Botanical Garden (2012) "Sachino" publishing, Tbilisi.

[8] Sharashidze Sh., Bagratishvili N. (1988) Batumi Botanical Garden of the Georgian Academy of Sciences, "Adjaria" publishing, Batumi, 179 p.

[9] Papunidze V., Bagratishvili N. (1998) "Batumi Botanical Garden of the Georgian Academy of Sciences", "Metsniereba" publishing, Batumi,155 p.

[10] Batumi Botanical Garden – 100 (2102) Published by „SACHINO" Ltd. Tbilisi, 239 p.

[11] Adjaria (1978) "Sabchota Adjara" Publishing 177: 3-46

[12] Rehder Arnold (1949) "Manual or cultivated trees and shrubs". The Macmillan Company, New York, 670 p.

[13] Takhtajyan A. (1978) Floristic Regions of the Earth, Leningrad, "Nauka" 349: 51-91

[14] Flora of North America. Vol. 3 www.erloras.org

[15] Tokin B. (1960) Microbe Destroyers, Phytoncides. Moscow, "Nauka" publishing, 160 p.

[16] Tokin B. (1975) On the Role of Phytoncides in Nature. Collection "Phytoncides, their Role in Nature", Leningrad University Publishing, 235: 5-21.

[17] Beideman I. (1974) "Methodology of the Study of Plants and Plant Communities Phenology", "Nauka" publishing, Novosibirsk

[18] Elagin, I., Lobanov A. (1979). "Atlas Guide of Plant Phenological Phases", "Nauka" publishing, Moscow, 95 p.

[19] Gocholashvili M., Zaldastanishvili Sh., Demetradze T. (1959), "Frost and Winter Resistance of Woody Subtropical Flower Varieties" (journal "Subtropical Cultures", №1, pp. 3-12.

[20] Methodological Guidebook for Plant Reproduction, Developed by Principal Botanic Garden (1974). Nauka" publishing, Moscow, 125 p.

[21] Khromova T. (1980) "Methodological Guidebook for Reproduction of Introduced Woody Plants with Grafts". ВАСХНИЛ" publishing. Moscow 140 p.

[22] An Update of the Angiosperm Phylogeny Group Classification for the Orders and Families of Flowering Plants: APG II (2003) Botanical journal of the Linnean Society, 141: 399-436.

[23] V. Kholyavko, Globa-Mikhaylenko D. (1976) "Valuable Woody Varieties along the Black Sea Littoral", "Lesnaya Promishlennost" publishing, 340 p.

[24] www.IUCNredlist.org

[25] http://www.cbd.int/doc/publications/plant-conservation-report-en.pdf

[26] Nevton A., Oldfield S., Fragoso G., Paul M., Miles L., Edwards M. (2003) "Towards a Global Tree Conservation Atlas. Mapping the status and distribution of the world's threatened tree species", World Conservation Monitoring Centre, Cambridge, United Kingdom.

[27] Tskhoidze T., Tchaidze F., Kontselidze N., Jakeli J. (2014) Bioecology of tulip Trees at the Adjara Black Sea Coast. „ Modern Phytomorphology." Vol. 6, Lvov, pp.189-194. www.phytomorphology.org

A New Approach and Tools for Perfecting Phytoremediation Technology

Tamar Varazi[1], Maritsa Kurashvili[1], Marina Pruidze[1], Gia Khatisashvili[1], Nino Gagelidze[1], George Adamia[1], George Zaalishvili[1], Marlen Gordeziani[1], Mark Sutton[2]

[1]Durmisidze Institute of Biochemistry and Biotechnology of Agricultural University of Georgia, David Agmashenebli Alley 240, Tbilisi, Georgia
[2]Chemical Sciences Division, Lawrence Livermore National Laboratory, California, USA

Email address:
t.varazi@agruni.edu.ge (T. Varazi), m.kurashvili@agruni.edu.ge (M. Kurashvili), m.pruidze@agruni.edu.ge (M. Pruidze), g.khatisashvili@agruni.edu.ge (G. Khatisashvili), n.gagelidze@agruni.edu.ge (N. Gagelidze), g.adamia@agruni.edu.ge (G. Adamia), g.zaalishvili@agruni.edu.ge (G. Zaalishvili), m.gordeziani@agruni.edu.ge (M. Gordeziani),sutton18@llnl.gov (M. Sutton)

Abstract: The development of methods for soil remediation for removal of chemical contaminants is a large and challenging problem. A new phytoremediation technology, given in the present work, is based on joint application of natural sorbents, microorganisms and plants. The method aims at prevention of chemical contamination of soils. It has the potential to restore a polluted environment and prevent dissemination of toxic compounds from hotbeds of pollution for a considerable length of time. The presented complex biotechnology is important from the viewpoint of prevention. This technology can be applied to soils contaminated with heavy metals, oil hydrocarbons, explosives, pesticides and other pollutants. The main purpose of the carried out work is the development of a novel approach to the provision of ecological safety. The approach is based on using natural minerals composites which are comprised of natural mineral rocks, microorganism strains with high detoxification abilities and plants-phytoremediators. In this composite material, the function of a sorbent is to uptake and to trap pollutants thus restraining their emission in the environment. The role of the microorganisms is to accomplish the first stage of biodegradation of organic contaminants and then to apply phytoremediation as a unique cleanup strategy. Natural sorbent can be applied during the initial stage of cleaning up of contaminated soil, followed by application of a phytoremediation technology through purposeful planting of selected plants. This results in the total assimilation and complete mineralization of pollutants and/or their intermediates or in their partial transformation by microbes.Following contaminants were used in the present work: explosive – 2,4,6 trinitrotoluene (TNT), oil hydrocarbon (hexadecane) and Wax (long chain hydrocarbons). Special attention was paid to enhancement of soil fertility, which improves microorganism and plant growth conditions and is important for bioremediation processes providing total rehabilitation of soil.

Keywords: Phytoremediation, Bioremediation, Contaminated Soils

1. Introduction

Cleansing of polluted environment, its absolute restoration and long-term protection can be achieved effectively and cheaply via application of phytoremediation technologies [1]. In this case the ecological balance is maintained: molecules of chemical toxicants are totally assimilated by plants and the atoms, forming toxic structure of xenobiotics participating in the natural metabolism as harmless compounds.

Phytoremediation is a unique cleanup strategy. Application of phytoremediation technologies implies the planting of contaminated area with one or more specific, previously selected species of plants having the potential to extract contaminants from the soil [2-7]. The treatment continues by harvesting the plants and composting or incinerating them. To create a truly effective phytoremediation system all components of the system should be thoroughly analyzed. The major constitutive component of such a system is obviously the plant. Plant selection implies choosing a certain plant species with appropriate characteristics. A survey of the site vegetation should be undertaken in order to determine the species of plants that will grow best on the contaminated site, taking into account the ability of the plants to accumulate and degrade the contaminants.

Phytoremediation has great advantage over other nonbiological and biological technologies. It is envisaged by many private or state companies while projecting and accomplishing remediation or prevention activities.

Some limitations to phytoremediation should be pointed out – very high concentration of pollutants, e.g. accidental oil spills, may significantly reduce plant growth. Therefore, dissemination of toxic compounds from the hotbeds of pollution in the environment (e.g. groundwater) occurs at a higher pace than in case plants provide their assimilation.

The main purpose of the carried out work, is to reduce limitations to phytoremediation, and to develop a new technological approach to the remediation of chemically contaminated soils using "biosorbents" - composites of natural mineral sorbents and microorganism strains with high detoxification abilities. In this composite material, the function of a sorbent is to uptake and to trap pollutants, thus restraining their emission in the environment. The role of the microorganisms is to accomplish the first stage of biodegradation of organic contaminants and then to apply phytoremediation as a unique cleanup strategy. Natural sorbent can be applied during the initial stage of cleaning up of contaminated soil, followed by application of a phitoremediation technology through purposeful planting of selected plants. This results in the total assimilation and complete mineralization of pollutants and/or their intermediates or in their partial transformation by microbes.

By combining specially selected microorganisms having high detoxification abilities with natural minerals having high sorption capacities, a new "biosorbent" technology provides restoration of the polluted environment, and prevents the dissemination of toxic compounds from hotspots of pollution for a considerable period of time. The method can be used to purify and decontaminate soils and water contaminated by a variety of eco-pollutants, including heavy metals, radionuclides, alkanes, aromatic and polycyclic hydrocarbons, pesticides, explosives, chlorinated solvents etc.

The purpose of the present project is to elaborate a new technological approach to chemically contaminated soils, based on simultaneous application of sorption properties of natural minerals and detoxificational abilities of microorganisms and plants.

2. Materials and Methods

2.1. Screening of Bacterial Strains

Bacterial cultures from the Collection of Microorganisms of Durmishidze Institute of Biochemistry and Biotechnology of the Georgian Agricultural University were used to determine the strains capable of degrading different organic pollutants, particularly crude oil, wax and 2,4,6-trinitrotoluene (TNT).

2.2. Plants

The following plant species were tested as vegetation to be used in purposeful planting during our investigation: maize (*Zea mays*), alfalfa (*Medicagosativa*), mung bean (*Vignaradiata*), ryegrass (*Loliummultiflorum*), soybean (*Glycinemax*), chickling vetch (*Lathyrussativum*), chickpea (*Cicerarietinum*) and chinabean (*Vignasinensis*).

2.3. Investigation of Sorbents

Standard methods were applied to define physical-chemical features (volumetric weight, porosity, displacement, filtration characteristics, oil absorption potential, absorption dependence on temperature, pH, etc.) of rocks [8].

2.4. TPH Determination

The sampling of soil was carried out according to standard methods. The content of total petroleum hydrocarbons (TPH) in the soil samples was measured gravimetrically after Soxhlet extraction as well as by gas chromatography with flame ionization (FID) detection [10-12].

2.5. TNT Determination

TNT content in the samples was determined according to the extinction of colorized TNT in high alkaline medium ($pH>12.2$) at 447 nm [9].

For determination of residual amounts of TNT, a sample of soil was taken and TNT was extracted by methanol (three times with 300 ml of ethanol on 50 g air-dried soil, mixed during 30 minutes). Extracts were concentrated to a volume of 5 ml on the vacuum-evaporator, then 2 ml of this extract was added to 2 ml of KOH solution (1 M) and after 8 minutes TNT content was determined spectrophotometrically, according to extinction at 447 nm. The amount of TNT determined at the beginning of the experiment was used as the control.

In case of experiments with [1-^{14}C] TNT, the air-dried soil was weighed (mass of sample 7 kg), 0.5 L of solution of 1.05 g [1-^{14}C] TNT (specific radioactivity 500 Bq/mg) in diethyl ether was added and was thoroughly mixed. The ether was evaporated during 2 days. After that 10% suspension of microorganisms was added to the samples and was dried to up to 130% of initial weight (according to the standards of soil investigations). Thereafter the soil was placed in special boxes for the experiments and the soybean seeds were sowed in samples (80 seeds on 7 kg of soil).

For determination of radioactivity of soil samples the methanol extract was evaporated until dry residue. The dry residue was dissolved in 5 ml of benzene; 1 ml sample was taken and was measured with the scintillation spectrometer SL-30 Rackbeta, with an efficiency of 95%.

Thin-layer chromatographic (TLC) analysis of methanol extracts from polluted soil samples was carried out. The plates were developed with mixture: benzene: dioxane : acetic acid (90 : 10 : 1). The chromatograms were exposed on X-ray film during 45 days for revelation of radioactive spots. After exposition the radioactive spots were extracted from chromatograms with benzene and their radioactivities were measured.

3. Results and Discussions

3.1. Sorbents

Application of mineral composites is important in case of all types of contamination, as, on the one hand, it is significant from the viewpoint of prevention, and, on the other hand, it creates favorable conditions for growth of vegetation. After the introduction of the composites into the upper part of the soil and as a result of subsequent watering, the composites become swollen and stimulate growth and development of vegetation at the corresponding stage of phytoremediation.

Three types of mineral rocks were studied for the potential to be used as biosorbent – 2 types of Bentonites and Zeolite. The absorption ability of minerals and their composites for oil and TNT were studied. In composite ratio of minerals for each contaminant was different. Composite content against contaminants was the following: for TNT – B1/Z3 in ratio 1/3, for oil – B1/Z3 in ratio 2/3. It has been stated, content of composites in soil - 20%.

3.2. Soil Contaminated with Crude Oil and Wax

3.2.1. Selection of Microorganisms

Screening of 200 strains of microorganisms was performed to study their abilities to effectively assimilate oil hydrocarbons during growth on media containing crude oil. As a result, optimal growth conditions (content of nutrient medium, pH, temperature) were established, resulting in 20 selected strains. The ability of these selected strains to degrade oil hydrocarbons was determined by their submerged cultivation in nutrient media with crude oil (3%) as the sole carbon source.

The gas-chromatographic analysis has shown that *Pseudomonas* and *Rhodococcus* strains were the most efficient combination of bacterial cultures for oil degradation. Chromatographic analysis also revealed that hydrocarbon fractions up to C_{17} are assimilated completely, while a small amount of heavier hydrocarbons (> C_{17}) remain in the incubation medium.

3.2.2. Plants

The following plant species: alfalfa (*Medicago sativa*), chickpea (*Cicerarietinum*), chickling vetch (*Lathyrussativum*), maize (*Zea mays*), pea (*Pisumsativum*), ryegrass (*Loliummultiflorum*), and soybean (*Glycine max*) were used as phytoremediators.

The model phytoremediation experiments for cleaning the soil artificially contaminated with crude oil were carried out according to the following scheme: the suspension of microorganisms was inoculated in the contaminated soil at the beginning of the experiment. After different incubation periods the plants were sowed in separate samples of soil. The conditions and details of experiments are presented in legends of Fig. 1.

As shown in Fig.1, bacterial consortia cause effective decreasing of oil content in soil in all model test variants (43-74% of initial contamination). Residual contamination of soil removes after plant using in phytoremediation processes. (55-90% of initial contamination). Most effective phytoremediator in presenting experiment is soybean.

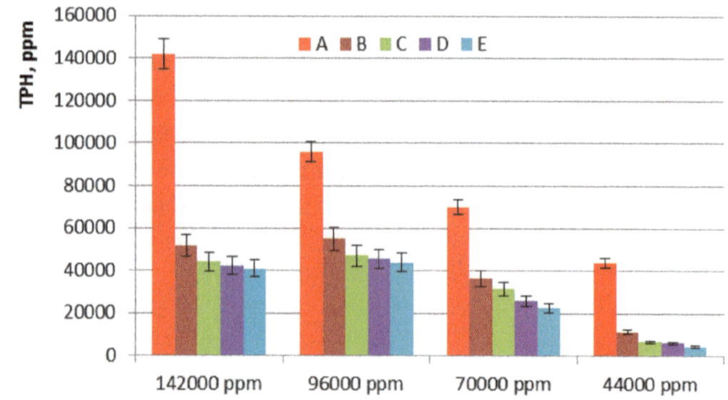

Figure 1. *TPH content in soils contaminated with oil hydrocarbons after being treated by bacterial consortium (Rhodococcus sp. MO227 and Pseudomonas sp. TP 335) and sowing of plants. The suspension of bacteria was inoculated in the soil at the beginning of the experiment. On 31th day of incubation the plants were sowed in separate samples of soil. Initial degree of contamination: 142 000; 96 000; 70 000 and 44 000 ppm of TPH; total time of incubation: 2 months; temperature: 20-2°C. Sample variants: A – Contaminated soil in the beginning of the experiment. B – Soil treated by bacterial consortium. C – Soil treated by bacterial consortium and sowed with ryegrass. D – Soil treated by bacterial consortium and sowed with alfalfa. E – Soil treated by bacterial consortium and sowed with soybean.*

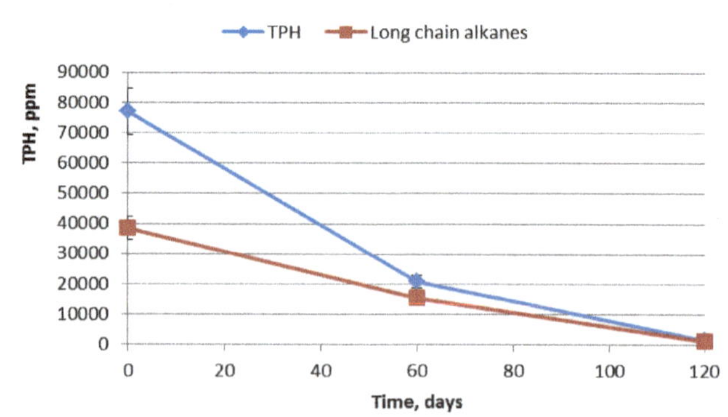

Figure 2. *Phytoremediation of soil artificially contaminated with crude oil and wax by using of Alfalfa, bacterial consortium (Pseudomonas sp. 6R67 + Bacillus spp.). Initial contamination of soil with TPH - 77 000 ppm, duration of experiment – 4 months.*

3.3. Soil Contaminated with TNT

3.3.1. Selection of Microorganisms

The screening of microorganisms for their ability to degrade TNT was performed. More than 200 bacterial strains from the Durmishidze Institute of Biochemistry and Biotechnology Collection of Microorganisms and Cultures were tested. These strains were isolated from the soils of different regions of Georgia, including the soil of proving grounds in the territory of Georgia.

The screening was carried out in the solid nutrient media containing different concentrations of TNT (0.01 mM, 0.1

mM and 1.0 mM) as the source of carbon and nitrogen. As a result of the screening 34 strains of genera of *Pseudomonas*, 1 strain of *Bacillus*, 8 strains of *Rhodococcus* and 5 strains of Mycobacterium were proven to be characterized by good growth on the solid nutrient medium containing 1 m MTNT, and were subsequently selected. For these strains optimal growth conditions (content of nutrient medium, pH, temperature) were established. In order to increase efficiency of selected strains towards TNT degradation, 7 different bacterial combinations were composed and tested under conditions of submerged cultivation containing 0.5 M of TNT. The cultivation was conducted during 3-7 days on a thermostatic shaker (180-200 rev. per min; at a temperature of 28-30°C) in a modified Czapek agar, containing 0.5 mM TNT. Glucose was added to the activation of strains.

3.3.2. Plants

High tolerance of soybean to TNT while comparing the growth parameters of testing plants should be mentioned. Growth parameters of this plant on the saturated solution of explosive (1.0 mM) decrease only by 5-10%. To summarize the data, it can be concluded that the tolerance of tested 8 plants to TNT decreases according to following orders:

Soybean >Mung bean > Ryegrass > Chickpea >Chickling vetch > Alfalfa > China bean > Maize

3.3.3. Model Experiments

For the purpose testing of selected consortia and plants, was carried out model experiments by using [1-^{14}C] TNT contaminated soils (conditions of the experiment are given in 2.5). The obtained results are present in Fig.3 and Fig.4.

According to the obtained results, presented inFig. 3, content of TNT in soil decrease, but radioactivity remaining in the soil. Results of TLC analyze indicates formation of TNT radioactive metabolites in soil (Fig. 4). After a short period of bioremediation (after 8 days) was accomplished purposeful planting of soils. The radioactive metabolites, formed as a result of bacterial transformation of [1-^{14}C] TNT, were effectively uptake from the soil by the plants. Fig. 4 evidences dynamic decrease of radioactivity and amount of metabolites with different R_f in soil extracts.

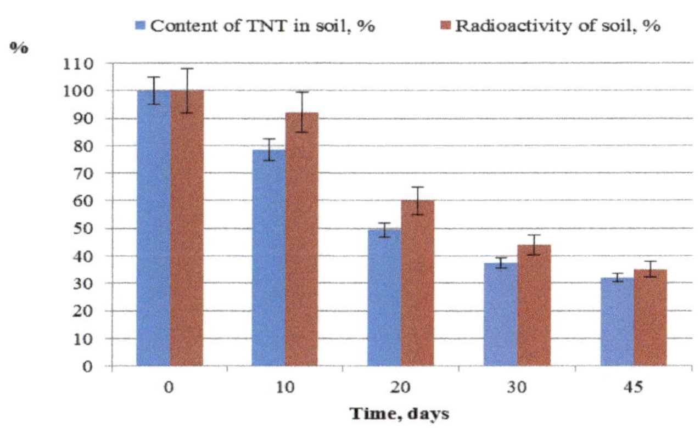

Figure 3. *The dynamics of changing of TNT content in soil contaminated with [1-^{14}C] TNT during cleaning by soybean and bacterial consortium containing strains of Pseudomonas sp. TNT-44 and Rhodococcus sp. TNT-74.*

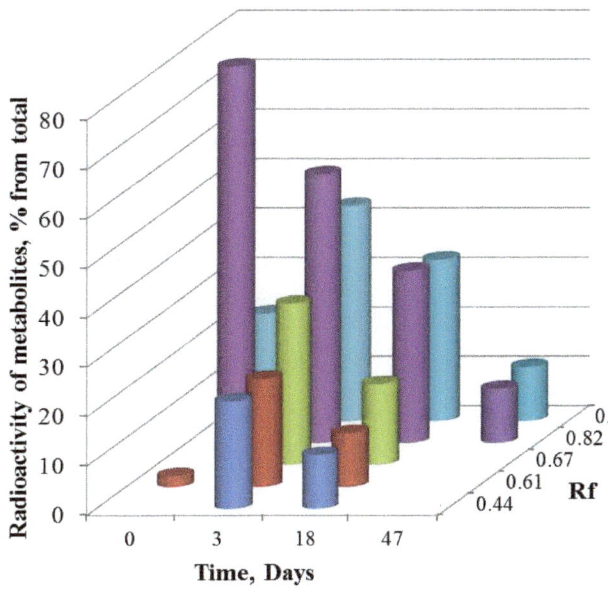

Figure 4. *TLC analyses of [1-^{14}C] TNT metabolites extracted from samples of polluted soil on 3rd, 18th and 47th days of the remediation process*

4. Conclusion

1. It has been stated, content of composites for each contaminants: For TNT – B1/Z3 in ratio 1/3, For oil – B1/Z3 in ratio 2/3.
2. According to carried out model experiments was stated:
   - For effective phytoremediation of crude oil contaminated soil, can be used bacterial consortia *Rhodococcus sp. MO227* and *Pseudomonas sp. TP 335* and Soybean.
   - For effective phytoremediation of wax contaminated soil, can be used bacterial consortia *Pseudomonas sp. 6R67 + Bacillus spp.* and Alfalfa.
   - For effective phytoremediation of TNT contaminated soil, can be used bacterial consortia *Pseudomonas sp. TNT-44* and *Rhodococcus sp. TNT-74* and Soybean.

Acknowledgements

This work was supported by STCU Partner Project # P499. In addition, Lawrence Livermore National Laboratory is operated by Lawrence Livermore National Security, LLC, for the U.S. Department of Energy, National Nuclear Security Administration under Contract DE-AC52-07NA27344.

References

[1] F. Korte, M. Behadir, W. Klein, J.P. Lay, H. Parlar, and I. Scheunert, "Lehrbuch der okologischen chemie," Grundlagen and Konzepte fur die Okologische Beureilung von Chemikalien.Suttgart, Georg Thieme Verlag, 1992.

[2] E.L. Arthur, and J.R. Coats, "Phytoremediation," In: Pesticide Remediation in Soils and Water, P. Kearney and T. Roberts. Eds. New York, Wiley, 1998.

[3] D.E. Salt, M. Blaylock, P.B.A. Nanda Kumar, V.P.Dushenkov, B.D. Ensley, I. Chet, and I. Raskin, "Phytoremediation: a novel strategy for the environment using plants," Biotechnology, vol. 13, 1998, pp. 468-474.

[4] D.T.Tsao, "Phytoremediation. Advances in Biochemical Engineering and Biotechnology," Berlin Heidelberg New York, Springer, 2003.

[5] G. Kvesitadze, G. Khatisashvili, T. Sadunishvili, and J.J.Ramsden, "Biochemical Mechanisms of Detoxification in Higher Plants," Basis of Phytoremediation, Berlin Heidelberg New York, Springer, 2006.

[6] AS 482.2 Guide to the sampling and investigation of potentially contaminated soils, 1999.

[7] Dee Schnoor, "Phytoremediation," Technology Evaluation Report TE-98-01, Ground-Water Remediation Technologies Analysis Center, Ser E. Iowa City, 1997.

[8] G.S.Fomin, and A.G.Fomin, "Soil. Control of quantity and ecological safety according to international standards," Moscow, BNII Standard, 2001.

[9] NWTPH-HCID "Hydrocarbon Identification Method for Soil and Water", 2001.

[10] US EPA Method3540C "Soxhlet Extraction;" Method 8572A. "Semivolatile organic compounds in soils and solid wastes using thermal extraction/gas chromatography/mass spectrometry", 1996.

[11] US EPA Method 418.1. "Total petroleum hydrocarbons," 1997.

[12] TNRCC Method 1006. "Characterization of C_6-C_{35} petroleum hydrocarbons in environmental samples," 2000.

An Integrated Model of Sustainability and Emission Control: The Concept of Society as a Super Organism That Lives by Consuming Its Own Waste Using Alternative Energy as Currency

Paul A. Comet

Comet Environmental Consulting, Houston, Texas, USA

Email address:
pcomet@prodigy.net

Abstract: There is a need for new political, social and economic theories, as neither present day capitalism nor communism are yet adequate in creating sustainable societies, or for reversing climate change. A rough "working" document for discussion purposes.

Keywords: Sustainability, Economic Theories, Alternative Energy, Super Organism

1. Introduction

"Strategies and actions can be pursued now which will move towards climate-resilient pathways for sustainable development, while at the same time helping to improve livelihoods, social and economic well-being and effective environmental management. In some cases, economic diversification can be an important element of such strategies. The effectiveness of integrated responses can be enhanced by relevant tools, suitable governance structures and adequate institutional and human capacity (medium confidence). Integrated responses are especially relevant to energy planning and implementation; interactions among water, food, energy and biological carbon sequestration; and urban planning, which provides substantial opportunities for enhanced resilience, reduced emissions and more sustainable development (medium confidence)" [1].

2. Concepts

A series of simple concepts are combined here for the creation of carbon- neutral population centers with sustainable economic systems.

The primary concept is the observable operation of an "ecosphere"; a small glass container, welded shut containing sea water, a bubble of air, a spring of algae and a few tiny shrimps. A small patch of basal gravel completes, the ensemble. Placed by a window the ecosphere can be indefinitely sustainable. It was developed by NASA. Somehow the shrimps "know" not to eat up all the algae by overbreeding. Perhaps the shrimps secrete an "overcrowding" pheromone.

A second concept is the idea of a "superorganism": this can mean many things from "Gaia" at the planetary level [2,3]; to bees, ants and other social organisms; which may (as in the case of ants and bees) have delegated their reproductive rights to a single individual (queen) or retained those rights (prairie dogs &, humans). At the smallest scale are symbiotic organelles that allow autotrophy within "animal like", otherwise heterotrophic, protists; these marvelous "wee beasties" can also be considered to "micro super organisms" which may or may not be colonial. Various protists come to mind. All of these examples should be indefinitely sustainable given enough water, air & sunlight.

The third concept is the concept of waste recycling; not just by the reuse or repairing of damaged, unwanted, surplus, or decayed commodities, packaging etc. but also by totally deconstructing materials by combustion (or consumption by detritovores) and using the recoverable energy, ash and gas for other purposes such as heating &fertilizer, Comet, P.A., 2006,

2007, 2010. [4-6]. This concept links closely to the ecosphere where everything is indefinitely recycled.

A fourth concept is the idea that money can be denominated in units of energy – joules or watts. Fawzi, M, 2009, [7].

3. Proposed Implementation

If each city could be self - reliant, using local crops as food (possibly "twinning" a rural town and larger cities) the city become reliant on urban areas for food and the rural town using effluent, as fertilizers and other recycled materials from the city, then a symbiotic arrangement might emerge. Symbiosis might be facilitated by local electrical grids which covers both rural country side and the large city. If alternative energy forms the basis of a single credit card "banking system", external to the present one, then a series of fully autonomous regions may be created. The waste stream then forms a resource of stored energy, much better than batteries, for night time, electricity or day peak demand. Similarly the incombustibles, such as steel and aluminum cans, could form the basis of local manufacturing. Other waste such as glass and ceramics could be used in decorative construction work such as sculpture. Plastics can be recycled for a multitude of purpose or used as fuel in a gasifier.

The value of superorganism theory is that it may potentially offer a predictive, integrated, holistic approach to understanding how a healthy society functions - economically, politically, sociologically, maybe even spiritually. Medical analogies then may become useful & the idea of societal ills being curable may become valid, Comet, P.A., 2015, [8]. The question then becomes what is the best biological analogy? Gaia, unicellular organisms, multicellular organisms, plants, animals, & symbiotic associations, particularly those that combine autotrophy and heterotrophy? Though the present model takes fragments of each of these analogies, it is initially the concept of a single photosynthetic cell, the basic functioning unit of any plant that seems to fit best to city & state, as the cell defines the basic parameters of self and non-self, chemical factories for creating the compounds of life , derived ultimately from sunshine, in the chloroplasts, partly analogous to "solar" or photovoltaic panels" that can place energy into storage as lipids & carbohydrates, cellular infrastructure such as walls, water reservoirs, pipelines between cells functioning by osmosis – the list is long. Many land plants (even cyanobacteria), need little but water, air, sunshine. However, nitrogen and phosphorus sources may require bacteria & fungi to collect these vital elements for the plant, using the plants stored photosynthate as "payment" to the fungus etc. Hence, the "currency" of this system is stored energy ultimately derived from the sun, it does not matter what the organism is. If one perceives that each town or city can be compared to heterotrophic nucleus, and the surrounding fields are photosynthetic chloroplasts; then roads and railways can be compared to veins, power lines and telecommunications to a neural system. However, it is the circulatory fluid (blood etc.) that can be compared to the modern banking system and the flow of money, as well as those resources that money can buy.

Though the above analogies are imperfect, they are useful in the sense that certain features of civilization, presently arcane, such as the novel identification of money as "energy tokens"; the health of a society or individuals , measured in not only in terms of conventional personal medical "wellness", but also in terms of social problems of unemployment, incarceration, substance abuse, domestic violence, divorce, homicide, child abuse, availability of clean water, healthy food, access to good medical care, degree of pollution, waste elimination, longevity etc.; all become group "symptoms" of the health (or lack thereof) of that society. If the group symptoms described above become overwhelming, then political metastasis ensues and the society collapses into war and anarchy. Essentially the "organs of state" can be compared to the organs of a body or the organelles of a single cell. If these malfunction, then sickness, death and disintegration ensues. The Middle East can be used as an example of what can go wrong.

Ultimately if the above models have any validity, then urban and rural areas can be reappraised in terms of a mutually beneficial economic "modular" structure, just using the waste stream and sunshine. By using some of these ideas, the first sustainable lunar base (or autonomous refugee camps, i.e. not requiring external charity, just land & water) may be viable.

Essentially we are advocating cooperative capitalism in which the municipal structure forms the basic administrative unit, this might be urban or rural; or best of all, both linked together by a common grid based on alternative energy, this system emphasizes the waste stream as stored energy, as well as solar, wind etc. for immediate use. During night time, or peak electricity demand, the energy stored in the organic materials of the waste stream can be accessed by fermentation, gasification, or combustion etc.; also utilizing cogeneration to also access the energy in "waste heat" produced by those processes. The electric grid might form a common bank of the 2 communities, in which personal credit cards monitor energy input and withdrawal from the grid.

If rural food production and urban manufacturing are correctly balanced, just by using the waste stream & alternative energy, then the population may become autonomous, with potentially higher levels of employment for those "left behind" in the present system. The existing system could be left intact. However the pressure to import raw materials from elsewhere would be much reduced. For instance, using the example of animal bones, a simple model can be proposed. These are the phosphate islands of the Pacific, for not until the last ton of phosphate be removed from these "disappearing" islands, will people routinely start recycling food bones as fertilizer. The concentrated phosphate deposits will always require less energy (& money) for extraction & processing, as compared to food bones which would need to be laboriously collected. Essentially the present system of "cheapest resource to be used first" based on fossil fuels, results in "material entropy" (a better term would be resource dilution or dispersal) and the eventual loss of everything (& everybody) to the landfill, cemetery, (dole or prison).The "rag & bone" man of Dickensian times maybe need be reintroduced. If the currency needed to pay the bone collector comes from the local grid, then the system becomes sustainable& the phosphate is indefinitely retained by the municipality. The money also should circulate indefinitely within the community, as long as the sun shines & as long as the credit card is not usable in other municipalities.

A difficulty to be overcome are that is that articles made for sale, may trade for considerably more elsewhere (outside the municipal grid),than they cost energy – wise to manufacture, also currency conversion between the existing system & the energy - based one proposed here, has not yet been solved. This will not be a problem immediately, as there is probably insufficient energy to be yet used in manufacturing. However, as new technology is introduced to better capture, use and store the sun's rays, there will eventually be a surplus of energy that can be used in local manufacturing. Customers from another grid could purchase items, or alternatively, replenish their own grid with more energy from the vending grid etc., that would continually revalidate the energy currency unit.

Once the idea of stored energy is introduced the model becomes much more complex. For an individual to purchase an item, for which the energy to manufacture it has been dispersed, a currency token would become a promissory note similar to those of the present system. Eventually the number of unbacked promissory notes could become overwhelming and the system collapses. However social organisms, plants etc. have solved this problem, for instance a tree that generates dead structural material (trunk) so that the cambium & leaves etc. can live & be supported in space. Termite mounds are similar, or virtually any "infrastructure" such as a skeleton, or shell of some kind. As noted above, the replenishment of the grid by new energy would continually validate circulating currency, but stored energy from biochar, batteries, living biomass & the waste stream might permit savings & investment.

By analogy with the ecosphere discussed earlier, a specific rate of energy incorporation and loss can be directly observed as one moves through the carbon cycle of photosynthesis, respiration and decomposition. Manufacturing of plant tissue, shrimp tissue and bacterial biomass is regulated by light energy and the continual regeneration of plant material from simple nutrients. Hence by transposing this ecological system to human economic systems, the amount of energy used or absorbed per unit of time should correspond to the number of currency units available. Essentially the cliché "money is power" would be validated.

So when photosynthesis, respiration and decomposition are in balance and confined to a specific area, such as in the ecosphere, the law of conservation of energy might be applied to the printing of "shrimp" currency. However, leaving the existing currency in place & just using an energy - based credit card for electricity addition / withdrawal, or for articles manufactured, or services provided, by alternative energy, might be easier. The amount of energy in the closed system can be expressed as a limited joule or watt currency as long as it bears a resemblance of amount of both free &"bound" energy available. The currency problem becomes much more difficult to solve when the system is open and may indeed correspond to "Laissez faire". Conventional capitalism automatically transfers wealth to the least expensive manufacturer or service provider. This problem might be avoided by a number of autonomous areas that are defined by either a gas or electrical grid that is used not only for running domestic appliances, light etc. but eventually in manufacturing & specifically oriented towards recycling in the waste stream by using it as a manufacturing & energy feedstock. The credit card used in one autonomous area would, initially, perhaps, not be valid in another area, except under highly controlled circumstances. Eventually manufacturing etc. in one place, that might complement manufacturing of different goods in other places, could justify use a common energy - based credit card.

If it becomes economically feasible to replace fossil with alternative fuels then, and if, this trend were to become widespread, carbon neutrality could be accomplished. The seas & boreal forests would continue to absorb 2 ppm per year of CO_2 (spring "blooms" on land and sea) as indicated from the Mauna Loa observatory "zig – zag" CO_2 record of atmospheric concentrations known as the Keeling curve [9]. However this absorption presently removes both CO_2 derived from decomposition of leaf litter & annual vegetation as well as from fossil fuels. Dead vegetation may have to be charred to fix at least half of this carbon on the land. A range of other techniques for carbon fixation will also be needed to move from carbon neutrality to carbon negativity.

4. Conclusion

The observation of the NASA "ecosphere" allows one to mentally model a carbon – neutral, sustainable society, that uses all its waste as a source of stored energy (where city & state replace the shrimps!) as well as sunlight, wind etc. The creation of an alternative energy – based currency "photonic dollar" is suggested that would gradually outcompete the petrodollar as domestic waste to energy, & other alternative energy devices are attached to a communal electric grid (that also functions as a bank) that gradually replace fossil fuel energy.

"Superorganism" theory is advocated as a way of distributing resources in an effective manner & as a way of understanding the complexities of how human society functions in a holistic manner. Societal ills then become diagnosable in terms of a malfunction of the entire social & economic system. The interposing of a separate economic "tier" specifically for alternative resources & energy is suggested as a non – revolutionary way of creating sustainable employment [10].

These ideas are proposed because neither the theories of Adam Smith nor Karl Marx address the waste problem, nor clearly identify the "energy nature" of capital. If this new combination of models is successful in solving climate change problems, then economic philosophical differences between competing nation states might prove to be resolvable.

References

[1] IPCC Climate Change Synthesis Report Summary for Policymakers, 2014, http://www.ipcc.ch/pdf /assessment-report/ar5/syr/AR5_SYR_FINAL_SPM.pdf

[2] J.E. Lovelock, "Gaia as seen through the atmosphere", Atmospheric Environment, 1972, pp. 579- 580.

[3] J.E. Lovelock, and L. Margulis, Atmospheric and homeostatic by and for the biosphere: the gain hypothesis, *Tellus* XXVI, 1974, vol 1-2, pp. 1-10.

[4] Comet, P.A "A plan for sustainability", C&EN News, vol. 85, issue 13, August 2007, pp.6, 8.

[5] P.A Comet, "Good use of waste". C&EN News, vol. 84, Numb. 41, October 2006, pp. 4-5,

[6] P.A Comet, "Biochar for CO2 Reduction", C&EN News, vol. 88, issue11, March 2010, p.4

[7] Fawzi, Marc., 2009, http://p2pfoundation.net/P2P_Energy _Economy.

[8] P.A. Comet, "A "lunatic" model of sustainability", 19th Annual Green Chemistry & Engineering Conference, July 14 to 16th 2015, Bethesda, Maryland, USA.

[9] Latest reading, October 2015. Mauna Loa Observatory, https://scripps.ucsd.edu/programs/keelingcurve/

[10] P.A. Comet, "Planetary regulation of greenhouse gases by changing municipal waste management practice: a commonsense approach. Presentation at American Chemical Society", Green Chemistry Symposium" July 16, 2012.

Genetic Characteristics of Brown Forest Soils on the Middle Urals

Samofalova Iraida

Perm State Agricultural Academy, The Faculty of Soil Science, Agrochemistry, Ecology and Commodity research, Perm, Russia

Email address:
samofalovairaida@mail.ru

Abstract: Feature of mountain soil formation is that the soils on the mountain slopes are formed in different bioclimatic and orogeomorfological conditions. The purpose of research is to study genetic properties of the brown forest soils in the Middle Urals. Features of the morphological structure: truncated profile (35-75 cm), weakly expressed in the differentiation of the soil profile into individual Horizons, detritus (20-65%), loamy fine earth, the signs of podzolization in the soil profile is not found. Feature granulometric size distribution of the soil - a gradual increase in weight on the profile into Sandy Loam to Silty Clay and Clay, the dominant factions are either big Silt or Sand. Soils are characterized by a very acidic environment and high hydrolytic acidity. The soils are enriched in organic matter and humus profile characterized by prolixity. Group composition humus shows mobility, high degree of humification of organic substances to fulvic acids. Gross composition shows that the processes of soil formation on Mount, North Basegi aren't leading to a distinctly different profile. Coefficients of geochemical accumulation, subsurface weathering, eluviation, oxidation calculated and helped define the features of the gross composition of soils. Character of distribution in profile and correlation of forms of iron help to diagnose physical processes of weathering and soil formation (burozemic pedogenesis et al.); signs Podzolization not revealed. Weathering and soil formation processes that occur with varying intensity, creating a diversity of soil cover and the spatial heterogeneity of soils, even within the same type. Thus, in the mountains of the Middle Urals brown forest soils form a number of subtypes: raw organic (900 m) - ferruginized (655 m) - metamorphosed (590 m) - clay-illuvial (577 m) - eluvial (565 m) - clay-illuvial (315 m).

Keywords: Brown Forest Soils, Properties, Genetic Horizons, Soil Profile, Soil Formation, Weathering Processes

1. Introduction

Mountain areas represent about 20 % of Earth's land area (excluding the areas covered by ice). The largest area is occupied by mountainous taiga zone (one-third), and mountain forest and brown soil zones (almost one quarter) [1-8]. In Russia, mountain areas occupy 34 % of the total area. Problems in studying of mountain soils are associated with many factors. These include soil-forming processes and constant weathering, special climate and vegetation. It is also natural that soils differ in different mountains depending on their location, height, composition of the material, etc. [4, 5, 9-18].

Scientists have established that brown forest soils are formed under broad-leaved forests in the warm-temperate and humid areas of subboreal belt [19-24]. It took a number of years for this type of soil to be recognized, yet only in the southern mountains of the Crimea and the Caucasus, and at the Far East. It was believed that the mountain brown forest soils are common in the Caucasus, the Crimea, the Carpathians and the Sikhote-Alin [20, 25, 26]. Brown forest soils have always been of great interest for researchers, causing numerous disputes and differences of opinions regarding their distribution and formation. It took some time for soil scientists come to a consensus on the type of brown forest soils, so the matter remained unclear for a long time [27-32].

Active studies of the genesis of brown forest soils in different mountain systems in Russia began in the middle of 50s-60s of 20 centuries [33-39]. A considerable number of papers on forest soils showed that the area of the brown forest soils is much broader than it was believed to be. As a result of development of soil studies in Siberia, the question arose if some forest soils of the Urals, Sayan and Baikal can be classified as brown forest soils. Brown forest soils have been allocated under the dark-coniferous forests, not only in the Caucasus and the Carpathians, but also in more continental areas, such as the Altai and Kuznetsk Alatau and the Urals

[40-42]. The assumption on the distribution of brown forest soils in colder regions of Russia was confirmed later.

Thus, the brown forest soils have no clear affinity to any particular type of vegetation, as they develop both under broadleaved and mixed coniferous-broad leaved forests, and even under coniferous forests. Knowledge of the brown forest soils was generalized. Due to numerous studies, brown soils have been allocated almost everywhere, even in colder climatic conditions. As a result, brown forest soils and brown soils are now considered the most common type of forest soils.

Rozanov B.G. [4] described the almost ubiquitous brown-colored soils, poorly differentiated on the horizons, having different types in different bioclimatic zones, located on the slopes of humid, sub-humid and subarid areas of the world, as one of the features of the soil covering of mountain systems of the world. Mountain brown soil is the main form of soil-forming in humid mountain slopes regardless of the climatic characteristics of the mountains. The scientist indicated that the presence of brown soils in the mountains is a definite historical stage in the development of the soil maintained for a long period in conditions of mountain slope soil-forming. Thus, similarity of mountain soils is noted in mountains worldwide in senile soils, and the diversity increases in mature soils.

On the background of large-scale studies of mountain ecosystems, the studies of Subpolar, Northern, and Middle Urals soils began to appear [43-57]. There are opinions that in the Middle Urals, under the middle-taiga forests, the formation of brown forest soils having its own characteristics is possible.

The purpose of the research is to study the genetic characteristics of brown forest soils in the Middle Urals.

2. Material and Method

The object of research is the brown forest soils of North Basegi, which is a part of the Basegi crest overlying in the West of the Urals watershed and is located between 58°50' and 60° N. Basegi is a meridionally elongated range of three mountains: North Basegi (951.9 m), Middle Basegi (994.7 m), and South Basegi (851 m). The ridge is a part of "Basegi" State Natural Reserve. The lowest point in the reserve is located in the mouth of the Korostelevka river on the altitude of314 m. The territory of the reserve is a part of the Urals mountain band, which is composed of metamorphic rocks [58]. The territory belongs to the field of ridge-remnant low mountains of the Middle Urals [59]. The climate is cold and wet with features of continental (precipitation of 496 mm to 1,071 mm; the average temperature of the coldest month, January, is -17,9 ° C, and the warmest, July, +13,3 ° C). By zonal distribution of vegetation cover, the territory is located in the middle taiga subzone of the boreal forest zone.

In the Middle Urals, there are mountain-forest belt, subgoltsy (subalpine) belt, and mountain-tundra (alpine) belt [60]. In the mountain-forest belt to a height of 450-600 m above sea level the cover is dark coniferous taiga with a sufficiently dense grass cover. Subalpine belt comprises three sub-belts (park open forest, subalpine meadows, and crooked forest). Park open forest belt (sparse, with the small closeness of undergrowth and tall grasses) is gradually transformed to the crooked forest with altitude increasing. Subalpine meadows are located at the same altitude as the crooked forest, often mixing with it. Meadow communities are common almost up to rocky places. At an altitude of 800 m or more, rocky, suffruticous, and grass-moss tundra is found.

In 2010-2012 soil cross sections on North Basegi Mountain were made in different altitudes and vegetation zones, namely in mountain forest, park open forest, under the sub-alpine meadows, in a crooked forest, and mountain tundra belts. In total, 12 soil cross sections were made for the study of morphological characters of soils on the slopes of different exposures. Soil samples were collected throughout the profile from genetic soil horizons.

The following parameters were evaluated at samples: gritty consistency, granulometric texture with pipette method, version of N.A. Kaczynski (with the pyrophosphate method of soil preparation for the analysis), accelerated gross analysis, exchange acidity with potential measurement method, hydrolytic acidity with Kappen method, total exchangeable bases by Kappen-Gilkovits, group composition of humus in mineral soils by M.M. Kononova and N.P. Belchikova [61], content of organic substances by the method of Turin in Antonova's modification [62]. Analytical studies were performed at the Department of Soil Science of the Perm State Agricultural Academy, and in the laboratory of soils physical chemistry of Soil Institute named after V.V. Dockuchaev. The following indicators were defined: total iron content by "Respect" XRF analyzer, the content of non-silicate and amorphous iron by methods of Tamm and Mer-Jackson with atomic-absorption ending. Based on acquiring data, the iron content in the silicate compounds (Fe_c) and crystallized compounds ($Fe_{окр}$) as well as Shvertmann ratio (the ratio of $Fe_{окр}/Fe_{нс}$) were calculated [63].

Russian soil classification as of the year 2004 was used [64].

Statistical analysis was performed in the "Data Analysis" software in Microsoft Excel and the STATISTICA 6.0 software.

3. Results and Discussion

Original physiographic position of the "Basegi" Reserve in the Middle Urals and a variety of environmental conditions influencing the formation of soil cover determine the presence of a broad range of soils, represented mainly by brown soils. The spatial distribution of the soils is subject to the laws related to the mountain nature of the area: exposure and slope angles, altitude, and composition of biocenosis. The main factors of classification differentiation of the soil cover are soil position in the landscape, vegetation diversity, and the nature of deposits.

The pedogenesis processes are closely related to the dynamics and functioning of the landscape. Brown forest soils of North Basegi are formed at different altitudinal positions and in different vegetation belts (table 1). Changes in vegetation formations can cause adifferent combination of different (by orientation) soil-forming processes: sod, alfehumus, gley, and cryogenic.

Table 1. *Conditions of formation of brown forest soils in the North Basegi Mountain*

№ of the cross section, Altitude, m asl	Belt, sub-belt	Exposure, graient	Relief form	Vegetation
№ 30, 900	Goltsy-subgoltsy	South, 30°	Drained steep part at the top of the mountain slope	Birch crooked forest with blueberry and green moss
№ 32, 655	Crooked forest	Southeast, 20°	Drained open forest gentle part in the middle of the slope	Birch crooked forest with large fern and groups of meadows
№ 17, 590	Park open forest	East, 5°	Drained gentle part at the middle of the mountain slope	Spruce and fir forest with large fern
№ 27, 590	Park open forest	South, 5°	Poorly drained gentle part in the intermountain	Birch and fir aconitic forest
№ 15, 577	Mountain-forest	East,5°	Poorly drained gentle part at the middle of the mountain slope	Spruce and fir aconitic forest
№ 19, 565	Mountain-forest	West,5°	Drained gentle part	Spruce and fir forest with large fern
№ 26, 315	Mountain-forest	North-west, 3°	Drained flat surface at the foot of the mountain	Birch forest with meadowsweet and mixed herbs (riverbank)

In the upper part of the southern slope of the studied mountain range, at an altitude of 800-900 m, in subalpine vegetation, a complex of brown (Cambisoils, FAO/UNESCO) earth has been formed, which is completely not typical of this high-altitude belt. In a small hollow in the crooked forest belt with bistort cover in a periodic waterlogging, ferruginated brown soils have been formed (c. 8, *AY-BM-BM*$_f$*-CLM*). Raw-organic dark-humus brown soils (c. 30, H_{ao}-AU_1-AU_2-*BM-CLM*) have been formed on a drained steep part of the hill terraces under the birch crooked forest with blueberry and green moss communities. The soils are 60 to 70 cm deep. The alternation of forest and meadow landscapes is probably due to the migration of the forest belt during the late Pleistocene and Holocene up and down the slope, which has already been noted in other mountain systems [65].

Conditions suitable for forming brown soils can mostly be found in the park open forest belt at an altitude of 500 m. At the bottom of the southern slope (closed intermountain saddle, altitude of 590 m) under the birch and fir aconitic forest in a higher humidification due to significant surface and lateral runoff, gleyey ferruginated brown soils are formed (c. 27, AY_{an}-AY_g-BM_g-$CLM_{f,g}$). A characteristic feature of the soil is the presence of the clearest signs of gleying in the profile, namely the rusty patches, compact constitution, long-term water saturation.

On the east macroslope, brown soils were discovered at the bottom of the slope at an altitude of 577-580 m. These are poorly drained flat areas (gradient 5°), occupied by spruce and fir aconitic forests on raw-organic eluviated brown soils (c. 16, *AO*-AY_{el}-BM_1-BM_2-*CLM*) and dark humus clay-illuviated brown soils (c. 15, *AU-BM*-BM_i-*CLM*). Above 590 m, aconitic associations are replaced by large ferns which are also located on dark humus brown soils (c. 17, *AU-BM*-BM_i-*CLM*). Thickness of the brown soil profile varies from 54 to 70 cm, naturally decreasing with area altitude increasing (r= -0.9). There is also a positive correlation between humus layer and absolute altitude (r = 0.8).

The next series of genetically conjugated soils is located on the western slope at the same relief altitude (557-590 m) with similar vegetation. Under the aconite, the short-profile dark humus eluviated brown soils have been formed (c. 19, *AU*-BM_{el}-*BM-CLM*; c. 21, AY_{el}-*AU-BM-CLM*) as well as dark-profile brown soils (p. 20, *AU*-BM_u-*BM*); under the large fern communities, a more deep dark humus eluviated brown soils are formed (p. 22, *AU*-BM_{el}-*BM-CLM*). There was an inverse mean correlation between the average depth of the *BM* horizon (r = -0,6) and the total depth of soil profiles (r = -0,5) and the area altitude.

On the shady northern slope at the bottom of the mountain slope (400-430 m) on a gentle slope section of 3° gradient under spruce forest with sorrel and small fern, eluviated brown soils have been formed (c. 9, *AY*-BM_1-BM_2-BM_{el}-*CLM* and c. 10, *AO-AY*-AY_{el}-BM_{el}-*BM- CLM*). Brown soils profile depth is 70-104 cm, moist, the structural and metamorphic horizon is lighter at a depth of 30-70 cm. At 300 m from the Usva river (altitude above sea level of 315 m), where drainage conditions are improving locally, vegetation gives way to a birch forest with meadow-sweet and mixed herbs, the clay-illuviated brown soils with a depth of 67 cm is formed (c. 26, *AY*-BM_1-BM_2-BM_i). It should be noted that increasing of absolute altitude is followed by increasing of thickness of structural metamorphic horizon *BM* (r = 0.9) in the soils, and the correlation with the total thickness of the profile is practically absent.

Subgoltsy belt soils profile thickness varies significantly depending on the slope angle and slope exposure, as at the slopes of different exposure an unequal distribution of heat and moisture is created, and it results in a different degree of erosion and soil-forming processes. These differences affect the structure of the soils profile. So, on the shady slopes of the northern and eastern exposure brown soils are formed, the thickness of which varies from 45-50 cm to 100-110 cm. On the slopes of the southern and western exposure, more rubbly and short (no more than 30-45 cm) soils are formed. Thickness of individual horizons also varies significantly: top humus 2 to 7 cm, humus horizon 3 to 16 cm. Humus horizon color varies from black and dark gray through shades of gray to dark brown and gray-black. There is a relationship between the position of the soil on the slope (altitude) and sod thickness (r = 0,5).

In the mountain-forest belt, brown soils of different subtypes occupy a dominant position both under deciduous and coniferous forests. They occupy middle parts of the gentle slopes with an absolute altitude of 315-500 m (c. 26, 9, 10). The morphological appearance of brown soils in mountain and forest belt is similar to that of brown soils under the park vegetation. Thus, the general morphological characteristics of mountain brown forest soils are the following: truncated profile, mild differentiation in separate horizons, high rubbility (20-65%). No morfological signs of podzolity were found in the soils profile despite the presence of fir and spruce forests and acid forest litter.

Granulometric texture of the fine soils is loamy and varies from medium loamy to clayey. The soils feature is the gradual increment in weight of granulometric texture (by physical clay content), and the dominant factions are either large silt or sand. Distribution of elementary particles in the soil profile indicates soils poligenecity, which is manifested in the presence of the barrier, on which there is a sharp decline or a sharp increase in granulometric particles content.

Morphologically and analytically, the following basic processes of common geochemical features typical of soil-forming and weathering in cold humid areas are clearly distinguished for the brown forest soils of the Middle Urals – ferrugination and aluminizing of the thickness of soil and eluvium combined with desilication of the soil profile, which are fixated by different types of differentiation of lithophile elements in the process of pedogenesis; staging of accumulation and the contrast in the profile distribution profile-forming elements (iron, aluminum, silicon) (table 2).

Table 2. *The bulk composition of the oxides in the brown forest soils on the North Basegi mountain*

Cut	Depth, cm	Horizon	SiO_2	Al_2O_3	Fe_2O_3	K_2O	Na_2O	CaO	MgO	TiO_2	MnO	P_2O_5
30	13-23	AU_1	68.67	11.45	3.40	1.74	0.62	0.18	0.41	1.16	0.02	0.29
	23-32	AU_2	66.18	12.41	5.09	1.82	0.52	0.25	0.71	1.04	0.03	0.23
	32-50	BM	69.69	13.28	4.11	1.77	0.68	0.29	1.30	0.94	0.04	0.11
	50-75	CLM	74.28	13.64	3.84	1.81	1.12	0.31	1.45	0.91	0.04	0.10
32	5-9	AY_1	61.33	14.67	8.02	2.67	0.59	0.16	0.96	1.68	0.10	0.61
	9-22	AY_2	61.51	15.34	8.23	2.66	0.38	0.17	0.82	1.55	0.10	0.43
	22-37	AYf	61.7	16.24	8.53	2.7	0.33	0.18	1.16	1.53	0.12	0.38
	37-58	BM	63.56	16.05	8.19	2.60	0.62	0.18	1.21	1.49	0.09	0.32
	58...	CLM	65.11	16.84	8.01	2.69	0.66	0.22	1.49	1.48	0.09	0.25
27	4-12	AYan	63.10	16.11	6.14	2.29	1.14	1.11	1.99	1.33	0.04	0.50
	12-26	AYg	66.03	15.52	6.65	2.41	0.82	0.99	1.99	1.49	0.06	0.31
	26-57	BMg	66.74	13.98	7.13	2.32	2.49	1.16	2.63	1.56	0.09	0.18
	57...	CLMf,g	66.38	15.07	7.59	2.45	1.77	1.20	2.52	1.38	0.08	0.14
17	4-13	AU	58.82	13.70	5.82	2.08	0.16	0.36	1.13	1.09	0.10	0.72
	13-23	AUm	59.38	14.37	6.40	2.17	0.45	0.30	1.46	1.09	0.08	0.49
	23-32	BM_1	67.66	14.91	5.69	2.29	0.74	0.38	1.58	1.11	0.06	0.16
	32-46	BM_2	69.56	14.57	5.41	2.28	0.19	0.40	1.63	1.05	0.05	0.06
	46...	CLM	68.57	15.56	5.62	2.41	0.86	0.39	1.86	1.07	0.06	0.12
15	10-14	AU	71.45	12.51	5.02	2.06	1.01	0.51	1.04	1.13	0.15	0.17
	14-21	BM	69.64	14.51	5.98	2.08	1.03	0.43	1.76	1.13	0.05	0.07
	21-43	BMi	67.55	15.59	6.39	2.15	0.98	0.42	2.03	1.10	0.06	0.06
	43-70	CLM	69.77	14.28	6.13	2.16	0.78	0.42	1.59	1.06	0.06	0.02
19	6-15	AU	46.31	7.874	7.81	1.30	1.10	0.30	0.84	1.09	0.06	0.17
	15-30	BMel	64.71	15.63	7.63	2.25	1.03	0.58	2.16	1.64	0.07	0.30
	30-72	BM	65.92	14.95	7.32	2.22	1.32	0.67	2.49	1.52	0.07	0.11
	72...	CLM	65.58	15.16	7.59	2.43	1.59	1.24	2.52	1.43	0.10	0.19
26	5-22	AU	65.62	14.66	5.60	2.17	1.09	0.42	1.66	1.02	0.10	0.16
	22-33	AY	65.23	15.88	5.88	2.14	0.66	0.36	1.37	1.05	0.08	0.16
	33-57	BM	69.68	14.29	5.59	2.13	0.81	0.39	1.65	1.06	0.07	0.05
	57...	BMi	71.40	14.50	5.45	2.27	0.98	0.43	1.51	1.06	0.07	0.03

For brown forest soils formed in a mountain-tundra belt (cross section 30), the seasonal permafrost plays a significant profile-forming role as hydrothermal factor and geochemical barrier.

Types of distribution of various oxides in soils vary depending on the role in soil formation. Thus, distribution of SiO_2, Al_2O_3, Fe_2O_3 is mildly differentiated throughout the profile, and only in some soils, it is closer to the eluvial-illuvial type with various manifestations of eluviality and illuviality. Accumulative eluvial-illuvial distribution characterized by accumulation on the surface was noted for CaO, while MgO is mainly characterized by progressive eluvial and eluvial-illuvial types of distribution. TiO_2 distribution in the soils is not differentiated. Regressively or uniformly accumulative distribution is noted for phosphorus oxide and sulfur. These substances distribution subtypes are of accumulative type of distribution that characterizes profiles with a maximum accumulation of substances from the surface with a gradual decrease of their content with depth. To determine the relationship between type of substances distribution and the altitude, correlation coefficients were calculated. Based on these, it can be said that the correlation between the types of substances distribution and the altitude is low ($r < 0,3$).

Based on data on the content of macroelements, their ratio was calculated, and various geochemical factors were defined that allow detection of changes in soil properties and determination of their geochemical characteristics. Detailed characterization of geochemical factors illustrating the process of soil-forming is presented in [57]. Geochemical factors show that soils are poligenetic and intergenerational judging by time of exposure and manifestation of various paedogenesis processes in their properties, which indicates a change in the conditions while forming these soil profiles.

The ratio of the various forms of iron compounds in the profile, and even within the same horizon, performs diagnostic role for the detection of elementary processes of soil formation, which can be controlled by accumulation or redistribution of different forms of iron [39].

Thus, in the brown soils at an altitude of 900 m asl the iron content through the profile is differential and is < 5 %. In crooked forest brown soils the maximum iron content is noted, that is more than 8 % with a maximum in the middle of the profile. In the brown soils of mountain-forest belt, the iron content is also more than Clark's and varies from 5.02 to 7.81 %.

The group composition of iron compounds shows that in brown soils at different altitudes ratios of iron compound forms may vary. In brown soils, the content of non-silicate iron is less (27-40 %) than silicate (49-80 %), especially in soils on the slopes of the southern and western exposure.

The ratio of amorphous and crystallized iron in the brown soils is also interesting. Thus, in the brown forest dark humus soils in the mountain forest belt and the transition sub-belt of park open forest (cross sections 15, 17, 19) there are more crystallized forms of iron compounds than amorphous. In brown forest clayey illuviated soil (crosssection 26) in the lower border of the mountain-forest belt, the content of amorphous iron is slightly higher than that of crystallized, with a maximum in structural and metamorphic profile horizons.

In brown soil profile in crooked forest and mountain tundra, there is an alternation in the prevalence of crystallized and amorphous forms of iron compounds. Chemogenic differentiation processes manifested in the soil profile indicate a change in ecological and landscape conditions at altitudes above 500 m asl. Influenced by various factors, the main soil-forming processes can be transformed, flow with different intensity and superimpose each other, thereby defining various modifications of the structure and properties of diagnostic horizons.

The results of the bulk composition allow to diagnose the following processes in the studied soils by the 2004 classification: humus accumulation, decilitization, argillization, pedogenic weathering, chemogenic differentiation, lessivage, metomorfization, ferrugination, illuviation.

The content of organic matter in humus horizons of soils ranges from 3.2 to 4.1 % per the weight of the soil. With the depth increment, wamount of organic substance gradually decreases (Table 3).

The soils have a very strong acidic reaction which depends on altitude and varies from 3.01 to 3.97. The soils are characterized by a rather high value of hydrolytic acidity (Ng) (in the upper horizons it varies between 10.3-25.2 mEq/100 g). The highest Ng rates are found in organogenic accumulative horizons. Hydrolytic acidity decreases down through the profile. There was noted a clear spatial pattern in the distribution of the magnitude of the potential acidity: the highest values have been observed in soils of north-western and western slopes, as well as at the top of the North Basegi. No regularities in change of acidity depending on altitudinal belts were identified. It has been established that the soils are depleted in exchangeable bases (0.5 to 22.3 mEq/100 g of soil). There is a gradual decrease in the cation exchange capacity down through the profile, and in some soils the content of exchangeable cations increases in the rock and the transitional horizon.

Based on the analysis of group composition of humus in brown forest soils of North Basegi (Table 4), it should be noted that humus in the soil is very mobile, since more than a half of humic substances goes into pyrophosphate extraction. Humic substances in brown soils of park open forests and mountain-forest belt have the highest solubility (95.79 and 78.62 %, respectively).

Table 3. *Physical and chemical properties of brown forest soils on the Middle Urals*

Cut	Slope exposure	Altitude at sea level, m	Depth, cm	Horizon	pH_{KCl}	pH_{H2O}	Organic matter, %	S	Hr	T	V,%
								mg-ekv/100 g of soil			
30	West	900	13-23	AU$_1$	3.01	4.07	4.1	2.7	20.7	23.4	6.4
			23-32	AU$_2$	3.45	4.16	3.8	0.7	22.2	22.9	1.7
			32-50	BM	3.91	4.48	1.5	0.5	13.3	13.8	1.2
			50-75	CLM	3.95	4.58	1.6	1.9	12.4	14.3	4.5
32	South–East	691	5-9	AY$_1$	3.50	4.01	4.0	2.5	16.2	18.7	6.0
			9-22	AY$_2$	3.77	4.18	2.6	1.1	13.9	15	2.6
			22-37	AYf	3.80	4.48	2.3	2.7	13.2	15.9	6.4
			37-58	BM	3.84	4.60	1.5	1.5	13.2	14.7	3.6
			58...	CLM	3.97	4.80	0.4	1.7	11.0	12.7	4.0
27	South	590	4-12	AYan	3.80	5.09	3.7	17.8	10.3	28.1	42.4
			12-26	AYg	3.71	5.39	1.8	15.3	9.7	25	36.4
			26-57	BMg	3.83	5.60	0.2	16.9	3.8	20.7	40.2
			57...	CLMf,g	3.84	5.67	0.1	22.3	3.8	26.1	53.1
15	East	577	10-14	AU	3.44	4.63	3.5	8.5	15.0	23.5	20.2
			14-21	BM	3.47	4.81	1.6	-	17.3	17.3	-
			21-43	BMi	3.54	4.92	0.8	6.6	19.8	26.4	15.7

Cut	Slope exposure	Altitude at sea level, m	Depth, cm	Horizon	pH_{KCl}	pH_{H2O}	Organic matter, %	S	Hr	T	V,%
								mg-ekv/100 g of soil			
			43-70	CLM	3.72	5.14	1.0	5.4	14.1	19.5	12.9
19	West	565	6-15	AU	3.32	4.10	3.2	5.0	18.0	23	11.9
			15-30	BMel	3.83	4.47	1.5	–	14.2	14.2	–
			30-72	BM	3.78	4.61	0.9	2.5	13.2	15.7	6.0
			72...	CLM	3.94	5.66	0.2	21.9	3.6	25.5	52.1
26	North–West	315	5-22	AU	3.33	4.66	3.5	–	25.2	25.2	–
			22-33	AY	3.37	3.85	2.7	0.2	25.2	25.4	0.8
			33-57	BM	3.54	4.66	0.6	–	20.4	20.4	–
			57...	BMi	3.48	4.83	0.5	2.5	17.5	20	6.0

Table 4. *Group composition of humus in the humus horizons of brown forest soils of North Basegi (%)*

Altitude sub-zone	Altitude asl, m	Cut	C_{total}	$C_{extract}$	$C_{humic\ acid}$	C_{fulvic}	NR	$C_{humic}:C_{fulvic}$	Degree of humification
Crooked forest	900	30	3.85	1.95	0.13	1.82	1.90	0.1	51
				50.65	3.40	47.27	49.35		
	691	32	3.63	1.94	0.37	1.57	11.69	0.2	53
				53.44	10.20	43.25	46.56		
Park open forest	590	17	3.80	3.64	0.66	2.98	0.82	0.2	96
				95.79	17.30	78.42	4.21		
Mountain-forest	565	19	2.81	2.11	0.37	1.74	0.70	0.2	75
				75.09	13.10	61.92	24.91		
	315	26	2.76	2.17	0.69	1.48	0.59	0.5	79
				78.62	25.00	53.62	21.38		

Note: NR – nonhydrolyzable residue

$C_{humic}:C_{fulvic}$ ratios reflect the structural features of humic acids, which allow us to estimate the degree of hydrolyzability of humic acids and the properties of the humic acid itself, its looseness [66].

In the brown forest soils of the Middle Urals, humic acids are rather loose ($C_{humic}:C_{fulvic}$ varies between 0.1 and 0.5) and able to hydrolysis, that means that they are not very persistent and therefore easy passing to solution. These indicators show that the degree of hydrolyzability of mountain soils is high, and the possibility of the formation of double bonds is very low. It is believed that fulvic acids are the split product of humic acids, i.e. fulvic acids are the derivants of the humic acids. The degree of humification calculated by the formula [66] gives a clearer picture of humification of plant residues. About half of the incoming plant residues are humified in the soils, with a greater extent in a mountain forest belt. Nonhydrolyzable residue characterizes conservative, stable part of humus. Nonhydrolyzable residue content in soils ranges from very low (4.2 %) to medium. This once again confirms the high mobility of humic substances.

Thus, the features of the humus condition of brown forest soils are recency and immaturity of humic substances, staying at the first stage of humification, which manifests in the predominance of fulvic acids in the humus and the high proportion of soluble humic substances ($C_{extract}$.).

4. Conclusions

Physical and geographical conditions of the Middle Urals cause the following genetic characteristics of brown forest soils.

1. Formation at an altitude of 315-900 m above sea level under different plant biocenoses on well-drained parts of the slopes.
2. Soils profile is poorly differentiated into genetic horizons.
3. Granulometric varieties of soil are loamy and clayey.
4. Distribution of the gross forms of oxides in the soil profile depends on a combination of processes in soils, regardless of an area altitude. The high content of *Ti, Si, S, P*, as well as *Al* and *Fe* was also revealed within the whole soils profile not having anthropogenic nature.
5. The bulk composition reveals the homogeneity of all soil horizons, suggesting the lack of podzolization processes; high content of total iron (6 % average for all soils) followed by a predominance of silicate compounds in the composition allows to diagnose brown soil forming process.
6. Brown forest soils in the Middle Urals have a very acidic environment and actively manifest all forms of soil acidity.
7. Characteristics of group composition of humus showed that mountain soils of the Middle Urals have the following specific features: significant spatial variation of parameters of qualitative composition of humus; stretched humus profile; acidic and immature nature of humic substances; there was also noted the impact of altitude and slope exposure to conditions of humus formation.
8. Based on diversified studies of the properties of brown forest soils, there has been made an assumption of soils polygeneticity and the presence of paleohorizons in the profile, which indicates a change in environmental conditions during the formation of mountain soils in the Middle Urals.

Integrity of mountain ecosystems allows us to observe the

development of soil in time and space and find a whole range of transition conditions and stages of soil formation in a small area. Soils located below the slope are constantly interact and contact with soils of overlying high-altitude belt and adjacent areas through gravitational movement of weathering products. Simultaneousness of processes of soil formation, weathering and denudation, having varying intensity, cause the diversity of soil cover so that even within a single altitude belt a spatial heterogeneity of soils can be observed. Spatial heterogeneity and evolution of the soils in the mountains is evident even within a soil type.

By the nature of the main and associated processes, brown soils at hypsometric levels of the Northern Basegi mountain form the next evolutionary series and are divided into the following subtypes: raw-organic dark humus brown soils (900 m), ferruginated brown soils (800, 655 m), ferruginated gleyey brown soils, metamorphosed dark humus brown soils (590 m), clayey illuviated dark humus brown soils (577 m), eluviated dark humus brown soils (565 m), eluviated brown soils (430, 400 m), raw-organic humic brown soils (383 m), clayey illuviated brown soils (315 m).

Acknowledgements

I thank for your assistance in determining the chemical composition of soils Soil Science Institute staff. V.V. Dokuchaev - head. Laboratory of Physical Chemistry of the soil Cand.Chem.Sci O.B. Rogova and a tester N.A. Plotnikova.

References

[1] Kovda V.A. Similarities and differences in the history of the soils of the continents // Trans. 8th Int. Congr. Soil Science. Bucharest, 1964. vol. 1.

[2] FAO/UNESCO Soil Map of the World, 1:5,000,000. UNESCO, Paris, 1971-1975.

[3] Buol S.W., Hole F.D., Cracken R.J. Soil genesis and classification // The Lowa State Univ. Press, Amer. 1973.

[4] Rozanov B.G. The soil cover of the globe. M.: MGU, 1977. 248 p.

[5] Urushadze T.F. Some aspects of the soil formation in the mountainous regions // Soil science. 1979. № 1. P. 131-143.

[6] Soil Taxonomy. A Basic system of Soil Classification for Making and Interpreting Soil Surveys. Second edition. USDA. Washington D. C. 1999. 869 p.

[7] World Reference Base for Soil Resources 2006. A framework for international classification, correlation and communication. Food and Agricultural Organization of the United Nations. Rome. 2006. 128 p.

[8] http://polyera.ru/geografiya-pochv/1119-pochvennye-zony-gornyh-territoriy-chast-8.html

[9] Baumler R., Zech W., Heuberger H., Weber-Diefenbach K. Investigations on the intensity of weathering jf soils developed from glacial and fluvioglacial deposits and their relationship with the history of the landscape in the Mt. Everest region // Geoderma, 1991. 48. P. 223-243.

[10] Grosswald M.G., Kuhle M., Fastook J.L. Wurm Glaciation of Lake Issyk-Kul Area, Nian Shan Mts.: A Case Stady in Glacial History of Central Asia // GeoJournal. 1994. 33.2/3. P. 273-310.

[11] Brooks P.D., Schmidt S.K., Williams M.W. Winter production of CO_2 and N_2O from alpine tundra: environmental controls and relationship to inter-system C and N fiuxes // Gecologia. 1997. Vol.110. N. 3. P. 403-413.

[12] Farley R.A., Fitter A.H. Temporal and spatial variation in soil resources in a deciduous woodland // Journal of Ecology. 1999. Vol. 87. № 4. P. 688-696.

[13] Lee B. D., Sears S. K., Graham R. C., Amrhein C., and Vali H. Secondary mineral genesis from chlorite and serpentine in an ultramafic soil toposeqnence // *Soil Science Society of America Journal*, 2003, vol. 67, no. 4, P. 1309-1317.

[14] Wilkinson M.T. et al. Soil production in heath and forest Blue Mountains, Australia: influence of lithology and palaeoclimate // Earth Surface Processes And Landforms, 2005. 30: P. 923-934.

[15] Humphreys G.S., Wilkinson M.T. The soil production function: A brief history and its rediscovery // Geoderma 139. 2007. P. 73-78.

[16] Heimsath A.M., Hancock G.R. and Fink D. The 'humped' soil production function: Eroding Arnhem Land, Australia // Earth Surface Processes & Landforms, 34: 2009. P. 1674-1684.

[17] Fu G., Shen Z., Zhang X., Yu C., Zhou Y. and Yang P. Response of ecosystem respiration to experimental warming and clipping at daily time scale in an alpine meadow of Tibet // *Journal of Mountain Science*, 2013. vol. 10, P. 455-463.

[18] Yanhong L., Huimei W., Wenjie W., Lei Y., Yuangang Z. Ectomycorrhizal Influence on Particle Size, Surface Structure, Mineral Crystallinity, Functional Groups, and Elemental Composition of Soil Colloids from Different Soil Origins // The Scientific World Journal Volume, 2013. P. 1-13. http://dx.doi.org/10.1155/2013/698752

[19] Rammann E. Soil and climatic zones of Europe // Soil science. 1901. N 1. p. 5-18.

[20] Prasolov L.I. Mountain-forest soils of the Caucasus // Tr. Soil institute. 1947. T. 25. P. 5-28.

[21] Antipov-Karataev I.N. On the brown forest and brown soils // Soil science. 1947. № 12. P. 697-703.

[22] Liverovskii J.A., Kornblum E.A. Zonation of soils in submontane areas // Publishing house of USSR Academy of Sciences. Ser. geograph. 1960. № 3. P. 34-41.

[23] Stefanovits P. Broun forest soils of Hungary // Academiai Kiado, Budapest, 1971.

[24] Urushadze T.F. Soils of Georgia mountain forests. Tbilisi: Mentsiereba, 1987. 243 p.

[25] Zakharov S.A. Soils of mountain regions of the USSR // Soil science. 1937. № 6 P. 810-848.

[26] Zakharov S.A. On some controversial issues of mountain soil science // Soil Science. 1948. № 6. P. 347-356.

[27] Zonn S.V. Mountain forest soils of the Northwest Caucasus. M.: Publishing House of USSR Academy of Sciences, 1950. 333 p.

[28] Friedland V.M. Brown forest soils of the Caucasus // Soil science. 1953. № 12. P. 28-44.

[29] Retzer J.L. Soil formation and classification of forested mountain lands in the United States // Soil Science, 1963. vol. 96. N. 1.

[30] Stase H. a.o. A handbook of Australian soils. – Adelaide. 1968.

[31] Prehtel L.V. Silvicultural properties of brown forest and brown sod soils // FESU Scientific notes. 1969. T. 20. P. 62-79.

[32] Ohtsuka T., Hirota M., Zhang X. et al. Soil organic carbon pools in alpine to nival zones along an altitudinal gradient (4400–5300 m) on the Tibetan Plateau. // Polar Science, 2008. vol. 2. №.4. P. 277-285.

[33] Petrov B.F. Soils of Southern Sakhalin // Soil science. 1947. № 11. P. 641-651.

[34] Glazovskaya M.A. Soils of mountain regions of Kazakhstan // Publishing house of Academy of Sciences of the Kazakh SSR, soil science series. 1949. Issue 14. № 52. P. 191-220.

[35] Gradoboev N.D. Soils of larch forests of Siberia (Altai, Kuznetsk Alatau and Tuva) // Works on Siberia Forestry. 1955. Issue 2. – Novosibirsk.

[36] Ivanov G.I. Soils of mountain coniferous and deciduous forests of South Primorye // Komarovski readings. Vladivostok, 1957. Issue 6. P. 55-87.

[37] Gerasimov I.P. Brown forest soils in the USSR, European countries and the USA // Soil science. 1959. № 7. P. 61-68.

[38] Karpachevskii L.O. Some features of the soil forming in Kamchatka // Soil science. 1965. № 11. P. 1-10.

[39] Zonn S.V. Features of allitic soil-forming on the islands of Primorye and the Far East // Studying and development of the natural environment. M.: Science, 1976. P. 125-137.

[40] Firsova V.P., Gorjacheva T.A., Prokopovich E.V. Comparative characteristic of properties of mountain soils of the Middle Urals // Soil science. № 5. 1963. P. 16-25.

[41] Firsova V.P. On the problem of the distribution and properties of brown forest soils in the Urals // The collection Forest and soil: Tr. of All-Union Scientific Conference. Krasnoyarsk, 1968. P. 200-203.

[42] Firsova V.P. The soils of the taiga zone of the Urals and Trans-Urals. M.: Science, 1977. 175 p.

[43] Ivanova E.N. Mountain forest soils of the Middle Urals // Proceedings of the Soil Institute of the USSR Academy of Sciences, 1949. T. 30. P. 168-193.

[44] Mikhailova R.P. Micromorphological and chemical properties of brown raw-organic soils of central mountainous zone of the Middle Urals // Soil science. 1976. № 5. P. 18-28.

[45] Firsova V.P. Brown mountain forest soils of the Urals // Soil science. 1991. № 4. P. 47-58.

[46] Karpachevsky M.L., Shevchenko E.M. The ratio of lithogenic and cenogenetic factors in the formation of brown forest soils of the Middle Urals // Soil science. 1997. № 1 P. 22-30.

[47] Semikolennykh A.A., Bovkunov A.D., Aleynikov A.A. Soils and soil cover of the taiga belt of the Northern Urals (the Pechora river headwaters) // Soil science. 2013. № 8. P. 911-923.

[48] Rusanova G.V., Kuhri P. Soil on the border of the forest and alpine tundra of the Polar Urals // Soil science. 2001. № 4. P. 409-417.

[49] Dymov A.A., Zhangurov E.V. Morphological and genetic features of soilf of the Enganepe ridge (Polar Urals) // Soil science. 2011. № 5. P. 515-524.

[50] Dymov A.A., Dubrovsky Y.A., Gabov D.N. Pyrogenic change in illuvial ferruginous podzols (middle taiga, Komi Republic) // Soil science. 2014. № 2. P. 144-154.

[51] Dymov A.A., Zhangurov E.V. Diversity and genetic characteristics of soils of the Polar Urals // Perm Agricultural Gazette. 2014. № 3 (7). P. 45-51.

[52] Samofalova I., Luzyanina O., Maulina E., Kulkova L. Features soil mountain-taiga zone the middle urals // Igdir university jurnal of the institure of science and technology. 2(2EK: A): 2012. P.93-100.

[53] Samofalova I.A., Luzyanina O.A. Ecological and genetic characteristics of soils of mountain forest zone in the Middle Urals // Proceedings of the Samara Scientific Center of the Russian Academy of Sciences. 2013. T. 15. № 3(4). P. 1426-1431.

[54] Samofalova I.A., Luzyanina O.A., Kondratieff M.A., Mamontov, N.V. The elemental composition of soils in undisturbed ecosystems in the Middle Urals // Bulletin of the Altai State Agrarian University. 2014. № 5 (115). P. 67-74.

[55] Samofalova I.A., Luzyanina O.A. Sokolova N.V. Morphological and genetic features of soils in the subalpine zone (Middle Urals) // Actual problems of the humanitarian and the natural sciences. 2014. № 1 (60). Pt. I. P. 24-28.

[56] Samofalova I.A., Luzyanina O.A. "Basegi" reserve soils and their classification // Perm Agricultural Gazette. 2014. № 1 (5). P. 50-60.

[57] Samofalova I. Geochemical features of the elemental composition of soils in undisturbed ecosystems in the Middle Urals (for example the Reserve "Basegi") // French Journal of Scientific and Educational Research. No.2. (12), July-December, 2014. Vol. III. "Paris University Press". 2014. P. 156-170.

[58] Sofronitsky P.A. Geological outline // Proceedings of the fifth meeting on the chemical water geography and hydrogeochemistry of Perm region. PSU, Perm wepartment of geographical Society of the USSR, Institute of Karst and Speleology. Issue 4 (5). Perm, 1967. P. 26-41.

[59] Voskresensky S.S., Leontiev O.K., Spiridonov A.I. [et al.]. Geomorphological zoning of the USSR and adjacent areas: tutorial. M.: Higher School, 1980. 343 p.

[60] Gorchakovskii P.L. The flora of the alpine Urals. M.: "Science", 1975. P. 13-67.

[61] Theory and practice of chemical analysis of soils. / Ed. by L.A. Vorobeva. M.: GEOS, 2006. – P.344.

[62] Antonova Z.P., Skalaban V.D., Suchilkina L.G. Determination of humus in soils // Soil Science. 1984. № 11. P. 130-133.

[63] Schwertmann U. Relations between iron oxides, soil color, and soil formation // Soil color. SSSA Special Publication no 31. Madison, USA, 1993. P. 51-69.

[64] Classification and diagnosis of soils of USSR / comp. V.V. Egorov, V.M. Friedland. M.: Kolos, 1977. 225 p.

[65] Samofalova I.A., Luzyanina O.A. Mountain soils of the Middle Urals (for example, SNR "Basegi"). Perm State Agricultural Academy. Perm: Publishing House of HPE Perm State Agricultural Academy, 2014. 154 p.

[66] Orlov D.S., Sadovniko L.K. Soils chemistry. M.: Higher School, 2005. 558 p.

Permissions

All chapters in this book were first published in AJEP, by Science Publishing Group; hereby published with permission under the Creative Commons Attribution License or equivalent. Every chapter published in this book has been scrutinized by our experts. Their significance has been extensively debated. The topics covered herein carry significant findings which will fuel the growth of the discipline. They may even be implemented as practical applications or may be referred to as a beginning point for another development.

The contributors of this book come from diverse backgrounds, making this book a truly international effort. This book will bring forth new frontiers with its revolutionizing research information and detailed analysis of the nascent developments around the world.

We would like to thank all the contributing authors for lending their expertise to make the book truly unique. They have played a crucial role in the development of this book. Without their invaluable contributions this book wouldn't have been possible. They have made vital efforts to compile up to date information on the varied aspects of this subject to make this book a valuable addition to the collection of many professionals and students.

This book was conceptualized with the vision of imparting up-to-date information and advanced data in this field. To ensure the same, a matchless editorial board was set up. Every individual on the board went through rigorous rounds of assessment to prove their worth. After which they invested a large part of their time researching and compiling the most relevant data for our readers.

The editorial board has been involved in producing this book since its inception. They have spent rigorous hours researching and exploring the diverse topics which have resulted in the successful publishing of this book. They have passed on their knowledge of decades through this book. To expedite this challenging task, the publisher supported the team at every step. A small team of assistant editors was also appointed to further simplify the editing procedure and attain best results for the readers.

Apart from the editorial board, the designing team has also invested a significant amount of their time in understanding the subject and creating the most relevant covers. They scrutinized every image to scout for the most suitable representation of the subject and create an appropriate cover for the book.

The publishing team has been an ardent support to the editorial, designing and production team. Their endless efforts to recruit the best for this project, has resulted in the accomplishment of this book. They are a veteran in the field of academics and their pool of knowledge is as vast as their experience in printing. Their expertise and guidance has proved useful at every step. Their uncompromising quality standards have made this book an exceptional effort. Their encouragement from time to time has been an inspiration for everyone.

The publisher and the editorial board hope that this book will prove to be a valuable piece of knowledge for researchers, students, practitioners and scholars across the globe.

List of Contributors

Sandra Martinovic, Rejhana Muhamedagic and Azrudin Husika
Department of Energetics/Faculty of Mechanical Engineering, University of Sarajevo, Sarajevo, Bosnia and Herzegovina

Otar Abdaladze, Ketevan Batsatsashvili, Khatuna Gigauri and Tamar Jolokhava
Alpine Ecosystems Research Program, Institute of Ecology, Ilia State University, Tbilisi, Georgia

Gia Nakhutsrishvili and George Mikeladze
Department of Plant Systematic, Institute of Botany, Ilia State University, Tbilisi, Georgia

Tsisana Basilashvili
Institute of Hydrometeorology, Georgian Technical University, Tbilisi, Georgia

Melor Alpenidze
Sokhumi State University, Department of Geography, Tbilisi, Georgia

Zurab Seperteladze, Eter Davitaia and Tamar Aleksidze
Ivane Javakhishvili Tbilisi State University, Department of Geography,Tbilisi, Georgia

George Gaprindashvili
National Environmental Agency, Department of Geology, Tbilisi, Georgia

Satbay Nurashov and Elmira Sametova
Institute of Botany & Phytointroduction Ministry of Education and Science, laboratory of Mycology and Algology, Almaty, Republic Kazakhstan

Lar Uriah, Yenne Ezekiel and Jibo Jesse
Department of Geology, University of Jos, P.M.B. 3084, Jos, Nigeria

Ozoji Toochukwu
Department of Geology and Mining, Ibrahim Babangida University Lappai, Niger State, Nigeria

I. Kupradze and A. Jorjadze
Institute of Botany of IliaStateUniversity, Tbilisi, Georgia

A. Arabidze, T. Beltadze and K. Batsatsashvili
Institute of Ecology of Ilia State University, Tbilisi, Georgia

N. Y. Paniagua Zambrana
Herbario Nacional de Bolivia, Universidad Mayor de San Andres, La Paz, Bolivia

R. W. Bussmann
William L. Brown Center, Missouri Botanical Garden, St. Louis, USA

Sophia Barinova
Institute of Evolution, University of Haifa, Mount Carmel, Haifa, Israel

Viktor Gabyshev
Institute for Biological Problems of Cryolithozone SB RAS, Yakutsk, Russia

Mariyo Boboev
Khatlon Scientific Center Academy of Sciences of the Republic of Tajikistan, Kulob-town, Tajikistan

Lali Kukhaleishvili
Tbilisi Botanical Garden and Institute of Botany, Tbilisi, Georgia

Olena Bilous
Institute of Hydrobiology of NAS of Ukraine, Kiev, Ukraine

N. G. Gemejiyeva and Zh. Zh. Karzhaubekova
Institute of Botany & Phytointroduction Ministry of Education and Science, Almaty, Kazakhstan

Hambaryan L. R.
Department of Ecology and Nature Protection, Yerevan State University, Yerevan, Armenia

Nalbandyan M. A. and Poghosyan A. M.
Institute of Geological Sciences of the NAS Yerevan, Armenia

Achille Lutumba Suika and Zhang Liqiu
College of Environmental Science and Engineering, Beijing Forestry University, Beijing, China

Anne Marie Nsaka Kabunda
College of international law, Zhongnan University of Economics and Law, Wuhan, China

Gulnara Sitpayeva, Tansara Murzatayeva, Saniya Inerbayeva, Karina Makhmudova
Institute of Botany and Phytointroduction, Almaty, Kazakhstan

Nino Kezevadze
Directorate, ltd "Georgian Water and Power", Tbilisi, Georgia

Sallau Adamu
The Carter Centre, Tudun Wada, P.O. Box 7772 Jos, Plateau State Nigeria

Mangs Ayuba, Adamu Murtala and Lar Alexander Uriah
Department of Geology, University of Jos, Jos, Nigeria

Arnold Gegechkori
Department of Biodiversity, Faculty of Exact and Natural Sciences, Iv.Javakhishvili Tbilisi State University, Tbilisi, Georgia

Eter Didmanidze
Georgian National Museum, Tbilisi, Georgia

David Kereselidze, Vazha Trapaidze, Giorgi Bregvadze and Zaal Gulashvili
Department of Hydrometeorology, Faculty of Exact & Natural Sciences, Ivane Javakhishvili Tbilisi State University, Tbilisi, Georgia

Lia Matchavariani
Department of Soil Geography, Faculty of Exact & Natural Sciences, Ivane Javakhishvili Tbilisi State University, Tbilisi, Georgia

Lamzira Lagidze
Department of Nature Use Management, Faculty of Exact & Natural Sciences, Ivane Javakhishvili Tbilisi State University, Tbilisi, Georgia

David Svanadze
Department of Geography, Faculty of Exact & Natural Sciences, Ivane Javakhishvili Tbilisi State University, Tbilisi, Georgia

G. M. Kudabayeva, P. V. Vesselova, M. P. Danilov and B. M. Sultanova
Institute of Botany &Phytointroduction Ministry of Education and Science, Almaty, Kazakhstan

Ahmed Boucherit and Mohamed Wahib Naceur
Department of Chemical Engineering, University of Blida, Blida, Algeria

Djamel Ghernaout
Department of Chemical Engineering, University of Blida, Blida, Algeria
Binladin Research Chair on Quality and Productivity Improvement in the Construction Industry, College of Engineering, University of Hail, Ha'il, Saudi Arabia

Abdulaziz Ibraheem Al-Ghonamy, Ammar Abdallah Mahjoubi, Noureddine Ali Elboughdiri, Noureddine Ait Messaoudene and Mohamed Aichouni
Binladin Research Chair on Quality and Productivity Improvement in the Construction Industry, College of Engineering, University of Hail, Ha'il, Saudi Arabia

Badiaa Ghernaout
Department of Mechanical Engineering, University of Laghouat, Laghouat, Algeria

Lia Matchavariani
Department of Soil Geography, Faculty of Exact & Natural Sciences, Iv Javakhishvili Tbilisi State University, Tbilisi, Georgia

Dali Nikolaishvili
Department of Geomorphology & Cartography, Faculty of Exact & Natural Sciences, Iv Javakhishvili Tbilisi State University, Tbilisi, Georgia

Rusudan Khukhunaishvili, Marina Koridze, Marina Nagervadze, Ciuri Khizrevanidze and Shorena Gabaidze
Department, of Biology, Faculty of Natural Sciences and Health Care, Batumi Shota Rustaveli State University, Batumi, Georgia

Richard O. Claus
NanoSonic Inc. Pembroke, Virginia, USA

Tamara V. Tulaikova
Wave Research Center at Prokhorov General Physics Institute, Moscow, Russia

Dimeyeva Liliya
Institute of Botany & Phytointroduction /Ministry of Education & Science, Almaty, Republic of Kazakhstan

Mohamed Wahib Naceur and Ahmed Boucherit
Department of Chemical Engineering, University of Blida, Blida, Algeria

Djamel Ghernaout
Department of Chemical Engineering, University of Blida, Blida, Algeria
Binladin Research Chair on Quality and Productivity Improvement in the Construction Industry, College of Engineering, University of Ha'il, Ha'il, Saudi Arabia

Abdulaziz Ibraheem Al-Ghonamy, Noureddine Ait Messaoudene, Mohamed Aichouni, Ammar Abdallah Mahjoubi and Noureddine Ali Elboughdiri
Binladin Research Chair on Quality and Productivity Improvement in the Construction Industry, College of Engineering, University of Ha'il, Ha'il, Saudi Arabia

Giorgi Khomeriki
Vakhushti Bagrationi Institute of Geography at Iv.Javakhishvili Tbilisi State University, Tbilisi, Georgia

Giorgi Meladze
Dep. of Geography, Iv. Javakhishvili Tbilisi State University, Tbilisi, Georgia

Abdeljabar Haboub, Meryem Hamlich, Souad Harkati, Anouzla Abdelkader and Mohamed Safi
Laboratoire Materials, Catalysis & Natural Resource Development, URAC University of Hassan II Mohammedia-Casablanca, BP 146, Mohammedia, Morocco

Yassine Riadi, Rachid Slimani and Saïd Lazar
Laboratory Biochemistry, Environment & Agri URAC University of Hassan II Mohammedia-Casablanca, BP 146, Mohammedia, Morocco

Mina Aadil
Laboratoire of Physical Chemistry and Bio-organic Chemistry, University of Hassan II Mohammedia-Casablanca, BP 146, Mohammedia, Morocco

Giorgi Meladze and Maia Meladze
Institute of Hydrometeorology at the Georgian Technical University, Tbilisi, Georgia

Gulnara Sitpayeva
Institute of botany and phytointroduction, Almaty, Kazakhstan

Younes Abrouki, Abdelkader Anouzla, Hayat Loukili, Rabiaa Lotfi, Ahmed Rayadh, My Abdellah Bahlaoui, Driss Zakarya, Mohamed Zahouily and Adil Mandil
Department of Research, Faculty of Science and Technology, University Hassan II, Mohammedia, Morocco

Rachid Mamouni
Laboratory Organic Chemistry, Team Bio-Organic Chemistry Applied, Ibn Zuhr University, Agadir, Morocco

Saïd Sebti
Department of Chemistry, Faculty of Science Ben M'Sik, University Hassan II, Casablanca, Morocco

Mariam Metreveli
Batumi Shota Rustaveli State University, Biology Department, Faculty of Natural Sciences and Health Care, Batumi, Georgia
Batumi Botanical Garden, Batumi, Georgia

Avtandil Meskhidze and Dali Beridze
Batumi Shota Rustaveli State University, Biology Department, Faculty of Natural Sciences and Health Care, Batumi, Georgia

Feride Tchaidze and Julieta Jakeli
Batumi Botanical Garden, Batumi, Georgia

Tamar Varazi, Maritsa Kurashvili, Marina Pruidze, Gia Khatisashvili, Nino Gagelidze, George Adamia, George Zaalishvili and Marlen Gordeziani
Durmisidze Institute of Biochemistry and Biotechnology of Agricultural University of Georgia, David Agmashenebli Alley 240, Tbilisi, Georgia

Mark Sutton
Chemical Sciences Division, Lawrence Livermore National Laboratory, California, USA

Paul A. Comet
Comet Environmental Consulting, Houston, Texas, USA

Samofalova Iraida
Perm State Agricultural Academy, The Faculty of Soil Science, Agrochemistry, Ecology and Commodity research, Perm, Russia

Index